AP
3A

Rediscovering Darwin: Evolutionary Theory and Archeological Explanation

C. Michael Barton & Geoffrey A. Clark, Editors

Contributions by

Douglas B. Bamforth
C. Michael Barton
Robert Bettinger
Douglas W. Bird
Peter Bleed
Geoffrey A. Clark
Michael W. Diehl
Jelmer Eerkens
John A. Gowlett
Daniel O. Larson
R. Lee Lyman
Steven Mithen
Hector Neff
Fraser Neiman
Michael J. O'Brien
David Rindos
Michael B. Schiffer
Michael Shott
Esmée Webb
Polly Wiessner

1997
Archeological Papers of the American Anthropological Association No. 7

About the editors . . .

C. Michael Barton was born in Wichita, Kansas. He received his BA from the University of Kansas and his PhD in anthropology and geosciences from the University of Arizona. His research focuses on prehistoric foragers and horticulturalists in the Mediterranean and the American Southwest, and on late Pleistocene and early Holocene lithic technology and settlement. His interests also include landuse, the origins of food production, geoarcheology and, of course, evolutionary theory.

Geoffrey A. Clark was born in Philadelphia and received his PhD from the University of Chicago. An Old World prehistorian and a paleoanthropologist, he has published extensively on human biological and cultural evolution in 'deep time'—the past three million years. Current interests turn on the logic of inference in the various aspects of the modern human origins debate. Clark is the coeditor, with C. M. Willermet, of *Conceptual Issues in Modern Human Origins Research* (Aldine de Gruyter, 1997).

The cover artwork for this book is an original drawing by Tom Dunne © *American Scientist* 85(2): 105 (March-April, 1997). Modified by G. A. Clark; used with the artist's permission.

Manufactured in the U.S.A.

Library of Congress Cataloging-in-Publication Data
Rediscovering Darwin : evolutionary theory and archeological explanation / C. Michael Barton & Geoffrey A. Clark, editors ; contributions by Douglas B. Bamforth . . . [et al.].
p. cm. -- (Archeological papers of the American Anthropological Association ; no. 7)
Includes bibliographical references.
ISBN 0-913167-87-8
1. Darwin, Charles, 1809–1882. 2. Human evolution. 3. Archaeology—Methodology. I. Barton, C. Michael. II. Clark, Geoffrey A. III. Bamforth, Douglas B. IV. American Anthropological Association.
V. Series.
GN281.R412 1997
599.93′8—dc21 97-47355

Rediscovering Darwin: Evolutionary Theory in Archeological Explanation

C. Michael Barton & G. A. Clark, Editors

Table of Contents

Part I - Introduction

Part II - General Theory & Method in Evolutionary Archeology

Part III - Material Culture, Behavior & Middle Range Theory

Part IV - Case Studies & Applications in Paleoanthropology and Archaeology

Part V - Epilogue

Preface

As is the case with all scholarly endeavors, there are individuals who played a major role in helping make *Rediscovering Darwin* a reality, but who-for one reason or another-are not listed in any bibliography. My own realization of the applicability of neo-Darwinian evolutionary theory was inspired by a lecture I attended as a graduate student at the University of Arizona in the Fall of 1978. Given jointly by Robert Netting and Keith Basso, the lecture was an assessment of the state of anthropology as a unified discipline. Their conclusion was a rather disheartening prognostication of the potential fragmentation of the field due to the diverse interests of its practitioners and the lack of a general theory of human behavior. Unwilling to accept this pronouncement, it came to me later that evening that Darwinian evolution might provide a much more robust and satisfactory explanatory framework than the 'evolutionary' (i.e., Spencer's, Tylor's, White's, Service's) and other theories Netting and Basso so effectively critiqued. As is so often the case with personal revelations, however, none of my colleagues seemed to appreciate this 'flash of insight'. Several years later, after listening to one of my occasional tirades about the unappreciated potential of Darwinian theory for understanding human behavior, my wife Margaret (also an archaeology graduate student, but at a different university) suggested that I read a book by Robert Dunnell. However, with a title of *Systematics in Prehistory* (Dunnell 1971), it didn't seem to have much to do with evolutionary theory nor did it seem relevant to my interests in paleolithic prehistory and Pleistocene environments. As I have long since learned, however, I should have heeded her advice. As it was, I finally read an article by Dunnell in 1987 and realized (with a mixture of disappointment, relief, and a little embarrassment) that I was neither the only one nor the first to recognize the potential of Darwinian theory in archaeology.

Rediscovering Darwin had its birth in something of the same way. It was first conceived as a Society for American Archaeology symposium (eventually held in 1994). The idea of a symposium was triggered by my reading of Elizabeth Brumfiel's Distinguished Lecture in Archaeology to the American Anthropological Association (Brumfiel 1992). In this article, Brumfiel criticized the 'ecosystem approach' for its lack of attention to subordinate social classes, gender differences, and political factionalism. In her critique, she seemed to equate the ecosystem approach with Spencerian progressivist theory, on the one hand, by citing White and Steward (1992: 551), and with Darwinian evolutionary theory on the other, by discussing processes like natural selection (e.g., 1992: 552). Interestingly, many of her most telling criticisms of the ecosystem approach were aimed at misapplications of Darwinian theory. I felt that Brumfiel's essay portrayed well the status of Darwinian theory in archaeology-seldom invoked explicitly, widely used in piecemeal fashion, and often misunderstood. At the same time, a small but significant number of programmatic works on Darwinian theory were beginning to appear in the mainstream archaeological literature. I thought that an SAA symposium might be a way to initiate a broader appraisal of the potential for Darwinian theory in archaeology. Never having organized such an endeavor, I asked my colleague Geof Clark if he would be willing to collaborate with me. Fortunately, he enthusiastically agreed. Geof has been the ideal collaborator in this undertaking. He has long been interested in epistemology in archaeology (e.g., Clark 1991, Clark & Willermet 1997) and in evolutionary theory as it applies to humans and, as an Old World archaeologist trained in a New World research tradition, he has a very good sense of the state of the discipline. Added to this, he had the experience I lacked in organizing symposia and in editing the proceedings.

Rather than organize a symposium that showcased only the few Darwinian theorists actively and explicitly publishing in archaeology, we decided to take another tack. In keeping with our assessment of the widespread use and misunderstanding of Darwinian theory in archaeology, Geof and I invited active archaeological practitioners who made good use of Darwinian theory in their research programs, regardless of whether they thought of themselves as Darwinians. We asked them to explicitly discuss this aspect of their research. The objective of the symposium was twofold. One goal was to show that anglophone archaeologists widely employed Darwinian theory in practice, regardless of the programmatic literature; the second was to encourage these archaeologists to confront more explicitly the theoretical implications of this practice. We sought to invite individuals who employed Darwinian theory in exceptionally diverse ways. The symposium played to a standing-room-only audience at the Anaheim meetings, and was very successful in achieving its objectives. Geof and I want to take this opportunity to thank Robert Dunnell and C. Loring Brace for serving as discussants. Their insightful comments helped to emphasize the theoretical commonalities among the participants and to counterbalance the diverse ways in which they made use of Darwinian theory.

In deciding to publish the proceedings of the 1994 symposium, Geof and I followed the same strategy of soliciting essays from practitioners employing Darwinian theory in diverse ways. We have done this by expanding the number of papers beyond those contributed by the original 1994 participants. Since several of the symposium participants were unable to contribute papers to the book, we took advantage of this opportunity to further increase the diversity represented in the volume. We think this diversity of applications is a particular strength of *Rediscovering Darwin*. An interesting outcome of this is that some participants-sharply critical of other 'Darwinian archaeologies'-have had the opportunity to take a broader perspective and perhaps recognize the potential of applying a Darwinian conceptual framework to an exceptionally diverse array of archaeological issues and problems. While disagreements are still to be found in the pages of *Rediscovering Darwin* (and are to be expected in our view), they are considerably more constructive than they might have been (this without intervention or editorial license on the part of Geof or myself). In this somewhat ecumenical spirit, I want to thank Mike Schiffer for writing the Foreword to *Rediscovering Darwin*. In a recent article (1996), Mike has called for a rapprochement between selectionist and behavioralist approaches to the archaeological record, pointing out that both are doing much the same thing within a broadly defined Darwinian theoretical structure. Similar sentiments are echoed in many of the papers published here.

Geof and I want to thank the contributors to *Rediscovering Darwin*. All have put considerable thought and effort into their chapters. They have also been wonderfully cooperative at meeting the often tight schedules for drafts, revisions, and proofs. This has made our jobs as editors much less stressful than they would otherwise have been. We also want to thank Marilynn Bubb for typesetting *Rediscovering Darwin*. Her knowledge and expertise, and her editing skills are clearly evident, and she did it all in addition to a more than full-time job helping to run the ASU anthropology department.

Some of the contributors to *Rediscovering Darwin* did not appear to think of themselves as employing a neo-Darwinian conceptual framework-much less as Darwinian archaeologists-prior to being asked to participate in the book. Geof and I think this is typical of the anglo-american research traditions. Recent programmatic works have called on archaeology to try Darwinian theory. In our view, however, archaeologists already are employing Darwinian theory and have been doing so increasingly for the past two decades. However, they often use it, or elements of it, implicitly, in a hit or miss fashion, and hence not very rigorously. To employ it more effectively, archaeologists need to be explicit about their use of Darwinian theory and associated processes. This means using the appropriate Darwinian terminology and concepts to describe evolutionary processes rather than the many circumlocutions archaeologists have devised to avoid being branded an 'evolutionist'. This means understanding how Darwinian processes actually operate so as not to misapply them. O'Brien and Holland's (1992) discussion of *post hoc* adaptationist argument is particularly apropros. More important, archaeologists need to take greater account of epistemology-to consider the theoretical implications of their research and the implications of theory for their research. If articles in mainstream journals and the recent appearance of several other compilations of papers on Darwinian theory in archaeology are any indication (e.g., Maschner 1996, O'Brien 1996, Teltser 1995), archaeologists are doing this to an increasing extent. Geof and I hope this trend continues.

This is not to claim, of course, that all archaeologists are operating to a greater or lesser extent within a Darwinian theoretical framework. But many are. Furthermore, if archaeology is to continue to develop as a science (and Geof and I emphatically believe that it should), it badly needs well-grounded theory. After over a century and a half, Darwinian theory remains the most powerful explanatory framework ever developed in western science to account for variation and change over time. As Daniel Dennett pointed out recently. archaeologists are scarcely alone in repeatedly attempting to devise alternative explanations for change to replace "Darwin's dangerous idea" (1995). None have yet succeeded in doing so. Rather than continuing to do this, archaeologists would do better to build on the solid foundation of Darwinian theory and seek to use it more effectively to better account for behavioral change.

Michael Barton
October 1997

REFERENCES CITED

Brumfiel, Elizabeth M.
1992 Distinguished lecture in archaeology: breaking and entering the ecosystem-gender, class, and faction steal the show. *American Anthropologist* 94: 551-567.

Clark, G.A. (editor)
1991 *Perspectives on the Past: Theoretical Bias in Circum-Mediterranean Hunter-Gatherer Research*. Philadelphia: University of Pensylvania Press.

Clark, G.A. and C.M. Willermet (editors)
1997 *Conceptual Issues in Modern Human Origins Research*. New York: Aldine de Gruyter.

Dennett, Daniel C.
1995 *Darwin's Dangerous Idea: Evolution and the Meanings of Life*. New York: Simon & Schuster.

Dunnell, Robert C.
1971 *Systematics in Prehistory*. New York: The Free Press.

Maschner, Herbert D.G. (editor)
1996 *Darwinian Archaeologies*. New York: Plenum Press.

O'Brien, Michael J. (editor)
1996 *Evolutionary Archaeology: Theory and Application*. Salt Lake City: University of Utah Press.

O'Brien, Michael J. & Thomas D. Holland
1992 The role of adaptation in archaeological explanation. *American Antiquity* 57: 36-59.

Schiffer, Michael B.
1996 Some relationships between behavioral and evolutionary archaeologies. *American Antiquity* 61: 643-662.

Teltser, Patrice A. (editor)
1995 *Evolutionary Archaeology: Methodological Issues*. Tucson: University of Arizona Press.

Foreword

As the late Fred Plog was fond of remarking throughout his brief but intense career, the archaeological record is above all else a record of change. If that is so, then archaeologists should assign a high priority to constructing and borrowing theories appropriate for explaining behavioral change. Despite his own considerable efforts (e.g., *The Study of Prehistoric Change*, 1974), Plog did not succeed in furnishing the theories needed to account *convincingly* for the sequences of behavioral change inferred from the archaeological record. Nor, for that matter, has anyone else.

And it is not for lack of effort. Indeed, since the tumultuous sixties, which gave us both Woodstock and processual archaeology, archaeologists have tried out countless theories, usually borrowed from other disciplines. Most of the flirtations with theories exogenous to archaeology have had few long-term effects. Does anyone, for example, remember "systems" theory?

Although the vast majority of imported theories have left few traces on contemporary archaeological practice (not *praxis*), there are some conspicuous exceptions. In particular, since the late 19th century, biology has furnished formulations-not always formal theories-that have inspired archaeologists to view change as the outcome of evolutionary processes. From the so-called unilinear evolutionists, through the Boasians and neo-evolutionists, to the selectionists of recent decades, evolutionary thinking in one guise or another has influenced archaeological explanations of behavioral change.

At the present time, Anglo-American archaeology is undergoing another infusion of evolutionary ideas, acquired both from biologists directly and from other social and behavioral scientists influenced by biologists. The most vocal proponents of evolutionary renewal have been the selectionists, led by Robert C. Dunnell, whose program seems, to some investigators, unduly narrow if not altogether misguided. Other persistent champions of evolutionary renewal are to be found among those, like James F. O'Connell, who advocate the thoroughgoing application of behavioral ecology to human behavior. In recent years, less easily characterized evolutionary variants have cropped up in the archaeological literature. It appears that what began as modest renewal is rapidly becoming a veritable-and highly visible-renaissance of evolutionary thinking.

The present volume, which I have the privilege of introducing, brings together a number of thoughtful essays and case studies that exhibit the range of evolutionary ideas that contemporary archaeologists are endeavoring to apply to behavioral change inferred from the archaeological record. Clearly, the discipline owes a considerable debt to Mike Barton and Geoff Clark for assembling this fine collection of papers and making it widely available to the profession.

The overwhelming impression that one gets from these well crafted contributions is that evolutionary archaeology is enjoying a spurt of variability unfettered by selectionist-or any other-theoretical orthodoxy. Indeed, one could properly speak of evolutionary archaeolog*ies* rather than a unitary evolutionary archaeology. This is as it should be. After all, biological evolution exhibits enormous theoretical variability itself (it is, for example, still trying to digest punctuated equilibrium, complexity, and chaos theories), and so it would be unwise for archaeologists to claim that only one variant of evolutionary theory-or one interpretation of a specific evolutionary theory-can be viably applied to human behavior.

Thus, the essays in this and other recent volumes allow archaeologists to assess the fit between various evolutionary theories and archaeological phenomena. The expectation is that, in an evolutionary process described by David L. Hull in *Science as a Process* (1988), the best theories-those with the most convincing applications-will enjoy higher rates of replication. If this volume simply encourages more archaeologists to assess the merits of evolutionary theories for explaining long-term behavioral change, then it will have succeeded brilliantly.

Today is a particularly opportune time for scientific archaeologists to renew and reconfigure their commitment to evolutionary theory. Across the sciences-biological and physical, social and behavioral-countless investigators are exploring the implications of evolutionary theories for their own problems, projects, and disciplines. This period of remarkable intellectual vitality perhaps presages a new evolutionary synthesis that can comfortably encompass changes in human behavior and all other natural phenomena. Whether such a Herculean synthesis will require a latter-day Darwin remains to be seen. What is already clear, however, is that the formulations and case studies of archaeologists must come to play a pivotal role in fashioning the new synthesis, for only archaeologists have access to the record of behavioral change that evolutionary theories purport to explain. As a result of books like this, when symposia are put together to celebrate the sesquicentennial of *The Origin* in 2009, I expect to see many archaeologists taking their place alongside scientists from other disciplines, contributing significantly to the shape and content of the new evolutionary synthesis.

Michael Brian Schiffer
University of Arizona

In Memoriam

This volume is dedicated to the memory of David Rindos (1947-1997), one of the pioneers whose work led to the rediscovery of Darwin by the discipline of archaeology.

Part I
Introduction

1

Evolutionary Theory in Archaeological Explanation

C. Michael Barton & G.A.Clark
Arizona State University

ABSTRACT

In spite of the the fact that archaeology embodies the study of change, it is only within the last few years that neo-Darwinian theory has begun to have an impact on archaeological explanation. This lack of acceptance of neo-Darwinian theory in archaeology is in part the result of a long history of archaeologists misunderstanding the processes and expectations of evolutionary theory. This volume attempts to illustrate the applicability of neo-Darwinian theory in archaeology by compiling studies that use this theoretical basis to resolve archaeological problems at varying degrees of temporal depth, at varying scales of social complexity, and employing varying metholodologies. Although a broad range of topics are covered in this volume, a number of issues remain to be addressed, including: links between evolutionary units and archaeological explanation; processes involved with the origin of behavioral variability; processes involved with the transmission of behavior; evaluating behavioral fitness; and the role of 'non-Darwinian' processes in behavioral change.

If I were to give an award for the single best idea anyone has ever had, I'd give it to Darwin...In a single stroke, the idea of evolution by natural selection unifies the realm of life, meaning, and purpose with the realm of space and time, cause and effect, mechanism and physical law...[N]ot only does Darwin's dangerous idea apply to us directly and at many levels, but the proper application of Darwinian thinking to human issues...illuminates them in ways that have always eluded the traditional approaches, recasting ancient problems and pointing to their solutions.
(Dennett 1995: 21, 23)

Theoretically (in both the scientific and colloquial meaning of the word) archaeology is a field focused on change. From its beginnings as a scholarly discipline in the early years of the last century, archaeology has played an important role in the shift in western thought from perceiving the world as unchanging–designed by and maintained by the constant hand of the divine watchmaker–to a dynamic place where nature makes itself and the only thing certain is change (Eiseley 1958). In fact, the realization of the pervasiveness of change in human cultural systems led 19th century archaeological theoreticians to postulate that all human societies inevitably progressed through a series of stages; only when something went terribly wrong did a society become stalled and fail to change. Change was taken as a given. In the climate of the industrial revolution, the root cause of change was seen as the innate human drive to improve–obvious in the growing cities, faster ships and trains, more productive factories, and the Horatio Algers of the world (see also Chapter 2).

Biologists did not have such easy explanations for an equally dynamic natural world. Although Lamarck and others proposed innate drives to explain biological change, they met with less widespread acceptance. (An innate drive for increasing neck length wasn't too difficult to attribute to a 'proto-giraffe', but imagining an analogous drive for blue-green algae stretched the imagination too much for many.) Fortunately, Charles Darwin and Alfred Russel Wallace proposed an alternative explanation of why and how life changes. Darwinian evolutionary theory transformed natural philosophy into scientific biology.

Progressivist social theory was less successful in this regard. By the early 20th century, it was apparent that it was inadequate to explain much of the recently discovered human behavioral diversity, and was abandoned by anthropologists and archaeologists. This left a theoretical vacuum in archaeology that has been filled only slowly and recently (see Chapter 2). Only with the calls of the 'New Archaeology' of the 1960s for a more scientific archaeology, with 'covering laws', did the discipline again seriously consider accounting for change rather than simply documenting it. Although most of the efforts at discovering general laws met with varying degrees of failure, some began to reconsider the ability of Darwinian evolutionary theory to explain change and the language of evolutionary biology began to creep into the archaeological vocabulary.

In most cases only selected aspects (the concept of adaptation, for example) were applied to human societies, often in loose or even inappropriate ways. In the 1970s a few individuals, Robert Dunnell and David Rindos among the most notable, began to call for more inclusive and explicit use of Darwinian evolutionary theory in the explanation of behavioral change (O'Brien 1996). Although initially strongly criticized by many as inappropriate for explaining 'culture' (e.g., see Rindos 1980 and commentary), the explanatory power of Darwinian theory has proven increasingly attractive to archaeologists attempting to take a scientific approach to past human societies and social change, and small but growing number of archaeologists began to echo this call for expanding Darwinian theory to human social systems (e.g., O'Brien & Holland 1990, 1992; Leonard & Jones 1987; see also Teltser 1995a).

By the early 1990s, it was apparent to us that neo-Darwinian evolutionary processes were increasingly employed to account for change in anglophone archaeology. Nevertheless, the history of evolutionary thought in archaeology–including the excesses of 19th century social Darwinists, the misapplication of Spencerian evolutionary theory, the failure of unilineal progressivist schemes, and the problems with recent sociobiological approaches–led many in the discipline to be wary of explicit ties to 'evolution' of any kind. This has retarded overt discussions of how Darwinian theory can be extended from biology to behavior and resulted in many cases in which evolutionary concepts have been misconstrued and misapplied (see O'Brien & Holland 1992, Rindos 1989).

In 1994 we organized a symposium at the Society for American Archaeology meetings to recognize the strong current of Darwinian theory that seemed to underlie a significant–possibly a major–portion of the archaeology actually being done in the anglophone tradition. To do this, we asked a diverse set of practitioners to explicitly address the role of Darwinian evolutionary theory in their research. The strongly positive response on the part of the participants and the audience supported our assessment of the growing importance of evolutionary theory in late-20th century archaeology. In the wake of the symposium, we felt that a compilation of papers highlighting the applicability of neo-Darwinian evolutionary theory in an even wider range of archaeological research programs would be of interest and value to the discipline. This volume is the result. It does not seek to be a manual for the employment of evolutionary theory in archaeology. What we have attempted to do is to portray a wide cross-section of the discipline to illustrate the pervasiveness, versatility, and utility of evolutionary theory in archaeological explanation. We do this with the hope of stimulating anew explicit discussion of general theory of social change and promoting the applicability of neo-Darwinian processes in the scientific study of human behavior and society.

THE APPLICATION OF EVOLUTIONARY THEORY TO CULTURAL SYSTEMS

To introduce this volume, some general remarks about the applicability of neo-Darwinian evolutionary theory (shortened to simply evolutionary theory in the following discussion) to cultural systems are in order. What is evolutionary theory? In the broadest sense, it is a theory to explain change. Evolution simply refers to change over time. No one questions that human behavioral systems (that is, 'cultures') change. But how do we explain this change, as opposed to simply recording its occurrence?

Evolutionary theory postulates that culture change results from the differential persistence, through time, of behavioral variation, and that this persistence can be explained through the application of a limited number of universal processes. How these processes operate and interact with each other in any particular case is complex and, at one level, always unique because their result is conditioned by environmental and historical circumstances. Nevertheless, their universality means that generalizations can be made for similar sets of circumstances, allowing us to explain patterns of variation in cultural systems. While few would question the application of evolutionary theory to biological systems, including humans as a biological species, why should it also apply to human behavioral systems? We can propose at least two logical reasons. The first has to do with the relationship between biology and behavior.

As anthropologists, we accept the importance of the relationship between biology and behavior for humans, but we often fail to fully appreciate its implications. It is well understood that the phenotype results from the interaction of the genotype and the environment, and that evolutionary processes (except for mutation) do not work directly on the genotype, but rather on the phenotype. However, these forces only affect the phenotype via the interaction between the phenotype and the environment. This phenotype/environment interface is behavior in its broadest sense. Seen in this light, behavior is the place where evolutionary forces have an effect. These forces affect the phenotype according to the behavior it exhibits. Furthermore, they mediate the differential transmission of behavior to others—regardless of how it is accomplished.

Another way to look at the applicability of evolutionary theory is from a systems perspective. There are fundamental similarities between biological and behavioral systems that cause evolutionary processes to affect these systems in similar ways. Both can be characterized as information-transmitting, self-replicating, negentropic, open systems. Information about both biological and cultural systems is transmitted from component to component within each system. This includes information about the operation and structure (i.e., relationships among components) of the system. Information transmission is not perfect, however, introducing variation. In this regard, imperfect transmission is as vital for evolution as information transmission itself.

Both biological and cultural systems maintain temporal continuity through the replication of their components and structure. Information about the system is transmitted from component to component during self-replication, making the current state of a system strongly conditioned by its previous state.

The negentropic and open characters of biological and cultural systems are closely related. These systems transform energy into more complex, structured forms in a variety of ways. That is, they operate to counter entropy. Results of this are the highly structured nature of these systems and their dynamic temporal continuity through self-replication.

Another outcome is that both biological and cultural systems tend toward expansion. To accomplish this, these systems must continuously take in energy. In the real world, however, the availability of energy is limited, and varies in both time and space. This encourages competition for available energy needed to maintain these systems, and the differential persistence of variants over time according to their abilities to successfully capture energy sufficient for continued information transfer and replication. The end result of the interaction of these various characteristics is vectored (that is, historical) change in these systems.

Regardless of how compelling such logical reasoning may or may not be, the archaeological community will not be convinced of the applicability of evolutionary theory to cultural systems on the strength of logical arguments alone. Nor will it respond to continual criticism of its inability to understand or utilize evolutionary theory. In the end, evolutionary theory must be better theory to be accepted broadly across the field. It must satisfy a performance standard. That is, it must better explain the archaeological record by accounting for more phenomena and by doing so more parsimoniously.

MISCONCEPTIONS ABOUT EVOLUTIONARY THEORY

As noted above, a review of current archaeological literature shows that many, perhaps most, anglophone archaeologists regularly utilize some version or components of evolutionary theory in explanations of culture change. Examples include concepts of adaptation and diffusion, and seriations that depend on the stochastic effects of drift. There are a number of misconceptions about evolutionary theory and its role in archaeological explanation, however, that prevent its wider application.

Back to Nature

The first of these perceives evolutionary theory as dealing solely with the interaction between humans and the natural environment. The primary problem here is a misconstrual of the concept of environment in evolutionary theory; it is not limited to the natural world, but refers to the complete—and often complex—context in which behaviors take place. The social environment is important in biology (in sexual selection, for example); it is at least as important as the 'natural' environment for humans, if not more so. A related 'back to nature' misconception is that evolutionary theory only applies to economic behavior. In truth, evolutionary processes are as applicable to ideational aspects of culture as they are to the more material ones. Recent examples include Neiman's work with style and social interaction (1995), Barton, Clark, and Cohen's (1994; Clark, Barton & Cohen 1996) model for European paleolithic art, Graves and Ladefoged's (1995) study of Polynesian ceremonial architecture, and a number of the contributions to this volume. Worse is the misconception that evolutionary theory is uncivilized. That is, it only applies to hunter/gatherers. As many of this volume's contributors show, general theory to explain culture change is as useful and necessary for complex as for simple societies.

Spencerisms and Social Darwinism

If 'back to nature' misconceptions relate to the application of evolutionary theory to archaeological questions, 'Spencerisms' involve fundamental misunderstandings of evolutionary theory in general—especially the concepts of selection and vectored change. Evolutionary theory is not a revived 'social Darwinism'. The specialization implied by social Darwinist 'survival of the fittest' scenarios may have appeared successful in the short-term. However, a better understanding of evolution shows that the maintenance of diversity provides a greater potential to deal with inevitable environmental change, and seems a better strategy for long-term success. This seems to be characteristic of humans as a species.

Thomas Malthus long ago pointed out the relationships between mortality-producing agents such as disease, starvation, and war on human population. Accepting these relationships does not mean, as 19th-century social Darwinists proposed, that we should condone them as inevitable. On the other hand, saying that we do not condone such agents does not make them untrue.

Related to the above is the misconception that evolution equals selection. Certainly selection is important. But non-Darwinian processes, such as drift, are clearly as important in cultural systems as they are in biological systems. Moreover, they probably played an especially important role in the small, semi-isolated social groups that comprised such a large part of the human past.

Finally, evolutionary theory is not progressivist. As repeatedly pointed out (e.g., Rindos 1989) evolution is non-directional in the sense that there is no progression toward any cultural stages or levels of social complexity. To the uncritical appraisal, evolutionary change may *appear* to be so because it is vectored through time (though in no predetermined direction) and changes are cumulative, building on the past. This, of course, does not preclude patterning in human systems or similarities in adaptive postures among societies in similar contexts, nor does it require such similarities.

Exceptions to the Rule

'Exceptions to the rule' are often in the form of cautionary tales. These are indeed useful in avoiding overly facile application of general theory. Gould and Lewontin's (1978) classic argument against overly-simplistic application of the concept of adaption (see Chapters 5 and 12), and O'Brien and Holland's (1992) revisiting of this issue in archaeology are good examples. Such 'exceptions' also have been portrayed, however, as invalidating general theory. While caution is useful, these again often result from misconceptions about expanatory theory in general and evolutionary theory in particular. We mention two examples here.

The first is the existence of unique culture histories that supposedly invalidates the applicability of general explanatory theory (see papers in Nitecki & Nitecki 1992, Clark 1993). All culture histories—like the histories of each biological species—are indeed unique. In fact, from a neo-Darwinian perspective, undirected change and the complex interaction of environmental and historical circumstances virtually assures uniqueness in cultural trajectories (Rindos 1989). However, the uniqueness of histories does not preclude the generality of the processes that govern them. One should not confuse the two.

The second exception we mention is the existence of apparently maladaptive behaviors. First, we must be aware of the possibility of ethnocentrism, and recognize the importance of context in evaluating fitness. One society's maladaptation could be another's success. Furthermore, many behaviors are, for all practical purposes, selectively neutral. The significance of non-Darwinian processes such as drift has been mentioned above. The occurrence of selective neutrality in behavior, and its effect on the archaeological record is clearly important. Without it, the seriations upon which so many archaeologists depend would not work very well (Chapter 8, Teltser 1995b). Furthermore, individual behaviors do not occur in isolation, but comprise a tightly and complexly integrated set that we call culture. Apparently maladaptive behaviors may be 'packaged' along with those that are strongly favored by selection. This concept, also called 'sorting' (Ramenofsky 1995, see also Chapter 4) is also recognized for biological structures.

With respect to apparently maladaptive behaviors, we should also keep in mind what we call the 'good enough' principal, recently discussed for human behavior by O'Brien and Holland (1992). That is, a behavior does not have to be optimal to be selected for; it only has to be good enough—that is, it only has to be a little better than competing behaviors.

Person to Person

The last group of misconceptions we want to discuss concerns the idea that evolution is dehumanizing. That is, evolutionary theory focuses on mechanistic forces (often seen as 'external' to human society) and ignores the very aspects that make us human—emotions, desires, and especially intent. This is due in a large part to misunderstanding the nature of evolutionary processes and the role of the environment. Evolutionary processes are not some sort of external 'forces' that direct human destiny, but an integral part of the operation of human systems. Evolutionary change is not imposed from without, but the result of human choice. Over time, individuals perform 'behavior A' less and less frequently, and perform alternative 'behavior B' more and more frequently. This is the essence of change seen from an evolutionary perspective. Why B becomes more frequent at the expense of A is the fundamental subject of inquiry from this same perspective. All human behavior and behavioral change takes place in a context. This is the environment. In one context, A may replace B; in another B may replace A. Does the environment determine behavior? No, but the outcome of any behavior, and its effect on whether or not people continue to perform it is often affected in some way by the context in which it takes place.

In evolutionary theory, the role of the individual is not precluded *a priori*. In fact, when change is viewed as the differential persistence of behavioral variation, the individual assumes a much more important role in evolutionary theory than in many other explanations that see change in terms of transformation at the level of 'culture'. As in biological systems, evolutionary processes operate at the level of the individual, even though their effects are manifest at the population level. Granted, the individual is difficult to resolve in prehistory. But this is a problem with the record and our abilities to interpret it, not an indictment of evolutionary theory. As seen in Chapters 4 and 9, evolutionary theory may help to better resolve individuals in prehistory.

Emotions, desires, intent, and other aspects of human consciousness also are not excluded *a priori* from interest in an evolutionary perspective. However, they too are often difficult or impossible to access in the archaeological record. As a scientific discipline, archaeology seeks general, testable explanations for human phenomena. While human feelings often may not serve to adequately account for the archaeological record, this does not negate their existence nor potential importance. Especially contentious is disagreement over the role of intent, although it is not always clear what is meant by intent. People intend to do many things, but this explains little. As O'Brien and Holland point out, "Many early aviators must have leapt from cliffs, propelled by hopeful inventions and the intent of flying. Ultimately, it was the ability to overcome gravity—not intent—that determined which aviators survived to pass their genes and inspiration on to others." (1995a: 180). An evolutionary approach focuses on why some are more successful and some are less so—regardless of their intent.

Along the same lines, evolutionary approaches are somehow felt to ignore human inventiveness and problem solving abilities. This misconstrual is more understandable given discussions in some of the theoretical literature (e.g., O'Brien & Holland 1990, 1995a, Rindos 1989). Nevertheless, the important point made in these and other somewhat less rigid applications of evolutionary theory is the conceptual separation between the origin of variation and its differential persistence through time. Only the latter process effects long-term change. Human problem solving introduces new behavioral variation into a cultural system. This variation is then subject to selection, drift, sorting/packaging, or other evolutionary processes. A new solution may indeed spread because it is somehow 'better' than existing alternatives—i.e., selection—but this is not guaranteed. It may also spread for other reasons (possibly due to selection, but also due to drift, packaging, or other processes), even at the expense of alternative 'better' solutions. Again, humans are constantly attempting to solve problems and human inventiveness regularly introduces new behavioral variants. However, the invention of a new solution does not mean that it will spread (as the majority of inventors filing patents learn). Separating the differential persistence of variants from the creation of variation is one of the conceptual milestones of Darwinian theory and a fundamental reason for its success in explaining change (see Chapter 5 for additional discussion of this point).

Because change is a result of the differential persistence of variation, most evolutionary theorists have focused primarily on the processes responsible for sifting variation and have paid less attention to those responsible for creating it. However, as discussed below, we should probably pay more attention to the origin of variation in human systems.

SCALES OF APPLICABILITY

An important aspect of evolutionary theory is that it operates and can be invoked in explanation at a wide variety of different temporal, socioeconomic, and operational scales. Hence, we should expect that research programs based in an evolutionary framework should similarly approach the archaeological record at different scales. The papers in this volume exemplify such diversity, ranging from studies of the earliest hominids to modern societies, from small-sized egalitarian groups to complex civilizations, and from macroevolutionary approaches to evolutionary ecology to human cognition.

Time Depth

The entire suite of papers show the applicability of evolutionary theory at various time depths and, hence, at varying degrees of archaeological resolution. At one extreme are the chapters by Geoff Clark (Chapter 12) and John Gowlett (Chapter 3) dealing with Pliocene and Pleistocene hominids. Gowlett reviews a wide variety of evolutionary models for dealing with human biocultural evolution through the Pliocene and Pleistocene, differentiating those that employ direct archaeological evidence and those that use comparative, uniformitarian analogy with modern humans and other primates. These approaches are exemplified in Clark's paper. He uses evolutionary theory to structure and support the use of modern human and primate behavior in order to reconstruct aspects of Pliocene hominid sociality that may not be visible in the archaeological record.

At the other temporal extreme are the chapters by Polly Weissner and Douglas Bird (9 and 16, respectively). Wiessner uses her ongoing study of style among the modern !Kung San to explore the evolutionary bases for aspects of cognition. She focuses on social identification via comparison, cultural transmission, and communication through style. Bird establishes foraging theory within a Darwinian framework, using shellfish collecting by the Meriam of Melanesia as a case study. He then examines the relationships between foraging behavior and archaeological residues.

Socioeconomic Complexity

Evolutionary theory also scales across the range of social complexity and economic systems found in human cultural systems. The chapters just mentioned focus on a variety of relatively small-scale forager societies, as do several others. Others address issues at more complex scales of socioeconomic organization.

Michael Shott (Chapter 11) examines the effects of community size and complexity on cultural transmission processes and lithic morphology in the context of a Mississippian settlement in the central United States. Michael Diehl (Chapter 14) uses a behavioral ecological approach and optimality modeling to evaluate models for incipient agriculture among semi-sedentary late Archaic populations of the Tucson Basin, in the southwestern United States. Hector Neff and Daniel Larson (Chapter 5) compare the spatial and temporal distributions for ceramic design and composition to assess the development of region-level craft and subsistence specialization in coastal Guatemala and the Colorado Plateau of southwestern North America. And Fraser Neiman (Chapter 15) develops an evolutionary basis for the linkage between political power and wasteful consumption. He applies this to Maya stelae construction and terminal monument dates, and uses statistical modeling to clarify the timing and causes of the Maya collapse.

Methodological Diversity

In archaeology, as in biology, there are significant theoretical and methodological differences between those whose interest lies in the operation of formal Darwinian processes in a historical context and those who focus more on the dynamics of relationships among individuals, societies, and the environment writ large. In general the former macroevolutionary approaches share several general objectives. They attempt to identify the effects of macroevolutionary processes like selection or drift on the spatial and temporal distribution of behaviors and their material residues. They also often focus on ways to recognize and distinguish the process(es) responsible for these effects. Such work helps to clarify the meaning of different dimensions of variability in the archaeological record, and to provide general explanations for broad patterns. It also helps to define and formulate questions about human behavior and map out ways to address them. Lee Lyman and Michael O'Brien (Chapter 2) take this approach to outline the history of evolutionary thought in American archaeology over the past century. A formal Darwinian perspective also characterizes the above mentioned chapters by Neff and Larson and by Neiman

Still, as the above authors would agree, to simply state that a phenomenon is the result of selection, for example, isn't a very complete or useful explanation in that we don't learn very much about human behavior (see, e.g., Jones, Abbott & Leonard 1995; O'Brien & Holland 1995). Selection, or other macroevolutionary processes, is not a giant hand from on high sorting behaviors to preserve and to discard. We need to understand how and why macroevolutionary processes operate, both in general and in particular cases. How do behaviors get selected? Under what conditions does drift operate? This is an important role of evolutionary ecology and related studies. Douglas Bamforth and Peter Bleed (Chapter 7) discuss the potential value of this ecological perspective on evolutionary change. They propose a more explicit definition for the concept of risk, within a more general evolutionary framework, noting that risk involves assessments of both the possibility and cost of failure. They use several test cases to examine human technological behavior in the context of variation in the nature of risk along these dimensions.

Esmée Webb and David Rindos (Chapter 13) also employ the perspective of evolutionary ecology to assess the ways in which landscapes are initially colonized by humans. They model differences in realized carrying capacity resulting from the degree to which colonizers are preadapted to the environments of uninhabited territory (due to its similarity or difference from their homeland) and its effects on the mode and tempo of colonization.

They test this model for Pleistocene Sahul and North America. The chapters by Bird and Diehl, mentioned above, are also strongly ecological in perspective.

On the other hand, these studies still need to operate within a general evolutionary framework for the development and evaluation of ecological models, and to make them relevant to the long-term patterns of change and stability seen in the archaeological record. It is clear that while all foraging may be optimal in some sense, some foraging is more optimal than others (with apologies to George Orwell). That is, optimality is not a universal law, like those of thermodynamics; it is a behavioral tendency that is widely observed in the archaeological (and biological) record. Both the widespread appearance of this tendency, and those circumstances under which it does not appear to hold, still call for explanation. To do this, we must seek recourse in more formal and mechanistic evolutionary processes, points cogently made in the chapters by Neiman and Mithen.

Mithen (Chapter 4) makes the case for long-term selective pressure affecting the evolution of human cognition, and underlying many of the behavioral regularities observed by evolutionary ecologists. He also points out that some important domains of human behavioral systems (e.g., religion) may be evolutionary 'spandrels' (cf. Gould & Lewontin 1978). That is, they are simply neutral consequences of selection acting on other characteristics and hence might not be explainable by formal Darwinian selection or by ecological models of optimality. This is discussed further below in the context of 'sorting'.

Biologists have been arguing the relative merits of these positions for decades (see summaries in Chapters 5 and 7) and, in recent years, archaeologists have entered the fray. However, the merits of each approach are often closely tied to the types of questions asked (proximate versus ultimate causality), the time frame of interest (synchronic versus diachronic), and the geography of applicability (local versus regional). Both approaches are very useful and should better be considered as conceptual tools to apply evolutionary theory at different scales appropriate to the focus of research in particular problem domains.

In addition to approaches that directly parallel those of evolutionary biology, several others are represented in the papers collected here. Drawing from evolutionary psychology, ethology, and to some extent from sociobiology, the previously mentioned chapters by Steven Mithen and Polly Wiessner (4 and 9) explore the evolutionary basis of human cognition and its effects on behavior manifest the archaeological record. Both studies employ aspects of macroevolutionary and ecological models for developing explanations of human behavioral systems.

In something of an intermediate position are chapters that deal with the relationships between human behavioral systems and the material residues that comprise the archaeological record–an area that has often been considered under of the rubric of 'middle range theory'. While most of the papers deal with this interface to varying extents, several do so in more detail. Robert Bettinger and Jelmer Eerkens (Chapter 10) and Michael Shott (Chapter 11) closely examine the effects of different processes of cultural transmission on the morphology and variability of Great Basin projectile points. Michael Barton (Chapter 8) employs Darwinian and non-Darwinian processes to construct inference chains between stone tool morphology, technological behavior, and social organization–especially social group membership. At a more general level, Peter Bleed (Chapter 6) develops theory, with evolutionary underpinnings, for human technology. Of importance to archaeologists, his goal too is to forge explicit and theoretically sound links between technological 'content' (including standards, information, and behavior) and 'results' (including material culture, environmental modification, and social organization).

While the essays collected here intentionally represent an extremely wide array of interests, covering much of the territory represented by the discipline as a whole, all the contributors share a common interest in applying neo-Darwinian evolutionary theory to human behavioral systems. Nevertheless, they differ (sometimes strongly) as to how this should best be accomplished. We think that this diversity is a strength of the volume, and a positive and encouraging sign. Only through overt discussion of theory–including marked disagreement at times–accompanied by the explicit application of theory to archaeological problems and rigorous testing of theory against pattern manifest in archaeological records can we begin to build a better, more scientific archaeology.

ISSUES TO BE ADDRESSED

Rediscovering Darwin attempts to portray the current state of the discipline with regards to the application of Darwinian evolutionary theory to archaeological problems. In so doing, it brings together a diversity of approaches and solutions. Most evolutionary theorists in archaeology have called for an expansion of current Darwinian theory to encompass behavioral systems as well as biological systems. As shown in this volume, this has been initiated in a number of dimensions. We feel that the discipline has matured considerably in the development of general theory of behavioral change and its use in archaeological explanation. It is also clear from this book and from other current literature, however, that several important areas need to be more intensively investigated. We briefly address several of these issues with the hope of stimulating further thought and discussion.

Explanation and Evolutionary Units

There is a need for serious discussion about what is being explained in archaeology. Is it the archaeological record *per se* or the behavior that produced it? Recent discussions by Robert Dunnell (1995) and Michael Schiffer (1996) are pertinent in this regard, as is the Neff and Larson chapter. It seems that archaeology is more satisfying to most practitioners and more valuable to non-archaeologists if it includes the latter, as well as the former. However, claiming that we are explaining behavior alone is overly simplistic. Material culture undeniably comprises an important component of the human phenotype. It predates the appearance of the genus *Homo* and is as fundamental to being human (and surviving as humans) as our bipedalism. Material culture also clearly varies across space and time, and explaining this phenotypic variation is essential to understanding the human past. Furthermore, regardless of our interest in past behaviors, the material record is all that remains. Fundamentally, *all* archaeological explanation must explain the archaeological record.

Nevertheless, we cannot lose sight of the fact that the archaeological record is but an irregularly preserved residue of a much richer set of human behaviors. The items that remain for archaeological study participated in complex, integrated human behavioral systems that are no longer directly visible. Failure to appreciate this complexity can lead to overly atomistic explanations and naïve adaptationist 'just so' stories (see Clark, Chapter 12).

Evolutionary processes often did not operate directly on the material culture that forms the present archaeological record, but on the behaviors that originally produced it and led to its incorporation in the record. In many respects, the archaeological record is the result of evolutionary processes acting on human behavior. In Chapter 5, Neff and Larson cogently argue that the inference chains linking the material record with past human behaviors are most reliable if they are short and explicit. While explicit inferences are always desirable, however, short chains are not always possible. The fact that material residues are all that remains of past human behaviors does not mean that our explanations cannot go beyond these residues, as is shown in Clark's chapter. The same is true for most other branches of science-especially the life sciences where explanations for evolutionary change must often come to grips with very incomplete remains of organisms that were unlike any modern creatures and have no living descendants (e.g., Erwin, Valentine & Jablonski 1997).

Origin of Behavioral Variability

As noted, a fundamentally important aspect of neo-Darwinian theory is the distinction between the origin of variation and the differential persistence of variation. The latter process, because it directly causes behavioral change, has generally been the focus of most theoretical discussions in archaeology. Closely paralleling evolutionary thought in biology, the origin of behavioral variation has generally been treated as potentially interesting from a humanist viewpoint, but inconsequential with regard to accounting for change. The general disregard of the origin of variation is understandable from the biologist's point of view. Because organisms have no control over the nature or frequency of mutations, their occurrence often can reasonably be treated as essentially a random process, although even at a molecular level this is not entirely accurate.

Nevertheless, the processes responsible for change, like selection and drift, are themselves controlled in part by the amount of variability available, the nature of that variability (e.g., its relative neutrality or fitness coefficient), and the rate at which new variability arises. Furthermore, it is not clear that the origin of *behavioral* variability can be treated as an effectively random process with the same impunity as can genetic mutation.

Dunnell (1978) has argued that the rate at which behavioral variants arise might be linked to absolute population size, and that the rate of new variants per individual may be relatively stable. This may be an overly simplistic view, however. For example, population density–in addition to absolute numbers–may also play a compounding role. More people in a social group means more opportunity for the exchange of ideas and their recombination in novel ways. It also means that each individual has a better chance to learn about–and avoid–'unsuccessful' behaviors. Increased population density might act to amplify variability, as sexual reproduction does for biological systems, increasing the rate at which variation arises and spreads and, consequently, the rate of evolutionary change. This *appears* to be the case with the rise of complex society, for example, and some form of empirical testing seems warranted.

Other factors may also affect the rate at which new variation appears. Situations of extreme stress, for example, may increase the rate of behavioral variant creation. An admittedly vague but useful concept, stress can be generalized to refer to the severely reduced effectiveness or failure of a set of previously adaptive behaviors. This is generally due to a change in contextual parameters (i.e., the physical or social environment) such that behavioral outcomes exceed acceptable tolerance limits. More obvious examples include social systems that become unable to adequately resolve interpersonal problems and maintain social cohesion, or technological systems that can no longer provide sufficient subsistence resources. Such stress may increase the rate at which novel behaviors are generated. An increase in the amount of variation, in turn, increases the chance that a successful alternative will be produced upon which selection might act. Such a response to stress may itself be the product of long-term selection—a successful adaptation to environmental change that has allowed humans to populate diverse environments across the globe.

In addition to changes in the rate with which variety is generated, it may not be appropriate to treat the actual variants produced as random in all cases. One effect of language is that human knowledge is potentially cumulative. Of course, all evolutionary change is cumulative in some sense, but this effect may play out differently in human behavioral systems than it does in biological systems. A single individual has access, through oral or written tradition, to the experience of many others—including those to whom he or she is unrelated in a genetic sense. This means that many unsuccessful behaviors can be avoided and successful ones can be elaborated when new behaviors are created. This is not to say that all behaviors that appear successful or unsuccessful are ultimately so. 'Success' is conditional—dependent upon temporal and spatial scale. However, language may well increase the percentage of potentially successful new variants for selection to work on and reduce potentially deleterious effects of variation reduction due to drift. In this way, the type and amount of information accessible to individuals may non-randomly affect the rate and direction of evolutionary change.

Individuals may also use such cumulative knowledge to make reasonably accurate estimates of the probability of successful transmission prior to introducing a new variant, or they may manipulate those probabilities in favor of the variant they are introducing. In effect, this can give a new behavioral variant a better chance of being favored by selection than would have otherwise been the case (see Dennett 1995: 370-383). In this sense, the origin of new variation might not be completely undirected, as Rindos (1989) has argued.

In an historical example of technological change, Thomas Edison correctly surmised that an incandescent light bulb might successfully replace extant gas and arc lighting. He announced to the press that his research lab was going to develop one. His team not only developed a functional bulb, but also an electrical system infrastructure (i.e., wiring, switching, and generating equipment) that would make it practical and commercially viable. Together with the advanced publicity and the prestige of his laboratory, these combined to encourage the rapid adoption of incandescent lighting in general, and Edison's version of it in particular. As much an entrepreneur as a creative genius, he 'stacked the deck' to give his new behavioral variant the best competitive position possible. Similar examples have been proposed for the rise of chiefdoms, in specific, and social complexity, in general. Intent doesn't always equate with success, however, as Edison later learned when he tried to push direct over alternating current. However, such deck stacking may have two somewhat different effects. In some cases it can result in selection favoring a variant that is less beneficial over the long term than others. Given the temporally vectored (albeit non-progressive) nature of evolutionary change, this kind of selection can profoundly affect the direction of change. Over the long-term, it may also increase the probability of the introduction and rapid adoption of potentially successful (or more successful) variants, giving individuals and groups who engage in such behaviors a competitive edge over those who don't.

This is not to claim that behavioral change is wholly Lamarckian, as some have done. That is, the *origin* of behavioral variation and the processes that control its *transmission* and *persistence* are indeed distinct, and change is still the result of differential persistence of variation, not directed variation. However, those processes responsible for the origin of behavioral variation should be treated within an evolutionary framework. They can have significant effects on the operation of selection and drift and can, in turn, be affected by such evolutionary processes.

The Transmission of Behavior

A little-explored issue central to applications of evolutionary process to human behavioral systems, is the transmission of behavior. It is clearly not genetic (or at least not *directly* genetic) for the majority of human behaviors (but see Chapters 4, 9 and 12), and this has numerous implications, some of which are discussed below. The fact that genetic information is almost exclusively transmitted via mitosis and meiosis (closely related in their operation) may be the primary reason for the restricted number of evolutionary processes identified for biological systems. Non-genetic transmission of behavior is much more varied and flexible, and may involve more processes and more variability in the operation of those processes than now recognized. Cavalli-Sforza and Feldman (1981), and Boyd and Richerson (1985) have opened the systematic treatment of this subject, but considerable work remains to clarify both the mechanisms of behavioral transmission in human societies and their implications for the application of evolutionary processes. Most evolutionary theorists refer to these studies, often questioning the need for the distinct sets of biological and cultural evolutionary processes, proposed in both works, for dealing with human behavior (e.g., Dunnell 1989, Leonard & Jones 1987, O'Brien & Holland 1990, 1992). However, these workers have done little to either apply them or to suggest alternatives. Notable exceptions include work by Neiman (1995) and the chapters by Bettinger and Eerkens and by Shott in this volume which test Boyd and Richerson's models for cultural transmission against different data sets.

Evaluating Behavioral Fitness

The concept of fitness, as applied to human behavioral systems, remains difficult to define or apply. Fitness—a measure of potential fertility and realized survivorship—is a cornerstone of evolutionary theory. In the operation of evolutionary processes, it is essential to be able to evaluate the relative benefits of alternative expressions of a trait. In biology, this is accomplished by application of the reproductive standard. Defined in terms of differential reproductive success, fitness is a universal, independent (in theory, at least), and empirical measure of the differential survival (both potential and realized) of variability that results from the operation of fundamental evolutionary processes such as selection and drift. Even biologists who employ concepts like group fitness, inclusive fitness, indirect benefits, and kin selection in sophisticated treatments of animal social behavior rely ultimately on the reproductive standard for measuring fitness (e.g., Emlen et al. 1995, Ruse 1987). The same reproductive yardstick is applied, regardless of the trait in question or the evolutionary process involved. The fact that it is not possible to actually observe reproductive success in many real-world situations (e.g., in the fossil record) does not diminish its utility or value to evolutionary theory.

Applying the reproductive standard for fitness to human behavioral systems is problematic for several reasons, however. Primary among these is that there is no demonstrable, direct, genetic basis for most human behavior. This does not mean that behavior does not affect nor is affected by reproductive success, simply that such behavior does not seem to be generated directly by genetic mechanisms (see Chapter 4). Human language is a well-known case-in-point. It has been known for over a century that there are physical structures in the brain (e.g., Broca's area, Wernicke's area) that are directly related to the human capacity for the development of language and its expression in speech. The information necessary for the development of these physical structures is coded in human DNA, and transmitted from one generation to the next via sexual reproduction. However, it is equally well-known that, if an individual is not exposed to language in a context of social learning prior to early adolescence (ca. 12 or 13 years of age), it is unlikely that he or she will ever be able to develop functional language. Hence, the information that codes for the phenotypically expressed behavior that we call language is passed on *both* in the human genome and via social learning.

Arguably, many human behaviors do not have even the partial genetic basis known to exist for language. This suggests that the reproductive standard, as it is now conceptualized, may be an inappropriate fitness measure for human behavior. On the other hand, behavior undeniably *can* affect reproductive success. Also, in the small groups typical of much of the human past, most social learning took place in the context of the biological family. Hence, reproductive success *can* affect the transmission of behavioral information via social learning. Nevertheless, the linkage is indirect and can be 'horizontal', rather than only via the transmission of genetic material 'vertically', from one generation to the next. In fact, it is easy to propose factors other than reproductive success that would affect the transmission of behavioral information even in small social groups comprised primarily of extended families. For example, individuals who live long lives will have a greater opportunity to pass on behaviors to others via social learning, even if they have few or no offspring, than individuals who have many offspring but die young.

This has left evolutionary archaeologists in a bit of a conundrum with respect to explaining the mechanisms of selection or identifying selective neutrality. One answer has been to tie behavior eventually to reproduction, in a manner not unlike sociobiological approaches to human behavioral systems. For example, if two alternative behaviors are postulated to be differentially 'efficient', the more efficient behavior is thought to increase the time available for other behaviors, including those directly related to reproductive success (e.g., Rindos 1984, 1989; Neiman 1995). Similar arguments have been made by sociobiologists (see Dunnell 1989, Ruse 1987). Others have utilized a more paleontological approach to fitness. Leonard and Jones (1987) propose an alternative behavioral fitness measure they term 'replicative success'. Cavalli-Sforza and Feldman (1981: 15) refer to this same concept as 'cultural fitness' (see also Stoltis, Boyd & Richerson 1995, and commentary). Simply put, the fitness of a given behavior can be measured in terms of how it is replicated over time relative to alternatives, in a way somewhat analogous to realized fitness in biology. Leonard and Jones go on to point out that, while all behaviors will show variable replicative success, some will also affect Darwinian fitness (i.e., reproductive success). They imply that these latter will be subject to selection, while those that do not affect reproductive success might better be explained by stochastic processes such as drift (O'Brien and Holland [1992] explore additional implications of replicative success). Although both approaches make the case that behavior can affect reproductive success, neither deal with the way in which reproductive success can affect the transmission of behavior.

A reasonable solution to this dilemma is to return to the idea that, as archaeologists, we need to expand current evolutionary theory to account for behavioral systems rather than simply import the package from biology (e.g., Dunnell 1989, Jones, Leonard & Abbott 1995). One way to do this for fitness is to build on suggestions made by Neff (1992, 1993, Chapter 5), Dunnell (1989), and Dawkins (1989) and treat the transmission of *information* responsible for behavior as the critical variable, rather than the mechanism by which it is transmitted. This approach requires expanding the concept of genotype to include *all* information needed to produce a phenotype, regardless of transmission mechanism, much like the expanded phenotype concept proposed by some archaeological theorists which includes human behaviors and material culture as well as biological structures (e.g., Dunnell 1989, O'Brien & Holland 1992, 1995b; Jones, Leonard & Abbott 1995; Neff 1993, see also Dawkins 1990).

In other words, fitness should be defined and measured in terms of successful information transmission—both potential and realized—rather than reproduction (or even 'replication' *sensu* Leonard and Jones [1987]). This applies to both biology and behavior. It is often forgotten (or the implications overlooked) that biological *traits* are not passed from one generation to the next; in reality, it is the *information* needed to reproduce these traits (in the form of DNA codons) that is transmitted. In evaluating fitness, reproductive success is simply a more easily measured surrogate for the transmission of genetic information. In the same way, *behaviors* are not transmitted from individual to individual in a social context, but rather the *information* that an individual needs to produce those behaviors that is transmitted in social learning (see also Dennett 1995: 335-369).

An example of this approach considers the set of biological and behavioral mechanisms required to make stone tools. The information needed for the phenotypic expression of grasping hands, precision and power grips is coded in human DNA, and genetically transmitted along descent lines. The information needed to make and use stone tools is stored and coded cognitively, and transmitted from individual to individual via social learning. Both sets of phenotypic characteristics are heritable—that is, they can be passed from individual to individual with relatively 'high fidelity'. The frequencies of the expression of both sets of phenotypic characteristics over time are conditioned by the probabilities that the requisite information will be transmitted from individual to individual. If the grasping abilities of the human hand positively or negatively affects reproductive success (the way in which genetic information is transmitted), then its frequency will change over time accordingly. The same is true of a set of technological behaviors. If they affect the social learning process positively or negatively, their frequencies will be affected accordingly. Clearly, whether technological behavior has a positive or negative impact on the biological survival of the social unit in which learning takes place will affect the replication of that behavior. However, the reason is not due to a *genetic* basis for lithic technology, and changes in the social transmission of such behavior are not are not simply a result of differential reproductive success.

The point is that we would do better to measure *potential fitness* according the probability that information needed to code for human phenotypes is transmitted from individual to individual—including the extended phenotypes of learned behavior and associated material culture, as well as biological characteristics—and *realized fitness* according to the extent that such information is transmitted. The way in which such information is stored—DNA, cognitive structure, the written word, or optical disk—and means by which such information is transmitted—through sexual reproduction, oral tradition, libraries, or a computer screen—is irrelevant at a conceptual level (although it *is* relevant in terms of the rate, direction, and other mechanics of transmission [see also Dunnell 1989]).

Decoupling fitness from biological reproduction and redefining it in terms of information transmission more accurately represents the evolutionary process, and has the potential of stimulating better (i.e., more inclusive, more parsimonious) explanations of human behavior and its change. It is especially useful for the many behaviors that have little direct impact—or negative impact—on reproductive success, but which are still replicated. Along these lines, such an expanded fitness concept could be equally valuable in ethology and sociobiology given the importance of learned behavior in vertebrates, and especially in mammals and birds.

Non-Darwinian Processes

Although Dunnell noted the potential importance of non-Darwinian processes such as drift two decades ago (1978), most evolutionary archaeological research—both theoretical and applied—has focused on selection.

Neiman's work (1995) is an exception, and Barton's chapter in this volume focuses on non-Darwinian processes. This perspective has the potential for considerable advancement in our understanding of cultural systems and the vectored changes that characterize them.

Most selectionist literature has a strongly gradualist tone (e.g., Braun 1987, Rindos 1980). Although many are aware of the potential for non-Darwinian processes, they tend to see selection as the overriding force in behavioral change (e.g., Jones, Leonard & Abbott 1995; O'Brien & Holland 1995a). Nevertheless, throughout most of the human past, behavior took place and was transmitted in the context of very small groups. This means that drift and 'behavioral bottlenecks' likely played important roles in behavioral change. For this reason, along with humans' wide geographical distribution and potential for long-distance movement, much of human behavioral change can more accurately be characterized by a punctuated equilibrium model rather than a gradualist one.

Sorting or packaging also probably played an important role in behavioral change, but has yet to be explored. Packaging refers to the fact that traits–including human behaviors–are rarely transmitted as discrete units, but rather as parts of an integrated set (Ramenofsky 1995). For example, use of a particular type of temper in ceramic production can be transmitted as part of a set of pottery making behaviors that included vessel construction, vessel form, surface decoration, and firing conditions. If one of the trait set (e.g., temper) was strongly selected for, the entire behavior set might be favored over another set, even if other specific traits (e.g., those related to vessel form) in the first set had a lower potential fitness than alternative traits in the second. In the 1994 symposium, the presentation by Michael Jochim (1994) addressed the potential for such processes to account for apparent anomalies in an archaeological record. In the present volume, Steven Mithen explores this idea further, focusing on religious practices. He suggests that such processes may be important for understanding the evolution of complex behavioral systems, such as food production.

EXPLANATION IN ARCHAEOLOGY

The human experience is diverse and complex, and can be approached in many ways. Of those that have more time depth than a generation, religion, art, folklore, and history exemplify some of the very different perspectives we can adopt to try to make sense of ourselves and our past. Of the many fields of study deal with the human experience, only archaeology and its sister discipline of physical anthropology employ a scientific paradigm to access the human past. And only archaeology focuses on the behavioral systems we call 'culture' that make us unique among the living organisms of this planet.

The scientific paradigm does not inherently warrant archaeology with a better claim to 'truth' about the human experience, nor does it imply that the meaning of humanness is somehow less accessible to archaeology. However, given the considerable success of the materialist world view of western science in accounting for many aspects of reality, applying it to ourselves should provide equally valuable insights. While there are archaeologists who feel that a scientific paradigm is an inappropriate one for the discipline, we wish to point out that there is no other discipline that employs a scientific approach to the study of the human past. In the century of so of its existence, archaeologists have learned more about our past and how we came to be the way we are than in all the preceding millennia for which we have written records–and this despite a continuous interest in our origins. To abandon the only scientific conceptual framework for the study of our heritage would be as much a loss to our self-understanding as to abandon historical or artistic inquiry.

Archaeology cannot, nor should it, encompass all avenues to an understanding of the human past. If it is unable to account for aesthetic genius or the depths of the human spirit, this should not be taken as a failure of the discipline. Other scholarly endeavors may be better equipped to cope with these questions. If archaeology fails to do good science, however, this is a more serious matter. Archaeology has claimed a scientific approach since its beginnings as a formal field of inquiry in the last century. Despite well-intentioned self-critique as to the adequacy of science for studying the past, most practitioners and those outside the field consider archaeology to be a scientific endeavor. The success of the discipline, both in terms of its internal goals and from the standpoint of the non-archaeological public, hinges on its ability to do good science.

An overt aim of this volume is to promote better scientific study of the human past on the part of the discipline. As others have argued before us, a powerful, unifying body of general theory is fundamental to the successful application of the scientific paradigm. We think that neo-Darwinian evolutionary theory, expanded to encompass human behavioral systems, is the best candidate for this role. However, as Robert Bettinger and Jelmer Eerkens point out in their chapter there is, "...an unfortunate parallel between evolutionary theory and the weather: everyone talks about it but no one does anything about it". We hope this volume will contribute to efforts to improve this situation.

REFERENCES CITED

Barton, C. Michael, G.A. Clark & A. Cohen
1994 Art as information: explaining Upper Palaeolithic art in western Europe *World Archaeology* 26: 185-206.

Boyd, R. & P.J. Richerson
1985 *Culture and the Evolutionary Process.* Chicago: University of Chicago Press.

Braun, David P.
1987 Coevolution of sedentism, pottery technology, and horticulture in the central Midwest, 200 B.C. - A.D. 600. *Emergent Hortocultural Economies of the Eastern Woodlands* (W.R. Keegan, ed.), pp. 153-181. Carbondale: SIU Center for Archaeological Investigations, Occasional Paper No. 7.

Cavalli-Sforza, L. L. & M. W. Feldman
1981 *Cultural Transmission and Evolution: a Quantitative Approach.* Princeton: Princeton University Press.

Clark, G.A.
1993 Review: *History and Evolution* (Nitecki & Nitecki, eds.). *The American Journal of Physical Anthropology* 90:129-130.

Clark, G.A., C. Michael Barton & Alison Cohen
1996 Explaining art in the Franco-Cantabrian refugium: an information exchange model. In *Debating Complexity* (D. A. Meyer, P.C. Dawson and D.T. Hanna, eds.), pp 241-253. Calgary: Archaeological Association of the University of Calgary.

Dawkins, Richard
1989 *The Selfish Gene.* Oxford: Oxford University Press.
1990 *The Extended Phenotype: the Long Reach of the Gene.* Oxford: Oxford University Press.

Dennett, Daniel C.
1995 *Darwin's Dangerous Idea: Evolution and the Meanings of Life.* New York: Simon & Schuster.

Dunnell, Robert C.
1978 Style and function: a fundamental dichotomy. *American Antiquity* 43: 192-202.
1989 Aspects of the application of evolutionary theory in archaeology. *Archaeological Thought in America* (C.C. Lamborg-Karlovsky, ed.), pp. 35-49. Cambridge: Cambridge University Press.
1995 What is that actually evolves?. *Evolutionary Archaeology* (P.A. Teltser, ed.), pp. 33-50. Tucson: University of Arizona Press.

Eiseley, Loren C.
1958 *Darwin's Century.* New York: Doubleday & Company.

Emlen, Stephen T., Peter H. Wrege & Natalie J. Demong
1995 Making decisions in the family: an evolutionary perspective. *American Scientist* 83: 148-157.

Erwin, Douglas, James Valentine, & David Jablonski
1997 The origin of animal body plans. *American Scientist.* 85: 126-167.

Gould, Stephen J. & Richard Lewontin
1978 The spandrels of San Marco and the panglossian paradigm: a critique of the adaptationist programme. *Proceedings of the Royal Society* 205: 581-598.

Graves, Michael & Thegn Ladefoged
1995 The evolutionary significance of ceremonial architecture in Polynesia. In *Evolutionary Archaeology* (P.A. Teltser, ed.), pp. 149-174. Tucson: University of Arizona Press.

Jochim, Michael
1994 Chance, History, and Context in Evolutionary Change. Paper presented at the annual meetings of the Society for American Archaeology, Anaheim, California.

Jones, G.T., A.L. Abbott & R.D. Leonard
1995 The structure of selectionist explanations in archaeology. In *Evolutionary Archaeology* (P.A. Teltser, ed.), pp. 13-32. Tucson: University of Arizona Press.

Leonard, R.D. & G.T. Jones
1987 Elements of an inclusive evolutionary model for archaeology. *Journal of Anthropological Archaeology* 6: 199-219.

Neff, Hector
1992 Ceramics and evolution. *Archaeological Method and Theory* 4: 141-193.
1993 Theory, sampling, and analytical technique in the archaeological study of prehistoric ceramics. *American Antiquity* 58: 23-44.

Neiman, Fraser
1995 Stylistic variation in evolutionary perspective: inferences from decorative diversity and interassemblage distance in Illinois Woodland ceramic assemblages. *American Antiquity* 60: 7-36.

Nitecki, M. & D. Nitecki (eds.)
1992 *History and Evolution.* Albany: State University of NY Press.

O'Brien, Michael J. (editor)
1996 *Evolutionary Archaeology: Theory and Application.* Salt Lake City: University of Utah Press.

O'Brien, Michael J. & Thomas D. Holland
1990 Variation, selection, and the archaeological record. *Archaeological Method and Theory* 2: 31-79.
1992 The role of adaptation in archaeological explanation. *American Antiquity* 57: 36-59.
1995a The nature and premise of a selection-based archaeology. *Evolutionary Archaeology* (P.A. Teltser, ed.), pp. 175-200. Tucson: University of Arizona Press.
1995b Behavioral archaeology and the extended phenotype. *Reconstruction Theory: a Behavioral Approach to the Archaeological Record* (A.E. Nielsen, J.M. Skibo, and W.H. Walker, eds.). Salt Lake City: University of Utah Press.

Ramenofsky, Ann F.
1995 Evolutionary theory and Native American artifact change in the postcontact period. In *Evolutionary Archaeology* (P.A. Teltser, ed.), pp. 129-148. Tucson: University of Arizona Press.

Rindos, D.
1980 Symbiosis, instability, and the origins and spread of agriculture: a new model. *Current Anthropology* 21: 751-772.
1984 *The Origins of Agriculture: an Evolutionary Perspective.* New York: Academic Press.
1989 Undirected variation and the Darwinian explanation of cultural change. *Archaeological Method and Theory* 1: 1-45.

Ruse, Michael
1987 Evolutionary models and social theory. *Prospects and Problems: Evolutionary Theory in Social Science* (M. Schmidt & F.M. Wuketits, eds.), pp. 23-48. Dordrecht: D. Reidel.

Schiffer, Michael B.
1996 Some relationships between behavioral and evolutionary archaeologies. *American Antiquity* 61: 643-662.

Stoltis, J., R. Boyd, & P.J. Richerson
1995 Can group-functional behaviors evolve by cultural group selection? An empirical test. *Current Anthropology* 36: 473-494.

Teltser, Patrice A. (editor)
1995a *Evolutionary Archaeology: Methodological Issues.* Tucson: University of Arizona Press.
1995b Culture history, evolutionary theory, and frequency seriation. In *Evolutionary Archaeology* (P.A. Teltser, ed.), pp. 151-68. Tucson: University of Arizona Press.

Part II

General Theory & Method in Evolutionary Archaeology

Introduction

Part II consists of five essays that deal with the conceptual framework of Darwinian evolutionary theory, and with how various construals of evolutionary theory in the life sciences have affected anthropology and archaeology, in the rather different contexts of the American and British research traditions. In an historical *tour de force*, R. Lee Lyman and Michael O'Brien (Chapter 2) examine the impact of evolutionary thinking on significant figures in American anthropology (Boas, Wissler, Spier, Kroeber, White, Steward, Sahlins, Service) and archaeology (Kidder, McKern, the Gladwins, Brew, Colton and Hargrave, Haag). They outline distinctions between Darwinian evolution, grounded in materialist biases and preconceptions, and Spencerian (or cultural) evolution, with its strongly essentialist character, and discuss some of the factors that led to the rise of cultural evolutionary paradigms in the 1940s and 1950s and how they became incorporated into the 'new archaeology' of the 1960s. Lost in the shuffle is the fact that some 'early' American archaeologists made a serious effort to incorporate elements of Darwinian theory into their work. Modern attempts to make archaeology a materialist discipline must contend with the fact that the Darwinian paradigm fell from favor in the U.S. after World War II, supposedly because of the misconception that humans, as cultural animals, were somehow 'unique', and that two kinds of evolutionary theory - one biological, the other cultural - were therefore required to account for change in organisms and cultures respectively.

Archaeologists in the United Kingdom did not have to contend with orthogenetic notions of progress, human intent, essentialist stage typologies and the like, and essays by Gowlett (Chapter 3) and Mithen (Chapter 4) are grounded firmly in the Darwinist camp. Focusing on the evolution of early hominids, John Gowlett makes a distinction between 'direct' models, using archaeological and paleontological data, and 'comparative' models, grounded in primate ethology. Taking note of the fact that morphological hominids have been around for at least 5 million years, whereas archaeological evidence has existed for only about half that long, he argues that our most productive inferences are likely to come from the interface between data relating to the past itself, and data derived from hominoid sociality studies, the latter having implications for the former. Of particular importance is 'Machiavellian intelligence' - the idea that the big brains found in Middle Pleistocene hominids might have evolved for dealing with the complexities of the social environment in the context of groups larger than those of extant hominoids. It can be shown empirically that there are strong correlations between neocortex ratios and group sizes across the primates, and this allows for retrodiction of fossil hominid group sizes. The question is what do these correlations imply for the evolution of human intelligence?

Steven Mithen's essay is also about human evolution in 'deep time'. He argues that early humans had domain-specific intelligence, whereas modern humans have evolved better integrated mental functions. He labels these domains 'technical', 'social' and 'natural history' intelligences, and suggests that early hominids could have operated effectively in each of them, but only modern humans were capable of cross-linking, exchanging and reworking information amongst the domains easily and effectively. He argues that Darwinian archaeologists should move toward a more explicitly cognitive approach to past behaviour based on some aspects of evolutionary psychology, illustrating this conceptual framework with the cultural transmission of religious ideas. Notions of group adaptation, cultural selection, memes and cultural viruses should be rejected. The most profitable way in which evolutionary theory can be used in archaeology, he argues, is for an explicitly individualistic and cognitive stance to be adopted.

Back on 'our' side of the Atlantic again, and complementing the historical perspective provided by Lyman and O'Brien, Hector Neff and Daniel Larson (Chapter 5) focus on 'the Dunnell era' (the 1980s), and the origins of the methodology of comparison in the evolutionary archaeologies of the 1990s. Dunnell argued that cultural evolution had little in common with scientific evolution because it provided no mechanisms with which to explain history, but a great deal in common with the non-scientific, progressive, goal-directed evolutionary philosophy of Darwin's contemporary, Herbert Spencer. Taking as a point of departure Dunnell's categorical rejection of cultural evolution, the authors develop a reasoned critique of 'the adaptationist program' in biology and archaeology, arguing that the 'Panglossian hyperselectionism' so harshly criticized by Gould and Lewontin (1978) is little more than a caricature of a more complex reality, that rejection of the concept of adaptation in archaeology is often capricious and unwarranted, and that comparative methods - basic to cultural evolution, cultural ecology, the adaptationist program and behavioural reconstruction - are discarded prematurely by many workers because of the impact of Dunnell's programmatic essays. They go on to explicate a methodology of comparison that can be used to test hypotheses generated from evolutionary theory, illustrating its utility with an adaptive design model of how human movement of goods through space (in this case, southwestern and Mesoamerican ceramics) is shaped by selection associated with greater or lesser degrees of environmental predictability.

The last paper in this section deals with technology, and with the evolution of technological systems. Peter Bleed (Chapter 6) argues that, despite nominal acknowledgment of the importance of technology by anthropologists and archaeologists, technology remains poorly understood because its material and non-material components are conflated and/or treated in isolation by many workers. He makes a fundamental distinction between the content of technology - the source of variability in technological systems - and the results of technology, upon which selection operates. The link between the two is complex and dynamic, he maintains, and neither is passive. Each reflects the other, and both are actively involved in shaping the evolution of technological systems. The essay defines the contents and results of technology in behavioural terms in an effort to clarify the relationship between its material and non-material aspects.

REFERENCES

Gould, S. & R. Lewontin

1978 The spandrels of San Marco and the Panglossian paradigm: a critique of the adaptationist programme. *Proceedings of the Royal Society* 205: 581-598.

2

The Concept of Evolution in Early Twentieth-Century Americanist Archaeology

R. Lee Lyman & Michael J. O'Brien
University of Missouri

ABSTRACT

American archaeologists who founded the culture-history paradigm early in the 20th century held a view of the evolution of cultures commensurate with Darwin's notions of continuous, gradual change. Initially, in their view, artifacts varied continuously across space and through time. This materialist metaphysic was, however, short-lived. Its fall from favor began in the 1920s, when it was noted that cultural evolution was reticulate whereas biological evolution was only branching. When it was argued in the 1940s that cultures were, after all, not organisms and that inanimate objects did not interbreed, any hope of adopting a Darwinian version of evolution in archaeology was abandoned. Rather, archaeologists adopted cultural evolutionism founded by Herbert Spencer in the 19th century and popularized by Leslie White in the 20th century. Franz Boas and his students flatly rejected this version in part because of its essentialist metaphysic. A. L. Kroeber had some significant insights into how to construct historical lineages, but he and his contemporaries did not know how to implement Darwin's version of evolutionism in anthropology and archaeology. Archaeologists of the 1950s thus came to view culture change largely in terms of discontinuous stages. Such a view gained legitimacy because biologists in the 1950s claimed the evolution of cultures was unlike the evolution of organisms. After some forty years, it is time to try a Darwinian version of evolution.

When cultural evolution is considered a natural extension of organic evolution, our thinking about the total process of evolution will be greatly clarified (W. G. Haag 1961: 440).

INTRODUCTION

Over the past decade or so, an increasing number of archaeologists have begun to show interest in employing Darwinian evolutionary theory to explain variation in the material record. Although Darwinian evolutionary archaeology has not enjoyed the same degree of popularity as some other approaches have over the years, a now-sizable literature attests to the fact that archaeologists are serious about incorporating the tenets of Darwinism into their work (see O'Brien 1996b, Schiffer 1996 and references therein). Given our personal perspectives, we are pleased with the careful attention archaeologists are giving Darwinian archaeology, but we also realize that, regardless of the attention it receives, its success can only be measured in terms of its own performance in successfully explaining the archaeological record.

Clearly, Darwinian evolution is a different kind of theory from the ones with which it competes, being rooted as it is in materialism as opposed to typological strategies. Because of this, most of the available data that routinely are used in archaeological analysis are inappropriate for the kinds of investigation that derive from a Darwinian perspective–an issue that is as pertinent in archaeology (Dunnell 1980, 1982, 1985, 1986, 1988; O'Brien & Holland 1990, 1992; O'Brien et al. 1994) as it is in biology (Dobzhansky 1951; Ghiselin 1966, 1974, 1981; Mayr 1959, 1963, 1973, 1977, 1987; Sober 1980, 1984). This is one issue with which scientific evolutionism must contend, but at this point it might not be the most critical issue. One of the risks involved in persuading a discipline that a particular approach has much to offer is that practitioners will view it as simply another in a long string of approaches being sold on the street corner–something seemingly exciting and appealing that quickly disappears after its original advocates find a new product to hype. In addition, there is a danger in the increase in visibility attached to a new approach. As one of us pointed out recently (O'Brien 1996a), history shows that archaeologists have been quick to jump on bandwagons without the slightest notion of why they are doing so and certainly without the background necessary to understand the nuances of the approaches they begin advocating. The lessons learned from Americanist archaeology of the 1960s and '70s should teach us that seemingly new approaches are not necessarily all they are purported to be, nor are they necessarily improvements on what came before.

One of the things missing in recent attempts to persuade the discipline that Darwinian evolutionism offers the best means of explaining the archaeological record is an examination of the roots of that approach in Americanist archaeology. Historical reviews by Dunnell (1980) and O'Brien and Holland (1990), for example, make it sound as if archaeologists all but ignored Darwinian evolution until an enlightened few made it accessible sometime after 1970. None of those involved in the renaissance of Darwinian evolution in archaeology actually believes such is the case, though it is clear in retrospect that their treatments of the historical roots of such evolutionism in archaeology are lacking in detail. Most such treatments spend considerable time on the differences between Darwinian evolution and cultural evolution, the latter of which made its way into archaeology during the 1950s and soon became a cornerstone of the so-called new movement in the discipline a decade later. These two kinds of evolutionism most assuredly have little in common, and hence discussion of the differences are warranted, but lost in the shuffle is the fact that Americanist archaeologists earlier this century made serious efforts to incorporate elements of Darwinian evolutionism in their work. Those committed today to such an evolutionary archaeology need to understand not only the arguments put forward by these earlier investigators but also why, when it appeared there was a growing consensus toward making archaeology a materialistic discipline, Darwinian evolutionism fell from favor. It is toward these two key issues that we address this paper.

CULTURE HISTORY: THE BACKBONE OF THE DISCIPLINE

Interpreting culture change—alternately referred to as *cultural development* by archaeologists early in the 20th century—was the central theme of Americanist archaeology beginning with the birth of the culture-history paradigm in the second decade of the 20th century (Lyman et al. 1997a, 1997b). As Clark Wissler (1917b: 100) remarked in what might be thought of as the birth announcement, Americanist archaeology not only now had some important questions to answer, but it also, for the first time, had a way to produce answers:

> [H]ow long has man been in America, whence did he come, and what has been his history since his arrival? . . . [T]he archaeologist finds in the ground the story of man and his achievements. The new, or the real archaeology is the study of these traces and the formulation of the story they tell. . . . [The archaeologist] must actually dissect section after section of our old Mother Earth for the empirical data upon which to base his answers. It is not merely the findings of things that counts; it is the conditions and interassociations that really tell the story.

This had been the goal of archaeology since the beginning of the 20th century. As Franz Boas (1904: 521-522) remarked,

> . . . the sequence of types of culture as determined by the artifacts of each period [is among] the fundamental problems with which archaeology is concerned. The results obtained have the most immediate bearing upon the general question of the *evolution* of culture, since the ideal aim of archaeology practically coincides with this general problem, the solution of which would be contained in a knowledge of the chronological development of culture (emphasis added).

The techniques for addressing the issues raised by Wissler and Boas—the latter having recognized that "In the study of American archaeology we are compelled to apply methods somewhat different from those used in the archaeology of the Old World" (Boas 1902: 1)—involved studying the fluctuation of frequencies of artifact *types* through time, with the passage of time initially confirmed and later established by the superposed positions of the actual artifacts themselves (Lyman et al. 1997b).

That the goals of culture history never varied for the next several decades is clear from the remarks of one of the parents of the paradigm, A. V. Kidder:

> Archaeologists, noting that modern biology has mounted above the plane of pure taxonomy [that is, classification], have attempted to follow that science into the more alluring fields of philosophic interpretation, forgetting that the conclusions of the biologist are based on the sound foundation of scientifically marshalled facts gathered during the past century by an army of painstaking observers. This groundwork

> we utterly fail to possess. Nor will it be easy for us to lay, because the products of human hands, being unregulated by the more rigid genetic laws which control the development of animals and plants, are infinitely variable. But that is no reason for evading the attempt. It has got eventually to be done, and the sooner we roll up our sleeves and begin comparative studies of axes and arrowheads and bone tools, make classifications, prepare accurate descriptions, draw distribution maps and, in general, persuade ourselves to do a vast deal of painstaking, unspectacular work, the sooner shall we be in position to approach the problems of cultural evolution, the solving of which is, I take it, our ultimate goal (Kidder 1932: 8).

Kidder's remarks are important not only because they outline the goals of the culture-history paradigm but because they also indicate how culture historians conceived of the phenomena they were studying. Cultures *evolved;* a historically documented culture had a developmental heritage or lineage, and it was the job of the culture historian to describe that lineage and to determine why it had the form that it did. But Kidder correctly indicated that archaeology lacked both the basic data and a *theory* consisting of cultural processes parallel to the biological ones of genetic inheritance and natural selection to help explain a culture's lineage in evolutionary terms. To those interested in incorporating Darwinian evolutionary theory in archaeology, Kidder's words couldn't have been better put because he focused on the issue that underlies any evolutionary study–namely, the documentation of variation. As we point out later, without the means to document variation and then to link that variation to applicable theory (which should dictate *how* we measure the variation), so-called explanations of how and why things change–the epitome of evolutionism–will be nothing more than a series of stories.

Kidder knew this, but most of his contemporaries either didn't appreciate the point or failed to pay it much heed. Compiling the data and building the theory evidently would take some work, Kidder suspected, and in the long run, only the former has been met with any sort of empirically verifiable success. Chronologies of artifact types and larger units variously termed cultures, phases, complexes, and the like have been constructed, tested, refined, empirically verified, and are now available for many areas of the Americas. Explanations of those sequences, particularly in the sort of evolutionary terms Kidder and many of his contemporaries envisioned, are, however, notably lacking.

Some Americanist archaeologists would, we suspect, argue long and loudly that explanations of an evolutionary sort *are* available and are made every year in professional journals and books and at professional meetings (see the various chapters and references in Maschner 1996). In one sense they would be correct, but the sort of evolution that usually is presented and discussed has nothing to do with the sort that Kidder had in mind. To fully understand the significance of what Kidder envisioned and why his vision has not yet been attained, we begin with a consideration of why many so-called evolutionary explanations are non-Darwinian in nature.

WHAT *IS* EVOLUTION?

Anyone familiar with the history of biology will know that Charles Darwin's version of evolution was not the first or the only version available in the theoretical marketplace of the 19th century. For example, Jean Baptiste de Lamarck had a version, as did Georges Cuvier, and both versions were, at one time or another, as popular with segments of the scientific community as Darwin's came to be. But in the end, it was Darwin's version that took hold, in the process relegating its competitors to obscurity. Relative to biological evolution, our concern here is only with Darwinian evolution. There also was a theory of evolution that concerned nonbiological phenomena, including such things as human society and technology. This was the version of evolution that Herbert Spencer was selling in the mid-19th century (O'Dowd 1982). Throughout the following discussion, we use the term *cultural,* or *Spencerian, evolution,* to denote the kind of evolution espoused by Spencer and his intellectual descendants, including such luminaries as Lewis Henry Morgan and Edward Burnett Tylor during the 19th century and Leslie White, Julian Steward, Marshall Sahlins, Elman Service, Morton Fried, and Robert Carneiro during the 20th century

To denote the kind of evolution introduced by Darwin, we use the terms *biological evolution,* or *Darwinism.* In this respect, it is important to note, following biologist Ernst Mayr (1972, 1982), that Darwinism was not some monolithic theory that swept through biological science in the 1860s and set it on an unwavering course. Just as with its biological subjects, Darwinian theory evolved. Darwinian evolution of the 1980s was not the same theory that it was in the 1880s or that it was in the 1930s. Understanding both the evolutionary history and the metaphysical underpinnings of Darwinian evolutionary theory is key to understanding why Americanist archaeology, after enjoying some success with that theory, rejected it and adopted cultural evolution in the 1950s.

Two Metaphysics

We use the unmodified term *evolution* to denote change, regardless of the mechanism, process, or form. As the quotations from Wissler and Kidder indicate, culture historians since the second decade of the 20th century were interested in documenting, studying, and explaining cultural change. Theirs was a historical science, in many ways similar to that of paleobiologists, whose business it is to document and explain the phylogenetic histories of organisms. But whereas paleobiologists were able to erect an explanatory theory of biological change after the synthesis of genetics and natural selection took place in the late 1930s and early '40s (Mayr 1972, 1982), archaeologists followed a different path and adopted a different method of explaining their subject matter. The most significant difference between the two fields of inquiry—paleobiology and Americanist archaeology—resides in the fact that by adopting different theories of evolution, the two fields adopted decidedly different metaphysics. Biologists in the early 19th century subscribed to an essentialist metaphysic, and it wasn't until the mid-1950s that they began to replace it with Darwin's materialist metaphysic (Mayr 1982; see also Gould 1986). Anthropologists and archaeologists of the early 20th century tended to subscribe to a notion of cultural evolution that was in many respects materialistic, but they replaced that notion with an essentialist one in the 1950s. What, then, are these two metaphysics?

Essentialism presumes the existence of discoverable, discrete kinds of things. Things are of the same kind because they share essential properties—their "essences"—and these essential properties dictate whether a specimen is of kind A or kind B. Essential properties define an ideal, or archetype, "to which actual objects [are] imperfect approximations" (Lewontin 1974: 5). This view renders nonessential variation between specimens as simply "annoying distraction" (Lewontin 1974: 5). An advantage of the essentialist metaphysic is that prediction is possible because the kinds are real and thus are always and everywhere of the same sort; they will therefore always interact in the same manner, and the same result will be produced by their interaction. Laws, in a philosophical sense, thus can be written because the interactions of things will always be the same. The things, as well as the interactions between things, will always, regardless of their positions in time and space, be the same because the essential properties of the things themselves never change. Specimens grouped within natural, essentialist kinds always, by definition, share essential properties. Not surprisingly then, ahistorical sciences such as chemistry employ an essentialist metaphysic, since what they are measuring is *difference* as opposed to *change* (Hull 1965, Mayr 1959, Sober 1980).

In fact, under essentialism, it is impossible to measure change. Only the difference between different kinds, or types, can be measured. For this reason, essentialism often is referred to as *typological thinking* (Mayr 1959, 1988). Mayr (1959: 2) notes that "Since there is no gradation between types, gradual evolution is basically a logical impossibility for the typologist. Evolution [change], if it occurs at all, has to proceed in steps or jumps." How could change be anything *but* transformational? If things have essences, the only way that things could evolve is by dropping one essence and adopting another. Thus a specimen of kind A is kind A in this time and place, but in another time and place it somehow has transformed into kind B. As Mayr (1982: 38–39) stated in regard to biological speciation, "Genuine change, according to essentialism, is possible only through the saltational origin of new essences [species]." Change is neither gradual nor continuous; it is jerky and punctuated.

In opposition to essentialism, materialism holds that phenomena cannot exist as bounded, discrete entities because they are always becoming something else. With specific reference to organisms, Mayr (1959: 2) pointed out that "All [things] are composed of unique features and can be described collectively only in statistical terms. Individuals . . . form populations of which we can determine an arithmetic mean and the statistics of variation. Averages are merely statistical abstractions, only the individuals of which the populations are composed have reality." As a direct result of its materialist metaphysic, a historical science can monitor *change* in phenomena: "For the [essentialist-thinking] typologist, the type is real and the variation an illusion, while for the [materialist-thinking] populationist the type (average) is an abstraction and only the variation is real" (Mayr 1959: 2). It is this variation between and among specimens that "is the cornerstone of [Darwinian evolutionary] theory" (Lewontin 1974: 5).

Classification Units

If only *variation* is real, how do we study it? The answer is, by constructing a set of units that allows *properties,* or attributes (*character states* to biologists; e.g., Szalay & Bock 1991), of phenomena to be measured. Archaeology is nothing if not a spawning ground for units, many of which-type, group, class, period, phase, and so on—are rarely defined explicitly. Here we use the term *measurement* to denote the assignment of a symbol—letter, number, word—to an observation made on a phenomenon according to a set of rules. The rule set includes specification of a scale (e.g., Stevens 1946) and the relation between the symbols and the scale. Observation and recording of attributes on specimens comprise measurement. But which attributes should we observe and record?

The analyst selects attributes relevant to some problem, and it is those attributes and their combinations that result in the sorting of specimens into internally homogeneous, externally heterogeneous piles. Importantly, specimens that share attributes or properties—those that end up together in one of the analyst's piles—do not have an essence (Mayr 1987: 155). They have been grouped together not because of some inherent, shared quality but rather because they hold in common some number of attributes *selected by the analyst.* Theory is the final arbiter of which attributes out of the almost infinite number that could be selected are actually chosen by the analyst. Theory also specifies the kinds of units to be measured. We might decide, based on theory, that the color of a stone tool is not related to its function, whereas the angle of the working edge is; thus, if we are interested in functional variation in stone tools, we choose as our attributes edge angles, traces of use wear (edge damage), and other attributes that *theoretically* are causally related to the property of analytical interest.

Because *we* decide on the attributes to be recorded, the resulting units—variously termed in archaeology types or classes—are *ideational;* they are not real in the sense that they can be seen or picked up and held. An edge angle—itself an ideational unit with different empirical manifestations—is measured, using a goniometer, in other ideational units known as degrees. In the simplest of terms, ideational units are tools used to measure or characterize real objects. An inch and a centimeter are used to measure length, just as a gram and an ounce are used to measure weight. Inches, centimeters, grams, and ounces do not exist empirically; they are units used as analytical tools to measure properties *of* empirical units. A writing pen is an empirical unit that can be placed in a set of things that are all six inches long, but only if the attribute distinguishing the set specifies that the things within it must be six inches long to be included; our theory, of course, tells us what length is and how to measure it, as well as how it differs from width, color, or any other dimension.

An *empirical unit* is a thing that has a real existence; it is phenomenological. We use the term *type* for units of unspecified kind; they may be either ideational or phenomenological. We note this is not always the meaning given to "type" by culture historians. We use the term *classification* in a general sense to denote an arrangement of units of whatever kind (see Lyman et al. 1997b for additional discussion).

What is Change?

Under the materialist view, things are continually in the process of becoming something else—not in a saltational, or jumpy, sense but in the sense of slow, gradual change (Figure 2.1). Depending on the scale at which we are operating, that change may be difficult or relatively easy to measure. Regardless, change must be measured as alterations in the frequencies of analytical kinds, or what we have termed ideational units. The classic model of the evolution of the modern horse (*Equus caballus*) from *Eohippus* to *Miohippus* to *Merychippus* and finally to *Equus* (Simpson 1967: 135) is merely a heuristic device that simplifies the phylogeny of equids into terms that allow discussion and description. In such a phylogeny, different forms, or types, of equid are given a (taxonomic) name and arranged against a time scale. Variation within the populations of horses that existed at different times and places is masked by such a procedure, but paleobiologists recognize this and use the names of various populations—each of a different ideational form—merely to discuss equid evolution. A realistic phylogeny explicitly constructed within the materialist metaphysic would include frequencies of those different forms plotted against time and space, with—and this is crucial—indications of the particular variant forms and specimens that represented particular taxa (e.g., Simpson 1951, Vrba 1980; see Gingerich 1979: 457 for a real example).

There is a critical question here that must be addressed: If two things are similar but also somewhat different in form and also different in age, do they indicate that change has somehow taken place? From a modern Darwinian viewpoint, they represent change only if they are phyletically (genetically) related, in which case the similarity of form and difference in age signifies inheritance and thus continuity—an ancestor—descendant lineage. How does one demonstrate a phyletic relation—that two phenomena are parts of a lineage? Paleobiologists accomplish this task by identifying homologous traits, or attributes, of the two phenomena. If the two phenomena share one or more such traits, they are by definition phyletically related. The analytical challenge, then, is to identify such traits.

Identifying homologous traits is a significant analytical hurdle (e.g., Bock 1977, Fisher 1994, Forey 1990, Simpson 1975, Smith 1994, Szalay & Bock 1991) because a trait that is shared by two phenomena may be analogous—the result of convergence. How are analogous and homologous traits to be distinguished? A. L. Kroeber (1931:151) suggested that "Where similarities are specific and structural and not merely superficial . . . has long been the accepted method in evolutionary and systematic biology." He was correct, for this was, and is, pre-

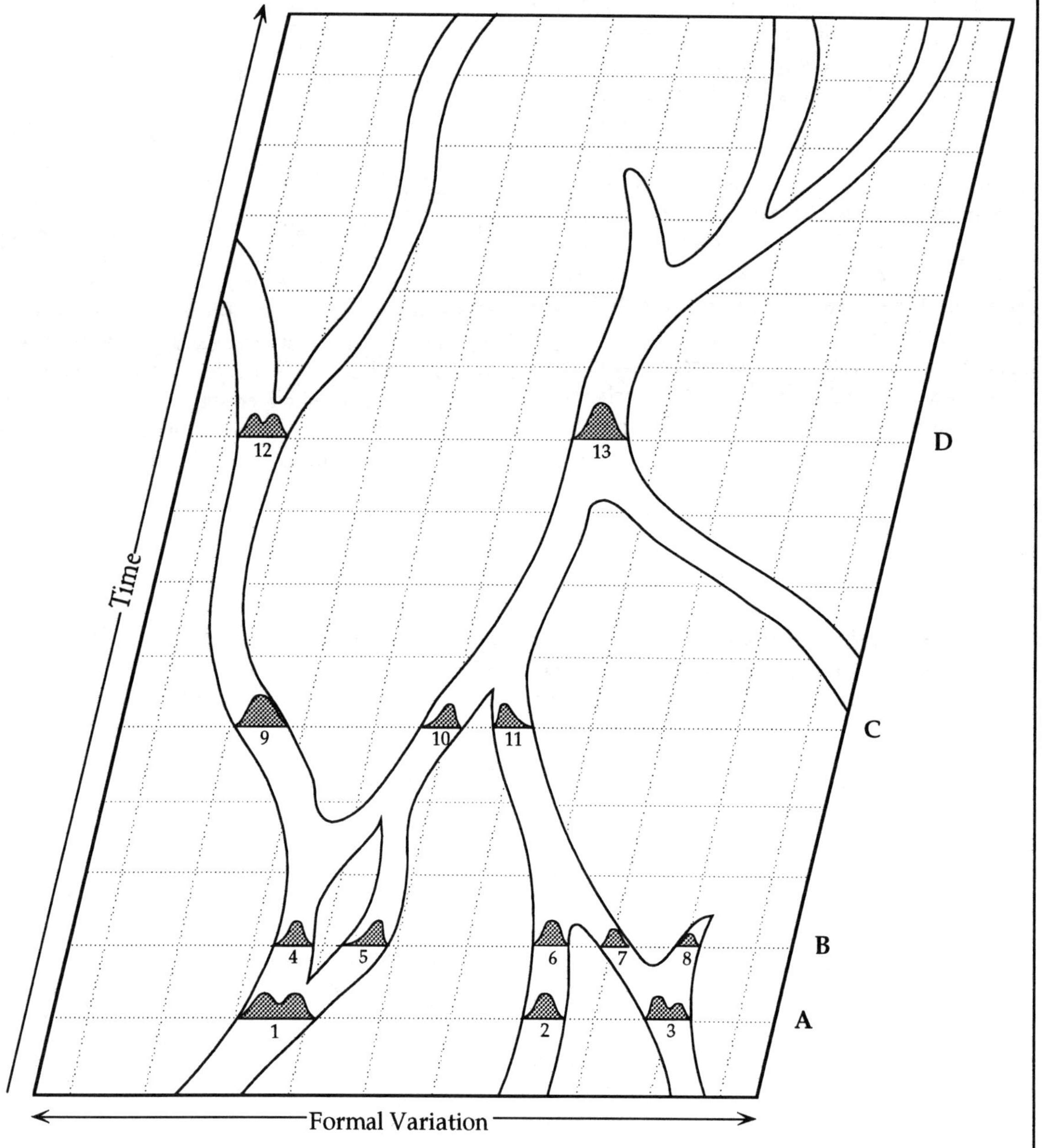

Figure 2.1. The braided (reticulate) continuum of Darwinian/materialist evolutionary change. Each shaded curve represents the frequency distribution of formal variants at a particular point in time. At time A, lineages 1 and 3 are diverging. At time B, lineage 1 has diverged into two (4 and 5), lineages 6 and 7 are converging, and lineage 8 is nearing extinction. At time C, lineages 10 and 11 are converging, hence the skewed frequency distributions of formal variants. At time D, lineage 12 is diverging. If each plotted frequency distribution represents an artifact assemblage or component (or set of fossils), and these distributions comprise the total known archaeological (fossil) record for an area, the analytical problem of culture history (paleobiology) is to determine their evolutionary relations, such as are indicated by the variously flowing, diverging, and converging lineages.

cisely how biologists distinguish between homologs and analogs (e.g., Rieger & Tyler 1979, Wagner 1989). The wings of eagles and those of crows are structurally as well as superficially similar; this is homologous similarity. The wings of eagles and those of bats are superficially, but not structurally, similar; this is analogous similarity. The important issue of distinguishing between analogous and homologous traits was only rarely mentioned and virtually never addressed in detail by culture historians, which was part of the reason the paradigm fell from favor in the second half of the twentieth century (Lyman et al. 1997b). But what is of most interest here is why culture historians abandoned a materialist conception of evolution in favor of an essentialist conception. By adopting the latter, the ability to distinguish between the two kinds of similarity became, in the minds of some, unnecessary.

EVOLUTION AT THE TURN OF THE CENTURY

Many archaeologists believe that "cultural evolution was generally anathema as late as the 1950s" but that it was revived by the end of that decade (Willey & Sabloff 1993: 220; see also Ehrich 1963: 21 [for an alternative view, see Trigger 1989: 295]). To be sure, cultural evolution—in the sense we have defined it here—did enjoy only minimal status within Americanist archaeology between about 1900 and the middle 1950s, faring only slightly better in anthropology as a whole. Willey and Sabloff's (1993: 306) contention that cultural evolution was dead during the first half of the century apparently was based in part on the fact that William Haag's (1959) review article "indicates only a limited number of evolutionary efforts in the field." Haag's article, however, published in the Anthropological Society of Washington's volume commemorating the centennial of Darwin's *On the Origin of Species* (Meggers 1959), was not intended to be an extensive review of the evolutionary-archaeological literature, nor was it written to adequately sample that literature. Gordon Willey's (1960) review, which was published at about the same time and in another centennial volume (Tax 1960), contained many more references, but most of them were to works that were published in the 1950s and represented recent syntheses that Willey was using to discuss the overall evolutionary trajectory of American cultural lineages.

We believe that there should be no disagreement at all over the alleged death and resurrection of cultural evolution because it is easy to demonstrate that the general notion of evolution—cultural or otherwise—played a *major* role in the growth and development of the culture-history paradigm in Americanist archaeology. Given the goal of that paradigm, how could it not have? As South (1955) indicated, its use was often covert and seldom explicit, but it was there nonetheless. Archaeology, after all, had one thing that cultural anthropology did not—access to an extensive temporal record. Everyone knew that evolution takes place over time. Once it was clear that measurable temporal differences existed in the archaeological record of the Americas, evolution of some kind was an obvious if not always explicit or well-developed explanation *of* that difference. More importantly, the *kind* of evolution used by archaeologists in the first several decades of the 20th century was at least conceptually of the Darwinian, or materialist, sort. However, this brand of evolution has been reborn only in the last twenty years (O'Brien 1996b). What seems to have happened is that in the 1940s and 1950s, an essentialist *cultural* evolution usurped the role previously played in Americanist archaeology by Darwinian materialist evolution. Understanding why cultural evolution replaced Darwinian evolution is important for those of us who wish to see the latter now replace the former. Knowing this history, various objections can be anticipated and defenses prepared. More importantly, we can learn from the mistakes of our Darwinian-leaning predecessors and avoid those errors in the future.

Evolution in Cultural Anthropology

Received wisdom holds that Boas killed any notion of cultural evolution at the end of the 19th century—an erroneous assumption attributable in large part to the grumblings of White (1943, 1945a, 1945b, 1947, 1959a, 1959b) but also to the reading of the death notice by others (e.g., Lesser 1952). White seemed to relish any opportunity to point out just how wrong-headed and closed-minded Boas and his students were. However, Boas and some other leading figures in anthropology at the time did not reject outright the basic notion of *evolution*, a point White (e.g., 1947) only occasionally acknowledged. Even today, Boas is often attributed with such a stance (e.g., Willey & Sabloff 1993). But Boas (1896: 904) merely was concerned with the fact that the theory of Spencerian *cultural evolution* resulted in a comparative anthropology founded on the interpretive "assumption that the same [cultural] features must always have developed from the same causes, [leading] to the conclusion that there is one grand system according to which mankind has developed everywhere; that all the occurring *variations* are no more than minor details in this grand uniform evolution" (emphasis added; see also Boas 1904: 519). Given that the key assumption—essentialist change was the only kind that had occurred—was demonstrably wrong, Boas (1896: 907) preferred the "historical method" and what he believed was its ability to discover the processes that resulted in the *particular* development of different cultures. The historical method was more in line with the materialist metaphysic and would ultimately reveal—inductively, in Boas's (1896, 1904) view—laws concerning how cultures evolved rather than assume them, as did the comparative method.

It was clear to Boas that each culture's heritage—its developmental lineage—was unique. Perhaps most importantly, Boas (1904: 516) noted that in the second half of the 19th century, the "regularity in the processes of [Spencerian] evolution became the center of attraction [to Morgan and Tylor and others] even before the processes of [cultural] evolution had been observed and understood." Thus, in Boas's view, as more data accumulated, Spencerian evolutionary theory came to be less and less tenable. As a result, Boas (1904: 522) indicated that instead of "a simple [universally applicable] line of evolution there appear a multiplicity of converging and diverging lines [read *cultural lineages*] which it is difficult to bring under one system." The important point here is that Boas did not reject evolution *per se.* As Bidney (1946: 297) pointed out, "Boas never rejected the concept of cultural evolution but merely the notion of uniform laws of cultural development and the *a priori* assumption that cultural development was always from a hypothetical simplicity to one of complexity." Similarly, some years later Naroll (1961:391) observed that "the reason Boas rejected 19th century [Spencerian] evolutionism was neither theoretical [n]or conceptual, but methodological. . . . Boas always maintained that the study of cultural evolution was a major task of cultural anthropology, but Boas was bitterly opposed to confusing speculation with scientific generalization." Thus, Boas rejected the interpretive assumption of Spencerian cultural evolution, and in so doing he rejected its essentialist metaphysic. Importantly, he retained the materialist notions of evolution as history and of variation as critical to any study of evolution.

Some of Boas's intellectual progeny followed his lead and made some rather astute observations that would greatly influence archaeological thinking. In the first edition of his classic *Anthropology*, Kroeber (1923: 3) used the term "tradition" to label the processes of cultural transmission and heredity, thus distinguishing them from the biological "force" of genetic heredity. Both biological evolution and cultural development involved transmission (as Boas [1904] had indicated), but because transmission within the latter could be between individuals who were not "blood descendants," the concept of biological evolution—that is, genetic transmission and change—was "ambiguous" in a cultural setting (Kroeber 1923: 7). Thus Kroeber's view paralleled that of Darwin. The latter avoided the term evolution—it was used by Spencer to label his social philosophy—and instead used "descent with modification." Morgan and Tylor had used the term evolution precisely in the Spencerian sense, and it was that sense of the term that Kroeber (and Boas) rejected, not the basic notion of evolution as change through time or descent with modification.

Kroeber failed to distinguish between what we have termed cultural evolution and Darwinian evolution. This was not critical to his discussion because not only did his comment that the term evolution was ambiguous signify his rejection of cultural evolution, it signified his rejection of Darwinism as well. Regarding the latter, cultural lineages were, after all, not the result of genetic transmission. As a result, in Kroeber's (1923: 8) view, "the designation of anthropology as 'the child of Darwin' is most misleading. Darwin's essential achievement was that he imagined . . . a mechanism [natural selection] through which organic evolution appeared to be taking place . . . [As a result, a] pure Darwinian anthropology would be largely misapplied biology." What had "greatly influenced anthropology," according to Kroeber (1923: 8), "has not been Darwinism, but the vague idea of evolution . . . It has been common practice in social anthropology to 'explain' any part of human civilization by arranging its several forms in an evolutionary sequence from lowest to highest and allowing each successive stage to flow spontaneously from the preceding—in other words, *without specific cause*" (emphasis added). This had been Boas's (1896, 1904) point too. Kroeber saw this Spencerian, essentialist notion as nonsense. As Bidney (1946: 295) argued a few years later, "*the concept of logical stages of cultural development presupposes the notion of the actual history of the cultural process* . . . [T]he stages of cultural development are but abstractions, useful to the student of culture, but not ultimately intelligible or explanatory of the dynamics of culture." To Kroeber (1923: 5), the development of a cultural lineage was historical, and historians "deal with the concrete, with the unique; for in a degree every historical event has something unparalleled about it . . . [Historians] do not lay down exact laws." This didn't mean that cultures do not evolve; it only meant that the Spencer-Morgan-Tylor model of essentialist cultural evolution was incorrect. Was there an alternative?

For Kroeber (1916c, 1917), culture was like a stream—it had a flow. It was heritable not in any genetic sense but rather was transmitted via learning (see also Boas 1904). Kroeber's notion of culture change was well expressed by one of his students, Nels Nelson (1932: 103), who wrote of "implements or mechanical inventions, i.e., material culture phenomena, as parts of a unique unfolding *process* which has much in common with that other process observed in the world of nature and generally called organic evolution" (emphasis in original).

Importantly, the evolution of culture was not only continuous, it also was gradual: "[A] study of the history of mechanization reveals few if any absolutely original contrivances that were not essentially the results of gradual transformation or combination of older inventions; that in reality spurts, mutants or leaps are as rare among artificial (intellectual) phenomena as among natural phenomena" (Nelson 1932: 122). Thus, there was a "mechanical culture stream" (Nelson 1932: 109).

As another student of his, Julian Steward, wrote, Kroeber "added time depth to the essentially synchronic ethnology of Boas" (Steward 1962: 203) and believed that the evolution of a cultural lineage involved a braided stream–what a Darwinist today would label a *reticulate* pattern of intersecting and diverging lineages–created by such processes as diffusion, trade, and migration. That this idea came from Boas is clear:

> There is one fundamental difference between biological and cultural data which makes it impossible to transfer the methods of the one science to the other. Animal forms develop in divergent directions, and an intermingling of species that have once become distinct is negligible in the whole developmental history. It is otherwise in the domain of culture. Human thoughts, institutions, activities spread from one social unit to another. As soon as two groups come into close contact their cultural traits will be disseminated from the one to the other (Boas 1932: 609).

Kroeber "constantly saw changes in styles as flows and continua, pulses, culminations and diminutions, convergences and divergences, divisions, blends and crosscurrents by which cultures develop and mutually influence one another" (Steward 1962: 206). He did not, however, attempt to determine causes of culture change, no doubt because his mentor (Boas) had indicated that the reticulate evolution of cultures "puts the most serious obstacles in the way of discovering the inner dynamic conditions of change. Before morphological comparison [read *evolutionary synthesis;* see Boas 1896] can be attempted the extraneous elements due to cultural diffusion must be eliminated" (Boas 1932: 609). Probably as a result, Nelson (1932: 122) wrote that "Final explanations of [culture change], as well as of the driving force and the ultimate goal of culture, may be left to the philosophers." White later would have something to say about this, as well as about the explanatory power of cultural evolution.

Genetic transmission was omitted from consideration of cultural development by Kroeber and his students for two reasons: (1) cultural ideas were not genetically transmitted, and (2) the transmission and inheritance of ideas could be up as well as down between generations and also within a generation. This prompted Kroeber (1923) to avoid adopting the biological version of evolution then current–a version that, after 1900, was dominated by genetics and its attendant mechanisms such as transmission, mutation, inheritance, and speciation. Natural selection, along with phenotypic variation, was relegated to the back room; these two critical concepts would reassume their proper place in biological theory only with the synthesis of the late 1930s and early 1940s (Mayr 1982). There was another reason not to transfer Darwinian evolution as an explanatory theory to culture–the evolution of culture was reticulate, whereas the evolution of biological organisms was simply and only branching. This was made clear in the second edition of Kroeber's (1948) *Anthropology,* where he illustrated the differences between the evolution of organisms and the evolution of cultures (Figure 2.2).

The reticulate form of the evolution of cultures was recognized prior to the publication of Kroeber's textbook. Kroeber himself, in an earlier, short article titled "Historical Reconstruction of Culture Growths and Organic Evolution," had written that while a "culture complex is 'polyphyletic' [and] a [biological] genus is, almost by definition, monophyletic. . . . the analogy does at least refer to the fact that culture elements [traits] like species represent the smallest units of material which the historical anthropologist and biologist respectively have to deal with" (Kroeber 1931: 149). The stumbling block was the implicit equation of a culture with an essentialist biological concept. Today, a biological species is a population that is reproductively isolated from other such populations; in 1930, the concept was more strongly founded in the essentialist metaphysic than in the materialist metaphysic (Mayr 1982: 272). Therefore, the problem was that Kroeber chose the wrong biological unit–species–to equate with cultures, no doubt because of the lack of theory. This problem of unit choice and construction would continue to plague the adoption of Darwinism by anthropologists through the 1950s (Birdsell 1957) and into the 1990s (Dunnell 1995).

In 1930, the conceived difference between the branching evolution of organisms and the reticulate evolution of cultures had considerable impact on Americanist archaeology's use of Darwinian evolution. But so too did the notion that evolution consisted only of genetic change. To illustrate this, we now turn attention to how archaeologists of the early 20th century used the concept of evolution.

Evolution in Archaeology

A Darwinian sort of evolution played a major role within culture history virtually since its birth as a viable

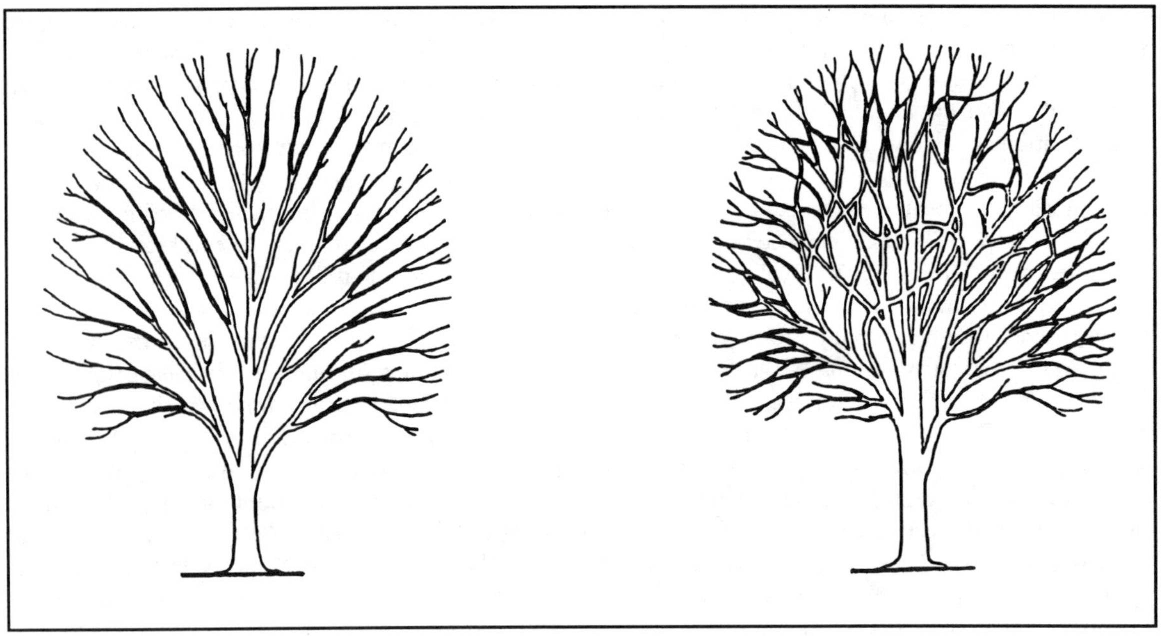

Figure 2.2. A. L. Kroeber's (1948:260, figure 18) tree of biological evolution (left) and tree of cultural evolution (right). Note the simple branching structure of the former and the reticulate (branching and intersecting) nature of the latter.

paradigm in Americanist archaeology. For example, when commenting on Max Uhle's (1907) stratigraphic excavations at the Emeryville Shellmound in San Francisco Bay, Kroeber (1909: 16) noted that most artifact types were found throughout the stratigraphic sequence and that although there was "some gradual elaboration and refinement of technical process . . . it was change of *degree* only" (emphasis added). Historians have tended to focus on the last phrase and have argued that Kroeber's vision of culture change was of the essentialist kind (Lyman et al. 1997b, Rowe 1962). Although this is correct, Kroeber clearly was aware of the fact that what he was looking at involved change of the materialist sort, though he did not recognize that his *conception* of culture change involved a particular metaphysic. He in fact measured just such materialist change a few years later when he invented frequency seriation (Kroeber 1916a, 1916b; Lyman et al. 1997b).

Two of Kroeber's students mimicked this materialist method of measuring the passage of time. Nelson's (1916) excavations in the Southwest and his use of percentage stratigraphy to show the waxing and waning of a pottery style's popularity was decidedly materialistic. Nelson's data consisted of the frequencies of ideational units termed types and showed continuous and gradual change through time. He had excavated precisely because he wanted to establish the chronological *continuity*–not just the relative temporal *positions*–of his pottery types (Lyman et al. 1997b). And he spoke, no doubt metaphorically, of types becoming extinct. Leslie Spier (1917a, 1917b) also used percentage stratigraphy and frequency seriation of ideational units to measure culture change in the Southwest. Although Nelson and Spier tended to avoid using the terms of Darwinian evolutionary theory in discussing what they documented, others were not so shy.

Kidder (1915: 453; 1917), for example, discussed different styles of pottery in terms of ancestral–descendent relationships. One pottery type might "father" another (Kidder 1915: 453), and, in wording similar to that of Nelson, a pottery type might become "extinct" (Kidder & Kidder 1917: 348). Kidder's types were of a kind that, hopefully, reflected the passage of time *as well as* evolutionary–that is, phyletic–relations, as is clearly implied in his seldom-mentioned 1917 paper on sequences, or "series," (Kidder 1917) of ceramic designs (Figure 2.3). Kidder (Kidder & Kidder 1917: 349) stated that "One's general impression is that [the types] are all successive phases of [particular pottery traits], and that each one of them developed [read *evolved*] from its predecessor." Kidder's word "phase" is a common-sense English designation for a portion of a continuum. To measure time, Kidder erected types that were both "more or less arbitrarily delimited chronological subdivisions of material" (Kidder 1936b: xxix) and "chronologically seriable" (Kidder 1936a: 625). Kidder's types were ideational units constructed to measure time: Conceptually, (1) variation was continuous–things were in the continuous process of becoming, and thus types had to be of the ideational sort to allow the measurement of change; and (2) change in

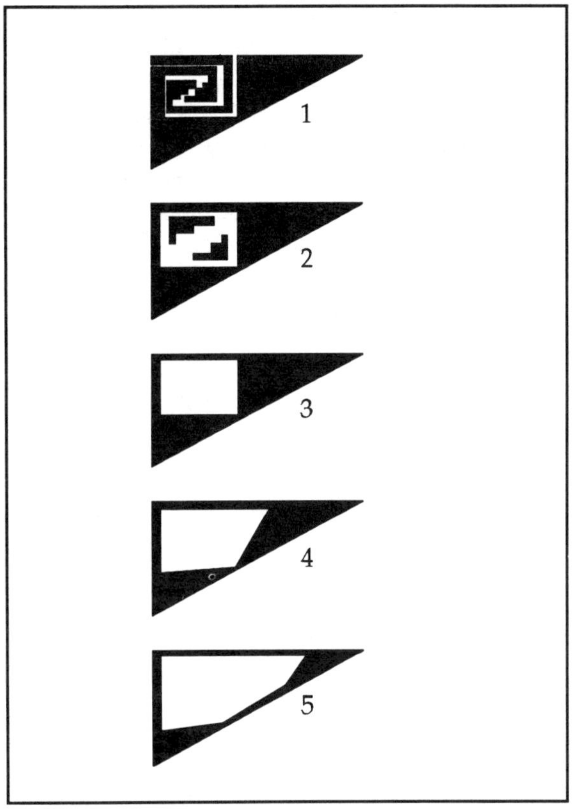

Figure 2.3. A. V. Kidder's (1917) example of a phyletic evolutionary "series" of Southwestern ceramic design elements.

the frequencies of types reflected the passage of time. This was, simply, the materialist metaphysic, but Kidder would abandon it in his Pecos volumes and elsewhere.

Numerous individuals (e.g., Dutton 1938, Ford 1935, Hawley 1934, Kniffen 1938, Martin 1936, Olson 1930, Reiter 1938b, Schmidt 1928, Strong 1925, Vaillant 1935) over the next two decades mimicked the methods and techniques of Kroeber, Nelson, Spier, and Kidder. Without explicitly recognizing it, they were using the materialist metaphysic and ideational units to document culture lineages. But they all also talked about *sections* of the materialistically measured cultural continuum–sections that were variously labeled cultures, complexes, periods, phases, and the like–as if they were real, phenomenological units (e.g., Kidder 1924, 1927). Such discussions represented what has been termed the materialist paradox (Dunnell 1995: 34). If cultures evolved, how could such units be real? Even Kroeber (1916a, 1916b), who surely must have recognized that his periods were arbitrary chunks of the continuum, treated his cultural units as if they were real, as did everyone else. What in reality were a few randomly selected frames of a reel of film–or arbitrarily chosen (they were accidents of sampling) one-foot sections of a mile-long cultural stream–conceptually came to have essentialist realities. The film had definite acts, and the stream had placid pools punctuated by brief but violent rapids.

Conflation of the materialist and essentialist metaphysics began with, among others, Kroeber (1916a, 1916b), who spoke of the sections of the continuum he distinguished in the Southwest as epochs, periods, eras, and the like. A few years later, Kroeber encountered the materialist paradox head-on. In his *Handbook of the Indians of California*, Kroeber (1925: 926) indicated that he was interested in the documentation that archaeology could provide on "the development of culture" in the state. By that time, it was clear to Kroeber (1925: 927) and other culture historians that "the correct [excavation] procedure [was] to follow lines of deposition in instituting comparisons; but this is not practical, stratification being confined to limited areas and often wholly imperceptible." After examining the shifting frequencies of various artifact types across selected vertical columns of sediment, Kroeber (1925: 926, 930) concluded that geographic subdivisions, or provinces, of the California culture area "were determined a long time ago and have ever since maintained themselves with relatively little change"; that "the basis of culture remained identical during the whole of the shell-mound period"; and that in California, "civilization, such as it was, remained immutable in all fundamentals." That his artifact types were variously functional or simply descriptive and not historical went unnoticed.

Most importantly, Kroeber (1925: 931) indicated, first, that types "must, of course, be interpreted as periods"; second, differences in types represented differences in culture; and third, cultures differed because the human groups that bore the cultures and made the artifacts differed. In short, Kroeber's types of artifacts and types of cultures were in some sense real, essentialist, units. Spier (1917a, 1917b, 1918, 1919) had interpreted his pottery data from the Southwest in just such terms, as had others (e.g., Kidder 1924, 1927; Schmidt 1928, Vaillant 1935). Kidder, for one, recognized the problem almost immediately, as revealed in the following statement:

> The division of the Glaze ware of Pecos into six chronologically sequent types is a very convenient and, superficially, satisfactory arrangement. For some time I was very proud of it, so much so, in fact, that I came to think and write about the types as if they were definite and describable entities. They are, of course, nothing of the sort, being merely useful cross-sections of a constantly changing cultural trait. Most types, in reality, grew one from the other by stages well-nigh imperceptible. My groupings therefore amount to a selection of six recognizable nodes of individuality; and a forcing into

association with the most strongly marked or "peak" material of many actually older and younger transitional specimens. . . . This pottery did not stand still; through some three centuries it underwent a slow, usually subtle, but never ceasing metamorphosis (Kidder 1936b: xx).

Kidder's comment concerned the materialist paradox—that is, the conceived materialistic "slow, usually subtle, but never ceasing metamorphosis" of artifact forms through time—was monitored using typological (in the sense of Mayr 1959), essentialist, "*recognizable* nodes of individuality." The important point is that despite this early recognition of the paradox, the equation of types of artifacts with particular *cultures* became the interpretive algorithm for culture history. Such an equation represented the first step toward adoption of the essentialist metaphysic and would eventually lead to the rejection of any form of materialist evolution. The conceived reality of the sections of the temporal continuum was reinforced by the perceived discontinuity of the stratigraphic column represented by discrete strata because the set of artifacts within each stratum was conceived as representative of a discrete cultural occupation (e.g., Fowke 1922, Thompson 1956, Wissler 1917a). As a result, the use of strata (or arbitrary levels) as artifact-collection units not only was beneficial to Americanist archaeology (because it allowed time to be measured) but came to have a serious cost as well.

Darwinism's Last Hurrah

The basic notion of Darwinian evolution affected the interpretations of early culture historians, but it was unclear to them how to transfer such a purely genetic process—as biological evolution was then thought to be—to a nongenetic phenomenon such as culture. Nor was it clear how the merely branching pattern of biological evolution could be applied to the clearly reticulate evolution of culture. Kroeber continued to grapple with these problems and managed early in the 1930s to identify one of the significant aspects of biological evolution that should have been employed by culture historians. Kroeber (1931: 151) pointed out that the "fundamentally different evidential value of homologous and analogous similarities for determination of historical relationship, that is, genuine systematic or genetic relationship, has long been an axiom in biological science. The distinction has been much less clearly made in anthropology, and rarely explicitly, but holds with equal force." Kroeber (1931: 151) went on to imply that a "true homology" denoted "genetic unity," and he argued that

. . . there are cases in which it is not a simple matter to decide whether the totality of traits points to a true [homologous] relationship or to secondary [analogous, functional] convergence . . . Yet few biologists would doubt that sufficiently intensive analysis of structure will ultimately solve such problems of descent . . . There seems no reason why on the whole the same cautious optimism should not prevail in the field of culture; why homologies should not be positively distinguishable from analogies when analysis of the whole of the phenomena in question has become truly intensive . . . [S]uch analysis has often been lacking but judgments have nevertheless been rendered (Kroeber 1931: 152–153).

Kroeber's (1931) remarks were largely ignored, leading him to lament over a decade later that anthropology was still "backward" with regard to distinguishing between analogous and homologous similarities (Kroeber 1943: 108). Instead of implementing Kroeber's suggestions, archaeologists adopted the rather more easily conceived and implemented dictum that "typological similarity is [an] indicator of cultural relatedness" (Willey 1953: 363). A not-so-liberal interpretation of this notion is found in the morphological species concept of early 20th-century biologists who still adhered to the essentialist metaphysic: Morphologically similar species were members of the same taxon or type and thus were phylogenetically related (Mayr 1982: 270). This notion was axiomatic in culture history (Willey 1953: 363), and in both contexts it put the cart before the horse. As George Gaylord Simpson (1961: 69) pointed out some years later, "individuals do not belong in the same taxon because they are similar, but they are similar *because they belong to the same taxon*" (emphasis added). In other words, archaeologists noted similarities between artifact types, assemblages of artifacts, and the like, but those similarities might not be of the homologous sort requisite to determining phylogenetic relations and writing phylogenetic histories (e.g., Szalay & Bock 1991). That some archaeologists were misled by Willey's axiom is clear. Not everyone was misled, but, as we discuss later, the objections of this minority ultimately led to the rejection of any potential archaeological utility of Darwinian evolution.

Ceramic and Cultural Phylogenetics

Shortly after Kroeber's (1931) paper was published, the Darwinian notion of evolution took on a form within Americanist archaeology that prompted several commentators to reiterate Kroeber's reasons for rejecting Darwinism as a viable theoretical model for the explanation of culture change. Harold S. Gladwin, working in the Southwest, created not only a binomial system of pot-

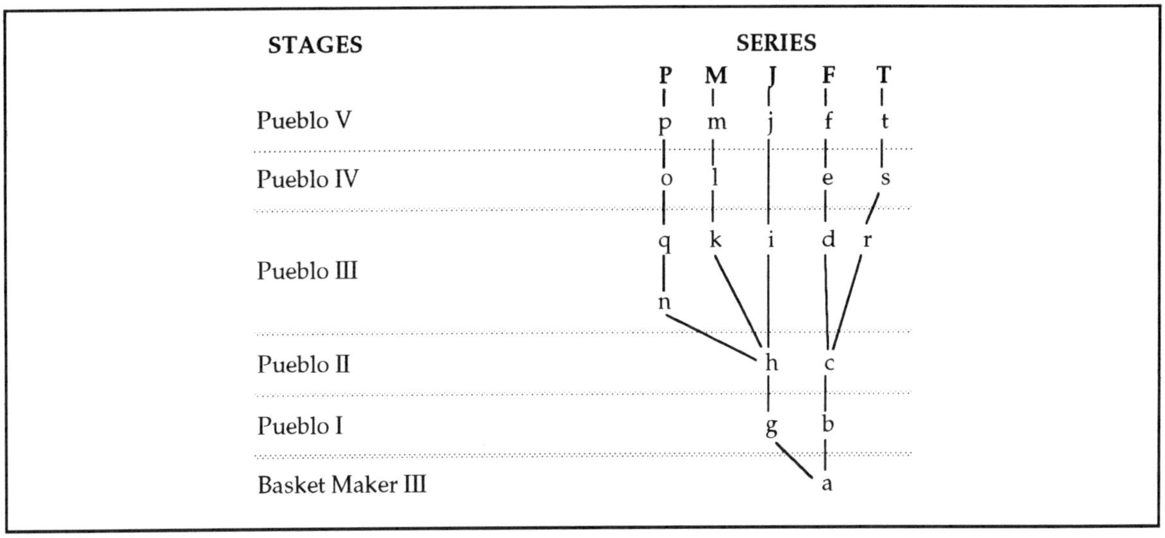

Figure 2.4. Colton and Hargraves's (1937:4 figure 1) model of the phylogenetic evolution of pottery. Each upper-case letter represents a ceramic "series" in the sense of Kidder (1917; see Figure 2.3); each lower-case letter represents a type; phylogenetic relations of types are indicated by the vertical lines connecting lower-case letters. Colton and Hargraves's figure caption reads, in part, "type c is collateral to types d and r, derivative [descended] from type b, and ancestral to types f and t . . . types q and n are both ancestral to type o, but collateral to each other, and derivative from types h, g, and a."

tery classification explicitly modeled on the biological genus-and-species concept (Gladwin & Gladwin 1930), but he also created a hierarchical structure for organizing archaeological units variously termed roots, stems, branches, and phases (Gladwin & Gladwin 1934). The latter were meant specifically to assist in the working out of "a comprehensive scheme by which relationships and relative chronology could be expressed" (Gladwin & Gladwin 1934: 8-9). In a few short years, Gladwin (1936) abandoned the scheme. His statement to that effect is revealing:

> My original suggestion . . . of using a generic [read *genus*] and a specific [read *species*] name for pottery types implied a biological analogy which I now think was a mistake. The idea is being carried too far along biological or zoological lines, and men do not realize the profound differences which exist between zoological species and the things which have been made by men and women.
>
> Zoological species do not cross and intergrade; evolution is so slow as to be hardly distinguishable. The evolution of culture . . . was stepped up to almost incredible speed, and on every side we find evidence of merging and cross-influences (Gladwin 1936: 158).

The individuals who were carrying the idea too far no doubt included Harold S. Colton, a professional biologist by training who left a professorship of biology to pursue archaeological interests. Along with Lyndon L. Hargrave, Colton published a major statement on the phylogenetic implications of pottery types. A *type* was "a group of pottery vessels which are alike in every important characteristic except (possibly) form," and a *series* was "a group of pottery types within a single ware in which each type bears a genetic relation to each other" (Colton & Hargrave 1937: 2-3). Thus, Colton and Hargrave's series was identical to Kidder's (1915, 1917). But Colton and Hargrave carried the biological analogy further than their predecessors had and distinguished among derived, collateral, and ancestral types and graphed the relations among them (Figure 2.4). Although Colton's knowledge of Darwinian evolution no doubt underpinned their scheme, the key to it was in the supposition that related forms were related because they were similar. As we pointed out above, it should be the other way around—similar forms are similar because they are related. The problem was that Colton and Hargrave offered no argument—nor did anyone else at the time—for the belief that the similarities described were of the homologous sort.

Kidder's (1915, 1917) simpler phyletic scheme (Figure 2.3) denoted some of the same sorts of relations between types, but his contemporaries did not heap criticism on him the way they did on Colton and Hargrave. In his review of their work, James Ford (1940: 265) noted that Colton and Hargrave had ignored the problem of selecting "a class of features [attributes] which will best reflect cultural influences [e.g., transmission via contact or heredity], and which in their various forms will be mutually exclusive, to serve as guides in the process" of determining ancestral–descendent relations. Ford was, without using the term, speaking of homologous similarity, but he failed to make this clear. In another review, Paul Reiter (1938a: 490) noted that he "was unable to find a single instance of proof of [the genetic relations of pottery types]." Colton and Hargrave (1937: 2–3, 5) indicated that genetic relations among types were "obvious" and "clearly revealed" and that "definite evolutionary characters were recognized," but they provided no list of them, nor did they provide guidance as to how such characters were to be analytically identified. They apparently were unaware of Kroeber's (1931) guidance in this regard.

Reiter (1938a: 490) also noted that Colton and Hargrave's pottery typology tended to ignore variation, and he insisted that "variation tendencies cannot be overlooked if genetic or chronologic emphasis is strong." Reiter's comments were on the mark–they underscored that a materialist conception of change was requisite to what Colton and Hargrave were attempting. But Colton and Hargrave's types were not ideational units that could be used to measure variation. The types were empirical, as evidenced by Colton and Hargrave's (1937: 30–31) suggestions that the members of a type "will not [always] fit the [type] description perfectly" and that "there are sherds that are intermediate between types."

Colton and Hargrave's classification scheme was hierarchical, but that structure was void of explanatory content. In another monograph, Colton (1939) employed the terms then becoming popular within the Midwestern Taxonomic Method, an organizational method explicitly built to be void of explanatory content (McKern 1934, 1937, 1939). As the name of the method implied, the structure of the system was hierarchical. In combination with the use of the Midwestern Taxonomic Method's terminology, then, Colton's (1939) phylogenetic interpretations, not surprisingly, were criticized. Reviewers focused on Colton's interpretations rather than on his methods. Erik Reed (1940:190) thought the scheme was a reasonable one and remarked that Colton's "genetic and temporal approach seems more desirable in a region such as this where chronology is relatively well-known." Chronological control is surely a requisite of determining a phylogenetic history, but it is only one of several requirements. Reed's failure to note that Colton had not established that the typological similarities he discussed were of the homologous sort was, however, typical rather than anomalous.

Not everyone overlooked the point, however. Julian Steward rejected Colton's analytical procedure and the results for the same reason that Kroeber had earlier rejected Darwinian evolution–it was nonapplicable to cultural phenomena:

> It is apparent from the cultural relationships shown in this scheme that strict adherence to a method drawn from biology inevitably fails to take into account the distinctively cultural and unbiological fact of blends and crosses between essentially unlike types . . . It is true that cultural streams often tend to be distinct, but they are never entirely unmixed and often approach a complete blend. A taxonomic scheme cannot indicate this fact without becoming mainly a list of exceptions. It must pigeon-hole . . . [T]he method employed inevitably distorts true cultural relationships (Steward 1941: 367).

Steward simply couldn't accept a taxonomic structure for any classification of cultural phenomena because such a structure implied the same thing to him that the Linnaean taxonomy tended to imply to some biologists–phylogenetic or branching evolution. This is clear in Steward's (1942, 1944) discussion of the Midwestern Taxonomic Method. Steward (1942: 339) offered the typical criticism of that method when he noted that it produced a "set of timeless and spaceless categories." In response, McKern (1942) protested that he had merely set those dimensions aside for the moment in favor of the formal dimension and had explicitly *not* discarded the time-space dimensions.

Steward's (1944: 99) rebuttal entailed two elements and is where the heart of the matter resides. He simply could not conceive of how taxonomic or hierarchical classification, which for him denoted branching evolution, could be forced onto cultural phenomena that not only branched or diverged through time but that also converged to create a reticulate form of evolutionary descent. Steward's implicit equation of species and cultures–both conceived as essentialist units–is clear. Second, Steward could not find any utility in the method because it seemed to lack any reference to theory. The latter was one reason numerical phenetics, also termed numerical taxonomy, fell from favor in biology several decades later (Mayr 1969), but Steward's observation had minimal impact on archaeological thought. Interestingly, the most damaging criticism of what had begun as Kidder's (1917) phyletic series of ceramic designs and what eventually grew into Colton and Hargrave's (1937) collateral and descendent types resided elsewhere.

The Paradox

J. O. Brew (1946: 46) argued that classifications are arbitrary constructs of the analyst–"no typological system is actually inherent in the material"–and that a classification should fit the purpose of the investigation. He clearly recognized the interplay of theory and unit construction and argued that we should change our classifications "as our needs change and as our knowledge develops" (Brew 1946: 64). His conception of cultural change was the same as Kroeber's: "We are dealing with a constant stream of cultural development, not evolutionary in the genetic sense, but still a continuum of human activity" (Brew 1946: 63). In holding to such a materialist conception of culture change, Brew (1946: 65) argued, we "must ever be on guard against that peculiar paradox of anthropology which permits men to 'trace' a 'complex' of, let us say, physical type, pottery type, and religion over 10,000 miles of terrain and down through 10,000 years of history while in the same breath, or in the next lecture, the same men vigorously defend the theory of continuous change." The paradox emanates "from the belief that the manufactured groups [types] are realistic entities and the lack of realization that they are completely artificial. . . Implicit in [the belief] is a faith . . . in the existence of a 'true' or 'correct' classification for all object, cultures, etc., which completely ignores the fact that they are all part of a continuous stream of cultural events" (Brew 1946: 48).

Although the term was not in general use, no one noticed that Brew had described the materialist paradox. What *was* taken notice of was Brew's (1946: 53) observation that the evolutionary implications of the Gladwin–Colton scheme were unacceptable for the simple reason that "phylogenetic relationships do not exist between inanimate objects" such as pot sherds. This statement was repeated by Beals et al. (1945: 87), who saw Brew's manuscript before it was published. Ironically, they mimicked Kidder's (Figure 2.3) technique of constructing ceramic series, as did others (e.g., Wheat et al. 1958). Brew's discussion nonetheless had one apparent impact on Americanist archaeology—to drive the final nail into the coffin containing the Gladwin-Colton scheme; no one after that time discussed ceramic series in such flagrantly Darwinian terms. That this was the singular impact of his arguments probably resulted because Brew could offer no clearly articulated alternative. For example, he stated that the "only defense there can be for a classification of [artifacts] based upon phylogenetic theory is that the individual objects were made and used by man" (Brew 1946: 55), but he failed to make the conceptual leap to "replicative success" (Leonard & Jones 1987)—the cornerstone of a Darwinian archaeology—for two reasons.

First, to Brew, evolution involved only the processes of genetic transmission and genetic change. There was, then, only a weak correlation between an organism and the "artifacts" that that organism might produce, such as birds and egg shells or molluscs and mollusc shells, and no connection at all between people and their artifacts (Brew 1946: 55–56). Since artifacts were not genotypic phenomena, they were not subject to evolutionary forces. That it is the phenotype—of which artifacts are an expression—that undergoes selection escaped Brew's—and contemporary biologists'—attention for a number of years (Leonard & Jones 1987, Mayr 1982). Second, Brew quoted a single biologist—a geneticist—who argued that a phylogenetic (or phyletic) history did not *explain* organisms; hence, to Brew (1946: 56), it could hardly explain artifacts: "This is a most important point, and I wish to emphasize it here." Of course a phylogenetic history is *not* an explanation, but such a construct *is* a requirement of using Darwinian evolutionary theory to explain the diversity of forms of organisms and their distributions in time and space (Szalay & Bock 1991).

THE REBIRTH OF CULTURAL EVOLUTION

Brew's (1946) and Steward's (1941, 1942, 1944) devastating criticisms of any suggestion that the evolution of cultures might have some similarities to Darwinian evolution and be discussed using Darwinian terms essentially killed any such notions from being stated explicitly. Archaeologists still used Kidder's (1917) phyletic-seriation technique (e.g., Beals et al. 1945, Wheat et al. 1958), but they were not explicit about the Darwinian implications of the results. Cultures still came in contact—the evolution of cultures was reticulate—but biologists would not come to recognize how often Darwinian evolution took that form until the 1980s, when studies of plant evolution became more rigorous and studies of animal evolution became more detailed (e.g., Davis 1996, Szalay & Bock 1991, Weiner 1994). Rather than retooling materialist evolution into something that was applicable to archaeological phenomena and that was aligned with their materialistic conceptions of culture change—just as paleobiologists had done to make it applicable to the fossil record (fossil bones, like pots, don't interbreed)—culture historians did something else. They adopted both Willey's (1953: 363) axiom that "typological similarity is [an] indicator of cultural relatedness" and a reborn version of Spencerian cultural evolution. There were problems with both, but the adoption was not a long and painful one because there was no effective competition; the only potential competitor had been eliminated, and the winner was therefore chosen by default.

Typological Similarity and Homologous Similarity

Willey (1953) disliked the Gladwin-Colton scheme, and when he and Philip Phillips published "Method and Theory in American Archaeology: An Operational Basis for Culture-Historical Integration" in 1953, they explicitly stated that the archaeological "use of the organic evolutionary model is, we believe, specious" (Phillips & Willey 1953: 631). Their basic operative unit was a *phase*—an archaeological manifestation of a culture that had some ethnographic reality. They suggested that the use of *traditions* and *horizon styles* would reflect the braided stream of the evolution of cultures and allow one to correlate phases. As Irving Rouse (1954: 222) noted, a metaphor could be drawn between the use of horizons and traditions as integrative devices for archaeological materials distributed across an area and a rectangular piece of cloth, the side edges representing the geographical limits of the area and the top and bottom edges representing the temporal limits:

> The warp threads of the cloth consist of a series of regional traditions running from the bottom towards the top of the cloth, while the weft is composed of a number of horizon styles which extend from one side of the cloth towards the other. The cloth is decorated with a series of irregularly arranged rectangles, each representing a single culture [read *phase*], and these are so colored that they appear to form a series of horizontal bands (Rouse 1954: 222).

But Rouse (1955) was concerned that the use of Willey's dictum that typological similarity denoted cultural relatedness was too simplistic. Rouse wanted to determine the phylogenetic relations among phases rather than merely to track the distributions of a few artifact classes, as the horizon and tradition units did. To illustrate this, he distinguished three ways to correlate phases. First, one might use a Midwestern Taxonomic Method-like procedure to group phases that shared traits. Why the traits were shared was a separate issue. Second, one could note similarities in the time–space distributions of phases. Identical or adjacent distributions of two or more phases "establishes contemporaneity and contiguity, or lack thereof, and nothing else" (Rouse 1955: 717). To argue that contemporaneous phases were phylogenetically related "because they share a given horizon style . . . is on the genetic rather than the distributional level of interpretation, for it requires an assumption that the style has diffused from one phase to the others with little or no time lag" (Rouse 1955:718). Third, one might trace the "genetic" relations among phases by establishing that the phases had been in "contact" temporally and spatially by using horizons and traditions that comprised homologous types (Rouse 1955:719). One needed to distinguish between analogous and homologous similarity, the linchpin to this kind of comparison, to ensure the relations were "genetic." The modeled result of such analyses is shown in Figure 2.5.

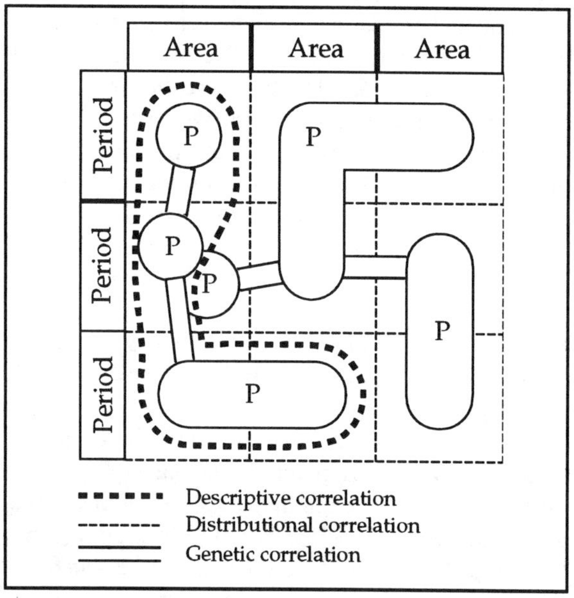

Figure 2.5. Irving Rouse's (1955:720, figure 4) model of the results of comparing phases (P) to show their relative contiguity in form, time, and space. Phases that were formally similar as well as contiguous in time and space had "genetic" relations.

Willey and Phillips (1958: 31) responded to Rouse (1955) by arguing that his "genetic" relations could only "be revealed and expressed by means of integrative concepts that are culturally determined." The phrase "culturally determined" was critical. Horizons and horizon styles by definition reflected cultural transmission or diffusion over space. A tradition was "a (primarily) temporal continuity represented by persistent configurations in single technologies or other systems of related forms" that operated at Rouse's (1955) "genetic level of interpretation" (Willey & Phillips 1958: 38) and reflected transmission or heredity across time. Thus, Willey and Phillips's conception of cultural development was well captured by the flowing braided-stream metaphor. Each trickle was a tradition that to varying degrees met and mixed with other trickles as denoted by horizons and horizon styles. Such a conception presumed that the typological similarities denoted by horizons and traditions were of the homologous sort.

For Willey and Phillips, horizons and traditions provided the empirical warrants for discussing the historical development of cultures. They were "integrative" units that denoted "some form of historical contact" *rather than* "implications of phylogeny" (Willey & Phillips 1958: 30). Culture history demanded "culturally determined" integrative concepts such as horizons and traditions, not phylogenetic ones (Willey & Phillips 1958: 30–31). Willey's (1953: 368) suggestion that "principles of continuity and change are expressed in the degrees of trait likeness and unlikeness which are the mechanisms of establishing the *genetic* lines binding the assemblages together" (emphasis added) thus was purely metaphorical. But cultural or historical "relatedness," when couched in a temporal framework aimed at studying the developmental lineages of cultures, such as that envisioned by Willey and Phillips, cannot fail to be phylogenetic in the sense of Darwinian evolution.

The generally discipline-wide abhorrence of anything Darwinian and/or genetic resulted in the contradiction internal to the Willey-Phillips scheme going unrecognized. Further, Kroeber's (1931, 1943) earlier critical point regarding the importance of the distinction between analogous and homologous similarity, even though repeated by Rouse (1955), was overlooked. It was, we suspect, overlooked at least in part because there was an alternative version of evolution that (1) did not require this critical distinction, rendering it pointless to consider; (2) did not entail genetic transmission, rendering further consideration of the mechanisms of heritability and thus consideration of continuity unnecessary; and (3) did not consider whether evolution was merely branching or reticulate, rendering this problem unworthy of further discussion. That version of evolution was Spencerian, or what we have termed *cultural*, and it had been lurking within anthropology since the early 1940s, precisely when Brew, Steward, and others were refuting the applicability of Darwinism to cultural phenomena.

LESLIE WHITE'S BRAND OF CULTURAL EVOLUTION

The 1940s witnessed the rebirth of cultural evolution within anthropology, initially at the hands of White (1943, 1945a, 1945b, 1947, 1949: 363–393) and later with contributions by Steward (1949, 1951, 1953). Not surprisingly, Kroeber was not impressed, and he and White exchanged broadsides into the late 1950s (Kroeber 1946, 1960; White 1945a, 1945b, 1947, 1959a, 1959b). A detailed analysis of their debate is beyond the scope of this paper, but several aspects of their exchange are significant in the present context. These do not entail the famous dictum of White's (later adopted by the processual archaeologists of the 1960s) that culture is humankind's extrasomatic means of adaptation, nor do they entail the statement by White (1959b: 30) that in his view history is ideographic and evolution nomothetic. Rather, the important points are that Kroeber's view of culture change was more in line with the materialist metaphysic, whereas White's was strongly within the essentialist camp. Detecting this critical difference is difficult because of the manner in which the two antagonists distinguished between history and evolutionism.

Although both conceived of cultural change as involving transmission and heritability and referred to it metaphorically as a flowing stream or continuum, White (1945b) and Kroeber (1946) viewed history and evolution very differently. White (1938) viewed specific events as varying formally—that is, structurally and/or functionally—with each event occupying a particular position in the time-space continuum. Kroeber (1946) apparently agreed. How these three dimensions—to borrow Spaulding's (1960) term—of form, space, and time were analytically interrelated is how White distinguished between history and evolution: "Events are related to each other spatially, and we may deal with [them] in terms of spatial, or formal, relationships, ignoring the aspect *time*" (White 1938: 375). White (1945b: 222) later termed this the "formal (functional) process, which presents phenomena in their non-temporal, structural, and functional aspects." Formal-functional aspects of events could be "repetitive," by which White (1945b: 229) meant different events as phenomena could have "generic likenesses." In our terms, events could be classified according to a set of ideational units, and thus while each event as a phenomenon occupied a unique time–space position, events as classes shared certain features in common. As classes or ideational units, events had distributions; that is, a class of event could occur in more than one time-space location.

In White's (1945b: 222) view, history concerned "non-repetitive" events: "History is that way of sciencing in which events are dealt with in terms of their temporal relationships alone. Each event is unique. The one thing that history never does is repeat itself" (White 1938: 374); the "temporal process [history] is a selective arrangement of events according to the principle of *time*" (White 1938: 376); the historic process is the one "in which specific and severally unique events take place in a purely temporal context" (White 1938: 380). White (1945b: 222) later referred to this as the "temporal process, being a chronological sequence of unique events, the study of which is history." In our terms, history concerned a set of empirical units—labeled events by White—arranged in a temporal sequence, each event in a particular spatial position. Because events were empirical, they had locations in the time–space continuum.

Evolution, in White's view, was distinct from history: "The temporal-spatial process is an evolutionary, or developmental process . . . Evolution is temporal-alteration-of-forms" (White 1938: 377). The "historic process [dimension] is merely temporal, the evolutionary process is formal as well: it is a *temporal-sequence-of-forms*" (White 1938: 379). The evolutionary process involves "new forms grow[ing] out of preceding forms" (White 1938: 380). White (1945b: 222) later described this as the "temporal–formal process, which presents phenomena as a temporal sequence of forms, the interpretation of which is evolutionism." Evolution was not a "chronological sequence of particular and unique events [this was history], but [rather] a general process of chronological change, a temporal-sequence-of-forms, with the growth of one form out of an earlier, into a later, form" (White 1945b: 224). Thus, evolution was different than history: "[T]he historic process and the evolutionist process are alike in that both involve temporal sequences. They differ, however, in that the historic process deals with events determined by specific time and space coördinates, in short with *unique* events [empirical units], whereas the evolutionist process is concerned with *classes* of events [ideational units] independent of specific time and place" (White 1945b: 230).

Kroeber (1946) simply couldn't understand White's distinction between history and evolution. His confusion no doubt arose from several sources, one being that White (1945b: 222) indicated both that evolution was "non-repetitive," suggesting that evolutionary events were unique, empirical units rather than ideational classes, and that evolution concerned "*classes* of events" (our ideational units) (White 1945b: 230, 238). Perhaps White merely misspoke here, because it is relatively straightforward to see in the bulk of his discussions that in the formal dimension events were ideational units, in the historical (temporal) dimension events were empirical units, and in the evolutionist view events were ideational. For example, in more than one place White (1945b: 239) indicated that the interests of the historian were in "the unique event at a specific time and place," whereas those of the evolutionist were in "a class of events."

To Kroeber (1946), history and evolution were one and the same—a point accepted by most biologists today (e.g., Gould 1986, Szalay & Bock 1991). In fact, the difference between Kroeber and White was their respective metaphysics. Kroeber (1946: 9) observed that apparently "what White means by evolution is a fixed, necessary, inherent, and predetermined process . . . White's evolution thus seems to be an unfolding of immanences." By the last we suspect Kroeber had in mind what we have termed essentialist units. This is important, for it comprises a much more significant source of confusion for Kroeber and source of disagreement between him and White. What White was speaking of is known in biology as the theory of *orthogenesis*—a point that was made by at least one anthropologist, Joseph Birdsell (1957), and an evolutionary biologist, Theodosius Dobzhansky (1957), at the time White was writing. Birdsell (1957: 399) noted that White was a "modern advocate of the orthogenetic evolution of culture." In the hands of Spencer, Morgan, and Tylor, the orthogenetic evolution of cultures was a "single inflexible and limited theory of culture change [that had] left scars on 20th century anthropology . . . [An] unreasonable amount of time and energy had been spent [by White and Steward] on beating [this] dead and specialized theory of evolution." Why was Birdsell so concerned?

Dobzhansky (1957: 382–383), in a companion article to Birdsell's, noted that orthogenetic evolution consisted of

> unfolding or manifestation of pre-existing rudiments"; there is in it nothing accidental or creative [no mutation], for evolution "proceeds in accordance with laws," through a predetermined sequence of stages or phases . . . Theories of orthogenesis represent evolution as unfolding of pre-existing but latent forms [Kroeber's immanences] . . . [An] idea popular among believers in orthogenesis is that the evolution of most phyletic lines tends towards evolutionary senility and extinction. If we were to accept this idea then all we can hope to do for our descendants is to postpone the inevitable.

When discussing his view of evolution, White repeatedly said that (1) "new forms grow out of preceding forms" (1938: 380, 1945b: 224, 1947: 175); (2) the evolutionary process was law-like (1949, 1959b); and (3) the sequence of stages was inevitable in the sense that all societies would eventually represent civilizations, whether they all were at one time chiefdoms or not (1947, 1959b).

It is not difficult to perceive essentialism in White's (1943) seminal discussion of his view of orthogenetic cultural evolution. It is even easier to see in his discussion of evolutionary stages: "For those who recognize that one form grows out of another, the concept of stages will be found useful as a descriptive, interpretative, and evaluative device . . . [Stages] serve to mark off steps in development . . . Stages are merely the succession of significant forms in the developmental process" (White 1947: 179). What made a stage "significant" was never made explicit by White, but it no doubt was the fact that metaphysically it was a real, essentialist category. Steward's (1955: 89) "cultural core" and "cultural type" were also essentialist units. Every anthropologist recognized a hunter-gatherer economy or tribal-level social organization or the like; these sorts of anthropological phenomena must therefore be real (recall Morgan's tripartite scheme of savagery, barbarism, and civilization). When such cultural traits occurred in particular combinations—and they seemed to covary in nonrandom fashion—they comprised a certain evolutionary stage—a point later shown to be fallacious by Leonard and Jones (1987; see also Rambo 1991). In short, this was an essentialistic view. Kroeber and other Boasians couldn't fathom what White was talking about because they held a rather more materialistic view, even though they seldom recognized that the materialist paradox tended to consistently thwart their explanatory efforts.

The cultural-evolutionary process was another matter of concern. Despite White's (1943: 339) disclaimer that he was not saying that "man deliberately set about to improve his culture," close reading of what he said indicates that he strongly believed all organisms, including humans, had an "urge" to improve and that *this* was the "motive force as well as the means of cultural evolution." In what we view as a maximal overstatement, White (1943: 339) proclaimed that the "culturologist knows more about cultural evolution than the biologist, even today, knows about biological evolution." White (1947: 177) also regularly indicated that he and other cultural evolutionists "did not identify evolution with progress [and that they] did not believe progress was inevitable."

But by default, cultural evolution was synonymous with progress: "[B]y and large, in the history of human culture, progress and evolution *have gone hand in hand*" (White 1943: 339). The key evolutionary mechanism–urge or necessity as a motive force–demanded absolutely no reference to a source of variation, to natural selection, or to the shape of lineages–all of which are of critical interest in Darwinian evolution. In White's view, every human invented new tools as necessary, and they were always better than the preceding ones because they allowed the procurement or exploitation of additional energy:

> The best single index [of progress] by which all cultures can be measured, is *amount of energy harnessed per capita per year.* This is the common denominator of all cultures . . . Culture advances as the amount of energy harnessed per capita increases. The criterion for the evaluation of cultures is thus an objective one. The measurements can be expressed in mathematical terms. The goal–security and survival–is likewise objective; it is the one that all species, man included, live by. Thus we are able to speak of cultural progress objectively and in a manner which enriches our understanding of the culture history of mankind tremendously. And finally, we can evaluate cultures and arrange them in a series from lower to higher. This follows, of course, from the establishment of a scientifically valid criterion of value and means of measurement (White 1947: 187).

If you don't think this sounds like the new, or processual, archaeology of the 1960s, then read the first thirty pages of White (1959b).

What gave White's evolution its distinctive punctuated form was his belief that change could occur only two ways: Either humans improve the efficiency of old tools, or they invent new tools. Evolution via the former was limited, however, as exemplified in White's (1943: 343) statement that the "extent to which man may harness natural forces [energy] in animal husbandry is limited" and his later statement that "some progress can of course be made by increasing the efficiency of the technological means of putting energy to work, but there is a limit to the extent of cultural advance on this basis" (White 1959b: 369). This is merely another expression of orthogenetic evolution, as the quote from Dobzhansky (1957) cited earlier makes clear.

It was the technological breakthroughs or revolutions that were important and resulted in "tremendous" changes, "extremely rapid" progress, and "great cultural advances" (White 1945b: 342, 344). Such breakthroughs gave cultural evolution its jerky, discontinuous appearance. The other sort of change–mere improvement of the efficiency of existing tools–produced "no fundamental difference" (White 1945b:344). To Kroeber, the evolution of a cultural lineage was continuous and gradual, like a gradually ascending ramp (Figure 2.6); to White, cultural evolution was not a ramp but a staircase, perhaps with each step at a slight incline to reflect the fact that existing technology was constantly being improved, but the risers were most significant because they represented the technological breakthroughs and the fundamental differences between what came to be referred to as stages (Figure 2.6).

Archaeologists opted for White's version, with its cultural stasis punctuated by relatively brief and abrupt change, because it showed that cultural change was discontinuous. This fit the discontinuous nature of change evidenced not only by stratigraphically superposed artifact assemblages (Lyman et al. 1997b) but also by the fact that, as Raymond Thompson (1956: 36) pointed out, "artifacts tend strongly to occur in spatial clusters." Hence, "the site provides a unit of artifact association for analysis and synthesis. . . . [F]orm classes of artifacts and a unit of artifact association provide the necessary data for quantitative description and comparison" (Thompson 1956:36). This fit well with White's (1947:175) notion that as a cultural evolutionist, an archaeologist "would begin, naturally, with the present, with what we have before us. Then we would arrange other forms in the series in accordance with their likeness or dissimilarity to the present form [Kidder's ceramic series]. . . . Stratigraphy is often involved here."

DISCUSSION

In 1953, Steward reiterated the typical objections to the application of Darwinian evolution to cultural phenomena:

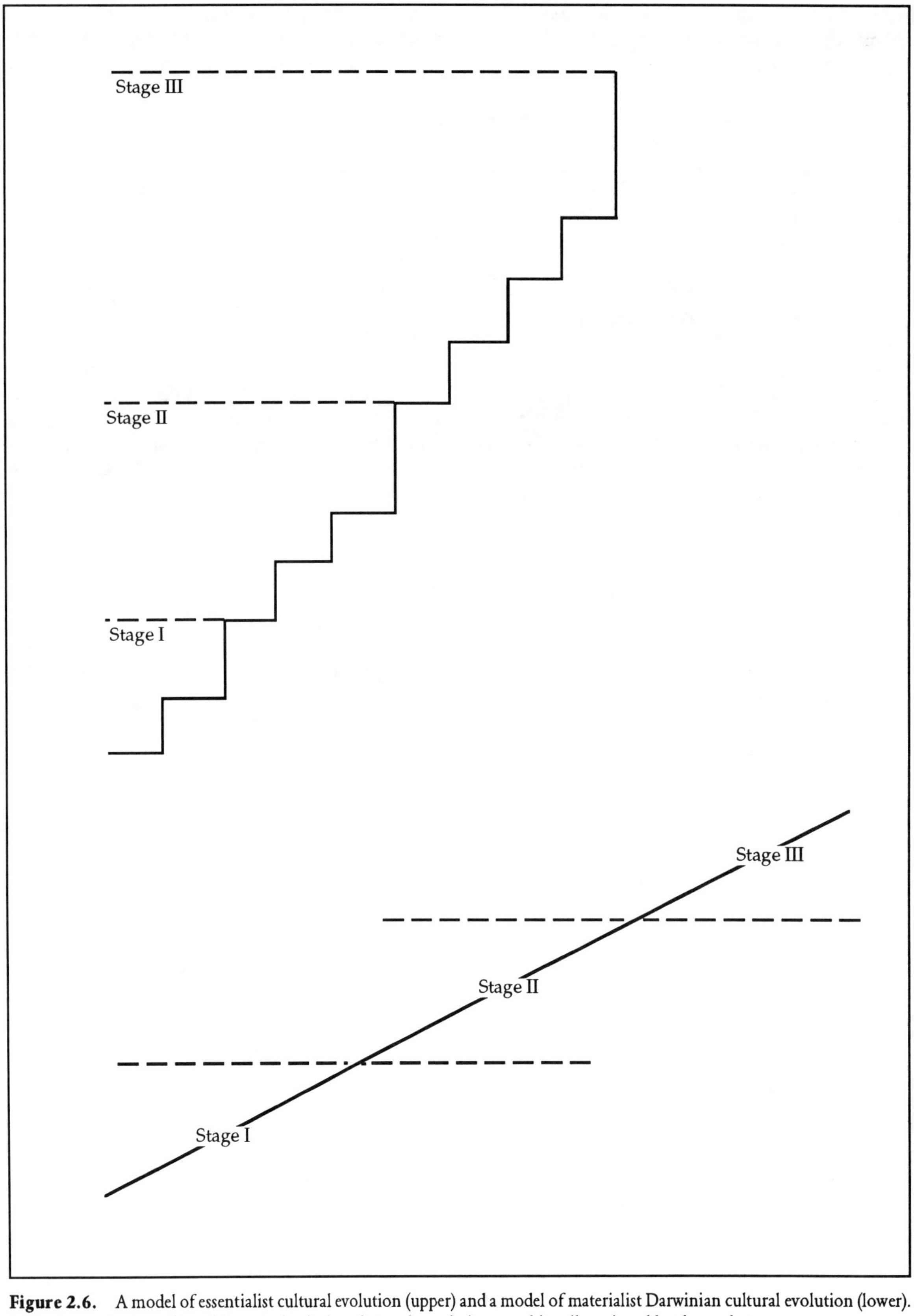

Figure 2.6. A model of essentialist cultural evolution (upper) and a model of materialist Darwinian cultural evolution (lower). In the latter, the term "stage" is metaphorical, and stage boundaries are arbitrarily assigned by the analyst.

> [C]ultural evolution is an extension of biological evolution only in a chronological sense. The nature of evolutionary schemes and of the developmental processes differs profoundly in biology and culture. In biological evolution it is assumed that all forms are genetically related and that their development is essentially divergent [branching] . . . In cultural evolution, on the other hand, it is assumed that patterns are genetically unrelated and yet pass through parallel and historically independent sequences (Steward 1953: 313).

Steward (1953: 315) went on to suggest that the use of cultural evolution as an explanatory model demanded two "vitally important assumptions. First, it [assumes] that genuine parallels of form and function develop in historically independent sequences or cultural traditions. Second, it explains those parallels by independent operation of identical causality in each case." Cultural evolution was concerned with generalities of processes and change—that is, with cross-cultural regularities or laws—and was therefore scientific (Steward 1953: 315; White 1945b, 1959b). These aspects of cultural evolution no doubt are what made it so attractive to the processual archaeologists of the 1960s and '70s. Culture historians had adopted it a decade earlier (Krieger 1953, South 1955, Spaulding 1955, Willey & Phillips 1955, 1958) because it was the only alternative, with Darwinian evolution having been discarded as a result of its conceived inapplicability to cultural phenomena.

1950s Cultural Anthropology and Evolution

The second "assumption" of cultural evolution—that causality was identical in case after case—was part of the orthogenetic version of evolution and denied any role for natural selection. Such a mechanism was unnecessary, as White's discussions indicated. Steward (1953, 1955, 1956) advocated multilinear evolution as distinct from what he characterized as the unilinear evolution of White, but there really was no significant difference, as White repeatedly (1945b, 1959a, 1959b) indicated. We suspect that Steward never had White's distinction of history and evolution clear in his mind, just as Kroeber hadn't. What is important here is that Steward, like White, saw "order" (Steward 1956: 73) in cultural evolution; this was typically glossed as "progress"—a notion that several individuals (e.g., Dunnell 1988, Mayr 1988, O'Dowd 1982) have shown to be inappropriate in the biological and cultural realms. Given such a connotation of the apparent order of evolutionary change, Steward, again like White, perceived no role for natural selection in determining the trajectory of a culture's lineage.

Such notions often were reinforced by some biologists writing in the mid-twentieth century. For example, Julian Huxley (1956), in an address delivered to anthropologists in 1951, implied that because cultural evolution was superorganic and involved the psychosocial realm, it could be directed by human intent. Paleontologist George Gaylord Simpson agreed (1949: 344–345). Huxley also reinforced White's views that (1) "every biological improvement appears eventually to reach a limit," such that coincident with the time the first culture-bearing hominids appeared, "biological evolution on this planet had reached the limit of its advance" (Huxley 1956: 6)—a patently orthogenetic statement; (2) "evolution includes advance, or improvement in organization" (Huxley 1956: 5)—a progressionist statement; and (c) in cultural evolution, just as in biological evolution, "major advance proceeds by large steps, each marked by the spread of the successful new type of organization" (Huxley 1956: 10)—a baldly essentialist stance. Thus evolution was jerky and consisted of one dominant type or group being "replaced" by another "related but improved type" (Huxley 1956: 6).

Kroeber (1960: 15) approved of how Huxley (1956) distinguished between biological and cultural evolution. Huxley (1956: 3) indicated that he did "not believe that any purely biological concepts and principles can be immediately applied or directly transferred to anthropology"—a notion that, as we have seen, was strong in Kroeber's (and Boas's and Steward's) thinking from the start. Kroeber also appreciated that Huxley made several other observations that were in line with his own thinking:

> The evolutionary approach in anthropology has been bedeviled by false starts and false premises—notably the erroneous idea that biological evolution could be represented by a single straight line of inevitable progress, [and this created] an evolutionary strait-jacket for culture (Huxley 1956: 15).

> Culture was not only that which was transmitted, it was also the mechanism of transmission and thus of reproduction (Huxley 1956: 9).

> [E]volution still works in man, but overwhelmingly as a cultural, not a biological process. . . . [C]ultural (psycho-social) evolution shows the same main features as biological evolution. Cultures advance, progress, diverge, and stabilize, but the mechanisms of change are different, and cultures tend to converge, whereas biological evolution involves only divergence (Huxley 1956: 23–24).

Huxley's ideas were, of course, expressed before explicit recognition within biology of the distinction between the materialist and essentialist metaphysics. Recognition of the effects of the essentialist metaphysic on biology came initially in the 1950s and was discussed at length by biologists during that decade and following ones (e.g., Hull 1965; Mayr 1959, 1972, 1982; Simpson 1961, Sober 1980). By the time the distinction had taken a fairly strong hold in biology, it was too late to have any influence within anthropology and archaeology. The distinction was made explicit in the latter two disciplines only in 1982 (Dunnell 1982). Despite the increasing number of attempts to apply an evolutionary perspective to anthropological and archaeological research (e.g., Boyd & Richardson 1985, Cavalli-Sforza & Feldman 1981, Durham 1991, Marks & Staski 1988), the critical role one's metaphysic plays in research remains generally unrecognized (Mithen 1989, Schiffer 1996, Tschauner 1994; although see Clark 1991, 1993, 1994; Clark & Lindly 1991). (The same has been argued for biology [Szalay & Bock 1991].)

1950s Archaeology and Evolution

Earlier, we suggested that culture historians had adopted cultural evolution as an explanatory framework in the 1950s. That suggestion is founded on papers such as Alex Krieger's "New World Culture History: Anglo-America" in which a set of four adaptational stages are described (1953). The abhorance of anything evolutionary, however, prompted Willey and Phillips (1958: 67) to label Krieger's (1953) set of stages and a similar one proposed a few years earlier by Steward (1949) as "historical-developmental schemes." They proposed their own developmental scheme in what would become–with the later substitution of PaleoIndian for Lithic–a well-known set of terms for five stages: Lithic, Archaic, Formative, Classic, and Postclassic. Noting that their scheme had been derived "from an inspection of archaeological sequences throughout the hemisphere," they concluded by noting that the stages were "abstractions which *describe* culture change through time in native America. The stages are not formulations which *explain* culture change" (Willey & Phillips 1958: 200).

With respect to Willey and Phillips's scheme, Swanson (1959: 121) observed that "no theory has been developed. What is *assumed* is an evolutionary theory about the nature of culture, though Willey and Phillips are shy about admitting this. Moreover, no history has been written." The parallels between Willey and Phillips's scheme and White's notions of cultural evolution are clear from both the assumed theory and the lack of history in White's sense of the term inherent in the former's scheme. Another parallel is found in the fact that Willey and Phillips characterize their developmental scheme of stages as descriptive rather than explanatory; recall that Bidney (1946: 295) characterized White's scheme as lacking explanatory power. Bidney may have been worried that White was unclear on whether energy capture was the cause or the effect of change; casting human intent as the catalyst–as White did–renders this point irrelevant. Alternatively, Bidney's concern may have been that change in energy capture might be construed as the *proximate cause* of cultural change; if so, this left the *ultimate cause* unidentified. As Mayr (1961: 1053) noted, the latter "are causes that have a history and that have been incorporated into the system through many thousands of generations of natural selection." White, as far as we know, never once mentioned *selection* as the ultimate cause of anything.

White's students took up his banner. The classic *Evolution and Culture,* edited by Marshall Sahlins and Elman Service (1960) sought to clarify, solidify, and expand various issues. Of most concern here is their distinction between general and specific evolution. The former was defined as "passage from less to greater energy transformation, lower to higher levels of integration, and less to greater all-around adaptability. Specific evolution is the phylogenetic, ramifying, historic passage of culture along its many lines, the adaptive modification of particular cultures" (Sahlins & Service 1960: 38). Ironically, given his earlier proposal of five developmental stages, Willey (1961: 442) thought the former kind of evolution was of little use because "its processes are obscure." Specific evolution was, however, useful because it combined "history plus explanation of process–the story of how a given culture, or culture continuum, changed through time by the processes of its adaptations to natural and superorganic environments" (Willey 1961: 442). Continuing, Willey (1961: 443) noted that "until the processes of this general evolution are better understood, particularly as these pertain to the way in which the many streams of specific evolution feed into the main one, I cannot appreciate the difference between a general universal evolution of culture and a general universal culture history."

Contrary to the processual archaeology that was to emerge a few years later, Willey (1961: 442) indicated that a central disagreement he had with Sahlins and Service (1960) resided in the latter's "insistence that the technico-economic realm must always be given priority in explaining the processes of cultural change." This notion, of course, resided in White's formulations: Culture is humankind's extrasomatic means of adaptation; culture is a system of interrelated variables; and energy capture is central to system stability and change. Despite the protests of Willey and others (e.g., Haag 1959), the cultural-evolutionary notions of White provided the basis for the emergence of what came to be known as processual archaeology. Definitions of key concepts, conceptions of how science was supposed to work, and those essentialist units known as bands, tribes, chiefdoms, and the like were carried over wholesale into that new paradigm (e.g., Binford 1969, 1972) because Darwinian evolution had already been discarded by the discipline.

CONCLUSION

On the eve of the birth of processual archaeology in the early 1960s, Haag perceptively observed that

> The usual reason for the rejection [by anthropologists and archaeologists] of the biological model [of evolution] is either, (a) that genetic mutations create "new materials" whereas human mutations (inventions) do not; or (b) that man can control and direct his evolution whereas animal mutations are random... Once the understanding of the nature of culture is gained, there is no longer any confusion between the genetic process in [biology] and the culture process in man (1959: 104).

Haag hit the nail squarely on the head as far as most anthropologists and archaeologists were concerned; there really *were* two kinds of evolution–one biological and one cultural–and different theory was needed for each. Haag's views of what cultural evolution entailed were no different than those of his contemporaries: a decidedly Spencerian and Whitean approach, complete with orthogenetic notions of progress, human intent, essentialist stages, and the like, and lacking any appeal to natural selection or a materialistic conception of variation. Regarding natural selection, Haag (1959: 103) made the observation that "[a]lthough an entirely different mechanism is involved in the adaptation of man to his environment, and thus a different mechanism controls his cultural evolution, adaptation is still the keyword." The biological mechanism was natural selection, the cultural mechanism was *integration*, defined by Haag (1959: 103) as "adjustment." Although Haag was vague, "adjustment" apparently meant addition to the pool of variant cultural traits; if a trait couldn't be adjusted to fit the existing pool, it wasn't integrated into (added to) the cultural system. *Why* it could not be integrated or adjusted was not addressed because a process comparable to natural selection was not included in Haag's rendition of cultural evolution, just as it wasn't in any of his contemporaries' views.

That Spencerian and Darwinian evolution are dissimilar is clear. Americanist archaeologists and cultural anthropologists of the early 20th century appear to have recognized at least some of the differences between the two, but they were insufficiently knowledgeable about Darwinism to figure out how to use it. Further, that theory was not in a form that could be readily adapted to the metaphor of culture change as a gradually, continuously flowing braided stream. Biologists were focusing on genetic change and essentialist categories termed species; typological thinking was rampant. Darwinism simply had little to offer anthropology and archaeology because (1) cultural evolution was reticulate, whereas biological evolution was branching; (2) cultural evolution did not involve the transmission of genes, whereas biological evolution did; and (3) people were not subject to the forces of natural selection and intentionally directed the evolution of their cultures, whereas biological evolution depended on the natural selection of random mutations. The option of Spencerian cultural evolution was attractive because it was not faced with these difficulties, and it concerned culture, not organisms.

The insights provided by Kroeber and several of his students have a somewhat unique and decidedly materialist feel to them, but the version of biological evolution then popular was not conducive to adoption for use in explaining the archaeological or cultural records. The alternative adopted–Spencerian evolution–has now been in active use for some five decades. It has granted us some significant insights to the past, and it has suggested some of the important variables with which we must contend. But it is in no sense a complete theory–some would argue it is not a theory at all but rather a model or philosophy (Dunnell 1980)–and thus its ability to provide testable explanations is limited. Its use of essentialist categories is one of its weaknesses; its orthogenetic view of evolutionary progress as inevitable is another. We suggest that it is appropriate to take up Kroeber's banner and try again to borrow from Darwinian evolution, the 1990s version, and see where it takes us. We believe the results will be significant.

ACKNOWLEDGMENTS

We thank E. J. O'Brien for his editorial advice and Dan Glover for producing the figures.

REFERENCES CITED

Beals, Ralph L., George W. Brainerd & Watson Smith
1945 Archaeological Studies in Northeast Arizona. *University of California, Publications in American Archaeology and Ethnology* 44: 1–236.

Bidney, David
1946 On the So-Called Anti-Evolutionary Fallacy: A Reply to Leslie A. White. *American Anthropologist* 48: 293–297.

Binford, Lewis R.
1969 Conceptual Problems in Dealing with Units and Rates of Cultural Evolution. *Anthropology UCLA* 1: 27–35.

1972 Comments on Evolution. In *An Archaeological Perspective*, pp. 105–113. New York: Academic Press.

Birdsell, Joseph B.
1957 On Methods of Evolutionary Biology and Anthropology, Part II. Anthropology. *American Scientist* 45: 393–400.

Boas, Franz
1896 The Limitations of the Comparative Method of Anthropology. *Science* 4: 901–908.
1902 Some Problems in North American Archaeology. *American Journal of Archaeology* 6: 1–6.
1904 The History of Anthropology. *Science* 20: 513–524.
1932 The Aims of Anthropological Research. *Science* 76: 605–613.

Bock, Walter J.
1977 Foundations and Methods of Evolutionary Classification. In *Major Patterns in Vertebrate Evolution* (Max K. Hecht, Peter C. Goody, & Bessie M. Hecht, eds.), pp. 851–895. New York: Plenum.

Boyd, Robert & Peter J. Richerson
1985 *Culture and the Evolutionary Process.* Chicago: University of Chicago Press.

Brew, John O.
1946 *Archaeology of Alkali Ridge, Southeastern Utah.* Cambridge: Peabody Museum of American Archaeology and Ethnology, Papers No. 21.

Cavalli-Sforza, L. L. & Marcus W. Feldman
1981 *Cultural Transmission and Evolution: A Quantitative Approach.* Princeton: Princeton Monographs in Population Biology No. 16.

Clark, G.A.
1991 Epilogue: paradigms, realism, adaptation and evolution. In *Perspectives on the Past: Theoretical Biases in Mediterranean Hunter-Gatherer Research* (G.A.Clark, ed.), pp. 411–439. Philadelphia: University of Pennsylvania Press.
1993 Paradigms in science and archaeology. *Journal of Archaeological Research* 1: 203-234.
1994 Migration as an explanatory concept in paleolithic archaeology. *Journal of Archaeological Method and Theory* 1: 305–343.

Clark, G.A. & J.M. Lindly
1991 Paradigmatic biases and paleolithic research traditions. *Current Anthropology* 32: 577–587.

Colton, Harold S.
1939 *Prehistoric Culture Units and Their Relationships in Northern Arizona.* Flagstaff: Museum of Northern Arizona Bulletin No. 17.

Colton, Harold S. & Lyndon L. Hargrave
1937 *Handbook of Northern Arizona Pottery Wares.* Flagstaff: Museum of Northern Arizona Bulletin No. 11.

Davis, Jerrold I.
1996 Phylogenetics, Molecular Variation, and Species Concepts. *BioScience* 46: 502–511.

Dobzhansky, Theodosius G.
1951 Mendelian Populations and Their Evolution. In *Genetics in the 20th Century* (Leslie C. Dunn, ed.), pp. 573–589. New York: Macmillan.
1957 On Methods of Evolutionary Biology and Anthropology, Part I. Biology. *American Scientist* 45: 381–392.

Dunnell, Robert C.
1980 Evolutionary Theory and Archaeology. In *Advances in Archaeological Method and Theory Vol. 3* (Michael B. Schiffer, ed.), pp. 35–99. New York: Academic Press.
1982 Science, Social Science, and Common Sense: The Agonizing Dilemma of Modern Archaeology. *Journal of Anthropological Research* 38: 1–25.
1985 Methodological Issues in Contemporary Americanist Archaeology. In *Proceedings of the 1984 Biennial Meeting of the Philosophy of Science Association Vol. 2* (Peter D. Asquith & Philip Kitcher, eds.), pp. 717–744. Philosophy of Science Association, East Lansing, Michigan.
1986 Methodological Issues in Americanist Artifact Classification. In *Advances in Archaeological Method and Theory Vol. 9* (Michael B. Schiffer, ed.), pp. 149–207. Orlando: Academic Press.
1988 The Concept of Progress in Cultural Evolution. In *Evolutionary Progress* (Matthew H. Nitecki, ed.), pp. 169–194. Chicago: University of Chicago Press.
1995 What Is It That Actually Evolves? In *Evolutionary Archaeology: Methodological Issues* (Patrice A. Teltser, ed.), pp. 33–50. Tucson: University of Arizona Press.

Durham, William H.
1991 *Coevolution: Genes, Culture, and Human Diversity.* Stanford: Stanford University Press.

Dutton, Bertha P.
1938 *Leyit Kin, A Small House Ruin, Chaco Canyon, New Mexico.* Albuquerque: University of New Mexico Bulletin, Monograph Series Vol. 1, No. 6.

Ehrich, Robert W.
1963 Further Reflections on Archeological Interpretation. *American Anthropologist* 65:16–31.

Fisher, Daniel C.
1994 Stratocladistics: Morphological and Temporal Patterns and Their Relation to Phylogenetic Process. In *Interpreting the Hierarchy of Nature* (Lance Grande & Olivier Rieppel, eds.), pp. 133–171. San Diego: Academic Press.

Ford, James A.
1935 *Ceramic Decoration Sequence at an Old Indian Village Site, near Sicily Island, Louisiana.* Baton Rouge: Louisiana State Geological Survey, Department of Conservation, Anthropological Study No. 1.
1940 Review of *Handbook of Northern Arizona Pottery Wares*, by Harold S. Colton & Lyndon L. Hargrave. *American Antiquity* 5: 263–266.

Forey, P. L.
1990 Cladistics. In *Palaeobiology: a Synthesis* (Derek E. G. Briggs & Peter R. Crowther, eds.), pp. 430–434. Oxford: Blackwell.

Fowke, Gerard
1922 *Archaeological Investigations.* Washington: Bureau of American Ethnology, Bulletin No. 76.

Ghiselin, Michael T.
1966 On Psychologism in the Logic of Taxonomic Controversies. *Systematic Zoology* 15: 207–215.
1974 A Radical Solution to the Species Problem. *Systematic Zoology* 25: 536–544.
1981 Categories, Life and Thinking. *Behavioral and Brain Sciences* 4: 269–283.

Gingerich, Philip D.
1979 Paleontology, Phylogeny, and Classification: An Example from the Mammalian Fossil Record. *Systematic Zoology* 28: 451–464.

Gladwin, Harold S.
1936 Editorials: Methodology in the Southwest. *American Antiquity* 1: 256–259.

Gladwin, Winifred & Harold S. Gladwin
1930 *A Method for the Designation of Southwestern Pottery Types.* Globe: Medallion Papers No. 7.
1934 *A Method for the Designation of Cultures and Their Variations.* Globe: Medallion Papers No. 15.

Gould, Stephen J.
1986 Evolution and the Triumph of Homology, or Why History Matters. *American Scientist* 74: 60–69.

Haag, William G.
1959 The Status of Evolutionary Theory in American History. In *Evolution and Anthropology: A Centennial Appraisal* (Betty J. Meggers, ed.), pp. 90–105. Washington: Anthropological Society of Washington.
1961 Review of *The Evolution of Man: Man, Culture and Society,* edited by Sol Tax. *American Antiquity* 26: 439–441.

Hawley, Florence M.
1934 *The Significance of the Dated Prehistory of Chetro Ketl.* Albuquerque: University of New Mexico Bulletin, Monograph Series No. 1.

Hull, David L.
1965 The Effect of Essentialism on Taxonomy-Two Thousand Years of Stasis. *British Journal for the Philosophy of Science* 15: 314–326; 16: 1–18.

Huxley, Julian S.
1956 Evolution, Cultural and Biological. In *Current Anthropology* (William L. Thomas, Jr., ed.), pp. 3–25. Chicago: University of Chicago Press.

Kidder, Alfred V.
1915 *Pottery of the Pajarito Plateau and of Some Adjacent Regions in New Mexico.* American Anthropological Association, Memoir 2: 407–462.
1917 A Design-Sequence from New Mexico. *Proceedings of the National Academy of Sciences* 3: 369–370.
1924 *An Introduction to the Study of Southwestern Archaeology, with a Preliminary Account of the Excavations at Pecos.* Papers of the Southwestern Expedition, Phillips Academy, No. 1. New Haven: Yale University Press.
1927 Southwestern Archaeological Conference. *Science* 66: 489–491.
1932 *The Artifacts of Pecos.* Papers of the Southwestern Expedition, Phillips Academy No. 6. New Haven: Yale University Press.
1936a Discussion. In *The Pottery of Pecos, Vol. II,* by Alfred V. Kidder & Anna O. Shepard, pp. 589–628. New Haven: Yale University Press.
1936b Introduction. In *The Pottery of Pecos, Vol. II,* by Alfred V. Kidder & Anna O. Shepard, pp. xvii–xxxi. New Haven: Yale University Press.

Kidder, Madeleine A. & Alfred V. Kidder
1917 Notes on the Pottery of Pecos. *American Anthropologist* 19: 325–360.

Kniffen, Fred B.
1938 The Indian Mounds of Iberville Parish. In *Reports on the Geology of Iberville and Ascension Parishes* (Henry V. Howe, ed.), pp. 189–207. Baton Rouge: Louisiana Geological Survey, Geological Bulletin No. 13.

Krieger, Alex D.
1953 New World Culture History: Anglo-America. In *Anthropology Today* (Alfred L. Kroeber, ed.), pp. 238–264. Chicago: University of Chicago Press.

Kroeber, Alfred L.
1909 The Archaeology of California. In *Putnam Anniversary Volume* (Franz Boas, ed.), pp. 1–42. New York: Stechert.
1916a Zuñi Culture Sequences. *Proceedings of the National Academy of Sciences* 2: 42–45.
1916b Zuñi Potsherds. *American Museum of Natural History, Anthropological Papers* 18: 1–37.
1916c Cause of the Belief in Use Inheritance. *American Naturalist* 50: 367–370.
1917 The Superorganic. *American Anthropologist* 19: 163–213.
1923 *Anthropology.* New York: Harcourt, Brace.
1925 *Handbook of the Indians of California.* Washington: Bureau of American Ethnology Bulletin No. 78.
1931 Historical Reconstruction of Culture Growths and Organic Evolution. *American Anthropologist* 33: 149–156.
1943 Structure, Function and Pattern in Biology and Anthropology. *Scientific Monthly* 56: 105-113.
1946 History and Evolution. *Southwestern Journal of Anthropology* 2: 1–15.
1948 *Anthropology* (revised edition). New York: Harcourt, Brace.
1960 Evolution, History, and Culture. In *Evolution after Darwin, Vol. II: The Evolution of Man* (Sol Tax, ed.), pp. 1–16. Chicago: University of Chicago Press.

Leonard, Robert D. & George T. Jones
1987 Elements of an Inclusive Evolutionary Model for Archaeology. *Journal of Anthropological Archaeology* 6: 199–219.

Lesser, Alexander
1952 Evolution in Social Anthropology. *Southwestern Journal of Anthropology* 8: 134–146.

Lewontin, Richard C.
1974 *The Genetic Basis of Evolutionary Change.* New York: Columbia University Press.

Lyman, R. Lee, Michael J. O'Brien & Robert C. Dunnell, Editors
1997a *Americanist Culture History: Fundamentals of Time, Space, and Form.* New York: Plenum.
1997b *The Rise and Fall of Culture History.* New York: Plenum.

Marks, Jonathan & Edward Staski
1988 Individuals and the Evolution of Biological and Cultural Systems. *Human Evolution* 3: 147–161.

Martin, Paul S.
1936 Lowry Ruin in Southwestern Colorado. *Fieldiana: Anthropology* 23: 1–216.

Maschner, Herbert D. G., Editor
1996 *Darwinian Archaeologies.* New York: Plenum.

Mayr, Ernst
1959 Darwin and the Evolutionary Theory in Biology. In *Evolution and Anthropology: A Centennial Appraisal* (Betty J. Meggers, ed.), pp. 1–10. Washington: Anthropological Society of Washington.
1961 Cause and Effect in Biology. *Science* 134: 1501–1506.
1963 *Animal Species and Evolution.* Cambridge: Harvard University Press.

1969 *Principles of Systematic Zoology.* New York: McGraw–Hill.

1972 The Nature of the Darwinian Revolution. *Science* 176: 981–989.

1973 *Populations, Species, and Evolution.* Cambridge: Harvard University Press.

1977 Darwin and Natural Selection. *American Scientist* 65: 321–337.

1982 *The Growth of Biological Thought: Diversity, Evolution, and Inheritance.* Cambridge: Harvard University Press.

1987 The Ontological Status of Species: Scientific Progress and Philosophical Terminology. *Biology and Philosophy* 2: 145–166.

1988 *Toward a New Philosophy of Biology: Observations of an Evolutionist.* Cambridge: Harvard University Press.

McKern, William C.

1934 *Certain Culture Classification Problems in Middle Western Archaeology.* Washington: National Research Council, Committee on State Archaeological Surveys.

1937 Certain Culture Classification Problems in Middle Western Archaeology. In *The Indianapolis Archaeological Conference,* pp. 70–82. Washington: National Research Council, Committee on State Archaeological Surveys, Circular No. 17.

1939 The Midwestern Taxonomic Method as an Aid to Archaeological Culture Study. *American Antiquity* 4: 301–313.

1942 Taxonomy and the Direct Historical Approach. *American Antiquity* 8: 170–172.

Meggers, Betty J., Editor

1959 *Evolution and Anthropology: A Centennial Appraisal.* Washington: Anthropological Society of Washington.

Mithen, Steven

1989 Evolutionary Theory and Post-Processual Archaeology. *Antiquity* 63: 483–494.

Naroll, Raoul

1961 Review of *Evolution and Archaeology,* Marshall D. Sahlins & Elman R. Service. *American Anthropologist* 63: 389–392.

Nelson, Nels C.

1916 Chronology of the Tano Ruins, New Mexico. *American Anthropologist* 18: 159–180.

1932 The Origin and Development of Material Culture. *Sigma Xi Quarterly* 20: 102–123.

O'Brien, Michael J.

1996a The Historical Development of an Evolutionary Archaeology. In *Darwinian Archaeologies* (Herbert D. G. Maschner, ed.), pp. 17–32. New York: Plenum.

1996b Editor. *Evolutionary Archaeology: Theory and Application.* Salt Lake City: University of Utah Press.

O'Brien, Michael J. & Thomas D. Holland

1990 Variation, Selection, and the Archaeological Record. In *Archaeological Method and Theory, Vol. 2* (Michael B. Schiffer, ed.), pp. 31–79. Tucson: University of Arizona Press.

1992 The Role of Adaptation in Archaeological Explanation. *American Antiquity* 57: 36–59.

O'Brien, Michael J., Thomas D. Holland, Robert J. Hoard & Gregory L. Fox

1994 Evolutionary Implications of Design and Performance Characteristics of Prehistoric Pottery. *Journal of Archaeological Method and Theory* 1: 259–304.

O'Dowd, M. C.

1982 The Relationship between Theories of Biological and Social Change. *South African Journal of Science* 78: 279–287.

Olson, Ronald L.

1930 Chumash Prehistory. *University of California, Publications in American Archaeology and Ethnology* 28: 1–21.

Phillips, Philip & Gordon R. Willey

1953 Method and Theory in American Archaeology: An Operational Basis for Culture-Historical Integration. *American Anthropologist* 55: 615–633.

Rambo, A. Terry

1991 The Study of Cultural Evolution. In *Profiles in Cultural Evolution* (A. Terry Rambo & Kathleen Gillogly, eds.), pp. 23–109. Ann Arbor: University of Michigan, Museum of Anthropology, Anthropological Papers No. 85.

Reed, Erik K.

1940 Review of *Prehistoric Culture Units and Their Relationships in Northern Arizona,* by Harold S. Colton. *American Antiquity* 6: 189–192.

Reiter, Paul

1938a Review of *Handbook of Northern Arizona Pottery Wares,* by Harold S. Colton & Lyndon L. Hargrave. *American Anthropologist* 40: 489–491.

1938b *The Jemez Pueblo of Unshagi, New Mexico.* Albuquerque: University of New Mexico Bulletin, Monograph Series Vol. 1, No. 5.

Rieger, Reinhard & Seth Tyler

1979 The Homology Theorem in Ultrastructural Research. *American Zoologist* 19: 655–664.

Rouse, Irving B.

1954 On the Use of the Concept of Area Co-tradition. *American Antiquity* 19: 221–225.

1955 On the Correlation of Phases of Culture. *American Anthropologist* 57: 713–722.

Rowe, John H.

1962 Alfred Louis Kroeber. *American Antiquity* 27: 395–415.

Sahlins, Marshall D. & Elman R. Service, Editors

1960 *Evolution and Culture.* Ann Arbor: University of Michigan Press.

Schiffer, Michael B.

1996 Some Relationships between Behavioral and Evolutionary Archaeologies. *American Antiquity* 61: 643–662.

Schmidt, Erich F.

1928 Time-Relations of Prehistoric Pottery Types in Southern Arizona. *American Museum of Natural History, Anthropological Papers* 30: 247–302.

Simpson, George G.

1949 *The Meaning of Evolution.* New York: Columbia University Press.

1951 The Species Concept. *Evolution* 5: 285–298.

1961 *Principles of Animal Taxonomy.* New York: Columbia University Press.

1967 *The Meaning of Evolution* (revised edition). New Haven: Yale University Press.

1975 Recent Advances in Methods of Phylogenetic Inference. In *Phylogeny of Primates: A Multidisciplinary Approach* (W. Patrick Luckett & Frederick S. Szalay, eds.), pp. 3–19. New York: Plenum.

Smith, Andrew B.
1994 *Systematics and the Fossil Record: Documenting Evolutionary Patterns.* London: Blackwell.

Sober, Elliott
1980 Evolution, Population Thinking, and Essentialism. *Philosophy of Science* 47: 350–383.
1984 *The Nature of Selection: Evolutionary Theory in Philosophical Focus.* Cambridge: MIT Press.

South, Stanley
1955 Evolutionary Theory in Archaeology. *Southern Indian Studies* 7: 10–32.

Spaulding, Albert C.
1955 Prehistoric Cultural Development in the Eastern United States. In *New Interpretations of Aboriginal American Culture History* (Betty J. Meggers & Clifford Evans, eds), pp. 12–27. Washington: Anthropological Society of Washington.
1960 The Dimensions of Archaeology. In *Essays in the Science of Culture in Honor of Leslie A. White* (Gertrude E. Dole & Robert L. Carneiro, eds.), pp. 437–456. New York: Crowell.

Spier, Leslie
1917a An Outline for a Chronology of Zuñi Ruins. *American Museum of Natural History, Anthropological Papers* 18: 207–331.
1917b Zuñi Chronology. *Proceedings of the National Academy of Sciences* 3: 280–283.
1918 Notes on Some Little Colorado Ruins. *American Museum of Natural History, Anthropological Papers* 18: 333–362.
1919 Ruins in the White Mountains, Arizona. *American Museum of Natural History, Anthropological Papers* 18: 363–388.

Stevens, S. S.
1946 On the Theory of Scales of Measurement. *Science* 103: 677–680.

Steward, Julian H.
1941 Review of *Prehistoric Culture Units and Their Relationships in Northern Arizona,* by Harold S. Colton. *American Antiquity* 6: 366–367.
1942 The Direct Historical Approach to Archaeology. *American Antiquity* 7: 337–343.
1944 Re: Archaeological Tools and Jobs. *American Antiquity* 10: 99–100.
1949 Cultural Causality and Law. A Trial Formulation of the Development of Early Civilizations. *American Anthropologist* 51: 1–27.
1951 Levels of Sociocultural Integration: An Operational Concept. *Southwestern Journal of Anthropology* 7: 374–390.
1953 Evolution and Process. In *Anthropology Today: An Encyclopedic Inventory* (Alfred L. Kroeber, ed.), pp. 313–326. Chicago: University of Chicago Press.
1955 *Theory of Culture Change: The Methodology of Multilinear Evolution.* Urbana: University of Illinois Press.
1956 Cultural Evolution. *Scientific American* 194: 69–80.
1962 Alfred Louis Kroeber, June 11, 1876–October 5, 1960. *National Academy of Sciences, Biographical Memoirs* 36: 192–253.

Strong, William D.
1925 The Uhle Pottery Collections from Ancon. *University of California, Publications in American Archaeology and Ethnology* 21: 135–190.

Swanson, Earl H., Jr.
1959 Theory and History in American Archaeology. *Southwestern Journal of Anthropology* 15: 120–124.

Szalay, Frederick S. & Walter J. Bock
1991 Evolutionary Theory and Systematics: Relationships between Process and Pattern. *Zeitschrift für Zoologische Systematik und Evolutionsforschung* 29: 1–39.

Tax, Sol, Editor
1960 *Evolution after Darwin, Vol. 2: The Evolution of Man.* Chicago: University of Chicago Press.

Thompson, Raymond H., Editor
1956 An Archaeological Approach to the Study of Cultural Stability. In *Seminars in Archaeology: 1955.* Society for American Archaeology, Memoirs 11: 31–57.

Trigger, Bruce G.
1989 *A History of Archaeological Thought.* Cambridge: Cambridge University Press.

Tschauner, Hartmut
1994 Archaeological Systematics and Cultural Evolution: Retrieving the Honour of Culture History. *Man* 29: 77–93.

Uhle, (Friedrich) Max
1907 The Emeryville Shellmound. *University of California, Publications in American Archaeology and Ethnology* 7: 1–107.

Vaillant, George C.
1935 Early Cultures of the Valley of Mexico: Results of the Stratigraphical Project of the American Museum of Natural History in the Valley of Mexico, 1928–1933. *American Museum of Natural History, Anthropological Papers* 35: 281–328.

Vrba, Elisabeth S.
1980 Evolution, Species and Fossils: How Does Life Evolve? *South African Journal of Science* 76: 61–84.

Wagner, G. P.
1989 The Origin of Morphological Characters and the Biological Basis of Homology. *Evolution* 43: 1157–1171.

Weiner, Jonathan
1994 *The Beak of the Finch.* New York: Vintage.

Wheat, Joe B., James C. Gifford & William W. Wasley
1958 Ceramic Variety, Type Cluster, and Ceramic System in Southwestern Pottery Analysis. *American Antiquity* 24: 34–47.

White, Leslie A.
1938 Science is Sciencing. *Philosophy of Science* 5: 369–389.
1943 Energy and the Evolution of Culture. *American Anthropologist* 45: 335–356.
1945a Diffusion vs. Evolution: An Anti-evolutionist Fallacy. *American Anthropologist* 47: 339–356.
1945b History, Evolutionism and Functionalism: Three Types of Interpretation of Culture. *Southwestern Journal of Anthropology* 1: 221–248.
1947 Evolutionary Stages, Progress, and the Evolution of Cultures. *Southwestern Journal of Anthropology* 3: 165–192.
1949 *The Science of Culture: A Study of Man and Civilization.* New York: Farrar, Straus and Giroux.
1959a The Concept of Evolution in Cultural Anthropology. In *Evolution and Anthropology: A Centennial Appraisal* (Betty J. Meggers, ed.), pp. 106–125. Washington: Anthropological Society of Washington.

1959b *The Evolution of Culture: The Development of Civilization to the Fall of Rome.* New York: McGraw Hill.

Willey, Gordon R.

1953 Archaeological Theories and Interpretation: New World. In *Anthropology Today* (Alfred L. Kroeber, ed.), pp. 361–385. Chicago: University of Chicago Presso.

1960 Historical Patterns and Evolution in Native New World Cultures. In *Evolution after Darwin, Vol. 2: The Evolution of Man* (Sol Tax, ed.), pp. 111–141. Chicago: University of Chicago Press.

1961 Review of *Evolution and Culture,* edited by Marshall D. Sahlins & Elman R. Service. *American Antiquity* 26: 441-443.

Willey, Gordon R. & Philip Phillips

1955 Method and Theory in American Archaeology, II: Historical-Developmental Interpretation. *American Anthropologist* 57: 723–819.

1958 *Method and Theory in American Archaeology.* Chicago: University of Chicago Press.

Willey, Gordon R. & Jeremy A. Sabloff

1993 *A History of American Archaeology* (3rd edition). New York: Freeman.

Wissler, Clark

1917a *The American Indian.* New York: McMurtrie.

1917b The New Archaeology. *American Museum Journal* 17: 100–101.

3

Why the Muddle in the Middle Matters: the Language of Comparative and Direct in Human Evolution

J.A.J. Gowlett
University of Liverpool

ABSTRACT

The two fundamental approaches to studying human evolution are those of using data relating to the past itself (i.e., archaeology, palaeontology), and those of establishing comparative frameworks, usually based on the study of extant related species. Work with primates has been important in establishing comparisons of social structure, frameworks of subsistence, and base levels of intelligence, and has recently been used to generate ideas about the origins of language. This chapter examines how archaeology can be used to address critically the application of comparative ideas, especially those which are based on the concept of 'Machiavellian intelligence', linking group size and neocortex ratio. Archaeology is able to map out the communication of cultural information in a number of ways. Its direct evidence establishes that early *Homo* had far greater ranges than those of living apes. Technology used across these ranges had to pay for itself in energetic terms. It is argued that such factors may be as important as group size in accounting for evolutionary change.

INTRODUCTION

'Human evolution': these two simple words are used to stake out a common ground, and have featured in the titles of innumerable text books. They weighted archaeology and physical anthropology differently, but data were scarce enough that the desired focus was obvious, and selection was chiefly imposed by huge gaps in the evidence. All that has changed in recent years through an explosion of data. Not only are practitioners forced to be specialists, but we actually have different aims according to our various disciplines. There are numerous ways of telling the story.

Regardless of discipline, the two principal approaches remain those of acquiring and analysing comparative and direct data. In a recent paper, I argued for the roughly equal merits of these (Gowlett 1996a). On the one hand, we can use the framework of archaeology (or hominid palaeontology) standing on its own to look at the past. On the other, using a comparative approach we set up a framework for evolution which can be far broader, and which may be based largely on data which do not come out of the past at all. It can include other primate species or carnivores, and their ecology; locomotion, dietary adaptation, and DNA studies and, of course, social behaviour.

A comparative approach was used by T.H. Huxley in his classic evolutionary analysis of the anatomy of humans, apes and monkeys, which first established the close structural similarities between humans and the African apes (Huxley 1863). Darwin's *Descent of Man* (1871) also adopted a comparative framework, and has often been regarded as a comparative psychology. In recent years, comparative approaches have been developed to provide a more formal comparative method, often involving studies of large numbers of related species, and using statistical methods to extract relationships.

The great limitation of a comparative approach is that it can only model across discontinuities in present evidence through inference projected into the past. For instance if two out of a group of related species possess a characteristic (e.g., knuckle walking, technology), it remains uncertain whether their common ancestor possessed this trait. For this reason, successful modern uses of comparative approaches tend to hang on pegs of direct past evidence. An example is provided by biomolecular studies, based on protein similarities or DNA evidence in living species. When the results are calibrated against well-dated fossil evidence which indicates a very early divergence of ape and Old World monkey ancestors, they provide limits for the divergence of apes and hominid ancestors - about 5-8 million years ago. These results gave a stronger framework than that of Miocene fossils, and helped to show that the fragmentary remains of the *Ramapithecus* had been misinterpreted as a hominid ancestor (Goodman et al. 1989, Chamberlain 1991). Even so, solid fossil evidence has amended interpretation again, since the direct record of early hominids now makes clear that 5 million years is an underestimate for the divergence.

In this chapter, I take a harder look at comparative and direct frameworks and approaches, to consider how far we as archaeologists can come to grips with and test the comparative. We have already seen that there is often an interplay between the approaches; I am not suggesting that in archaeology we do not practise any direct approach without using comparison at some level. Conversely, most comparative research–such as that on language–employs data from the past. Even an early stone tool is best interpreted in the context of modern replication. The point of my previous paper was to highlight the difficult interface where comparative and general meets specific and direct. But, those who study social behaviour, cognition or the origins of language are free to cover the whole scope of human evolution, and sometimes find that direct evidence is of little help. Where it is tedious or seems irrelevant, they work from other evidence, and concentrate on painting the story broad-brush (e.g., Deacon 1997, Dunbar 1996).

That position tends to become reinforced in human evolution because direct methods often fail to reach across the whole field of interest. Archaeology embraces only 2.8 million years of perhaps seven million years of hominid ancestry, and probably touches on a minority of hominid species. The hominid fossils themselves virtually fade out beyond four million years, so that comparative approaches reign supreme.

Although those approaches have a pedigree going back to Darwin, this position represents a change from one which was held for a generation. From 1959, when vital discoveries were made at Olduvai Gorge, archaeology had a two million year record, and so did the principal hominid finds (Leakey 1959). Beyond that, there was no more than a vague long span of apehood, ripe for only the most general of modelling. The discoveries of *Australopithecus afarensis* in the 1970s (Johanson & Taieb 1976) heralded the change, because they resurrected the challenge of inferring hominid behaviour without the aid of an archaeological record (previously a problem only in the South African cave sites). The collapse of *Ramapithecus*, a trickle of earlier hominid finds, and methods of biomolecular comparison (relying on direct fossil evidence for their calibration), have given us a seemingly better-defined framework of divergence, defining a period of 6 - 8 Ma for hominid origins (Goodman et al. 1989, Chamberlain 1991). Similarly, ethology, and especially primate field studies, have shown far more similarities with ape behaviour than had been evident earlier (Table 3.1).

What then remains the value of the direct evidence, especially that of archaeology? It is highly specific, but it may not tell us what we most want to know, and can also be highly redundant in what it does tell us. Nevertheless, it resolves points which a comparative approach cannot even touch. Only direct evidence can tell us about rates of change in lineages, and indeed about the variety of past life. The comparative method can examine extant humans and chimpanzees, but would tell us nothing of the existence of the extinct robust australopithecines - even though it can be applied to interpretation of the fossils once discovered. Its function must be to bring us into unavoidable contact with reality. Fossil hominids and stone tools do not talk very loudly, but they are there. They exist beyond theory, and constantly spring surprises on us which show the limits of theory. The 'muddle in the middle' was a good title for addressing the Middle Pleistocene (Isaac 1976), and refers to an area that still bristles with the trickiest problems of human evolution.

Table 3.1. Some elements of behaviour shared by chimpanzees and humans.

Tool use
Tool making
Reuse of stone tools at work sites
Processing of bones with tools
Hunting and eating of meat
Extracting and eating of insects
Scavenging of carcases from other predators
Food sharing
Prolonged period of maturation
Social learning
Close family relationships
Nonverbal communication
Symbol use (in captivity)

Sources: Goodall (1976), McGrew (1993), Savage-Rumbaugh (1986), Boesch (1993).

MODELS AND THEIR DIMENSIONS

This account is simplified, as are all our models. The comparative is favoured now, partly because more disciplines have come into play, but it has also caught the flavour of the age—we live in a time of *ésprits simplistes* (Lovejoy 1936).

Models can provide reductions in dimensions, thus may have a great appeal as well as add clarity. No formulation can show how much simplification is good, but there are two points to make here: first, comparative and direct approaches entail different kinds of simplification; and second, simplification tends to be implicit rather than explicit.

In human evolution the simplifications frequently operate as a restriction of the variables considered. Very often this is done intentionally, but without label. So if we see modern humans and chimpanzees as sibling species, descended from a common ancestor, this can be portrayed in a simple Y-diagram (Fig. 3.1). Probably the hominids are actually a species-bush (see below), but this does not make the diagram 'wrong'. Much of such simplification is implicit. Much is forced on us by the sheer weight of data. One common practice is to abandon earlier literature as outmoded—paradigm shift can make it conveniently irrelevant. Another is to let one area, even a single lake basin, stand as proxy for large areas, through long periods. Negative evidence may be ignored, even if it characterizes half the world. A third is to focus on 'early hominid behaviour' without reference to species variability. Much of this is forced on us by the pattern of data—abundant here, entirely absent there.

An approach favoured in palaeoecology aims to build up a short-term picture ('Pleistocene Park'), arguing that we must work out the many variables in ecological terms, before even attempting to construct longer term models. This admits complexity. Potts (1994) has called for greater emphasis on variability, rather than single solutions.

Where archaeology looks at the long term, it has tended to see simple directionality or stability, and can be taken to task for ignoring short-term fluctuations that are hard to resolve in the record (Bar-Yosef 1995, Gamble 1996b). Another habit of selection is to ignore the negative evidence, even when it occupies half the world. Again, we use one investigated region as proxy for huge areas that are unknown. This does not mean that archaeology is uncritical: indeed ideas or labels are challenged, and then it may be that we abandon the usage. Home-bases, culture, curation and style have all been threatened.

Much of this seems obvious, but it is basic to my theme of how we use comparative and direct evidence in modelling. With a little effort we can hope to perceive

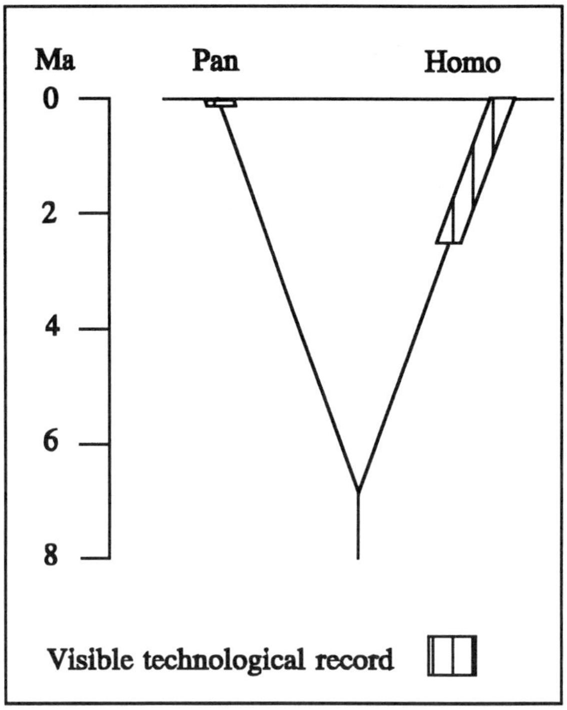

Figure 3.1. A simple model of divergence of hominids and chimpanzees from a common ancestor, attempting to chart the use of technology, but omitting many aspects of the record.

something of the nature of present biases (Clark 1991). Aspects of our own evolution determine how we think. I have argued that we have difficulty managing more than three variables at the same time; and that though we like the dynamic, we are better at static labels (Gowlett 1995). Even when we have a vague intuitive appreciation of a multivariate situation, and address it with appropriate techniques, we often aim to reduce the position to two new explanatory variables. Correspondingly, we seem able to manage only a similar number of levels of intentionality (Dunbar 1996). As everyday pressures of the Pleistocene have shaped the abilities which we now apply so esoterically, it is remarkable that any of us can even read and write.

MODELS OF HUMAN EVOLUTION

I would like now to probe further into some models, starting from those of the mainstream of human evolution (i.e., based on 'hard' evidence supported and informed by the comparative—analogy in some form or other), and then progress to those that are more purely comparative. This gives in synopsis the following frameworks:

(1) general biological models for hominisation: greatly influenced by direct anatomical evidence, but incor-

porating general evolutionary theory (c. 10 million years onwards).

(2) early hominid behaviour models, which incorporate archaeological evidence, and seek to explain the beginnings of technology in a social context (centering around 2 Ma).

(3) middle period models, which struggle with a million years of stasis, and still find that it is a muddle, even with better chronological resolution (about 1.5 to 0.5 Ma).

(4) symbolic explosion models(e.g., Duff et al. 1992, and many others) traditionally linked with the Middle/Upper Palaeolithic transition (chiefly last 100,000 years).

(5) comparative models, which seek to explain, say, changes in social structure, or the evolution of language, from non-fossil and non-archaeological evidence (all periods).

Does specialisation mean that no general models of human evolution are relevant to all of us? Scepticism seems fair over any new 'single shot' theory which seeks to bring everything together with one evolutionary motor. These are too simple. Even a cursory examination of the evidence shows the many scales of modelling that are possible and necessary (Fig. 3.2). The models themselves vary in power. Some aim to give a descriptive roughout, a basic analysis. Others attempt to explain the interaction of variables, and so are more functional. Others are simulations which can be put to work as computer programs (e.g., Wobst 1974, Potts 1988, Steele 1994, Lake 1996).

THE BIOLOGICAL FRAME: HOMINISATION AND EVOLUTIONARY TRENDS

Darwin lacked modern genetic knowledge, of course, but possessed the great insight to set out a basic model of human evolution impressively close in outline to those still accepted, and anticipating many discoveries of fossil evidence. Continuity of thought since Darwin wrote has been stressed by Washburn and Moore (1983). Darwin emphasised the primacy of bipedalism, and brought into play most of the major factors which still concern us. He did this almost entirely on comparative evidence in the lack of almost all hominid fossils.

Traditional versions of the model have emphasised fairly simple linkages: a primate forced to the ground; bipedalism freeing the hands, the upright position providing a better view; socialisation implied by the smaller canines; tool-making discovered, and leading to advantages which set off a cultural feedback loop. The social and economic aspects have been stressed differently by different writers (see below).

Darwin analysed many of these factors. Although his view is clearly pre-modern, he avoids many subsequent oversimplifications, including the persistent one that hominids alone were 'forced' to ground (debunked by Bishop 1976, Foley 1987). He even saw the locomotion of chimpanzees and gorillas as transitional. So, the fundamental outline of hominid evolution has been known for over a century, but following a long accumulation of finds, two major developments of perspective in recent years have recently come to the fore:

(1) living apes have far more overlap with hominids in behaviour than had been widely appreciated (Table 3.1); and

(2) early hominids are anatomically (and therefore perhaps behaviourally) far more apelike than was at first assumed.

The details of chimpanzee behaviour, including its cultural and symbolic aspects, are plainly of high relevance for human evolution. McGrew (1992) goes so far as to suggest that chimpanzees make a better analogue than modern humans for looking at early hominid behaviour. Other authors would argue that we should never 'see' one animal as another, but there seems little real danger of this. What has been gained is a comparative frame rich in specific detail (see Clark, Chapter 12).

The early stages of hominisation nevertheless remain a difficult area for comparative as well as direct approaches, with much of the most important information coming from the fossils themselves. One of the most striking features of hominid evolution is its directionality. First, there is the rapid evolution of bipedalism, and then the phenomenal evolution of the brain. Ideas about bipedalism and its origins have been reworked somewhat, both by ethological and metabolic studies, and as well as biomechanical studies based on hominid finds (Hunt 1994, Wheeler 1984, 1994; Li et al. 1996). I am going to skip almost entirely the early stages of bipedalism and canine reduction, just noting the consensus that they must be seen in a context of social structure and ecology, including feeding habits, as well as the mechanics of locomotion. Rather than explore these further, I will focus on the problems of studying and explaining the changes in the brain and its functions that have been fundamental to the subsequent course of human evolution.

ARCHAEOLOGICAL MODELS (1): EARLY HOMINID BEHAVIOR

If we start from the general biology, it is little more than a truism to say that the primary model is neo-Darwinian. This tells us that natural selection operating on genetic variation is important among hominids, but the same would be true for any other species. A rare challenge to this idea came from Groves (1989), who stresses the importance of non-Darwinian modes of evolution. His ideas are interesting because they are a radical alternative and offer a total framework. The *nomogenetic* evolution proposed by Groves, following the work of Berg, sees a bush, in which the ascending lineages are continually reborn through what Groves terms 'quasi-speciation events'. Groves suggests that this is the form of evolution which should ensue for the hominids. In this he seems to characterise human evolution successfully, offering heterochrony–especially neoteny–as an evolutionary mechanism, but not giving it a driving cause.

Groves is the champion of a particular evolutionary paradigm, which is reluctant to accept the force of natural selection (and hence adaptive explanations). Most mutations in genes are indeed likely to be deleterious, giving support to the idea that mutations altering rates of development are most likely to be successful (through leaving an organism that is not disastrously handicapped). Heterochrony is the general field (e.g., McNamara 1990), and neoteny the retention of juvenile characteristics in adults. Neoteny can be observed, and clearly is important, as is the great growth in brain size in humans. But for Groves (and for Blumenberg 1983) an adaptive explanation must be ruled out. Groves sees the large brain as secondary, an 'epiphenomenon' of neoteny, which he links with longevity and *K*-selection (following Gould 1977). Perhaps the most interesting feature of this argument is just how hard it is to dump an adaptive explanation without ending up with another argument which admits the advantage of such a brain (i.e., an exaptive explanation). *For Groves and Blumenberg, the large brain may be a useful thing, but it was beyond the scope of natural selection to forge it.*

Groves is influenced by the ideas of Stephen Jay Gould, although he points out antecedents to these ideas. Gould, together with Niles Eldredge, was largely responsible for the idea of 'punctuated equilibrium' which has greatly influenced palaeoanthropology through the application of ideas inherent in 'cladistics'. Most hominid palaeontologists now adopt a cladistic approach to analysis, because its rationale of distinguishing recently acquired and ancestral characters helps them in building classifications (see Clark 1988, 1989a; Trinkaus 1992 for reservations). They tend to start by grouping finds into species, and then postulate rapid intervening speciation events as necessary. The use of cladistics does not force this, but the approach tends to echo Gould's views (Brace 1989). For Gould (1990), inter-species competition is the most important element in evolution, but he admits the existence of trends, and is happy to cite Darwin: biotic competition must predominate, for the abiotic would be too random have a marked effect. Therefore trends will proceed in steps, with 'better' species succeeding one another. Simple progressional evolution, or 'anagenesis', tends to be regarded as outmoded.

Why is this discussion necessary? Ayala, Cain, Maynard-Smith and others have rebutted the arguments of macroevolution, demonstrated again the importance of adaptation, and concluded that the ideas of punctuated equilibrium add little if anything to classic evolutionary theory (Ayala 1983, Cain 1989, Maynard-Smith 1993). Groves does describe human evolution uncannily well, however. The species bush seems real, and in some aspects–including brain size growth–the record seems 'steppy'. If his theoretical explanation is in error, it is nevertheless highly thought-provoking in the search for alternatives.

Does all this affect our archaeological view? We are now dealing with a position in which palaeontologists have apparently found us far more species–at least eight within the timescale of technology. To an extent, Groves eases the position by the idea of 'quasi-speciation' which would allowing interbreeding, and presumably the transmission of cultural ideas. But this leaves the point moot of whether we might expect to see a series of *archaeological* discontinuities. Foley (1987, 1996) has aimed to test the idea that we might link industrial traditions and species, although most archaeologists feel there is too much local variability to support this (Clark 1989b).

If there truly were as many hominid species as are now posited, however, it emphasises the matter of different evolutionary responses to different environments. As Foley has said (1987: 273) "there is no single way of being a hominid." Rather than launching a general model, Foley concentrates on stating ecological principles, and then isolating individual 'problems' for treatment.

This discussion suggests that directionality of evolution may be perceived as a problem, and indeed that is the case for growth in brain size. Rensch (1959) talked of 'brain victories', but Barrow and Tipler (1986) emphasise that large brains are very rare in the biosphere, and they do not always win out (elephants, though large-brained, are not especially successful). Progression in one direction is far from inevitable, indeed species stability - and stabilising evolution - are very common (Ayala 1983, Dean 1992). Equally, a trend does not need to go on forever. There are many evolutionary cases where some faculty or other is lost, which was earlier laboriously gained. The nub of the problem is that if one character develops very strongly over a long period, the endproduct seems to imply the deficiences of the starting point. Yet evolution demands that every generation must be viable, and it is not easy to comprehend the

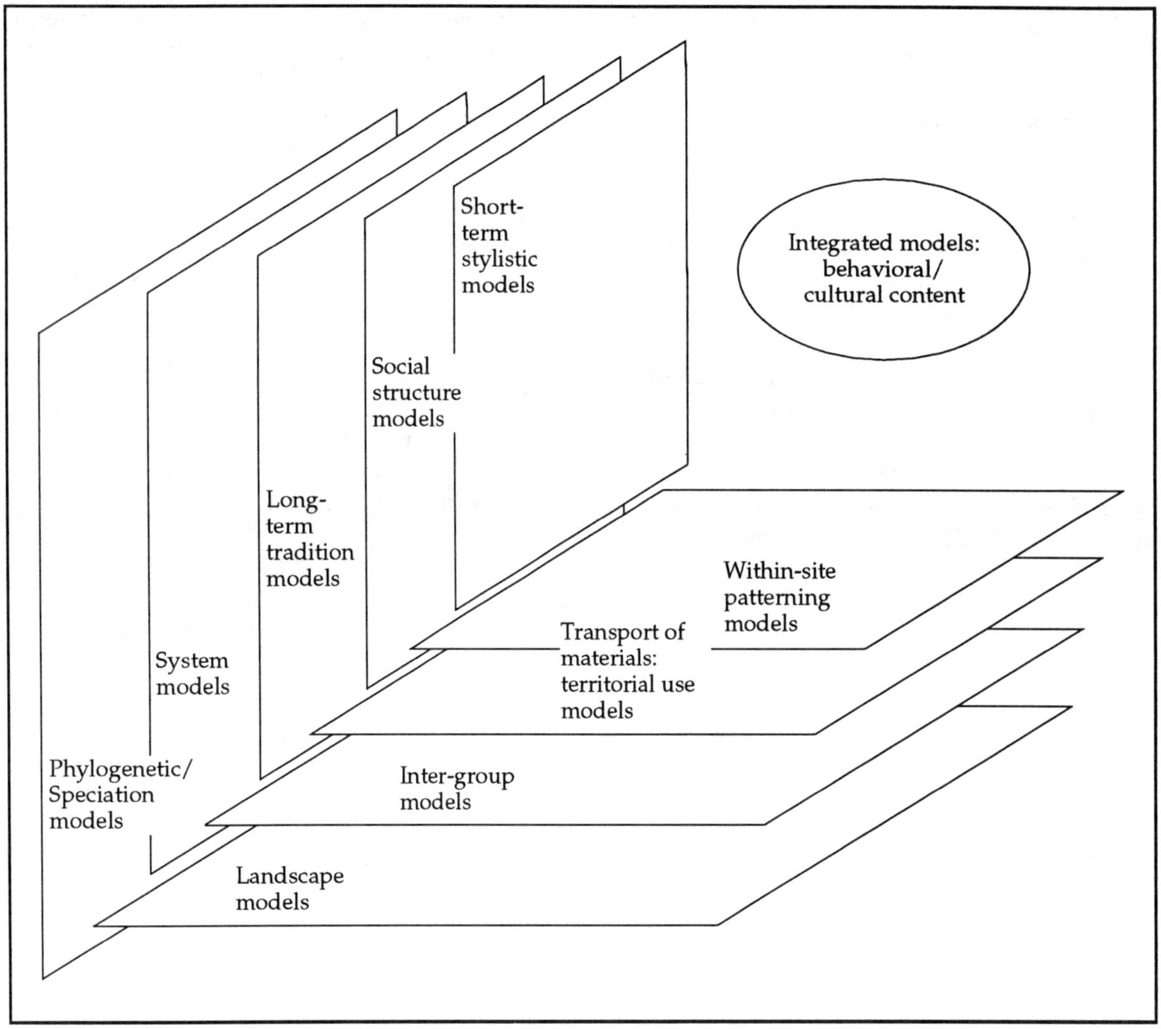

Figure 3.2. The varying scales of time (vertical axis) and space (horizontal axis) which are employed in models of human evolution.

circumstances under which all those generations would be subject to roughly equal evolutionary pressure, and have roughly equal scope to respond to it.

This may, however, come about through a feedback loop. Many authors have invoked feedback as the cause of a directional shift. Dobzhansky (1963) was among the first to bring this terminology to human evolution, linking the genetic component, culture and brain size. This is amplified by Washburn (1967), Tobias (1981,1991) and Wilson (1983), amongst others. The idea of a feedback loop, or autocatalysis, is useful because it automatically implies an instability, managed by evolution in a stable kind of way. Tobias (1981) points out the dramatic extent of growth in brain size, or encephalisation, which "was sustained for 2 Ma or more, carried the lineage across a generic boundary into *Homo* and continued in the three consecutive chronospecies *H. habilis, H. erectus* and *H. sapiens.*" He notes that it transcends ecological radiation, ethological variegation, and that it benefitted hominids through a bewildering range of lifestyles.

Recent biological work, however, emphasises that brain does not just 'happen'–it is expensive and must pay for itself (see e.g. Aiello 1996, Byrne 1996). Aiello distinguishes two periods of sharply increased growth in brain size, one at about 2.0 Ma in early *Homo*, the other at about 0.5 Ma. in archaic *H. sapiens/heidelbergensis.* These are isolated directly, empirically, from the fossil evidence, but her explanations combine physiological observations and inferred social factors. It is argued first that refined bipedalism requires a better brain; then increased group size is regarded as a prime factor, with perhaps increased range being another.

Can these steps be interpreted as evidence of punctuated equilibrium, with natural selection not being a main cause? That would undermine the idea of feedback loops. But it seems to me that Groves and Gould are only able to distance natural selection by one remove, and this both restricts the dimensions of their arguments and results in clumsier explanations. A great deal of modern study is preoccupied with the influences of climate, or of carnivore competitors (abiotic and biotic respectively), but another possibility is that the main source of competition comes from within. Sexual selection, which was a major concern of Darwin, at least offers an analogy, even though the main competition is between members of a sex (for recent views see, e.g., Hawkes 1996, Knight 1995). Social competition models, which are considered further below, offer more scope for explaining evolutionary change.

Early Hominid Behaviour and Archaeology

Most of the early-period models can be conceived as tandem: they make use of bipedalism, to create a stimulus for brain enlargement, which then subsequently accelerates. They invoke smaller canine size as evidence of early socialisation or changes in social structure. Then comes the feedback loop which involves technology, subsistence, and later more refined aspects of culture.

The foodsharing model of Isaac is now the best known integrated model, displacing simpler hunting models such as that advocated by Washburn (1967; see also Clark, Chapter 12). Although its elements were not entirely new, Isaac incorporated more aspects or dimensions, and integrated them more formally, also adding a sequence of events. In this structure he was influenced by the systems models of Clarke (1968) (for a recent version see Megarry 1995:124).

Lovejoy (1981) contributed a careful analysis of the benefits of bipedalism in provisioning, arguing that hominids avoided the reproductive trap into which the apes were falling. The hominid pattern helped to bring greater responsibilities to males, but this idea concentrates on ground-living and large ranges, rather than on home bases.

The importance of Isaac's model is emphasised by the criticisms which it has endured, and partly weathered. Newer statements have aimed to isolate non-controversial essentials, and to discard the more hypothetical aspects, such as the ideal of 'home', or division of labour (e.g. Schick 1991, Rose & Marshall 1996). Inevitably, this leads to a loss of explanatory power. The model did however have two potential Achilles heels. One was the very emphasis on food-sharing, which does occur in other animals, especially chimpanzees (McGrew 1992). Isaac claimed that hominid food-sharing was different, but that contention involved the circularity that if it was different because they were hominids, they had already acquired behaviour which the model sought to explain. The other was the linking of 'home bases' with actual archaeological patches, which has often been contested.

Binford (1981) took a somewhat popularized synopsis of the Isaac model (Isaac & Isaac 1975), and analysed it as "a speculation as to what the past was like", which involved no recognisable theoretical arguments. In terms of archaeological theory, this may have been strictly correct, since the account or model runs across the categories which I outlined above. It depends partly on direct evidence from the fossil record about the primacy of bipedalism and the size of hominid brains. It also makes empirical use of information about the archaeological record, its length and complexity. In constructing two tiers of development, it uses comparative evidence a good deal, and finally imagination,which the late David Clarke felt was a vital ingredient of archaeology (pers. comm.).

Lake (1996) regards Isaac's model as lying at the root of the 'current goals and inferential framework' of early hominid archaeology. There are of course many other partial models of events in this period, contributing other relevant variables, but otherwise tending to fall into descriptive lists. Tanner's theory of human evolution established the importance of women and of gathering, relying considerably on archaeological evidence (1981). The model of Holloway (1972) is perhaps the most complete of all, constructed on the basis of anatomical and physiological changes, and embodying three stages.

ARCHAEOLOGICAL MODELS (2): THE MIDDLE MODELS

The middle period of the Pleistocene (1.5 to 0.5 Ma) is usually seen as one of apparent stasis. I have tried to seek out general models for this period (and perhaps missed some), but I found that the models either start at c. 2 Ma, set up the brain-tools-feedback loop, and leave it to run; or they are focussed far later in time, and concerned with an ancient/modern transition, starting from a state of primitiveness (which for simplification could be regarded as 'primeval'). Stability of course is a model in itself, but it still poses questions: why was it entered; why was it left? And in any case, is it real or illusory?

The two tiers of Isaac's general model actually bracket this period. Previously, however, he had offered a series of Middle Pleistocene models. Isaac (1972a) considered the low levels of craft specialisation found in the Acheulean, and linked these with low levels of group boundaries. He inferred that there was not enough cultural content, or possibly linguistic vocabulary, for marked regional variations to occur. Rather, there occurred random variations in tool forms, within functionally-determined limits. Steele (1994) has developed these ideas and tested them through simulations of communication networks and dispersal patterns. Gamble (1993, 1996a) has also studied the spatial dimensions, and contributed his 'Local Hominid Networks' - the LHN models. These are a theoretical construct, depending partly on past archaeological models (e.g., of Higgs), partly on social and geographic models, and partly on the direct evidence of raw material transport distances of artefacts. Much other work done on artefacts confirms the many potential strands of evidence and supports the idea (Table 3.2). Gamble is among those who conclude that a single model is inappropriate in the face of Middle Pleistocene variability (1996b).

Colonisation of the temperate regions was one of the great achievements of this period. Gamble (1993) sees the factors listed above, including brains and technology, as preconditions rather than causes. His preferred explanation lies in the consistent selection pressure for calculated migration - which reduced risk and benefited the fitness of individuals who belonged to a more complex social unit. Hence, the emphasis on alliance which is an important thread in studies of later periods (Gamble 1986, Barton et al. 1994, Clark et al. 1996). Directly or indirectly, this places an emphasis on cognitive changes, which perhaps operated largely unseen in the rest of the archaeological record.

ARCHAEOLOGICAL MODELS (3): TOWARDS MODERNITY

A speeding-up of events is seen with the arrival of more modern-looking humans about 500,000 years ago (archaic *H. sapiens* or *heidelbergensis*) and the beginnings of Levallois technology, about 250,000 years ago. Symbolism is a major preoccupation of those who write about the beginnings of modern behaviour, and features in most models. The authors themselves sometimes seem uncertain about how far these models apply 'back' into earlier periods. Some, like Binford (1989), make a point of enhancing the distinctions between 'ancient' and 'modern'; others like Stringer and Gamble (1993) contrast these 'ancients' and 'moderns' as merely 'different' (cf. Clark & Willermet 1995).

The difficulty may be one common to all problems of 'emergence'. What we can see in later times is a cultural diversity and complexity which is not there earlier, so we can only look for its origin if we have an accepted way of analysing the later phenomenon into components. The 'before' and 'after' approach cannot achieve this, because it deals in presence and absence *en bloc.* Rigaud (1989) and Reynolds (1991) advocate an approach of scheduling–monitoring developments in many strands of evidence. Again, it is necessary to establish 'of what'. The older record may not present enough relevant threads for us to achieve this, but the range shown in Table 3.2 suggests that it might.

Writers are so much concerned with symbols precisely because these *do* inform us about complex behaviour, and more so because of the information that they might provide about language origins.

Some authors are willing to contemplate the possibility of symbolic behaviour at the earlier dates, but see no necessity for it to exist, and no necessity that we could recognise it in archaeological evidence even if it did exist (see, e.g., Chase & Dibble 1987, Chase 1991, Duff et al. 1992). Although these and papers on style (e.g., Sackett 1982, 1986) do not have a common viewpoint, they explore the difficulties of disentangling style and function, and show how hard it is to demonstrate the existence of symbolic intention in early times. Standardisation in the form of artefacts is taken as possible evidence, but only if there is no possibility that it is imposed by functional constraints.

Symbols, in any case, need not denote language. The best *a priori* argument that there might be a symbolic quality in early artefacts is its existence in the behaviour of chimpanzees (e.g., McGrew 1992). That might imply a long evolutionary history, but this is not accepted by all. Because of this and other difficulties language has had little place in recent archaeological models, although the issues have been explored (Davidson 1991, Davidson & Noble 1993). Graves (1994) felt that the equation of 'symbolic' artefacts with language missed the point of the social meanings which artefacts might have had for their makers.

Mithen (1996) has developed a challenging model which covers this period of transition, and which is not heavily dependent on language/symbols. He postulates that earlier humans have domain specific intelligence, whereas modern humans have achieved a much better integration of mental functions. He labels the domains as 'technical', 'social' and 'natural history' intelligences. Earlier hominids would be able to operate effectively in each of these domains, but only modern humans would be able to exchange, rework and crosslink information between them easily and effectively.

Although archaeology has much to do to test this model, some preliminary comments can be made. First, it implies that hominids earlier than the Neanderthals also had this dissociation. Can we discern this in the record? An alternative would be that new mental/brain structures had grown throughout the Pleistocene and were in need of this integration–i.e., that chimpanzees and earlier humans would not have had the difficulty. There is some reason to think that chimpanzees may have a problem through not having a natural symbolic language. For example, "if a chimpanzee learns to discriminate between a bright and a dim light, it cannot transfer this learning in order to discriminate between a loud and a soft sound" (Glucksberg 1988: 223).

One concern about the model is the way in which the domains are chosen. Mithen's categories appear to echo previous threefold divisions imposed by anthropologists (e.g., L. White) into technological, sociological and ideological/philosophical aspects (Dillingham & Carneiro 1987), or into aspects of material, social and mental culture by Osgood (cited in Alland 1969). Of course, Mithen's domains are not identical with those earlier divisions, but they have been isolated similarly on a logical basis. Mapping of the living brain is the best way to elucidate their validity in living humans, but this will not help us with Neanderthals. Gibson (1993) accepts the importance of the question about domains, but is very cautious about trying to answer it on present evidence. Sternberg also notes the trend towards seeing abilities as domain-specific, but believes that some fundamental processes are quite general (1988).

A more general approach to this problem is to suggest that as the neocortex has grown and its functions expanded, 'rewiring' may have restructured the operation of even basic functions. In modern humans, almost all these are under some cultural control or influence. For Alland, "The expansion of the cerebral cortex and the reordering of cortical structures are the basis of all high-order thinking, including language (1967)." For Holloway (1972), the later phase involved the elaboration of 'complexity management', defined as fineness of sensory discrimination, and associations between 'larger sets of past memories and skills'. The interactions of such factors in a 'biocultural brain' are discussed further by Gibson (1996), Aiello (1996) and Kien (1991). Modern individuals vary widely in their planning and associative abilities, perhaps supporting the idea that entire past populations were less gifted in these areas, apart from considerations of language and what it facilitates. All this work tends to stress complexity in the brain's operations - perhaps demanding language, but the difficulties of isolating this in the direct record have not yet been resolved.

COMPARATIVE MODELS

Suppose for a moment that we had all the material available to Darwin, and the information of modern palaeontology and genetics, but that no fossil hominid had ever been found, and that stone tools melted away. We would then be dependent on another set of comparative approaches, which would be largely independent of our direct record. These might reach towards social structure, social behaviour, and the origins of language. These are precisely the 'soft' tissues which our 'hard' evidence can reach only with the greatest difficulty.

They are the domains, in various measure of ethologists, behavioural ecologists, sociobiologists and comparative psychologists. Evolutionary ecology has worked towards a better understanding of human evolution over a long period (Thorpe 1979, Crook 1980). Some of the earlier sociobiology failed to appreciate the complexities of human behaviour, but this is not a fair criticism of later work, which makes a point of recognising human uniqueness (e.g., Alexander 1979, Crook 1980). Alexander, in particular, has sought to explain the special characteristics of the human species which led to its divergence and the extinction of near relatives. For him complex social competition is relevant not only to human intelligence, consciousness and foresight, but to such matters as parental care and sexual behaviour. Intergroup competition is the guiding force, but if a group grew in size, this would lead to intensified within-group competition. The parental care mentioned by Lovejoy (1981) would then be more important.

Here Alexander's ideas anticipate the notion of 'Machiavellian intelligence'. The main thrust of this notion is that the large brain has not evolved for dealing with an extra-species environment, but rather for dealing with the social environment within a large group (Byrne & Whiten 1988, Byrne 1996). Since many primates have large brains, and in this measure humans merely represent the extreme of the range, it can be argued that the need for it was primarily social rather than the management of technology. The needs of simple technology do not present a powerful argument, for it is used by chimpanzees, and would place demands on human brain power for only short periods at a time.

Dunbar has focused on the importance of the neocortex ratio - the ratio of the neocortex to the old limbic brain (e.g., Dunbar 1995, 1996). Examining variables pairwise, he has found a very high correlation between neocortex ratios and group size across the primates. He uses this information to read off natural group size in modern humans, producing the 'magic' number of 150. This is not taken to mean that humans go around in groups of this size, but that it is a relevant limit of community coinciding approximately, for example, with the numbers in a clan, or a military company. Reasonable estimates can be made of the neocortex ratio in fossil hominids so that–if the calculations are accepted–group sizes can also be predicted in fossil hominids (Aiello &

Table 3.2. Examples of lines of archaeological research in Lower Palaeolithic studies over the last generation.

Site spatial configuration	Howell & Clark (1963) Leakey (1971) Roberts (1986)
Definition of entities	Bishop & Clark (1967) Clarke (1968) Isaac (1977)
Regional/assemblage variation	Kleindienst (1962) Isaac (1977) Bar-Yosef & Goren-Inbar (1992) Wynn & Tierson (1990)
Artefacts - typology	Kleindienst (1962) Leakey (1971) Isaac (1977)
Artefacts - technology/replication	Roche (1980) Toth (1985) Bradley & Sampson (1986) Bar-Yosef & Goren-Inbar (1992)
Artefacts - morphology	Roe (1968) Isaac (1977) Wynn (1979) Wynn (1981, 1985) Wynn & Tierson (1990)
Artefacts - modalities in design	Gowlett (1988)
Artefacts - allometry	Crompton & Gowlett (1993) Gowlett & Crompton (1994)
Artefacts - transport	Hay (1976) Clark (1980) Wilson (1988) Bunn (1994)

Dunbar 1993). Aiello and Dunbar (1993) extended this approach to examining the issue of language, which is again treated comparatively. Dunbar's basic premise is that in humans language has replaced the grooming which is characteristic of other primates (Dunbar 1996). Grooming is the prime channel for servicing social relationships in non-human primates, but at some expanded group size, it would become too time-intensive to function effectively. Experimental observations in various contexts seem to give support to the hypothesis, since they show that language facilitates social interactions as much as communicating information.

These ideas are intuitively very appealing, so much so that it might be surmised that the statistical aspects are incidental. This is not so, however, because alternative hypotheses may be viable. Dunbar himself has carried out some testing and, for example, dismisses the extractive foraging hypothesis as an alternative explanation of relative neocortex size (Dunbar 1996). Byrne (1996) has also examined reasons for growth in brain-size, and has concluded independently that range size is not a factor of explanation. He does however find that tactical deceptive abilities also correlate with the neocortex ratio, whence 'Machiavellian' intelligence.

As testing continues, other contributing factors and qualifications will emerge. First, these are multivariate situations, but pairs of variables are being selected. It is often difficult to compile adequate data for broader testing, but Foley (1996) has considered other variables. Then, the clear straight line functions are the result of log/log transformations. This sort of fairly heavy treatment of the data is common in the life sciences, but it may over-

Table 3.3. Selected questions in archaeological research of the Palaeolithic

(1) Location and density of sites: can we accumulate evidence for land-use and ecology, especially in relation to water and vegetation zones?

(2) site integrity: to what extent do the artefacts and bone remain as deposited by hominids?

(3) site sizes: how big were they, how long occupied, and at what season(s)? Does any evidence relate to community size and organisation?

(4) internal structure: is there any evidence for structural features or separations of activities which can inform about social behaviour?

(5) resource transport: how are the sites localted in relation to material resources, and how far were these transported, for modification or use ?

(6) artefact dynamics: can we disentangle threads or chains of activity, and in particular examine to what extent artefacts were made elsewhere or discarded elsewhere?

(7) faunal potential: what relationships or overlaps can be traced, with either carnivores or possible prey species?

simplify complex relationships. Each species is represented as a point, rather than a range of values - safe for neocortex ratio, but less so for group size. Figures derived by Steele (1996), who has used alternative equations to predict group sizes, and finds that these vary from those of Aiello and Dunbar by a factor of up to two. Moreover, Steele cites Martins (1993), whose least squares regression on neocortex ratios predicts a human group size of about 43 from neocortex ratios, to show that the total range of predicted group size varies from about 43 to about 300. He then poses the question, which of these group sizes might be supported empirically from the archaeological literature? Steele sees raw material transport distances as the most promising archaeological avenue to determine this. Gamble (1996b) however, sees the 40 km range of his LIINs as fairly static over long time periods, implying less scope for analysis. Archaeological methods can be used to examine the residuals of variations in transport distance, but it seems unlikely that site evidence is good enough for archaeologists to recognise group sizes directly (although see Jewett & Clark 1987). We need to avoid ending up with imprecise estimates from one discipline being geared against imprecise estimates in another.

The language issue is also difficult to test from archaeological evidence. Steele and Shennan (1996: 33) cite Eibl-Eibesfeldt to the effect that almost 60% of !Kung conversations centre on food. Perhaps the ratio of information content to gossip would be higher in hunter-gatherer societies, with their large territorial ranges, than in the ordinary conversation of village and urban societies. But Dunbar's work does not minimise the importance of practical information exchange—he simply does not see it as the primary engine of language evolution. Just as technology is very important at certain moments, so it might be that specific information imparted by language is crucial at times. This would not conflict with Dunbar's hypothesis.

Dunbar (1996) implies that all archaeologists favour a late view of language origins, but we are actually badly divided. Some of us accept the possibility of very early language (following, e.g., Tobias [1991] who argues for anatomical evidence for language capability in *Homo habilis*); others subscribe to a late origin, coupling the lack of strong direct evidence in the archaeological record with alternative anatomical interpretations (e.g., Lieberman et al. 1992). Aiello and Dunbar's approach is as rewarding as anything that has yet come from archaeology, but we are free to challenge underlying models, and look for alternative formulations that are more testable by archaeological means.

ARCHAEOLOGICAL EVIDENCE: TOWARDS AN ALTERNATIVE MODEL

Through archaeology we can address some aspects of about 2.87 million years of human evolution. Archaeology is pivotal because it appears near the cross-over between bipedalism-dominated and brain-dominated models, is preserved on a much larger scale than any other direct evidence, and takes us closer to multiple-stranded patterns of behaviour. Archaeology has demonstrated that it can contribute vital data, but there is no necessity that these will be able to address the questions in other people's minds. To the detriment of 'high theory', archaeologists have learnt largely to focus on questions which might be answered.

A general series of archaeological research questions in palaeoanthropology was set out by Isaac in his papers of the 1970s (e.g., Isaac 1972a, b). Recently Potts (1994) has set out a new set of questions for palaeoecological investigation. Inspection of these shows that there is a good deal of concordance with issues outlined by Isaac (Gowlett 1996a). In recent work at Beeches Pit (a Middle Pleistocene site in Suffolk), we evaluated these schemes and abstracted a set of questions for study (Table 3.3; Andresen et al. 1997). It is striking that almost all our questions would be as relevant at 2.4 million as at 0.4 million years—and perhaps also at 40,000 years ago. The primary questions in the list are easier to tackle than the higher level issues which can be inferred from them (Lake 1996). Arguably we can get to an ideational level only through bringing in the questions of artefact form, style and symbolism, as has been mentioned above. But I already have considered the difficulty of interpreting these in early periods. I prefer to start with, and so opt for an alternative - at a lower point on the ladder of inference - that we can focus on charting *information content* in the early archaeological record, trying especially to elucidate the *instruction sets* that underlie what we see (Gowlett 1995b).

If we turn now to *direct archaeological* evidence, we find that by around 1.5 Ma, artefact transport distance is vastly greater than the size of habitual ape ranges - up to 100 km on occasion (Clark 1980, Gowlett 1995a). Therefore, *Homo* (*ergaster/erectus*) was certainly travelling and probably mapping mentally ranges 20-100 times greater than those of chimpanzees and gorillas. A larger brain might well pay for itself, if it greatly improved returns (i.e., if it saved energy through better understanding of the environment). Technology, in the form of stone, and perhaps other artefacts provides extractive efficiency: it concentrates energy like a lever (metaphorically and also sometimes literally). But it is cheaper to have no technology, than to use it wastefully. One expends energy going to raw material sources, shaping and carrying tools. Once acquired, there is a powerful impetus for technology to be self-selecting for relative efficiency. The use of energy has to operate under social control if, as we know, hundreds of kilograms—*or even tons* – of material were sometimes transported for several kilometers across a landscape. Such efforts must be only a small part of the total picture of energy management. I am convinced, therefore, that we must aim for a combined model, in which group size is only one component.

All the Machiavellian intelligence models, however, have a motor behind the correlations which they observe: that is, they postulate that brain size grew because group size grew, and Dunbar (1995) postulates that group size grew because of some unknown ecological variable. Group size had to accord with some features of the environment (*a primary cause*), so again brain size was *secondary* (cf. Groves 1989). Apart from Aiello (1996), authors tend to rule out range size, either as a variable correlated with neocortex ratio, or as a prime mover. Range size is, however, an important variable throughout primate life, probably correlated with many other variables, and arguably no number of bivariate, log-transformed comparisons are going to elicit its real contributions—or allow us to dismiss it. Nor is it clearly established that brain size grew in relation to managing the social complexity of life within large groups—rather than through dealing with an external world, and *inter-group* competition, which might well exert a higher selection pressure.

To sum up, natural selection is the main motor of change (despite recent argument). Hominids can respond to it through any part of their complex genome that can change (mutate) successfully. The very great changes in hominids therefore demonstrate intense selection pressures coupled with a great ability to respond (perhaps chiefly through heterochrony). The pressures can be partitioned as follows:

(1) abiotic—overriding changes in climate
(2) biotic—consequent local fluctuations in environment (floral, etc.)
(3) biotic—formation of new habitats (floral, etc.), the colonisation of which requires adaptation
(4) biotic/faunal—competition from other species, especially primates and carnivores
(5) biotic/within-species—inter-group competition
(6) biotic/within-species—intra-group competition.

Here are six factors, of which 'Machiavellian intelligence' magnifies only the last. This does not minimise the existence or importance of such intelligence, nor, for example, lead us to reject Aiello and Dunbar's idea that language has replaced grooming. It does emphasise that in a multivariate world models need to be multidimensional, and that archaeology may well be able to recover other dimensions that are equally important. Even if we are so far unable to interpret these in terms of language origins (which are much debated), we are plainly able to map out the communication of cultural information in a number of other ways.

CONCLUSIONS

This essay shows something of the great variety of models that we use in human evolution, and how far we are from sorting them out and integrating them. None of us can be specialists along the whole timescale, and we view the evidence from different perspectives. Comparative and direct evidence have fundamentally different resolutions. High resolution comes only from the direct, but the comparative often allows us to tackle problems from more angles and with greater sophistication.

Among the fundamental questions are, *Why did the brain keep increasing in size?* and *How did it become organised to do the things that modern humans do with it?* These are at best part-answered by the points that most palaeoanthropologists would agree on: the importance of language, and its influence in restructuring the mind.

The extent of disagreement over brain evolution is shown by the several groups into which scholars fall. A very basic issue, which I have begun to appreciate only through writing this paper, is the way we divide into those who

(1) favour a traditional archaeological model in which technology is the driving element in a feedback loop;
(2) those who see brain development as merely an evolutionary process (i.e., not driven by natural selection, but arrived at stochastically or secondarily);
(3) those who postulate ecological pressure forcing greater group size, and hence leading to greater brain size for the management of social relations; and
(4) those who no longer see technology as a prime mover, but who are not convinced by the social intelligence paradigm, and who would rather put faith in multiple influences operating in feedback loops.

Archaeology deals with a multitude of variables, even for early periods characterised as 'simple'. We try to winnow these in our simplifications, but our subject is such that they keep forcing their way back in. This is more a strength than a weakness. The record does not record what we desire most to know, yet hardly ever can we be sure that it does not record information which is useful. Language seems to have dropped out of the early archaeological record, but until we find it, I am happy to leave the last word to Glucksberg:

> *'. . it is the speech of the mind, not of the tongue, that matters' (1988).*

ACKNOWLEDGEMENTS

This paper was completed during my tenure as a British Academy/Leverhulme Trust Senior Research Fellow, which I gratefully acknowledge. I am also grateful to Clive Gamble and James Steele for discussion and thoughts concerning other paper drafts, which have proved useful here. I am indebted to Geof Clark for sorting out an extremely garbled e-mail transmitted manuscript. Figures 3.1 and 3.2 are drawn from my 'Social Frameworks' paper of 1996 (with modification); Table 3.3 is based on Andresen et al. (1997) but the issues (I believe) are substantially different.

REFERENCES CITED

Aiello, L.C.
1996 Hominine preadaptations for cognition and language. In *Modelling the Early Human Mind.* (Mellars, P. & Gibson, K., eds.), pp. 89-99. Cambridge: McDonald Institute.

Aiello, L.C. & Dunbar, R.I.M.
1993 Neocortex size, group size, and the evolution of language. *Current Anthropology* 34: 184-193.

Aiello, L.C. & Wheeler, P.
1995 The expensive tissue hypothesis. *Current Anthropology* 36: 199-221.

Alexander, R.D.
1979 *Darwinism and Human Affairs.* London: Pitman.

Alexander, R.D.
1989 Evolution of the human psyche. In *The Human Revolution: Behavioural and Biological Perspectives on the Origins of Modern Humans.* (P. Mellars & C. Stringer, eds.), pp. 455-513. Edinburgh: Edinburgh University Press.

Alland, A.
1969 *Evolution and Human Behaviour.* London: Tavistock Publications.

Andresen, S.A., Bell, D.A., Hallos, J., Pumphrey, T.R.J. & Gowlett, J.A.J.
1997 Approaches to the analysis of evidence from the Acheulean site of Beeches Pit, Suffolk, England. *Archaeological Sciences 1995,* pp. 389-394. Oxford: Oxbow Monographs No. 64.

Ayala, F. J.
1983 Microevolution and macroevolution. In *Evolution from Molecules to Men* (D.S. Bendall, ed.), pp. 387-402. Cambridge: Cambridge University Press.

Barrow, J.D. & Tipler, F.J.
1986 *The Cosmic Anthropological Principle.* Oxford: Oxford University Press.

Bar-Yosef, O.
1995 The Lower and Middle Palaeolithic in the Mediterranean Levant: chronology, and cultural entities. In *Man and Environment in the Palaeolithic'* (Ullrich, H., ed.) pp. 247-263. Liége: ERAUL No. 62.

Bar-Yosef, O. & Goren-Inbar, N.
1992 *The Lithic Assemblages of the Site of Ubeidiya, Jordan Valley.* Jerusalem: Qedem No. 34.

Barton, C.M., Clark, G.A. & Cohen, A.E.
1994 Art as information: explaining Upper Palaeolithic art in western Europe. *World Archaeology* 26: 185-207.

Binford, L.R.
1981 *Bones: Ancient Men and Modern Myths.* New York: Academic Press.

Binford, L.R.
1989 Isolating the transition to cultural adaptations: an organizational approach. In *The Emergence of Modern Humans: Biocultural Adaptations in the Later Pleistocene* (Trinkaus, E., ed.), pp. 18-141. Cambridge: Cambridge University Press, 18-41.

Bishop, W.W. & Clark, J.D. (eds.)
1967 *Background to Evolution in Africa*. Chicago: University of Chicago Press.

Bishop, W.W.
1976 Pliocene problems relating to human evolution. In *Human origins: Louis Leakey and the East African evidence* (G.Ll. Isaac & E. McCown, eds.), pp. 139-154. Menlo Park, California: Benjamin Cummings.

Boesch, C.
1993 Aspects of transmission and tool-use in wild chimpanzees. In *Tools, Language and Cognition in Human Evolution* (Gibson, K.R. & Ingold, T., eds.), pp. 171-183. Cambridge, Cambridge University Press.

Brace, C.L.
1989 Medieval thinking and the paradigms of paleoanthropology. *American Anthropologist* 91: 442-446.

Bradley, B. & Sampson, C.G.
1986 Analysis by replication of two Acheulian artefact assemblages from Caddington, England. In*Stone Age Prehistory: Studies in Memory of Charles McBurney* (Bailey, G.N. & Callow, P., eds.), pp. 29-45. Cambridge: Cambridge University Press.

Bunn, H.T.
1994 Early Pleistocene hominid foraging strategies along the ancestral Omo River at Koobi Fora, Kenya. *Journal of Human Evolution* 27: 247-266.

Butzer, K.W. & Isaac, G.Ll. (eds.)
1975 *After the Australopithecines: Stratigraphy, Ecology and Culture Change in the Middle Pleistocene*. The Hague: Mouton.

Byrne, R.W.
1996 Relating brain size to intelligence in primates. In *Modelling the Early Human Mind*. (Mellars, P. & Gibson, K., eds), pp. 49-56. Cambridge: McDonald Institute.

Byrne, R.W. & Whiten, A. (eds.)
1988 *Machiavellian Intelligence: Social Expertise and the Evolution of Intellect in Monkeys*. Oxford: Oxford University Press.

Cain, A.J.
1989 The true meaning of Darwinian evolution. In *Evolution and Its Influence* (Grafen, A., ed.), pp. 1-18. Oxford: Clarendon Press.

Chamberlain, A.
1991 A chronological framework for human origins. *World Archaeology* 23: 137-146.

Chase, P.
1991 Symbols and palaeolithic artefacts: style standardization, and the imposition of arbitrary form. *Journal of Anthropological Archaeology* 10: 193-214.

Chase, P. & Dibble, H.
1987 Middle Paleolithic symbolism: a review of current evidence and interpretations. *Journal of Anthropological Archaeology* 6: 263-296.

Clark, G.A.
1988 Some thoughts on the Black Skull: an archaeologist's assessment of WT-17000 (*A. boisei*) and systematics in human paleontology. *American Anthropologist* 90: 357-371.

1989a Paradigms and paradoxes in paleoanthropology: a reply to C. Loring Brace. *American Anthropologist* 91:446-450.

1989b Alternative models of Pleistocene biocultural evolution: a response to Foley. *Antiquity* 63: 153-162.

Clark, G.A. (editor)
1991 *Perspectives on the Past: Theoretical Biases in Mediterranean Hunter-Gatherer Research*. Philadelphia: University of Pennsylvania Press.

Clark, G.A., C.M. Barton & A. Cohen
1996 Explaining the art in the Franco-Cantabrian refugium: an information exchange model. In *Debating Complexity* (D. Meyer, D. Dawson & D. Hanna, eds.), pp. 241-253. Calgary: Archaeological Association of the University of Calgary.

Clark, G.A. & C.M. Willermet
1995 In search of the neanderthals: some conceptual issues with special reference to the Levant. *Cambridge Archaeological Journal* 5: 153-156.

Clark, J.D.
1980 The Plio-Pleistocene environmental and cultural sequence at Gadeb, northern Bale, Ethiopia. *Proceedings of the 7th Panafrican Congress of Prehistory and Quaternary Studies*. Nairobi (R.E. Leakey & B.A. Ogot, eds.), pp. 189-193. Nairobi: TILLMIAP.

Clarke, D.L.
1968 *Analytical Archaeology*. London: Methuen.

Crompton, R.H. & Gowlett, J.A.J.
1993 Allometry and multidimensional form in Acheulean bifaces from Kilombe, Kenya. *Journal of Human Evolution* 25: 175-199.

Crook, J.H.
1980 *The Evolution of Human Consciousness*. Oxford: Clarendon Press.

Darwin, C.
1871 *The Descent of Man and Selection in Relation to Sex*. London: John Murray.

Davidson, I.
1991 The archaeology of language origins: a review. *Antiquity* 65: 39-48.

Davidson, I. & Noble, W.
1993 Tools and language in human evolution. In *Tools, Language and Cognition in Human Evolution* (K.R. Gibson & T. Ingold, eds.), pp. 363-388. Cambridge: Cambridge University Press.

Deacon, T.
1997 *The Symbolic Species: the Co-evolution of Language and the Human Brain*. London: Allen Lane.

Dean, D.
1992 The interpenetration of dialectic and hierarchy theory: a hominoid paleobehavioral example. In *Levels of Social Behavior: Evolutionary and Genetic Aspects* (E. Tobach & G. Greenberg, eds.), pp. 33-49. Wichita: Wichita State University.

Dillingham, B. & Carneiro, R.L. (eds.)
1987 *Leslie A. White's Ethnological Essays*. Albuquerque: University of New Mexico Press.

Dobzhansky, T.
1963 Cultural direction of human evolution—a summation. *Human Biology* 35: 311-316.

Dunbar, R.I.M.
1995 Neocortex size and group size in primates: a test of the hypothesis. *Journal of Human Evolution* 28: 287-296.

Dunbar, R.
1996 *Grooming, Gossip and the Evolution of Language*. London: Faber and Faber.

Duff, A.I., Clark, G.A. & Chadderton, T.J.
1992 Symbolism in the early Palaeolithic: a conceptual Odyssey. *Cambridge Archaeological Journal* 2: 211-229.

Foley, R.
1987 *Another Unique Species*. London: Longman.
1996 Measuring cognition in extinct hominids. In *Modelling the Early Human Mind* (Mellars, P. & Gibson, K., eds.), pp. 57-65. Cambridge: McDonald Institute.

Gamble, C.S.
1986 *The Palaeolithic Settlement of Europe*. Cambridge: Cambridge University Press.
1993 *Timewalkers: the Prehistory of Global Colonization*. London: Alan Sutton.
1996a Making tracks: hominid networks and the evolution of the social landscape. In *The Archaeology of Human Ancestry* (Steele, J. & Shennan, S., eds.), pp. 253-277. London: Routledge.
1996b Raw materials, technology and variability in Middle Pleistocene Europe. In *Evolucion Humana en Europa y los Yacimientos de la Sierra de Atapuerca* (E. Carbonell et al., eds.), pp. 387-402. Tarragona: Universidad Rovira i Vergili.

Gibson, K.R.
1993 General introduction: animal minds, human minds. In *Tools, Language and Cognition in Human Evolution* (Gibson, K.R. & Ingold, T., eds.), pp. 3-19. Cambridge: Cambridge University Press.

Gibson, K.R.
1996 The biocultural human brain, seasonal migrations and the emergence of the Upper Palaeolithic. In *Modelling the Early Human Mind* (Mellars, P. & Gibson, K., eds.), pp. 33-46. Cambridge: McDonald Institute.

Glucksberg, S.
1988 Language and thought. In *The Psychology of Human Thought* (Sternberg, R.J. & Smith, E.E., eds.), pp. 214-241. Cambridge: Cambridge University Press.

Goodall, J.
1976 Continuities between chimpanzee and human behaviour. In *Human Origins: Louis Leakey and the East African Evidence* (G.Ll. Isaac & E. McCown, eds.), pp. 80-95. Menlo Park, California: Benjamin Cummings.

Goodman, M., Koop, B.F., Czelusniak, J., Fitch, D.H.A., Tagle, D.A. & Slighton, J.L.
1989 Molecular phylogeny of the family of apes and humans. *Genome* 31: 316-35.

Gould, S.J.
1983 Irrelevance, submission, and partnership: the changing role of palaeontology in Darwin's three centennials, and a modest proposal for macroevolution. In *Evolution from Molecules to Men* (Bendall, D.S., ed.), pp. 347-366. Cambridge: Cambridge University Press.
1990 Speciation and sorting as the source of evolutionary trends, or 'things are seldom what they seem'. In *Evolutionary Trends* (McNamara, K.J., ed), pp. 3-27. London: Belhaven Press.

Gowlett, J.A.J.
1988 A case of Developed Oldowan in the Acheulean? *World Archaeology* 20: 13-26.
1995a Psychological worlds within and without: human-environment relations in early parts of the Palaeolithic. In *Man and Environment in the Palaeolithic'* (Ullrich, H., ed.), pp. 29-42. Liége: ERAUL No. 62.
1995b A matter of form: instruction sets and the shaping of early technology. *Lithics* 16: 2-16.
1996a The frameworks of early hominid social systems. In *The Archaeology of Human Ancestry* (Steele, J. & Shennan, S., eds.), pp. 135-183. London: Routledge.
1996b Mental abilities of early *Homo*: elements of constraint and choice in rule systems. In *Modelling the Early Human Mind* (Mellars, P. & Gibson, K., eds.), pp. 191-215. Cambridge: McDonald Institute.

Gowlett, J.A.J. & Crompton, R.H.
1994 Kariandusi: Acheulean morphology and the question of allometry. *African Archaeological Review* 12: 1-40.

Graves, P.
1994 Flakes and ladders: what the archaeological record cannot tell us about the origins of language. *World Archaeology* 26: 158-171.

Groves, C.P.
1989 *A Theory of Human and Primate Evolution*. Oxford: Clarendon Press.

Hawkes, K.
1996 Foraging differences between men and women: behavioural ecology of the sexual division of labour. In *The Archaeology of Human Ancestry* (Steele, J. & Shennan, S., eds.), pp. 283-305. London: Routledge.

Hay, R.L.
1976 *Geology of the Olduvai Gorge*. Berkeley: University of California Press.

Hayden, B.
1993 The cultural capacities of Neandertals: a review and re-evaluation. *Journal of Human Evolution* 24: 113-146.

Holloway, R.L.
1972 Australopithecine endocasts, brain evolution in the Hominoidea, and a model of human evolution. In *The Functional and Evolutionary Biology of Primates* (R. Tuttle, ed.), pp. 185-203. Chicago: Aldine.

Howell, F.C. & Clark, J.D.
1963 Acheulian hunter-gatherers of sub-saharan Africa. In *African Ecology and Human Evolution* (F.C. Howell & F. Bourlière, eds.), pp. 458-533. Chicago: Aldine.

Hunt, K.D.
1994 The evolution of human bipedality: ecology and functional morphology. *Journal of Human Evolution* 26: 183-202.

Huxley, T.H.
1863 *Evidence as to Man's Place in Nature*. London: Williams & Norgate.

Isaac, G.Ll.
1972a Early phases of human behaviour: models in Lower Palaeolithic archaeology. In *Models in Archaeology* (D.L. Clarke, ed.), pp. 167-199. London: Methuen.
1972b Chronology and the tempo of cultural change during the Pleistocene. In *The Calibration of Hominoid Evolution* (W.W. Bishop & J.A. Miller, eds.), pp. 381-430. Edinburgh: Academic Press.
1975 Sorting out the muddle in the middle: an anthropologist's post-conference appraisal. In *After the Australopithecines: Stratigraphy, Ecology and Culture Change in the Middle Pleistocene* (Butzer, K.W. & Isaac, G.Ll. eds), pp. 875-877. The Hague: Mouton.

1976 The activities of early African hominids: a review of archaeological evidence from the time span two and a half to one million years ago. In *Human origins: Louis Leakey and the East African Evidence* (G.Ll. Isaac & E. McCown, eds.), pp. 483-514. Menlo Park, California: Benjamin Cummings.

1977 *Olorgesailie: Archaeological Studies of a Middle Pleistocene Lake Basin*. Chicago: University of Chicago Press.

Isaac, G.Ll. & Isaac, B.
1975 Africa. In *Varieties of Culture in the Old World* (R. Stigler, ed.), pp. 8-48. New York, St. Martin's.

Jewett, R.A. & G.A. Clark
1987 Observations on estimating local group size at Olduvai Gorge, Tanzania. In *Coasts, Plains and Deserts: Essays in Honor of Reynold J. Ruppé* (S. Gaines, ed.), pp. 117-127. Tempe: Arizona State University Anthropological Research Papers No. 38.

Johanson, D.C. & Taieb, M.
1976 Plio-Pleistocene hominid discoveries in Hadar, Ethiopia. *Nature* 260: 293-297.

Kien, J.
1991 The need for data reduction may have paved the way for the evolution of language ability in hominids. *Journal of Human Evolution* 20: 157-165.

Kleindienst, M.R.
1962 Components of the East African Acheulian assemblage: an analytic approach. In *Actes du IVe Congrès Panafricain de Préhistoire* (G. Mortelmans & J. Nenquin, eds.), pp. 81-105. Tervuren, Belgium.

Knight, C.
1996 Darwinism and collective representations. In *The Archaeology of Human Ancestry* (Steele, J. & Shennan, S., eds.), pp. 331-346. London: Routledge.

Lake, M.
1996 Archaeological inference and the explanation of hominid evolution. In *The Archaeology of Human Ancestry* (Steele, J. & Shennan, S., eds.), pp. 184-206. London: Routledge.

Leakey, L.S.B.
1959 A new fossil skull from Olduvai. *Nature* 184: 191-193.

Leakey, M.D.
1971 *Olduvai Gorge. Vol. III: Excavations in Beds I and II, 1960-1963*. Cambridge: Cambridge University Press.

Lieberman, P., Laitman, J.T., Reidenberg, J.S. & Gannon, P.J.
1992 The anatomy, physiology, acoustics and perception of speech: essential elements in analysis of the evolution of human speech. *Journal of Human Evolution* 23: 447-467.

Li, Y., Crompton, R.H., Gunther, M.N., Alexander, R.McN. & Wang, W.J.
1996 Characteristics of ground reaction forces in normal and chimpanzee-like bipedal walking by humans. *Folia Primatologica* 66: 137-159.

Lovejoy, A.O.
1936 *The Great Chain of Being*. Cambridge: Harvard University Press.

Lovejoy, C.O.
1981 The origin of man. *Science* 211: 341-350.

McGrew, W.R.
1992 *Chimpanzee Material Culture: Implications for Human Evolution*. Cambridge: Cambridge University Press.

McHenry, H.M.
1994 Behavioral ecological implications of early hominid body size. *Journal of Human Evolution* 27: 77-87.

McNamara, K.J.
1990 The role of heterochrony in evolutionary trends. In *Evolutionary Trends* (McNamara, K.J., ed.), pp. 59-74. London: Belhaven Press.

Martins, E.P.
1993 Comparative studies, phylogenies and predictions of co-evolutionary relationships. *Behavioural and Brain Sciences* 16: 714-716.

Megarry, T.
1995 *Society in Prehistory: the Origins of Human Culture*. London: Macmillan.

Mithen, S.
1996a Social learning and cultural tradition: interpreting early Palaeolithic technology. In *The Archaeology of Human Ancestry* (Steele, J. & Shennan, S., eds.), pp. 207-229. London: Routledge.

1996b Domain-specific intelligence and the Neanderthal mind. In *Modelling the Early Human Mind* (Mellars, P. & Gibson, K., eds.), pp. 217-229. Cambridge: McDonald Institute.

Potts, R.
1988 *Early Hominid Activities at Olduvai*. New York: Aldine de Gruyter.

1994 Variables versus models of early Pleistocene hominid land use. *Journal of Human Evolution* 27: 7-24.

Reynolds, T.E.G.
1991 Revolution or resolution? The archaeology of modern human origins. *World Archaeology* 23: 155-166.

Rigaud, J.-Ph.
1989 From the Middle to the Upper Palaeolithic: transition or convergence? In *The Emergence of Modern Humans: Biocultural Adaptations in the Later Pleistocene* (Trinkaus, E., ed.), pp. 142-153. Cambridge: Cambridge University Press.

Rightmire, G.P.
1996 The human cranium from Bodo, Ethiopia: evidence for speciation in the Middle Pleistocene? *Journal of Human Evolution* 31: 21-39.

Roberts, M.
1986 Excavation of the lower palaeolithic site at Amey's Eartham Pit, Boxgrove, West Sussex: a preliminary report. *Proceedings of the Prehistoric Society* 52: 215-245.

Roche, H.
1980 *Premièrs Outils Taillés d'Afrique*. Paris: Société d'Ethnographie.

Roe, D.A.
1968 British Lower and Middle Palaeolithic handaxe groups. *Proceedings of the Prehistoric Society* 34: 1-82.

Rose, L. & Marshall, F.
1996 Meat eating, hominid sociality and home bases revisited. *Current Anthropology* 37: 307-338.

Sackett, J.R.
1982 Approaches to style in lithic archaeology. *Journal of Anthropological Archaeology* 1: 59-122.

1986 Isochrestism and style: a clarification. *Journal of Anthropological Archaeology* 5: 266-277.

Savage-Rumbaugh, E.S.
1986 *Ape Language: from Conditioned Response to Symbol*. Oxford: Oxford University Press.

Schick, K.D.
1991 On making behavioral inferences for early sites. In *Approaches to Understanding Early Hominid Life-ways in the African Savanna* (J.D. Clark, ed.), pp. 79-107. Bonn: Rudolf Habelt.

Smith, J. Maynard.
1993 Introduction. In *The Theory of Evolution*. Cambridge: Cambridge University Press.

Steele, J.
1994 Communication networks and dispersal patterns in human evolution: a simple simulation model. *World Archaeology* 26: 126-143.
1996 On predicting hominid group sizes. In *The Archaeology of Human Ancestry* (Steele, J. & Shennan, S., eds.), pp.230-252. London: Routledge.

Steele, J. & Shennan, S.
1996 Introduction. In *The Archaeology of Human Ancestry* (Steele, J. & Shennan, S., eds.), pp. 1-41. London: Routledge.

Sternberg, R.J.
1988 Intelligence. In *The Psychology of Human Thought* (Sternberg, R.J. & Smith, E.E., eds.), pp. 267-308. Cambridge: Cambridge University Press.

Stringer, C. & Gamble, C.
1993 *In Search of the Neanderthals*. London: Thames & Hudson.

Tanner, N.M.
1981 *On Becoming Human*. Cambridge: Cambridge University Press.

Thorpe, W.H.
1979 *The Origins and Rise of Ethology*. London: Heinemann.

Tobias, P.V.
1981 The emergence of man in Africa and beyond. *Philosophical Transactions of the Royal Society of London, Series B* 292: 43-56.

Tobias, P.V.
1991 The emergence of spoken language in hominid evolution. In *Approaches to Understanding Early Hominid Life-ways in the African Savanna* (J.D. Clark, ed.), pp. 67-78. Bonn: Rudolf Habelt.

Toth, N.
1985 The Oldowan reassessed: a close look at early stone artefacts. *Journal of Archaeological Science* 12: 101-121.

Trinkaus, E.
1992 Cladistics and Later Pleistocene Human Evolution. In *Continuity or Replacement? Controversies in Homo sapiens Evolution* (Brauer, G. & Smith, F.H., eds.), pp. 1-7. Rotterdam: Balkema.

Washburn, S.L.
1967 Behaviour and the origin of man. *Proceedings of the Royal Anthropological Institute for 1967*: 21-27.

Wheeler, P.E.
1984 The evolution of bipedality and loss of functional body hair in hominids. *Journal of Human Evolution* 13: 91-98.
1994 The thermoregulatory advantages of heat storage and shade-seeking behaviour to hominids foraging in equatorial savannah environments. *Journal of Human Evolution* 26: 339-350.

Wilson, E.O.
1983 Sociobiology and the Darwinian approach to mind and culture. In *Evolution from Molecules to Men* (D.S. Bendall, ed.), pp. 545-553. Cambridge: Cambridge University Press.

Wilson, L.
1988 Petrography of the Lower Palaeolithic tool assemblage of the Caune de l'Arago (France). *World Archaeology* 19: 376-387.

Wobst, H.M.
1974 Boundary conditions for Palaeolithic social systems: a simulation approach. *American Antiquity* 39: 147-178.

Wynn, T.
1979 The intelligence of later Acheulian hominids. *Man* 14: 371-391.
1981 The intelligence of Oldowan hominids. *Journal of Human Evolution* 10: 529-541.
1985 Piaget, stone tools and the evolution of human intelligence. *World Archaeology* 17: 32-43.

Wynn, T. & Tierson, F.
1990 Regional comparison of the shapes of later Acheulean hand-axes. *American Anthropologist* 92: 73-84.

4

Cognitive Archaeology, Evolutionary Psychology and Cultural Transmission, with Particular Reference to Religious Ideas

Steven Mithen
University of Reading

ABSTRACT

Archaeologists have drawn on evolutionary theory for more than three decades resulting in a diversity of approaches that lay claim to being a type of Darwinian archaeology. These are briefly reviewed, and some suggested to be of limited value to the discipline. The paper argues that a valid evolutionary approach requires an explicitly cognitive perspective, and suggests that there will be considerable benefit from an integration between cognitive archaeology and evolutionary psychology. The result would be a non-functionalist Darwinian archaeology with potential to throw new light on cultural developments during the Palaeolithic and more recent times. To illustrate this, the paper discusses the cultural transmission of religious ideas.

INTRODUCTION

Charles Darwin's theory of evolution by natural selection has achieved remarkable success in explaining patterning and variation in the natural world. It provides the theoretical basis and unifying theme for academics working in all areas of the life sciences, ranging from molecular biology to community ecology. Few challenge the central tenants of Darwin's theory with debate concerning the relative role of historical contingency in evolutionary process and the value of adaptationism (e.g., Gould & Lewontin 1979, Gould 1989, Eldredge 1995, Dennett 1995, Orr 1996).

Without Darwinian theory, the diversity of the natural world would appear bewildering. As bewildering, perhaps, as the diversity of the cultural world of humans. If we document the range of human cultures in the modern world, and supplement that range with those inferred for the past, we find remarkable variation in human behaviour, notably in the material culture that people make and use. To what extent can Darwinian theory aid in the explanation of this diversity? Just as the diversity of life forms today are understood by evolution from some primordial single celled organism, is it possible to explain the evolution of modern material – such things as space shuttles and CD players – from the Oldowan tools of 2.5 million years ago by the application of the same theory? Archaeologists have struggled with such questions for at least two decades and tried to make use of evolutionary theory in a variety of ways. In this paper I argue that the only respectable evolutionary approach, and the only one that will make a contribution to explaining past behaviour, is one that adopts an explicitly cognitive approach.

DARWINIAN ARCHAEOLOGIES

A recent edited book on the use of evolutionary theory in archaeology was appropriately titled *Darwinian Archaeologies*, (Maschner 1996) rather than Darwinian Archaeology, as even among those who feel committed to a Darwinian approach, there is no consensus as to how evolutionary principles can be applied to human behaviour and changes in material culture. Such pluralism may be a fashionable feature of post-modernism – there is no single truth – but it is a crippling attitude to accept when trying to make progress in our understanding of the past or present. Some of the approaches currently being pedalled as Darwinian are fundamentally flawed and must be rejected.

The most obvious of these is a group adaptationist approach. It was common during the 1970s to read about hunter-gatherer groups being adapted to their natural environments. By this was meant that their way of life appeared appropriate for their environments—it was a way of life that appeared to use energy and information efficiently. This was indeed a good type of short hand description. But the term adaptation has no relevance to group behaviour: while the use of such group adaptationist terminology may have provided a veneer of science it provided no explanatory power. An individualistic approach is a prerequisite for any evolutionary approach in archaeology, or indeed any other domain of the human sciences (Mithen 1989, 1993).

A second manner in which supposedly Darwinian theory was adopted by archaeologists was for the development of cultural selectionist models (e.g., Dunnell 1980, Rindos 1984, O'Brien & Holland 1990). Although such approaches are variable, their basic premise is that the material culture that an individual adopts is as much a part of his/her phenotype as his/her physical appearance and behaviour. Consequently such material culture traits should be as subject to the process of natural selection as any other trait: those which increase the reproductive success of an individual will survive and spread within a population. Those which do not will disappear from the pool of material culture variants. As with the group adaptationist arguments, such ideas have an intuitive appeal and initially appear effective: we are indeed all followers of fashion and adopt material culture, either consciously or unconsciously, that appears most effective at solving the problems we face in the world. The problem with such ideas is that the material culture of an individual is not an expression of his/her genotype in a particular environment, as are true phenotypic traits. There is no reason why the descendants of a reproductively successful individual will necessarily have the same material culture of their ancestor, as in biological evolution. As with group adaptationism, cultural selectionist models are fatally flawed.

Models of cultural transmission (e.g., Cavalli-Sforza & Feldman 1981, Boyd & Richerson 1985) differ from those of cultural selectionism by avoiding a functionalist agenda and attempting to explain how maladaptive traits may survive within human populations. The basis of these models is that cultural transmission is analogous to genetic transmission. The most sophisticated versions are dual inheritance models, which recognise that to explain variation in human behaviour we need to consider both genetic and cultural transmission within a single model. But the failure of such models to engage with archaeological data and to address archaeological problems, let alone contribute to their solution, is worrying to say the least. After at least two decades of development, such models should have moved beyond the purely theoretical phase and be providing some respectable case studies. The fact that they have not suggests that there may be fundamental flaws in cultural transmission models—as indeed there are. One of these is that such models lack reference to people as intentional agents—an essential element of explanation in the human sciences. Another flaw will be exposed below.

One variant of cultural transmission models focuses not on material artefacts but on the underlying ideas, which are referred to as memes—an explicit analogy with genes. Associated with this is the notion that some ideas form 'cultural viruses' that are located in human minds and might spread through a population (e.g., Dawkins 1976, Cullen 1993, Dennett 1995). The notion of the 'meme' is notoriously ill-defined. Lake (1997) has discussed and analysed the concept in detail, exploring the analogy between memes and genes, and asking whether the idea of memes has any value for models of cultural transmission and archaeological explanation. He takes a sympathetic approach, believing that there may be mileage in the concept when applied to symboling behaviour. My own judgement is harsher: the concept of a 'meme' as a unit of replication and selection is simply fallacious.

HUMAN COGNITION AND HUMAN BEHAVIOUR

The ideas of memes and cultural viruses, along with that of cultural transmission in general, focus attention on human minds rather than human behaviour. As such, these models are aligned with a further variant of Darwinian archaeology and one I want to explore within this paper: an explicitly cognitive approach.

The premise of this approach is that the behaviour of individuals, of groups, the character of their material culture, and ultimately the long term patterns of culture change we see in the archaeological record, arise from the short term decisions of individuals (Mithen 1990). Some of these decisions, perhaps only a small subset, are made consciously and with intent. As such, this approach explicitly rejects ideas that there are long term, and as yet undefined, processes of culture change, as argued by Murrey and Walker (1988). Long term patterns in the archaeological record are no more than the accumulated results of many individual decisions and post-depositional processes concerned with site formation. Consequently to provide adequate explanations for variation in their data sets and inferred patterns of past behaviour—whether they are concerned with the artefacts of the earliest *Homo* or modern material culture—archaeologists have no choice but to focus on processes of decision making and learning by individuals. Some of these processes are concerned with social learning, others with learning from an individual's own experience. It is the cognitive processes for these tasks which have been moulded by evolutionary abilities (including natural selection), not behaviour itself, let alone material culture.

Humans are no different in this respect than other animals. When biologists study the foraging behaviour of animals ranging from blue tits to gorillas, they assume that this behaviour arises from decision making processes which were shaped by natural selection. Those individuals who could make decisions which reduced foraging time (and hence exposure to predators) or increased the rate of energy gain over expenditure, were reproductively more successful. And consequently the decision making processes they used spread within the population. There is no reason to adopt anything other than this conventional evolutionary approach to humans, either those living today or long dead in prehistory.

EVOLUTIONARY PSYCHOLOGY AND COGNITIVE ARCHAEOLOGY

As the realisation that archaeologists need a more explicitly cognitive approach to past behaviour emerged during the last decade, it was matched by the development of a more explicitly evolutionary approach in psychology. Indeed a distinct sub-discipline has emerged calling itself 'evolutionary psychology' (Cosmides & Tooby 1987, Barkow et al. 1992). Its basic premise is that the human mind-brain is a product of evolution in precisely the same way as any other organ of the human body. As such, to understand how it works today, we must understand its evolutionary history.

Unfortunately, psychologists adopted a very simplistic attitude to that evolutionary history, proposing the notion of the EEA, the 'Environment of Evolutionary Adaptedness' (Symonds 1979, Tooby & Cosmides 1992). This was argued to be the world of Pleistocene hunter-gatherers. Consequently all human minds are described in their work as the mind of a Pleistocene hunter-gatherer, whether that mind is possessed today by a city dweller, peasant farmer, industrialist or modern hunter-gatherer.

It is easy to criticise this idea. It is clear from even the most rudimentary study of comparative psychology that there can no such thing as a single EEA. It is readily evident that parts of our mental apparatus are shared with all primates, others with the great apes, others with early *Homo* while others are unique to our species alone. In other words these evolved in response to a succession of different environments and problems during the last 100 million years: there is no single, unitary EEA (Foley 1996). Moreover, even the most rudimentary knowledge of Pleistocene environments and societies indicates that these were characterised by immense diversity: the identification of a discrete set of adaptive problems beyond the trivial level of 'gaining food' and 'selecting mates' seems improbable.

Nevertheless there is much to be welcomed in the development of evolutionary psychology. It does, after all, stress the importance of palaeoanthropology to understanding the modern world–and soon evolutionary psychologists will realise that instead of simply making vague claims about life in the stone age, they will need to draw on the work of archaeologists. The basic idea that our minds are somehow adapted to a hunter-gatherer lifestyle does appear to have considerable value in understanding our behaviour today. Why, for instance, do so many people today crave foods high in fat and sugar which cause such serious health problems and consequently appears highly maladaptive behaviour? Well, possibly it is simply because these foods were in short supply in our ancestral environments and our desire for and pleasure in them was at that time a highly adaptive mental traits. But today, when such foods are in abundance, such mental traits are now seriously maladaptive as our physiology remains adapted to a low fat diet.

A wide range of issues concerned with human health today can indeed be illuminated by adopting such an evolutionary perspective (Eaton et al. 1988, Nesse & Williams 1995). Evolutionary psychologists are making many advances in our understanding of human behaviour today in the realms of inter-personal violence and cooperation, social interaction and mate choice by adopting this type of approach (e.g., Daly & Wilson 1988, Barkow et al. 1992, Buss 1994). But evolutionary psychologists, concerned with explaining human behaviour in the modern world, have much to learn from archaeologists who can provide the understanding about our hunter-gatherer past and the selective pressures on the human mind.

Conversely, archaeologists concerned with explaining past behaviour have much to learn from evolutionary psychologists as they can provide the understanding of decision making and learning that are required in a cognitive archaeology. One of their fundamental tenets is of particular importance to an evolutionary archaeology: that rather than there being a single learning process, there are likely to be multiple modules in the human mind, each attuned to learning and making decisions about a specific domain of behaviour.

MENTAL MODULARITY AND DECISION MAKING ALGORITHMS

To explain the notion of mental modularity, Leda Cosmides and John Tooby (1987) have adopted the metaphor of the mind as a 'Swiss army knife' of specialised mental adaptations. Each of these, they argue, is a mental device specifically for solving a problem faced by our ancestors in their EEA. They would include devices for selecting food stuffs, recognising predators, and choosing mates. Cosmides and Tooby (1994) provide powerful arguments for why the mind would have such a design, rather than having one or a small number of general purpose problem solving mechanisms. In essence, this is because each of these problems have a different structure and to be solved efficiently a unique type of decision making process is required: what one needs to know when making a choice about foodstuffs, and how one should process that information, are different from the knowledge required for choosing mates. An individual with a single decision making process applied to these problems would be out competed by one who has specialised mental algorithms for these different types of problems.

These mental algorithms facilitate the learning of complex tasks because they are content laden. In other words, a certain degree of knowledge about the world appears to be hard wired into our minds at birth. The classic example concerns language acquisition. It is simply impossible that in their relatively short exposure to spoken words, young children could learn all the rules of grammar that they have mastered by the age of three or four (Pinker 1994). Somehow, these rules are partly already encoded into their minds. Similarly, knowledge about the physical and natural worlds appears to be encoded as an 'intuitive physics' and 'intuitive biology', while children seem to have a vast intuitive understanding about other minds, as explored in current research on 'theory of mind' (Hirschfeld & Gelman 1994, Mithen 1996a, chapter 3). Because such mental algorithms are content laden, they make learning about their relevant domains relatively quick and efficient.

One of the cognitive domains for which we should expect hunter-gatherers to have specialised, content rich, mental models is that relating to the natural world, and more specifically for making foraging decisions. As I have discussed at length, there are good grounds for believing that early humans possessed a distinct domain of 'natural history intelligence'—a bundle of integrated modules used for learning and making decisions about the natural world (Mithen 1993, 1996a, 1996b). And even within modern human minds, in which natural history intelligence is integrated with that of other previously isolated intelligences, those content rich decision making rules for foraging continue to exist.

A consequence of this is that we should expect individuals to make foraging decisions which ultimately lead to an increase in their reproductive success. This may be achieved by reducing the time spent foraging, or improving the rate of energetic return, or avoiding the risk of shortfall in food supply or by some other proximate goal—our task is to identify what types of goals were adopted and the constraints under which they were achieved. As I have demonstrated elsewhere, with a case study exploring Mesolithic foraging in southern Scandinavia (Mithen 1990), it is possible for archaeologists to build models based on the foraging decisions of individuals and test these against the archaeological record. This involves modelling information exchange between individuals, the formation of the archaeological record as well as decision making processes of individuals themselves.

We can do such work because there are strong reasons for believing that human minds have evolved mechanisms for solving foraging problems. But of course the modern mind today, and for much of prehistory, does not possess a unique problem solving device for each problem it faces. Hunter-gatherers in the EEA did not need to make choices about buying cars or computers. Consequently learning and making decisions about these problems for which there are no specialised mental modules is notably less efficient, and involves much more conscious awareness of the decision making process. Compare, for instance, how children learn the rules of language and those of mathematics. The first are acquired effortlessly and with very limited awareness. But when children need to learn multiplication tables they struggle, needing to be formally taught and to invest great effort, even though this set of rules is remarkably simpler than those they use in every spoken utterance they make. Pleistocene hunter-gatherers are unlikely to have needed to know their seven times table, and the mind today lacks specialised mental adaptations for acquiring such knowledge.

The significance of this for a Darwinian archaeology is that it has serious consequences for models of social learning and cultural transmission. Indeed it exposes a fundamental flaw in such models. Within models such as those of Boyd and Richerson (1985), no reference is made to the actual content of what is being learnt. There is an assumption that content has no relevance to the nature of cultural transmission. Those models which have addressed the relative significance of individual or social learning in different conditions (e.g., Boyd & Richerson 1996) have failed to consider how the value of these vary as to what is actually being learnt. Yet the message from evolutionary psychology (and indeed that from our everyday experience) is clearly that some things are much easier to learn than others. The nature of cultural transmission for knowledge about social relationships – something which our minds seem to be well tuned to learning about—will be very different to that regarding the internet, a recent topic of knowledge about which communication and thought are far less efficient because we do not have evolved mental algorithms for this domain of activity.

ARTEFACTS AND THE CULTURAL TRANSMISSION OF RELIGIOUS IDEAS

While this might be a flaw in the abstract models of cultural transmission and one reason why they have made very limited impact on explaining variation in the archaeological record, the differential ease with which different types of information can be transmitted does in fact reveal a new approach to understanding many aspects of cultural variability. To illustrate this, I will briefly consider the cultural transmission of religious ideas.

Explaining the existence of religious ideas from a Darwinian perspective on human behaviour is clearly a challenging task (Dennett 1997). It is a characteristic feature of religious behaviour that people act in ways that appear clearly maladaptive for their own reproductive success. People have destroyed their own material possessions, sacrificed their own offspring, devoted themselves to celibacy in the cause of worshipping deities. Explaining why and how ideas about religious entities could arise in humans minds is one problem; explaining how such ideas can be so persistent is perhaps even more challenging. It seems remarkable that during the last two decades there has been such a rise of fundamentalism throughout the world, such that more than 50% of Americans apparently believe that humans were divinely created, when during that period the fossil, archaeological and genetic evidence for human evolution has become so detailed.

My own views about the origin of religious ideas have been described elsewhere (Mithen 1996a). In brief terms, I have argued that these are essentially a spandrel of a change in mental architecture (cf. Gould & Lewontin 1979) that arose during the last 100,000 years of human evolution. A spandrel is a 'necessary architectural by-product'. Gould and Lewontin argued that many aspects of animal biology and behaviour are spandrels–byproducts of other features that in themselves lack direct functional value, drawing an analogy between these and spandrels found in architectural design (see discussion of this by Dennett [1995] and Houston [1997]). To understand how religious ideas evolved as a spandrel, we must first consider the mental architecture of early humans, those whose minds appear to have lacked any religious concepts.

The early human mind–that of archaic *H. sapiens*, Neanderthals, *H. erectus*–was structured, I have argued, on the basis of four 'intelligences', or bundles of mental modules. Three of these, those concerning the social world, making and using artefacts, and interaction with the natural world, were essentially isolated from each other. This isolation explains the rather odd character of the early Palaeolithic record in which we see evidence for very complex and sophisticated behaviour within these domains, but very simple behaviour at the 'domain interfaces' (Mithen 1996a, 1996b). For instance, while Neanderthals clearly possessed great technical skill in producing artefacts such as levallois points, and to have survived in Pleistocene Europe must have had an intimate knowledge of the natural world, the design of their hunting weapons appears remarkably simple. The recent discovery of 400,000 year old hunting spears at Schöningen, Germany (Thieme 1997), further indicates the absence of technological innovation during the Lower and Middle Palaeolithic as these appear as well designed as anything produced by the Neanderthals. This lack of innovation and the absence of multi-component hunting weapons, notably projectiles, arises, I have argued, from an inability to integrate their knowledge of tool making with that of the natural world.

The substantial development in cultural behaviour that we see in the archaeological record which begins at c. 100,000 years ago, and becomes dramatic after 50,000 years ago derives from a new ability by *Homo* to integrate their intelligences, a capacity I have termed cognitive fluidity. This appears to be restricted to modern humans, although some traces of cognitive fluidity may be present within the minds of the last Neanderthals (Mithen 1996a: 209-210). This change in the nature of intelligence is, I suggest, related to changes in the nature of language and consciousness (see also the idea of off-line thinking in Bickerton [1996]). Such cognitive fluidity had enormous adaptive benefits. By being able to integrate technical and natural history knowledge, tools could be designed to markedly improve the efficiency of hunting, plant gathering and food processing; by being able to integrate technical and social intelligence, artefacts could be designed to mediate social relationships providing new means to manipulate other individuals to ones' advantage.

Yet other consequences of such cognitive fluidity have no clear adaptive benefits and can be thought of as spandrels, inevitable by-products of such adaptations. For instance by integrating social and natural history intelligence beliefs could arise that entities exist which are half human and half animal, as clearly evident in the first representational art. And by integrating technical and social intelligence, inert objects could be attributed with ideas, feelings and intentions, 'living' entities could exist which did not need to feed, which were not born and could not die. This mixing up of natural categories is the essence of a supernatural being (Guthrie 1993, Boyer 1994a, 1994b, 1996; Mithen 1996c).

The human mind does not, therefore, have an evolved module/domain for supernatural beings or indeed any types of religious knowledge. This inhibits the cultural transmission of religious ideas. When an individual is told about a social relationship between two other humans, that information is embedded into content rich modules about human social relationships allowing many inferences to be drawn from a limited amount of information (Boyer 1994a). If we are simply told that those individuals are a boy and a girl and that they are 'in love' we can accurately guess how they are likely to be interacting with each other, what they will be doing, their feelings, how they will react in a host of circumstances. But if we were told about someone who was 'in love' with a supernatural being, no such inferences could be drawn. Perhaps we are told that supernatural being is invisible but exists in all places, that he once took human form and walked on water, that he died and then came alive again. Well, such details are of little help. What can it mean to love such a being? How is a person supposed to communicate with or behave towards such a being?

So the cultural transmission of religious ideas is difficult when compared to ideas about an evolved domain of human behaviour, such as social interaction or (for hunter-gatherers) animal behaviour. Boyer (1994a) has discussed this and noted that those religious ideas which survive the rigours of cultural transmission are those which have a link to a domain of intuitive knowledge. As he has recognised, while concepts of supernatural beings have, by definition, elements which are 'super natural', such as abilities to be omnipotent, invisible or ever-lasting, they also frequently have human like features, such as suffering jealously and desires. The gods of ancients Greece provide a typical example – supernatural beings who quarrel and deceive each other in a very human-like manner. By having these human like qualities, concepts of supernatural beings can be more easily transmitted and understood, than if all features were supernatural and unable to be grasped by any domain of intuitive knowledge.

As archaeologists we can never reconstruct the specific ideas that past people held about their religious beings, although we may be able to invoke what appear to be universal features of religious beings to suggest what ideas may have been present (Mithen 1997). But the difficulty of transmitting religious ideas also has enormous implications for the archaeological record which cannot be fully understood without understanding the human mind as a product of evolution. There are two cultural means which are widely used to facilitate the transmission of religious ideas and which have major impacts on the archaeological record. First, religious ideas are often transmitted in a context of ritual–the rote repetition of movement and utterances in sequences that must be conformed to precisely. Such ritual is essential: religious ideas cannot be transmitted in an informal manner if people are to share religious concepts because there is no evolved domain of religious ideas within the human mind. Without ritual, religious ideas might exist within individual minds but a religious institution, based upon shared religious concepts, would be impossible.

A second means by which the cultural transmission of religious ideas is achieved is even more fundamental for our understanding of human behaviour: the use of material culture. The last century of Palaeolithic archaeology has supported Durkheim's (1915: 307) assertion that "the principle forms of art seem to have been born out of religious ideas" (here I reject the idea that there are 'art objects' prior to those of the Upper Palaeolithic, finding the arguments of Bednarik [1995] for 'concept mediated marks in the Lower Palaeolithic' unconvincing [Mithen 1996d]). Throughout human history religious behaviour has involved visual symbols. Why should there be such a close connection between the two? Well, as Leach (1976) argued, we transform religious ideas into material form so that we can perform operations on them which are beyond the capacity of the mind. The evolutionary understanding of the emergence of religious ideas that I have summarised above explains why this is necessary: religious symbols, and more particularly the images of religious beings, serve to anchor religious ideas within the mind. Ideas about social relationships, the natural world, and stone artefacts did not need anchoring in hunter-gatherer minds as each of these related directly to an evolved domain of mental architecture which made them easy to learn, understand and transmit. Religious ideas had no such domain and the archaeological record of modern humans is replete with religious symbols.

SUMMARY

I have chosen to focus on religious ideas and behaviour in this paper because this domain of human activity is the one which most clearly lacks an adaptationist explanation but is one which can only be understood from an evolutionary perspective. The ability for humans to create ideas about supernatural beings is a product of the emergence of cognitive fluidity between 100,000 and 40,000 years ago (Mithen 1996a). This cognitive fluidity, the ability to integrate ideas and ways of thinking from what had been isolated domains of thought, evolved due to the selective advantage it gave those individuals who could do things such as design better hunting weapons, or use material culture to mediate social relationships. But by having a cognitively fluid mind, a whole host of new ways of thought lacking in any functional significance were also made possible – such as believing in beings which were half human/half beast. Transmission of such ideas required cultural support, such as visual symbols and ritual to anchor them in the mind as they have no evolved domain of their own. Consequently, unless we understand the human mind as a product of evolution we will be unable to understand religious ideas and their transmission. And of course religious ideas are just one type of thought made possible by cognitive fluidity. Another is pursuit of pure science–investing time in discovering laws of mathematics or the origins of the universe or of modern humans–activities of no adaptive benefit but which can be understood as a spandrel of our evolved mental architecture.

Other types of ideas can be transmitted without cultural support as they directly relate to a domain of intuitive knowledge. If we are dealing with hunter-gatherers, ideas and information about social relationships, the natural world and technology are of this nature and their minds have content rich, mental modules for making decisions within these domains. We would expect such decisions to make adaptive sense and ultimately to improve the reproductive success of the individuals involved. This is fortunate for archaeologists as data relating to past subsistence behaviour is often the most readily available in the archaeological record and can provide a means to test models for decision making in foraging strategies.

All people, either in the past or present, those who lived by hunting and gathering, by agriculture or by industry, live in a social world and have an evolved domain of social intelligence. And consequently we should also expect decisions about social relationships, such as the choice of friends and lovers, to also make evolutionary sense. This seems to be born out by the cross-cultural studies of Buss (1994) on mate choice, and on homicide victims by Daly and Wilson (1988). Of course, evolutionary sense is not necessarily adaptive sense: our decision making apparatus evolved for life in small scale, highly mobile hunter-gatherer societies. We use the same apparatus when living in our urban societies today and consequently may make decisions of little adaptive sense but which can only be explained by an evolutionary understanding of the human mind.

In summary, evolutionary theory is of vital importance to archaeologists, and indeed to any one studying human behaviour. But notions of group adaptation, cultural selection, memes and cultural viruses should be rejected. Models of cultural transmission must take into account the fact that some types of information are easier to learn and transmit than others. The most profitable manner in which evolutionary theory can be used in archaeology is for an explicity individualistic and cognitive stance to be adopted. The human mind is a product of evolution and as a consequence for those domains of behaviour which were of significance during the evolution of the mind, we should expect people to make decisions which lead to behaviour that increases their reproductive success. This provides a means for building models of human behaviour which can be used to explain the variability and patterning in the archaeological record. Other domains of behaviour and thought are spandrels of our evolved mental architecture and will defy adaptationist explanations. Nevertheless their existence can only be understood from an evolutionary perspective and cultural transmission theory may be able to explain the persistence of ideas with no functional utility.

REFERENCES CITED

Barkow, J.H., Cosmides, L. & Tooby, J.
1992 *The Adapted Mind*. Oxford: Oxford University Press.

Bednarik, R.G.
1995 Concept mediated marking in the Lower Palaeolithic. *Current Anthropology* 36: 605-634.

Bickerton, D.
1996 *Human Behaviour and Language*. London: University College Press.

Boyer, P.
1994a *The Naturalness of Religious Ideas, A Cognitive Theory of Religion*. Berkeley: University of California Press.
1994b Cognitive constraints on cultural representations: Natural ontologies and religious ideas. In *Mapping the Mind: Domain Specificity in Cognition and Culture* (L.A. Hirschfeld & S.A. Gelman, eds.), pp. 391-411. Cambridge: Cambridge University Press.
1996 What makes anthropomorphism natural: intuitive ontology and cultural representation. *Journal of the Royal Anthropological Institute* 2: 83-97.

Boyd, R. & Richerson, P.
1985 *Culture and the Evolutionary Process*. Chicago: Chicago University Press.
1996 Why culture is common, but cultural evolution is rare. In *Evolution of Social Behaviour Patterns in Primates and Man* (W.G. Runciman, J. Maynard-Smith & R.I.M. Dunbar, eds.), pp. 77-83. Proceedings of the British Academy, Vol. 88.

Buss, D.
1994 *The Evolution of Desire: Strategies of Human Mating*. London: Harper Collins.

Cavalli-Sforza, L.L. & Feldman, M.W.
1981 *Cultural Transmission and Evolution: A Quantitative Approach*. Princeton: Princeton University Press.

Cosmides, L. & Tooby, J.
1987 From evolution to behaviour: evolutionary psychology as the missing link. In *The Latest on the Best: Essays on Evolution and Optimality* (J. Dupré, ed.), pp. 277-306. Cambridge: Cambridge University Press.
1994 Origins of domain specificity: The evolution of functional organization. In *Mapping the Mind: Domain Specificity in Cognition and Culture* (L.A. Hirschfeld & S.A. Gelman, eds.), pp. 85-116. Cambridge: Cambridge University Press.

Cullen, B.
1993 The Darwinian Resurgence and the cultural virus theory. *Cambridge Archaeological Journal* 3: 179-202.

Daly, M. & Wilson, M.
1988 *Homicide*. New York: Aldine de Gruyter.

Dawkins, R.
1976. *The Selfish Gene*. Oxford: Oxford University Press.

Dennett, D.
1995 *Darwin's Dangerous Idea*. New York: Simon and Schuster.
1997 Appraising Grace: What evolutionary good is God? *The Sciences* Jan/Feb: 39-45.

Dunnell, R.
1980 Evolutionary theory and archaeology. In *Advances in Archaeological Method and Theory*, vol. 3 (M.B. Schiffer, ed.), pp. 35-99. New York: Academic Press.

Durkheim, E.
1915 *The Elementary Forms of Religious Life*. London: Allen and Unwin.

Eaton, B. et al.
1988 Stone Agers in the fast lane. *American Journal of Medicine* 84: 739-49.

Eldredge, N.
1995 *Reinventing Darwin*.. New York: John Wiley and Sons.

Fodor, J.
1996 Deconstructing Dennett's Darwin. *Mind and Language* 11: 246-262.

Foley, R.
1996 An evolutionary and chronological framework for human social behaviour. In *Evolution of Social Behaviour Patterns in Primates and Man* (W.G. Runciman, J. Maynard-Smith & R.I.M. Dunbar, eds), pp. 95-143. London: Proceedings of the British Academy, Vol. 88.

Gould, S.J.
1989 *Wonderful Life: The Burgess Shale and the Nature of History*. Norton: New York.

Gould, S.J. & Lewontin, R.
1979 The spandrels of San Marco and the panglossian paradigm: A critique of the adaptationist programme. *Proceedings of the Royal Society* B205: 581-98.

Guthrie, S.
1993 *Faces in the Clouds : A New Theory of Religion*. Oxford: Oxford University Press.

Houston, A.
1997 Are the spandrels of San Marco really panglossian pendentives? *Ecology and Evolution* 12: 125.

Hirschfeld, L.A. & Gelman, S.A.
1994 *Mapping the Mind*. Cambridge: Cambridge University Press.

Lake, M.
1997 Digging for memes: The role of material objects in cultural evolution . In *The Archaeology of External Symbolic Storage* (C. Renfrew & C. Scarre, eds.). Cambridge: McDonald Institute for Archaeological Research (in press).

Maschner, H. (ed.)
1996 *Darwinian Archaeologies*. New York: Plenum Press.

Leach, E.
1976 *Culture and Communication*. Cambridge: Cambridge University Press.

Mithen, S.
1989 Evolutionary theory and post-processual archaeology. *Antiquity* 63: 483-494.
1990 *Thoughtful Foragers: A Study of Prehistoric Decision Making*. Cambridge: Cambridge University Press.
1993 Individuals, groups and the Palaeolithic record: A reply to Clark. *Proceedings of the Prehistoric Society* 59: 393-398.
1996a. *The Prehistory Of The Mind: A Search For The Origins Of Art, Science And Religion*. London: Thames and Hudson.
1996b Domain specific intelligence and the Neanderthal mind. In *The Early Human Mind* (P. Mellars & K.Gibson, eds.), pp. 217-229. Cambridge: McDonald Institute for Archaeological Research.
1996c Putting anthropomorphism into evolutionary context (a comment on Boyer 1996).*Journal of the Royal Anthropological Institute* 2: 717-719.
1996d Early Palaeolithic 'concept mediated marks', mental modularity and the origins of art. *Current Anthropology* 37: 666-670.
1997 The Supernatural Beings of Prehistory and the external storage of religious ideas. In *The Archaeology of External Symbolic Storage*. (C. Renfrew & C. Scarre, eds.). Cambridge: McDonald Institute for Archaeological Research (in press).

Murrey, T. & Walker, M.J.
1988 Like WHAT? A practical question of archaeological inference and archaeological meaningfulness. *Journal of Anthropological Archaeology* 7: 248-287.

Nesse, R. & Williams, G.
1995 *Evolution and Healing: The New Science of Darwinian Medicine*. London: Phoenix.

O'Brien, M. & Holland, T.
1990 Variation, natural selection and the archaeological record. In *Advances in Archaeological Method and Theory*, vol. 2 (M.B. Schiffer, ed.), pp. 31-79. Tuscon: University of Arizona Press.

Orr, H.A.
1996 Dennett's Dangerous Idea. *Evolution* 50: 467-72.

Pinker, S.
1994 *The Language Instinct*. New York: William Morrow and Company, Inc.

Rindos, D.
1984 *The Origins of Agriculture: An Evolutionary Perspective*. New York: Academic Press.

Symonds, D.
1979 *The Evolution of Human Sexuality*. Oxford: Oxford University Press.

Thieme, H.
1997 Lower Palaeolithic hunting spears from Germany. *Nature* 385: 807-810.

Tooby, J. & Cosmides, L.
1992 The psychological foundations of culture. In *The Adapted Mind* (J.H. Barkow, L. Cosmides & J. Tooby, eds.), pp. 19-136. New York: Oxford University Press.

5

Methodology of Comparison in Evolutionary Archaeology

Hector Neff
University of Missouri, Columbia

Daniel O. Larson
California State University, Long Beach

ABSTRACT

Evolutionary archaeology needs methods for recognizing how selective retention of cultural variation shaped the archaeological record. One approach that has not yet been exploited fully is to derive adaptive hypotheses from evolutionary design arguments, then to test the hypotheses' predictions by comparing two or more segments of the archaeological record. In order to encourage further development of comparative methods, we discuss aspects of their application within evolutionary archaeology. We also present an example in which comparative observations are used to test a hypothesis about the evolution of local productive specialization and circulation of materials through space.

INTRODUCTION

Comparative methods have been central to a number of different archaeological approaches. Culture historians compared trait occurrences and frequencies among assemblages for the purpose of building chronologies and defining culture areas. As interest shifted away from culture-historical relationships in the 1950s, archaeologists used comparisons involving both archaeological and ethnographic cases in a search for cultural regularities with broad distributions in time and space (Beardsley et al. 1955, Willey et al. 1955, Willey & Phillips 1958: 2-3). In the 1960s "new archaeologists" sought to test general theories of systemic cultural change by comparisons involving both extant and extinct cultures (Binford 1962: 224). And in the 1970s, behavioral archaeologists proposed to seek regularities in the relationships between human behavior and material culture in all times and places (Reid et al. 1975, Schiffer 1996).

One might expect comparative methods to have seen their most sophisticated and extensive development in evolutionary archaeology. After all, the long time spans necessary for the realization of evolutionary processes largely preclude experimental testing, and that leaves only observation and comparison as means for evaluation. As Harvey and Pagel (1991: v) point out, Darwin frequently resorted to comparison, and so have "almost all [other] right-thinking evolutionary biologists." In evolutionary archaeology, comparisons between two or more segments of the archaeological record could be used to generate data for testing and refining design arguments about how natural selection helped shaped cultural aspects of past human phenotypes; this suggestion has already appeared in the literature of evolutionary archaeology (e.g., Neff 1992, O'Brien & Holland 1990, 1992). But the potential of this simple formulation has not yet been exploited fully, in part, we believe, because several issues related to the measurement of phenotypic change in the archaeological record remain unresolved. We hope this paper makes a contribution toward removing this impediment.

We begin our discussion with some background observations on how comparative methods have fared in the literature of evolutionary archaeology that has grown up over the last decade and a half or so. In brief, there was a need initially to define evolutionary archaeology in terms of what it was not, and there was a long tradition in anthropological archaeology of atheoretical comparison that merited unequivocal rejection (Dunnell 1980). In subsequent sections of the paper, we turn to a general discussion of methods of evolutionary investigation, then to a more focused consideration of how comparative data of use to evolutionists can be collected from the archaeological record. In the final section, we present an example illustrating the role of comparative methods in evolutionary archaeology.

ORIGINS OF AMBIVALENCE REGARDING COMPARISON IN EVOLUTIONARY ARCHAEOLOGY

Although Darwinian evolution had had some influence in archaeology before 1980 (e.g., Meggers, Evans, & Estrada 1965, Meggers 1983; also see Lyman & O'Brien, this volume), the persuasive essays of Dunnell (esp. 1980) sparked a wave of interest that sustained a concerted effort to explore the theory's implications for archaeology. In light of their seminal role in evolutionary archaeology, Dunnell's essays demand special attention in a search for understanding of methodological trends and biases. Dunnell does not explicitly reject comparative methods, but neither does he mention any specific role for them. More to the point, he argues that several approaches which do incorporate comparative methods are incompatible with evolutionary archaeology.

Cultural Evolution

Dunnell's (1980) point of departure in advocating development of a scientific evolutionary theory for archaeology was a categorical rejection of "cultural evolution." Cultural evolution was integral to the new archaeology of the 1960s (e.g., Binford 1962, 1965), which drew from White (1949), Sahlins and Service (1960), Steward (1955), and, more remotely, the early anthropological evolutionists, Morgan and Tylor. Comparative ethnographic analysis is a crucial methodological component of all cultural evolution: for unilinear evolutionists (e.g., White 1949), comparisons among unrelated groups lead to recognition of universal stages, whereas for multilinear evolutionists (e.g., Steward 1955), comparisons reveal developmental regularities.

Dunnell (1980) argued that cultural evolution has little in common with scientific evolution but a great deal in common with the nonscientific, progressive evolutionary philosophy of Darwin's contemporary, Herbert Spencer. Unlike the scientific theory of evolution, cultural evolution provides no mechanisms with which to explain history (Dunnell 1980: 44), but instead yields plausible empirical generalizations that merely summarize the comparative ethnographic data. Further, the extent to which ethnographic observations can be generalized is highly debatable because they were compiled by representatives of an expanding European culture at a time when most of the subject groups were becoming extinct (Dunnell 1980: 46). Since (a) cultural evolution fails to provide explanatory mechanisms and (b) the empirical generalizations that it does provide are highly suspect outside of the very limited historical contexts in which ethnographic data were collected by European travelers and ethnographers, Dunnell concluded that cultural evolution is not a suitable basis on which to build a scientific archaeology.

Comparative methods emerge from Dunnell's discussion of cultural evolution in a bad light because they are central to the nonscientific version of evolution that had to be rejected decisively in order to clear the decks for construction of a scientific evolution. Dunnell (1980: 50) is explicit regarding the necessity to discard everything associated with cultural evolution: "the flaws in cultural evolution are fundamental and characterize all aspects of the approach; no simple refinement of its concepts will materially alter its structure."

Cultural Ecology and Adaptation

The association of comparative methods with cultural ecology and the concept of adaptation may also be seen as a liability in the context of Dunnell's initial programmatic statements on evolutionary archaeology. Cultural ecology, the methodological arm of Steward's multilinear cultural evolution (see above), was conceived as "a method for recognizing the ways in which culture change is induced by adaptation to environment" (Steward 1955: 5). It is an explicitly comparative method designed to document the occurrence of similar processes of adaptation in similar environments (Steward 1955: 39-42). The concept of adaptation was discussed extensively in both anthropology (e.g., Alland 1975) and archaeology (e.g., Binford 1962, 1964; Kirch 1980) during the 1960s through 1980s.

Several interrelated suspicions about the role of adaptation in evolution arose around the time that the first proposals for a selectionist archaeology were being advanced. For one thing, Dunnell (1980: 49-50; also see Dunnell 1987) suggested that cultural evolutionists like Steward perhaps had embraced "adaptation" rather than selection because adaptation kept causality internal to culture, within the realm of human problem solving and intention. In a slightly different but related vein, Rindos (1989: 35-37) has argued that cultural ecology embraces a non-Darwinian view of evolution as a process driven by directed variation that arises at the level of "culture" in order to bring about more perfect adaptation of human societies to ecological conditions. Linking cause to the intentions of the actors (or to "culture") whose behavior needs to be explained contrasts with science, in which theoretical propositions about *mechanisms* furnish causes (Dunnell 1992, Neiman 1995). In a scientific sense, then, "adaptation" explains nothing but rather itself needs to be explained (e.g., O'Brien & Holland 1992). Once "adaptation" is shown to be devoid of explanatory content, the rationale for using comparative methods to document similar adaptations becomes subject to debate.

Unlike anthropologists and archaeologists, biologists have long conceived of adaptation as a product of natural selection (e.g. Mayr 1982). Around 1980, however, the "adaptationist program" began drawing fire because of the propensity of its practitioners to advance untestable "just-so" stories to account for any observed pattern of phenotypic variation (Lewontin 1979, Gould & Lewontin 1979).[1] Gould and Lewontin (1979: 584) refer to this as the "Panglossian paradigm." They claim that its adherents tend to analyze individual traits of organisms in isolation and to explain each as an optimal design for some function. Suboptimality is accommodated by arguing that optimal design for the whole organism may require suboptimality of some parts, and "the notion that suboptimality might represent anything other than the immediate work of natural selection is not entertained" (Gould & Lewontin 1979: 585). Sociobiological explanations of human behavior are singled out as particularly egregious examples of this Panglossian paradigm (Lewontin 1979). The critique of adaptation within biology provided additional grounds for Dunnell to minimize its importance for evolutionary archaeology:

> . . . showing how something is adaptive is not equivalent to demonstrating why it exists in the first place, and it can lead to the ad hoc Darwinian explanations criticized by Lewontin (1977) in modern biology. Functional explanations and adaptation are valuable elements of evolution to be sure, but they are not identical with evolutionary explanations founded in natural selection (Dunnell 1980: 50).

With so little room for the concept of "adaptation" in evolutionary archaeology, it is no surprise that comparative methods, which are basic to the study of adaptation (see below), did not figure in Dunnell's initial methodological suggestions.

Discomfort with cultural ecology and the adaptationist program has continued to color discussions of how to fashion an evolutionary approach to the archaeological record. O'Brien (1996: 26), for instance, synopsizes his views by noting that "[s]illy adaptationist stories are as much a problem in archaeology "as they are in biology." O'Brien's point, which is developed in a number of places (O'Brien 1996, O'Brien & Holland 1990, 1992, 1995a, 1995b; O'Brien et al. 1994) is that consideration of adaptation within archaeology is misplaced without a focus on the only mechanism capable of generating adaptation, natural selection. We discuss this point in more detail later on.

Reconstruction

A third theme in Dunnell's writings that works against serious consideration of comparative methods is his opposition to reconstruction of behavior from archaeological data. The inference of human behavior from material traces is addressed most systematically by behavioral archaeology (e.g., Reid et al. 1975, Schiffer 1996, Walker et al. 1995). This approach is manifestly comparative, with observation of material–behavior relationships in the present (i.e., experimental archaeology and ethnoarchaeology) making up a key component. Dunnell (1978a, 1978b, 1980, 1982, 1989), however, has long rejected a role for behavioral inference in evolutionary archaeology on the grounds that the theory must be written in terms that are observable in the archaeological record, not in terms of behaviors that are connected to the record only by lengthy, ad hoc inferential chains. Furthermore, since behavior-material relations must be invariant in order for behavior to be inferred from material, reconstruction of behavior requires the essentialist and obviously incorrect assumption that human behavior never changes (Dunnell 1989: 43-44). Dunnell's rejection of behavioral inference does not necessarily close the door on all use of comparative data in evolutionary archaeology, but it does rule out the use of comparisons between the archaeological and ethnoarchaeological records (e.g., Walker et al. 1995: 8).

In summary, comparative methods are central to cultural evolution, cultural ecology, the adaptationist program, and behavioral reconstruction, all of which were singled out for criticism by Dunnell in his initial theoretical and methodological statements regarding evolutionary archaeology. In light of these criticisms, it is perhaps understandable that comparative methods have not loomed large within the programmatic literature of evolutionary archaeology. This is not to say that comparison has been neglected entirely; despite the implications of Dunnell's programmatic essays, evolutionary archaeologists have looked to comparisons as means for testing hypotheses generated from evolutionary theory. Later on, we return to the topic of how comparative archaeological data can be brought to bear in the testing of evolutionary hypotheses. First, however, we need to backtrack and say a few words about the derivation of the hypotheses that are the focus of evolutionary science.

METHODOLOGICAL ISSUES IN EVOLUTIONARY INVESTIGATION

Evolutionary archaeology is based on the idea that artifacts in the archaeological record are parts of past human phenotypes (Dunnell 1989, Leonard & Jones 1987, Neff 1992, 1993; O'Brien & Holland 1992, 1995a, 1995b; O'Brien et al. 1994). As with other phenotypic traits, the frequencies of traits recorded in assemblages of artifacts wax and wane through history, giving rise to what Darwin called "descent with modification." For cultural traits, this process is generated by (1) communication among individuals of information on how to make and use implements, i.e., cultural inheritance, and (2) variation in the success individuals achieve at communicating knowledge about how to make and use their implements.

Despite the apparent simplicity of the above conception and increasing acceptance of its usefulness among archaeologists, methods for constructing and testing evolutionary accounts of the archaeological record cannot yet be said to have materialized. What might such methods look like? For a historical science like archaeology, whose subject matter consists of a finite, unique material record of the human past, the conclusion seems inescapable that testing of evolutionary hypotheses will have to be based not on experimentation but on observation and comparison.[2] Before explicating a methodology of comparison, however, some preliminary theoretical issues related to the investigation of evolutionary phenomena need to be addressed. The literature of evolutionary biology and philosophy of evolutionary science treats these issues at length, so, inasmuch as artifactual evolution is produced by the same mechanisms as biological evolution, the theoretical contributions of evolutionary biologists and philosophers are directly relevant to the methodology of evolutionary archaeology; evolutionary archaeologists (e.g., Dunnell 1980, 1989; O'Brien & Holland 1990, 1992) have, of course, recognized this before.

Biologists tend to distinguish two different approaches to the investigation of evolution (Harvey & Pagel 1991, Lewontin 1979,Maynard Smith 1994[1978]). On one hand, population geneticists use formal mathematical descriptions of migration and innovation rates, spatial structure, sampling error, and selection coefficients to model changes in gene frequencies (e.g., Crow & Kimura 1970). In Hull's (1980; also see Dawkins 1976) terminology, this approach emphasizes replicators as opposed to interactors. In archaeology, such an approach is endorsed by Bettinger (1991, Bettinger et al. 1996), who adapts Boyd and Richerson's (1985) cultural transmission models for archaeological ends; by Neiman (1995), who has modeled the temporal dynamics of stylistic (selectively neutral) traits; and by Lipo et al. (n.d.), who have shown how neutral trait frequency distributions can be used to infer the spatial structure of past populations (see Chapters 10, 15).

An alternative to the investigation of evolution based on the equations of population genetics is one that focuses on the evolutionary entities that Hull (1980, 1988: 408; also see Williams 1992) calls 'interactors.' These are the cohesive units that interact with the environment in such a way that replicators may persist differentially. Whereas the population genetics approach originated with the neo-Darwinian synthesis of genetics and natural selection, the focus on interactors dates back to *The Origin of Species.* The basic assumption underlying this approach is that the observed adaptations of plants and animals in the present were created by natural selection based on differential adaptedness (fitness) within populations in past environments. This approach entails "optimality modeling," in which an optimality criterion and alternative strategy sets are specified, after which an optimal morphology or behavior is deduced based on the predicted fitness payoffs from each strategy (Harvey & Pagel 1991, Maynard Smith 1994[1978], Sober 1993, Williams 1992). Some assumption about heredity is also required, but Maynard Smith(1994[1978]) has suggested that a broad "like begets like" assumption (explicitly including an assumption of cultural inheritance) is sufficient for most evolutionary optimization models. As mentioned above, this basic approach to evolutionary investigations has been severely criticized as the "Panglossian paradigm" by Gould and Lewontin (1979).

Inasmuch as archaeologists have little or no access to transmission mechanisms (replicators) but a great deal of access to interactors in the form of material remains in the archaeological record, it would seem that an approach that focuses on adaptation and optimality modeling would have much to offer.[3] First, however, it is necessary to salvage the concept of adaptation from the criticisms leveled within both evolutionary biology and archaeology.

The Adaptationist Program

Defenders of the adaptationist program in biology note that its purpose is not to test the general concept of adaptation, as suggested by critics (e.g. Gould & Lewontin 1979, Lewontin 1979), but rather to test specific hypotheses about the diversity of life that can be deduced from the theory of natural selection (Maynard Smith 1994[1978], Mayr 1983). In essence, "adaptation" is a key component of a strategy for understanding the living world as the product of natural selection coupled with chance, historical constraints, and developmental constraints.

Specific criticisms of adaptationist research leveled by Gould and Lewontin (1979, Lewontin 1979) are argued to pertain to an extreme, caricatured version (the "Panglossian paradigm") of what is otherwise a valid, productive research strategy (e.g., Maynard Smith 1994[1978], Mayr 1983, Sober 1993). For instance, Mayr (1983) agrees that atomizing an organism into a collection of traits, which are then explained individually as optimal design solutions to some problem imagined ad hoc is a strategy deserving of Gould and Lewontin's (1979) derisive criticism. Following Dobzhansky, however, Mayr (1983: 328) argues that it is perfectly valid to view the totality of traits produced by a given genotype as adapted (i.e., shaped by natural selection) to life under a certain range of environmental conditions; in fact, this is a fundamental tenet of evolutionary theory. Similarly, to characterize natural selection as a mechanism that infallibly produces perfect adaptation is an untenable perversion of Darwinian theory, as Gould and Lewontin (1979) suggest, but such a stance is explicitly rejected in modern evolutionary biology (Mayr 1983: 327); on the contrary, natural selection works within the constraints of available heritable variation and does not produce absolutely perfect adaptation to the current environment but merely adequate adaptation to a series of past environments (Gans 1993, Maynard Smith 1994[1978]: 113, Mayr 1983: 331, Van Valen 1973).[4]

O'Brien and Holland (1992: 44) reject Gould and Lewontin's (1979) criticisms of biological adaptationist research, but they nonetheless maintain (O'Brien & Holland 1995a: 188-189) that the Panglossian paradigm is alive and well in anthropology and archaeology (see Chapter 2). Archaeological application of optimal foraging theory, which derives from evolutionary ecological approaches elaborated in biology and anthropology (e.g., Smith 1983, Smith & Winterhalder 1992) has been singled out recently by O'Brien (1996) to illustrate the acceptance of adaptationist "just-so" stories in archaeology.[5] Evolutionary ecologists (e.g, Smith & Winterhalder 1992: 27-28) and selectionists (e.g., O'Brien & Holland 1992) seem to agree, however, that all adaptations must be understood as they are understood in biology - not as perfect solutions to present environmental conditions but as products of natural selection acting in past environments. Presumably both would agree that a crucial task confronting evolutionary archaeologists is to develop methods that will permit the recognition of adaptationist "just-so" stories and their replacement by testable hypotheses about how natural selection and other processes shaped the content of the archaeological record.

Fitness

Since the operation of natural selection is predicated on the existence of fitness differentials among individuals in a population, hypotheses about how natural selection has shaped adaptations in the living world require a clear concept of fitness, or adaptedness. As a result, biologists have expended considerable effort to clarify the concept of fitness in order to warrant the testing of evolutionary hypotheses based on natural selection (e.g., Mills & Beatty 1979, Sober 1984, 1993). As O'Brien and Holland (e.g., 1990, 1992, 1995a) have recognized, these discussions have a direct bearing on the causes of differential perpetuation of culturally-inherited traits.

If 'fitness' is defined in terms of actual survival of offspring (or actual events of cultural inheritance), natural selection is left open to the charge that it is a circular theory: more fit individuals are identified by the fact that they transmit their genetic or cultural traits more frequently, yet more frequent transmission of traits is explained as the result of greater fitness. Biologists have answered the charge of explanatory circularity by proposing that fitness refers to a *propensity* to transmit traits more frequently (Mills & Beatty 1979; Sober 1984, 1993). Propensities can be assessed through mechanical or other design considerations; for example, running speed may be expected to confer fitness advantages for zebras subject to predation (Sober 1984, 1993), while dark coloration may be expected to confer fitness advantages for moths subject to predation in sooted, industrial areas of England (Mills & Beatty 1979, Sober 1993).

To the extent that evolution is the same theory in archaeology and biology, it would seem that evolutionary archaeology can avoid the charge of circularity by adopting something resembling the propensity definition of fitness (O'Brien & Holland 1992, O'Brien et al. 1994). As in biology, this will require description of the products of evolution (archaeological remains in this case) in terms of measurements that can be argued to correlate with variation in propensity to transmit traits more frequently by cultural means. The theory must connect differential replicative success of cultural traits (Leonard & Jones 1987) to variation in propensities specified by optimal design arguments.

THE COMPARATIVE METHOD IN EVOLUTIONARY INVESTIGATION

Natural selection is a mechanistic process capable of creating convergent solutions to similar problems in similar environments. Convergence arises between unrelated groups because, given sufficient variation, exposure to similar environmental constraints and opportunities yields similar patterns of differential propensity to leave descendants. Biological examples of convergence are innumerable and can be found in any introductory textbook. The *observation* of convergence constitutes some of the strongest evidence favoring the view that the diversity of life has been created in part by natural selection (Mayr 1988: 137-138), whereas the *prediction* of convergence helps make hypotheses about how adaptations were shaped by natural selection amenable to falsification in both biology (Mayr 1983) and archaeology (Kirch & Green 1987: 433, Neff 1992, 1993).

Adaptive hypotheses about biotic diversity can be argued to come from two conceptually distinct sources. On one hand, as discussed above, the propensity model of fitness uses engineering design analysis and optimization as a basis for predictions about variation in reproductive success; in effect, these are hypotheses about how natural selection shapes the form and behavior of organisms under specified environmental conditions. O'Brien et al. (1994) use the propensity model of fitness to justify a focus on engineering design in the evolutionary investigation of prehistoric ceramics. Alternatively, optimal design can be inferred ex post facto by looking at differences in survival and reproductive success within a population, a procedure that Sober (1984; also see O'Brien & Holland 1990: 48, 1992: 44-45) calls "bootstrapping." Whether derived from an engineering analysis or other means, however, these hypotheses are all design arguments: certain features are predicted, based on engineering principles or other criteria, to confer a reproductive advantage in certain environments. Thus, for example, mechanical design features that increase running speed confer an adaptive advantage for zebras subject to predation, while cryptic coloration confers an adaptive advantage for predated moths. With appropriate qualifications for historical and developmental constraints, design arguments permit the direction of change to be predicted: engineering features that increase running speed should increase in frequency in populations of zebras subject to predation, and darker color should increase in frequency in predated moths of industrial regions.

The strategies for testing hypotheses derived from evolutionary design models can be categorized in various ways (e.g., Maynard Smith 1994[1978], Harvey & Pagel 1991, Sober 1984), but a similar logic underlies all approaches: all deduce from design arguments a set of predictions about variation in traits in one or more groups of organisms. For example, selection favors dark coloration of predated moths in polluted, industrial areas, but lighter coloration that would minimize a moth's visibility on unsooted trees would be favored in unpolluted regions. Mayr (1983: 329) succinctly characterizes the comparative approach as follows:

> Basically, the methodology consists in establishing a tentative correlation between a trait and a feature of the environment, and then to analyze in a comparative study, other organisms exposed to the same feature of the environment and see whether they have acquired the same specialization. There are two possible explanations for a failure of confirmation of the correlation. Either the studied feature is not the result of a selection force or there are multiple pathways for achieving adaptedness.

In a similar spirit, Sober (1984: 82) suggests that design arguments obtained ex post facto through "bootstrapping" can be tested by looking at other populations within which similar features should determine fitness differentials. Sometimes precise, quantitative predictions can be derived from a design argument, but failure of quantitative data to match theoretical predictions only demonstrates that the selective pressures postulated by the model are not the only significant influences on the trait in question, and this finding does not by itself disprove an adaptive explanation that is grounded in persuasive qualitative observations (cf. Maynard Smith 1994[1978], Sober 1993).

Comparisons are crucial for testing design arguments in archaeology just as they are in biology. Design arguments specify predictions about variation in one or more artifact categories for one or more segments of the archaeological record. The predictions are then evaluated against appropriate empirical data. One example of an archaeological design argument that has been discussed extensively in evolutionary archaeology (e.g., Braun 1983, 1987; O'Brien & Holland 1990, O'Brien et al. 1994, Neff 1992) is the shift toward thinner vessel walls in Woodland pottery from the Midwest U.S. Braun (1983) constructed a model that relates vessel wall thickness to thermal stress resistance, which he suggests would have been favored by the Middle-Late Woodland trend toward increased dietary reliance on starchy seeds and consequent longer boiling times. Observations on Woodland assemblages from western Illinois agree with predictions based on thermal stress considerations (Braun 1983, 1987). Later, O'Brien and Hoard (1996; see also O'Brien et al. 1994: 280-281) documented a closely similar trend in vessel wall thickness at the Burkemper site in eastern Missouri. The design argument in this example originates from a combination of ceramic engineering considerations and ex post facto reasoning (Sober's 1984 "bootstrapping") based on observed trends. While the trend observed in the Illinois ceramics is convincing, finding a parallel trend in the Missouri data offers dramatic additional supporting evidence.

Comparisons may even straddle the gap between genetic and cultural inheritance. Rindos' (1984) coevolutionary theory of plant domestication and the origins of agriculture is largely an argument about the general conditions that favor evolution of mutualistic relationships between organisms. Specific examples include mutualism between organisms whose phenotypes are determined primarily by genes, such as ants and acacias, but also mutualism between various plants and organisms whose phenotypes are determined to an important extent by cultural inheritance (humans). In a further extension of this same basic design argument, Cackler and Neff (n.d.) have suggested recently that the history of Maya society can be viewed as a history of fluctuating mutualistic interaction between elites and commoners.

In summary, comparisons offer a means for testing adaptive hypotheses derived from optimality (design) arguments. The method of testing is to derive from the design argument predictions about how a cultural trait or constellation of traits should vary, and then to check the predictions against two or more segments of the archaeological record. The use of comparative methods in evolutionary archaeology is illustrated in more detail in the final section of this paper. Before proceeding, however, four issues regarding the implementation of comparative methods in evolutionary archaeology need to be clarified.

Cultural Variation

One of Gould and Lewontin's (1979) criticisms of the adaptationist program is that there is no reason to assume that the requisite genetic variation will be available in a population at the precise time when it is needed for natural selection to fashion an optimal solution to some set of current environmental conditions and opportunities. In another discussion of how history constrains optimal design, Gould and Vrba (1982) point out that traits designed by natural selection for one purpose (or evolved for no purpose at all) are often co-opted for other purposes altogether, thus becoming "exaptations." Maynard Smith (1994[1978]) takes a different view of historical constraints, arguing that genetic variance of *some* appropriate kind does usually exist in populations; existing variation may not be that required for selection to fashion the absolute optimal solution, but it will be adequate to fashion a better solution. Similarly, Sober (1993: 121) has observed that "natural selection will optimize with respect to *existing* variation, and it is reasonable to expect existing variation to be *rich*" (emphasis in original).

One of the distinguishing characteristics of cultural inheritance is the ease with which new variation is introduced through mechanisms such as accident, imperfect communication, and intentional problem solving (e.g., Neff 1992; Rindos 1980, 1989). New cultural variants also tend to diffuse rapidly between populations, thus further enhancing the available cultural variation on which selection may act. To paraphrase Sober, it is reasonable to expect existing cultural variation to be richer even than existing genetic variation, so natural selection has even greater latitude to shape adaptations out of existing cultural variation. Moreover, because of the richness of cultural variation, natural selection can be expected to create phenotypic convergence in response to similar constraints and opportunities even more frequently for traits inherited culturally than for traits inherited genetically. In short, because cultural variation is so abundant, historical constraints on optimal design may be less important than they are for genetically determined traits. This in turn suggests that comparison and other methods for testing evolutionary design arguments may have an even more obvious role to play in archaeology than they do in biology.

Parenthetically, the forgoing observations provide a counter-argument to the often-expressed view that evolutionary approaches in archaeology are flawed because they deny the importance of consciousness and intentionality in human behavior (e.g., Trigger 1989: 306). In fact, these human characteristics play a central role in evolutionary theory as applied to people. That role, however, is not to direct the course of evolution but to provide a constant supply of novel variation on which selection may act in the future (Rindos 1989). And, to repeat, the abundance of cultural variation created by problem solving and other expressions of human intentionality gives reason to expect adaptation and convergence to be frequent outcomes of selection acting on cultural aspects of the human phenotype.

Observational Language

In the two examples of biological 'fitness propensity' mentioned above (running speed in zebras and color in moths), the organisms are described in terms of measurement units (speed and color) that are independent of the specific phenomena on which they are observed. Similarly, Braun's (1983) argument about the design of Woodland ceramics from the Midwest US concerns measurable attributes of any ceramic vessel (vessel wall thickness), and Rindos' (1984) coevolutionary model of plant domestication invokes measures, such as per-capita yield and total yield, that are independent of any specific empirical entities.

While seemingly obvious, these observations highlight a fundamental distinction that is crucial not just to comparative methodology but to the overall goal of developing scientific evolutionary archaeology. The distinction is between what Dunnell (1971, 1986, 1995) calls "ideational," or "theoretical," units on the one hand and "phenomenological," or "empirical," units on the other hand. Ideational units, such as classes, define phenomena and are space-time invariant, whereas phenomenological units, such as groups, are manifest by the phenomena that they contain and thus are defined only for a unique set of coordinates in space and time. According to Dunnell (1995), this distinction resolves the "materialist paradox:" ideational units make up the terms in which laws of history (e.g., evolutionary theory) must be written, whereas empirical units subsume the individuals on which the historical laws act. This issue would probably not even be worth mentioning were it not for the fact that archaeologists still tend to view units such as ceramic "types" as if they measured some theoretical dimension of variation.[6] If we wish to make comparisons between evolutionary sequences observed in different places and times, it is clear that the evolutionary phenomena of interest (archaeological remains, which are empirical) must be described in terms of units that are ideational (or theoretical).

The importance of describing archaeological phenomena in terms of theoretical units rather than empirical units has not been lost on those seeking to develop methods for evolutionary archaeology. The engineering design analyses favored by O'Brien and Holland (e.g., 1990, 1992, 1995b; O'Brien et al. 1994) are cases in point: the variables they and others (e.g., Feathers 1989, Schiffer & Skibo 1987) use to describe ceramics, such as temper type and amount, ceramic strength and toughness, vessel-wall thickness, and others, are ideational and related to one another by an explicitly articulated model. If evolutionary archaeology is to escape the "materialist paradox," such theoretical units must be adopted as the observational language used to describe and explain change (Dunnell 1995: 35).

Ideational units suitable for construction of evolutionary archaeology can be derived from any kind of design analysis, not just the engineering analyses advocated by O'Brien and colleagues. For instance, in previous work, Neff (1989, 1990, 1992, 1995b) has suggested that miniaturization and decorative elaboration associated with the emergence of a highly distinctive Mesoamerican ceramic ware known as Tohil Plumbate are convergent with a number of modern instances of miniaturization and decorative elaboration among traditional potters faced with growing external demand for their products. Although the observational language in this case consists of simple measures of vessel size and extent of decorative modification, the language is clearly "ideational" in that any ceramic assemblage could be described in the same terms. The use of theoretical units that are measurable in any ceramic collection makes it possible to compare the Plumbate industry, which is known only from archaeological collections, with modern ceramic industries, which are documented through the much richer medium of ethnography or ethnoarchaeology. It is clear, then, that many measurements are space-time invariant or "ideational" and thus suitable for comparing two or more evolutionary sequences.

Empirical Sufficiency, Ethnoarchaeology and Experimentation

Perhaps the most explicit use of optimality modeling in anthropology and archaeology has been the application of evolutionary ecology (Smith & Winterhalder 1992), in the form of optimal foraging theory, to hunter-gatherer studies (e.g., Bettinger 1991, Hawkes & O'Connell 1992, O'Connell 1987, 1995). As already mentioned, selectionists (e.g., O'Brien 1996) have sometimes characterized optimal foraging as Panglossian, adaptationist story telling, a characterization which would surely be debated by the other side (e.g., Bird, Chapter 16). A more serious problem with optimal foraging applications in archaeology has been pointed out by one of its own practitioners: "[models from behavioral ecology] were designed to analyze the behavior of individual organisms directly observed in ecological time, not the behavior of aggregates indirectly reflected in material evidence accumulated and incompletely preserved in archaeological time" (O'Connell 1995: 237). The observational units of evolutionary ecology (e.g., travel time, processing time, patch choice, etc.) are ideational, as required for scientific evolutionary archaeology, but in practice these units are unobservable to archaeologists and instead must be "reconstructed" or inferred from the archaeological remains. The problem here is with what Dunnell (1980) calls "empirical sufficiency."

As mentioned previously, opposition to the "reconstruction" or "behavioral correlates" approach to archaeology is a consistent theme running through Dunnell's work (e.g., Dunnell 1978a, 1978b, 1980, 1982, 1989). In Dunnell's view, not only must the data language of evolutionary archaeology incorporate ideational (theoretical) rather than phenomenological (empirical) units, but the theoretical units must themselves be measurable directly in the record. If, as proposed by evolutionary ecologists, evolutionary archaeology were phrased in transactional (behavioral) terms that had to be inferred from the material record, then it would continue to produce untestable "just-so" stories (Dunnell 1982, O'Brien 1996). As noted above, at least one archaeologist working from an evolutionary ecological perspective (O'Connell 1995) explicitly acknowledges this liability; unfortunately, he provides no advice as to how it might be circumvented.

Schiffer (1996: 650-652) has recently argued that Dunnell's opposition to behavioral inference is itself incompatible with evolutionary archaeology. Schiffer argues, first, that measurements of interest in scientific investigation can only rarely be made directly in the phenomenological world, as Dunnell would require of an empirically sufficient archaeology. Second, he argues that evolutionary biology depends on paleoenvironmental reconstruction as well as inferences about the behavior of extinct organisms, so the opposition to reconstruction has no parallel in other evolutionary sciences. Third, he suggests that close inspection of evolutionary models in archaeology, even Dunnell's, reveals that they are actually expressed in behavioral terms. Fourth, Schiffer argues that behavioral inference does not require stasis, as Dunnell asserts, but rather establishes the boundary conditions under which certain behavioral processes will lead to certain material results.

To some extent, Dunnell and Schiffer are talking past one another. Certainly Dunnell would not assert that "empirical in the record" means that human senses can provide the only access to archaeological phenomena. Almost any scientific measurement entails theoretical arguments that connect what is actually measured with the dimensions of interest. In chemistry-based archaeological provenance research, for instance, the goal is to monitor geographic space (the location of a "source"), but the actual measurements made in the laboratory are parts per million or weight-percent of the determined oxides or elements; the archaeological units of interest (geographic coordinates) are connected to the analytical measurements by a theoretical warrant labeled the "provenance postulate" (Neff n.d., Weigand et al. 1977). As Schiffer (1996: 650) points out, Dunnell himself utilizes such inferences, as in his reference to "costly imports" in the Woodland mortuary cult of the Midwest U.S. (Dunnell 1989: 48). Schiffer is right to characterize this as an inference: "import" is an inference based on knowledge of the natural distribution of raw materials, and "costly" is an inference based on the distance to source areas of the imports; both inferences are based on measurements given theoretical warrant by the provenance postulate, and the compound characterization "costly import" then reflects only a deduction about what must have been necessary (long-distance movement by humans) to give the artifacts their observed archaeological distributions. In accord with O'Brien and Holland (1995b: 159-160), we would suggest that the issue is not whether evolutionary archaeology requires inference but instead whether it requires extensive inferential chains that lead by a circuitous path from measurable material properties (whether theoretically justified or not) to reconstructions of gender, kinship, political structure, and other nonobservables. It may be true that "every selectionist model can be shown to rest, implicitly or explicitly, on a network of behavioral inferences" (Schiffer 1996: 650), but the requirement of empirical sufficiency forces us to admit only the shortest, strongest possible inferential chains into our research programs.

The major implication that Schiffer (1996: 652) draws from his discussion of Dunnell's views on reconstruction and inference is that comparative strategies of actualistic and ethnoarchaeological research would enhance evolutionary archaeology if they were taken seriously. This is a point with which a number of evolutionists (e.g., Neff 1990, 1992: 179, 1995; O'Brien & Holland 1990: 48, 1992: 4) would agree. But this does not mean that we need to adopt an observational language filled with behavioral terms. Rather, actualistic and ethnoarchaeological studies must be carried out using terms that are both time-space invariant (ideational) and measurable in the archaeological record. Experimental studies cited by Schiffer (Dunnell & Feathers 1991, O'Brien et al. 1994) as well, arguably, as Schiffer's own work (e.g., Schiffer & Skibo 1997), exemplify these principles.

The Plumbate case introduced above exemplifies the importance of appropriate measurement scales and empirical sufficiency when ethnoarchaeological data are combined with archaeological data in a comparative study. Ceramic size and decorative coverage can be observed both archaeologically and ethnographically, and therefore a hypothesis of convergence can be formulated to account for the trends toward miniaturization and decorative elaboration observed both in the Plumbate industry and several ethnographically known ceramic industries. This hypothesis gains additional credence from the observations that, like the modern industries, which have experienced growing tourist demand during the 20th century, Plumbate potters were tied into an interregional exchange system that expanded rapidly at the beginning of the Early Postclassic period, at roughly the same time that elaborately decorated and miniaturized vessels began to appear (Neff 1989b, 1990, 1995).

Homology and History

We noted previously that abundant novel variation tends to reduce historical constraints on adaptive design of cultural traits. We wish to disavow emphatically, however, any claim that cultural aspects of human phenotypes are liberated from the constraints of history. As stressed by one of us elsewhere (Neff 1991), evolutionary archaeology is as much about history as it is about ecology and economics. To paraphrase Stephen J. Gould (1986: 74), the material reality of history flows through the design standards that are external to history, forming a set of contrasts that define evolutionary archaeology: homology and analogy, history and immanence, movement and stability.

History matters in the present discussion because use of comparative data to test hypotheses about cultural adaptations requires independence of the cases being compared (Mace & Pagel 1994; also see Bettinger et al. 1996: 153). The problem is parallel to that encountered in comparative biology (e.g., Harvey & Pagel 1991), in which comparisons are relevant to testing adaptive hypotheses only if a common origin for the trait(s) in question can be ruled out. In anthropology, the problem of distinguishing historical (homologous) from functional (analogous) associations is known as Galton's problem after Francis Galton, who first described it to the Royal Anthropological Institute in 1889 (Naroll 1961). Mace and Pagel (1994) suggest that the key to resolving Galton's problem lies in reconstructing cultural phylogenies on the basis of genetic, linguistic, historical, or archaeological evidence. Based on such phylogenies, the number of independent occurrences of some trait association can then be determined.

Verifying the independent origins of presumed adaptive traits is no less important in evolutionary archaeology than in evolutionary anthropology or evolutionary biology (Neff 1990; also see Kirch & Green 1987: 443). The methods are clearly distinct, however. Whereas anthropologists may have access to linguistic, genetic, and historical data, archaeologists most often have only the material record on which to base conclusions about phylogeny. Where evolutionary archaeologists have the advantage is in their access to a more complete, better-resolved historical record of the traits of interest: comparative biologists and anthropologists look only (or primarily) at the tips of a phylogenetic tree and must infer relationships based on parsimony and the pattern of contemporary trait occurrence; archaeologists, in contrast, can look at the histories of traits in one or more sequences. In the archaeological case, independent origin and the direction of evolutionary change are monitored by establishing the timing of introduction, fixation, and disappearance of the traits in the sequences being compared and by considering the evidence pertaining to common descent (Neff 1990, 1993).

Decorative elaboration and miniaturization in the Plumbate tradition again serves as a convenient example of how independent occurrence can be established in archaeological investigations that employ the comparative method. The Plumbate tradition, which apparently disappeared before the Late Postclassic period, is hypothesized to have undergone convergent evolution with several 20th century Mesoamerican ceramic traditions whose roots lie in late prehistoric or post-conquest ceramic traditions of their respective regions. Moreover, parallels in the selective pressures favoring miniaturization and decorative elaboration (increasing external demand) are rooted in entirely different historical circumstances, one associated with growing interregional interaction in Early Postclassic Mesoamerica and the other associated with growing tourism during the 20th century.

The continuous historical character of the archaeological record offers additional potential for the testing of adaptive hypotheses, as O'Brien and Holland (1990, 1992) have recognized. The argument rests on Dunnell's (1978a) distinction between stylistic (adaptively neutral) and functional (non-neutral) traits. As corroborated recently by Neiman's (1995) simulations, neutral traits show smooth, temporal distributions approximating the battleship curves on which convincing chronologies are based. Traits under the control of selection, in contrast, are expected to increase rapidly to fixation once they come under the control of selection. O'Brien and Holland (1990, 1992) view the rapidity with which shell temper replaced grit temper in the Late Woodland period of the Midwest as convincing evidence that several aspects of ceramic resource selection and pyrotechnology were shaped by directional selection into a new adaptation around 1000 A.D.[7] Similarly, Neff and Arroyo (1996) argue that the rapidity with which ceramics appeared soon after 2000 B.C. on the southern Mesoamerican coast indicates that their adoption was a selection-driven process. These examples illustrate how the high-resolution ratio-scale measurement of time possible with archaeological data permits comparison not only of the final outcomes of adaptive processes but of their temporal trajectories as well. Access both to the direction of evolutionary change and to its pace gives archaeology a decided advantage over most other evolutionary sciences: in archaeology, comparative methods can be applied not just to adaptive traits (the outcomes of selection) but to the selection-driven adaptive trends themselves.

EXAMPLE OF A DESIGN ARGUMENT AND ITS EVALUATION BY COMPARATIVE METHODS

Ultimately, methodological discussions like that presented in the previous sections should pay off in the form of enhanced understanding of how evolutionary mechanisms shaped specific segments of the archaeological record. Therefore, in this section we consider a model

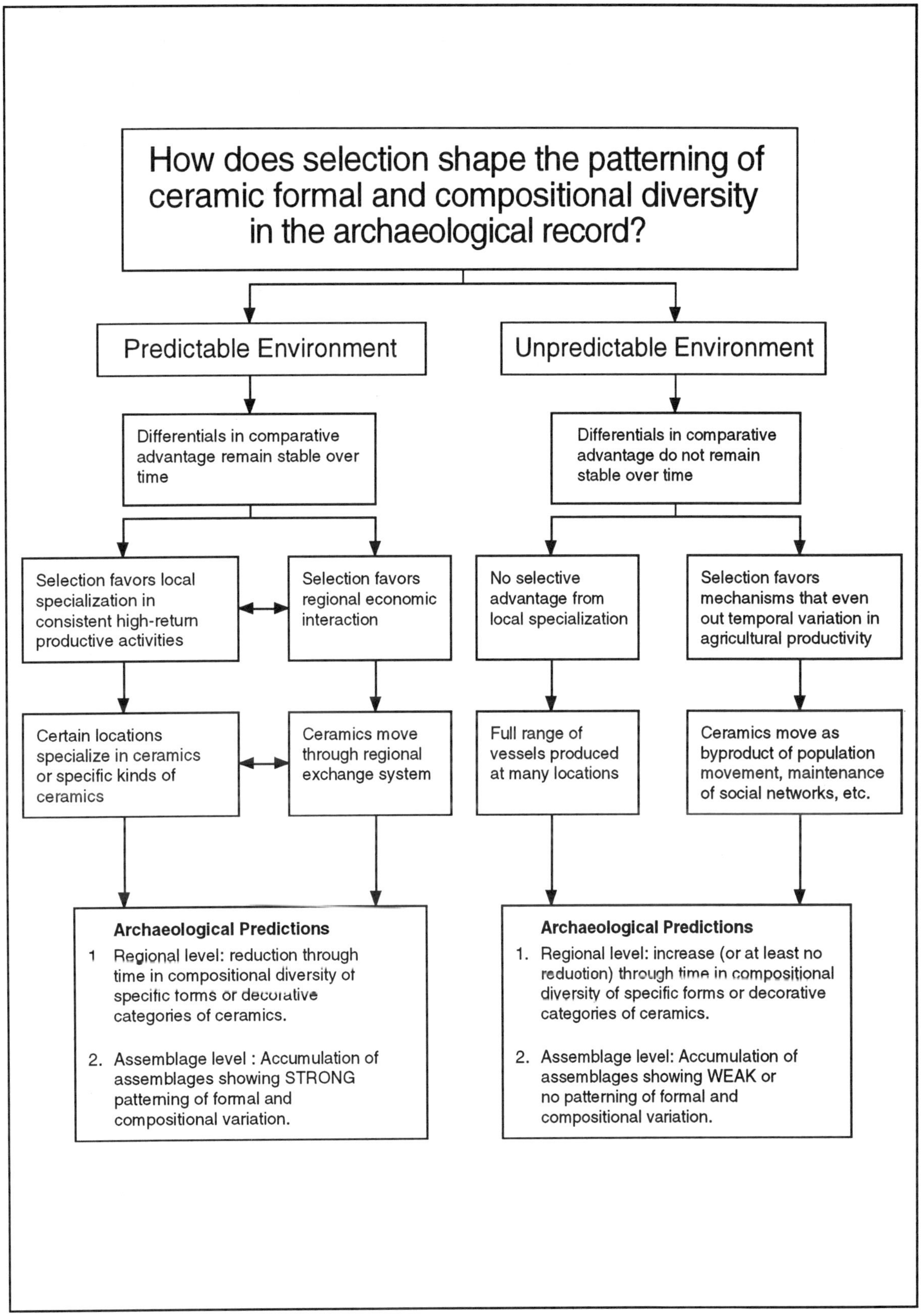

Figure 5.1 A model of how selection shapes the patterning of formal and compositional diversity in the archaeological record.

of how human movement of goods through space is shaped by selection associated with environmental predictability. The empirical details alluded to here are documented more extensively in other publications (Larson et al. 1996, Neff 1995a, Neff & Hodge n.d.,Neff & Larson n.d., Neff et al. 1997).

The Model

Movement of objects through space is a universal human propensity that entails significant investment of human energy and therefore must confer some substantial fitness benefit in order to persist. Because of the energy investment, one would also expect the specific ways in which humans move materials to be controlled by selection. We therefore propose that selection designs different kinds of mechanisms for moving objects through space in predictable vs. unpredictable environments: local specialization and economic interaction are selected for simultaneously in predictable environments, but exchange and other mechanisms of ceramic movement are favored *without* local specialization in unpredictable environments. This is an adaptive hypothesis that we would like to be able to test by evaluating how well its predictions hold under various historical circumstances. Fortuitously, the history of how humans have moved non-perishable objects (such as ceramics) through space in various times and places can be monitored with some confidence using methods of archaeological provenance investigation (Neff n.d.). As a result, we can connect the design argument of our model to specific predictions about the patterning of ceramic remains in the archaeological record (Figure 5.1; also see Neff & Larson n.d.).

Let us first consider an ideal predictable environment (the left half of Figure 5.1). Geographic differences in return from subsistence and other productive activities are consistent from year to year, and these consistent differentials in comparative advantage favor differentiation of productive strategies. Consider two locations, one in which biotic productivity is consistently high but ceramic resources are scarce or of low quality, and another in which biotic productivity is consistently low but ceramic resources are high-quality and plentiful. Other things being equal, everyone's fitness is maximized by obtaining sufficient food and sufficient pots for the lowest possible expenditure of time and energy: for people situated favorably with respect to ceramic resources, fitness is enhanced by any practice that reduces the costs of obtaining food; conversely, for people situated favorably with respect to subsistence resources, fitness is enhanced by any practice that lowers the costs of obtaining sufficient pots. Two alternative strategies for lowering fitness costs in each location are (1) steal the needed goods (pots or food) from someone else or (2) produce a slight excess of the goods that are easiest to produce locally and then seek to obtain the needed goods through exchange. The second strategy, economic interaction, will be favored if its fitness costs are lower than those associated with stealing (which may include, for example, the cost of planning and executing a raid, the cost of guarding one's own resources, the increased possibility of suffering a violent death, etc.). Thus, given the right balance of fitness costs and benefits, the selective pressures on populations distributed across a landscape of consistent differentials in comparative advantage will be complementary and will favor the growth of local specialization coupled with regional economic interaction. Increasing regional economic interaction permits individuals to capitalize on the potential for local specialization due to differential comparative advantage, and this reinforcing feedback favors increasingly complex patterns of local specialization, such as specialization in specific kinds of ceramic vessels. In short, local specialization and regional economic interaction are mutually reinforcing aspects of a single selective process arising from consistent differentials in comparative advantage. Arnold (1975, 1978) and Stark (1991) enumerate ethnographic examples of part-time specialist pottery making communities that apparently evolved through this process of niche diversification.

Archaeological expectations for predictable environments are shown at the bottom left in Figure 5.1. First, the selection-driven trend toward community specialization within a region is reflected in declining compositional diversity of specific forms or decorative classes of pottery, as some communities stop producing vessels that they increasingly obtain through economic interaction. Second, ceramic assemblages are expected to show a strong association between ceramic form/decoration on the one hand and raw material source on the other hand. One extreme possibility is that pottery production becomes the exclusive specialty of one community in a region, in which case the ceramics in the region's assemblages manifest a restricted range of compositions characteristic of raw materials near the specialized community. In more complex regional economic systems, several ceramic producing communities may specialize in distinctive shapes and/or decorative techniques, in which case the accumulated ceramic assemblages should show strong association between composition and formal/decorative modes.

Now consider the mechanisms that would move ceramics through space in an extremely unpredictable environment (the right side of Figure 5.1). In such an environment, differentials in comparative advantage are inconsistent from one year to the next, so selection for local specialization is relaxed or non-existent. Without local specialization, exchange is favored as but one of several mechanisms that permit differential allocation of subsistence effort across space and through time. Other mechanisms may include migration, short-term population movements, and maintenance of widespread networks of social interaction (Braun & Plog 1982, Leonard 1989, Leonard & Reed 1993, Plog 1983, 1984). Any one or combination of these behaviors confers a fitness benefit by increasing the likelihood that an adequate food supply will be available, despite the unpredictability of biotic productivity. Goods such as ceramics move across the landscape as a result of any or all of these behaviors, but the movement of ceramics can be seen as an incidental byproduct of selection for mobility and/or social interaction among people who happen to have made and used ceramic pots. Pots may be imported to a location as gifts in anticipation of reciprocity during a future shortfall, as containers used to transport foodstuffs during an actual shortfall, by an immigrant relocating in an attempt to enhance return from his/her agricultural labor, or by an immigrant relocating as a result of marriage. These behaviors are equivalent phenotypic expressions within the framework of the model. That is, the model predicts only that ceramics will move through space and makes no specific prediction about conditions that would favor one behavior over another. It connects observations that can be made on the archaeological record (the presence of a nonlocal pot) to selective pressures in an unpredictable environment

Expectations for ceramic assemblages accumulated in unpredictable environments are shown on the bottom right in Figure 5.1. First, because selection does not favor local specialization, regional compositional diversity is expected to be high (people living at many locations produce ceramics out of local raw materials). Second, because selection favors various activities that move vessels across the landscape, evolution is expected to bring *similar* ceramics from *different* sources together in the same assemblage. That is, in contrast to the expectation for a predictable environment, there should be weak or inconsistent association of decorative and formal attributes with composition. Whether ceramics moved through migration, residential mobility, or exchange, accumulation of a diverse yet weakly patterned ceramic assemblage is favored by selection associated with fluctuating environmental conditions and the attendant subsistence risk.

It is important to emphasize that the model diagrammed in Figure 5.1 is written in an ideational data language, as required for an evolutionary design argument (Dunnell 1995; see above). Resource procurement practices are monitored via provenance determination, and forming and decorative practices of potters are monitored via attributes of form and decoration observed on sherds. The specific predictions that we are testing have to do with the way that attributes of form and decoration pattern with respect to compositional group (source). In some cases, the formal attributes that we record may coincide with attributes used to identify "wares," "types," or similar empirical units that appear in culture-historical frameworks. But this does not mean that the formal attributes we are concerned with should be regarded as empirical rather than ideational, even if we use "type" or "ware" names to refer to a set of consistently associated formal attributes. What transforms the latter (which are ideational) into the empirical units of culture history is the hypothesis that common descent connects empirical entities on which certain formal attributes can be observed (Neff 1993, Tschauner 1994; also see footnote 3).[8]

Another important observation about the model diagrammed in Figure 5.1 is that it does not explicitly differentiate between ceramic traits that are functional (under the control of selection) and traits that are stylistic (selectively neutral). Instead, it concerns the control that selection exerts on the distribution of productive activities among populations; it contrasts the pattern of production and movement of goods expected in predictable vs. unpredictable environments. Thus, for example, given a sufficiently predictable environment, comparative advantage may favor local specialization in ceramic production in some community, even though the specific formal/decorative attributes that characterize pottery from that community reached high frequencies entirely by chance processes (drift); alternatively, their high frequencies may be a result of a selective process that shaped the technological and decorative practices of potters in the community. Our model specifies only that selection favors a pattern in which pottery-making practices (including whether or not pottery is made at all) are distributed heterogeneously among populations and therefore that a strong association will be detectable between formal/decorative attributes and composition (source).

What we wish to evaluate, then, is an adaptive design argument that predicts a contrast between the patterning of ceramic diversity in assemblages accumulated in predictable vs. unpredictable environments (Figure 5.1). In keeping with the principles discussed previously, the expectations about ceramic diversity are written in an ideational data language, and they are connected to the adaptive model by short, explicit inferential chains, as dictated by the requirement of empirical sufficiency.

Evolution of Ceramic Diversity in Predictable Environments

Humid tropical regions tend to have high and predictable available moisture, which translates into predictable variation in primary biotic productivity and, according to the model presented above, into strong selection in favor of coupled local specialization and exchange. Miriam Stark's (1991) survey of specialized pottery-making communities described in the ethnographic literature bears out the prediction that they should be most frequent in low-latitude regions. In Sober's (1984) terms, Stark's observations provide a warrant via "bootstrapping" for the part of the model depicted on the left side of Figure 5.1. Based on the theoretical arguments outlined above and Stark's ethnographic data, we predict that the archaeological records for humid tropical regions should show trends over time toward increasing local specialization and regional economic interaction.

Pacific Coastal Guatemala

One region where economic interaction has been examined in some detail via archaeological ceramic sourcing research is Pacific coastal Guatemala. With a mean of 200+ cm of rainfall annually and a coefficient of variation in rainfall of less than 15% in recent times (Mosiño Alemán & García 1974, Vivó Escoto 1964), the region epitomizes a predictable environment for agricultural production. Furthermore, the coastal plain itself receives runoff from the volcanic mountain chain to the north, where rainfall is even higher, so it seems unlikely that severe drought or flooding would have posed a severe threat to primary biotic productivity. Assuming that moisture in this region also was plentiful and predictable in the past, such things as soil type differences and drainage would have created consistent, long-term differentials in agricultural productivity that would have selected for both local specialization and regional exchange. In the absence of historical discontinuity, these features should become more prominent over time.

Ceramics first appeared on the southern Mesoamerican coast shortly after 2000 B.C. The initial ceramic assemblages are very similar across a wide area, and, while there is some slight evidence for vessel movement, there is no indication of local specialization (Neff & Arroyo 1996). On the central Guatemalan coast, the Late Formative through Early Classic periods (500 B.C. - 500 A.D.) saw the emergence of a tendency for vessels to flow west to east, toward sites located on the well-drained, extremely productive agricultural land of central Escuintla (Neff 1995a; Neff, Bove & Arroyo n.d.). The trend toward local specialization and regional exchange on the coastal plain culminates in the Late Classic period (A.D. 700 - 900), when potters exploiting a distinctive, volcanic ash-derived clay source located on poorly-drained land in northwestern Escuintla supplied serving vessels to coastal plain sites perhaps as far east as the Salvadoran border (Neff 1995a; Neff, Bove & Arroyo n.d.).

Other specialized ceramic industries evolved elsewhere on the Guatemalan Pacific slope during the Classic period. Of particular interest is the history of Flesh ware, which originates as Esperanza Flesh during the Early Classic period. There are at least four compositional variants of Esperanza Flesh, some linked to ceramic resources on the coast and some linked to resources in the highlands (Neff et al. 1994). Flesh ware occurs archaeologically throughout the central highlands and on the coastal plain as far east as the province of Santa Rosa, near the Salvadoran border. Amatle ware, the Late Classic successor to Esperanza Flesh, was made in only a single compositional variant, presumably within a much more restricted zone than the earlier Esperanza. Apparently, then, Flesh ware potters were exploiting opportunities presented by interregional demand for their products, but over the course of the Classic period, some potters (those utilizing the Amatle source) came to dominate this particular specialized pottery production niche. As predicted by our model, stable differentials in comparative advantage selected for increasing local specialization, and this is reflected in the trend toward decreased compositional diversity in Flesh ware during the Classic period.

The Southern Basin of Mexico

The history of ceramic production at the Aztec city of Chalco, in the southern Valley of Mexico (Hodge n.d.) also documents the evolutionary effect of consistent differentials in comparative advantage. The southern Valley of Mexico consistently receives more rainfall than the north (Tolstoy 1975), and as a result, the southern Valley was better able to produce an agricultural surplus. This is reflected, for instance, in the fact that the Tepaneca attempted raids on Chalco around A.D. 1400 in an attempt to acquire grain reserves (Chimalpahin 1965).

Chalco was probably founded in the 6th or 7th century A.D., during the period of turmoil that surrounded the fall of Teotihuacan (Hodge n.d.). At first, differentials in comparative advantage among the various Epiclassic communities established around this time would not have been manifest. Not surprisingly, similar Red-on-Cream and Red-on-Buff vessels were produced at Chalco and elsewhere in Central Mexico during the Epiclassic period (Neff & Hodge n.d.). As time passed, differentials in returns from various productive activities would have arisen gradually as a result of natural environmental differences (e.g., rainfall differences, differential access to ceramic resources, etc.) and chance historical differences (e.g., differential proficiency of local potters). At Chalco, a local serving vessel tradition continued into Early Aztec times, but, while other Black-on-

Orange production centers in the Basin of Mexico had begun to supply serving vessels through a growing regional economic system, Chalco supplied vessels primarily to a local, southern Basin market. In Late Aztec times, Chalco was incorporated into the Aztec empire centered at Tenochtitlan, an event that can be expected to have amplified disparities in returns from different productive activities, thus favoring even more strongly the activities (e.g., agricultural production) for which local returns were relatively high. As would be expected under our model, Aztec Black-on-Orange production ceased at Chalco in Late Aztec times, and the city imported Black-on-Orange serving vessels from other centers, such as Tenochtitlan and Texcoco. In sum, then, the 1000-year history of serving vessel production at Chalco follows a course that is highly predictable given the city's consistently advantageous situation with respect to agricultural production.

Evolution of Ceramic Diversity in Unpredictable Environments: the Southwest

Arid and temperate regions contrast with humid tropical regions in that they are marked by high temporal variability in the availability of moisture, which translates into high variability in primary biotic productivity. Floods and, especially, droughts can dramatically reduce supplies of wild or domestic food supplies, as Larson (1994) has demonstrated for the historic period in California. High-resolution dendroclimatic records available for several parts of the Southwest (Larson et al. 1996, Larson & Michaelson 1990) provide grounds for recognizing specific periods of extreme subsistence risk in several parts of this region. Selection for exchange in the absence of local specialization should be strongest during periods of extreme climatic variability (Larson et al. 1996).

Numerous ceramic sourcing studies in the Southwest have yielded evidence consistent with the view that exchange or other mechanisms of vessel movement evolved in the absence of local specialization. Several provenance investigations have recently been undertaken in the Grasshopper region of east central Arizona. At Chodistaas, a small Pueblo-III site abandoned around A.D. 1300, whole vessels of Corrugated, Plain, and Cibola White ware excavated from room floors were found to include local compositions along with several nonlocal compositions (Zedeño 1994). In accord with our model, Zedeño (1994: 102) suggests that environmental marginality favored maintenance of wide social networks, which in turn produced accumulations of non-local ceramics at sites such as Chodistaas. Triadan's (1994) study of White Mountain Redware at Grasshopper and other Pueblo IV sites in the Grasshopper region demonstrates that post-1300 assemblages contain both local and non-local Fourmile Polychrome. Migration, which may account for some of the non-local White Mountain Redware in east-central Arizona (Tridan 1994), is a particularly salient aspect of the later prehistory of the Southwest, especially after the Great Drought of the late 1200s (Larson et al. 1996).

The case for vessel movement in the absence of local specialization may be even stronger in areas to the north of the Mogollon Rim. In the Mesa Verde region, Glowacki (1995, Glowacki et al. 1995) has found that Black-on-White pottery in Pueblo-III assemblages includes locally made examples together with examples derived from non-local production zones. On Black Mesa, Deutchman (1980) found several compositional groups represented in Black-on-White assemblages and no statistical association between "type" and compositional group. Based on "low-tech" analyses, Kojo (1996) concurs with Deutchman's conclusion regarding widespread production of Tusayan White and Gray ware, but he suggests that Deutchman's estimate of the incidence of vessel movement may be somewhat inflated.

Our own analysis of a ceramic assemblage from a Pueblo-III site in south central Utah has revealed surprising compositional diversity that shows only slight patterning with formal diversity (Larson et al. 1996, Neff et al. n.d.). A two-level structure is evident in the chemical data for 206 analyzed sherds, with two broad but very distinct compositional groups represented and a total of seven compositional subgroups present within the two main groups. All five decorative (ware) categories present in the assemblage are represented in both of the main compositional groups, and all are present in more than one of the subgroups of the main groups. There seems little question that all decorative categories originate in more than one location. The best-sampled decorative class, Black-on-White, is represented almost evenly in the two major compositional groups and occurs in all but one of the seven subgroups.

The patterning of compositional and decorative diversity in the assemblage just described cannot possibly have been produced by exchange coupled with local specialization based on comparative advantage (Neff et al. n.d.). Although the pottery in this assemblage clearly originated in several distinct locations, there was a high degree of formal and decorative overlap in the output of the spatially distinct workshops, and most or all decorative classes may have been produced locally. Since movement of pottery cannot have been based on considerations of direct economic benefit, it must have been a byproduct of activities like small- and large-scale residential mobility and non-economic reciprocal exchanges (gift-giving). As specified by the model presented here (Figure 5.1), such activities enhance individual fitness in an unpredictable environment like the northern Southwest; incidentally, they move ceramics through space in the absence of local specialization and regional economic interaction.

CONCLUSION

In order to persist, energetically expensive behaviors must confer offsetting fitness benefits, i.e., they must be shaped and maintained by selection. If this is a valid assumption, then the signal of past selective processes must be strong in the material record accessible to archaeologists; every lithic flake, potsherd, and animal bone in an archaeological deposit records energy and time spent by someone making his/her way through life in the past.

Although we can be quite confident that selection has shaped the material record of the human past, so far we have made little progress toward perceiving the nature of its effect. Part of the solution, as argued by O'Brien and Holland (1992), is to adopt archaeological definitions of adaptation and adaptedness that parallel the definitions used in evolutionary biology: adaptations are traits that were shaped by selection in a series of past environments, and adaptedness (or fitness) is the propensity (which is assessed by design analysis) of a trait or set of traits to increase in frequency. Attention to these definitions can promote the formulation of plausible adaptive explanations. In archaeology as in biology, however, even the most ingenious adaptive design arguments by themselves represent wasted effort ("silly just-so stories") unless explicit methods of hypothesis testing are brought to bear. Comparative methods, which are widely advocated as key to the testing of adaptive hypotheses in evolutionary biology (e.g., Harvey & Pagel 1991, Maynard Smith 1994[1978], Mayr 1983; Sober 1984, 1993), would also seem to hold considerable potential for testing adaptive hypotheses in archaeology.

The unique properties of the archaeological record present both unique opportunities and special demands for archaeologists seeking to employ comparative methods. First, it must count as an opportunity that cultural variation is abundant and easily generated, and adaptation and convergence are therefore expected to arise commonly as the result of selective processes acting on cultural aspects of human phenotypes. On the other hand, because of the ease with which empirical archaeological units such as ceramic types can be reified, archaeologists are particularly vulnerable to the trap of the "materialist paradox" (Dunnell 1995); to avoid it, we must be careful to distinguish between empirical units, which are the things contained in the archaeological record, and theoretical (ideational) units in terms of which the record is described and comparisons between different segments of it are made. Another problem is unevaluated behavioral inference: while many statements about the archaeological record may involve inference, the requirement of empirical sufficiency demands that inferential chains be short and explicit. Finally, again on the positive side, the continuous historical character of the archaeological record ameliorates the problem of establishing independent origins of adaptive traits and extends the potential for comparison beyond adaptive traits, to include the direction and pace of the adaptive trends themselves.

We do not believe that comparative methods should be abandoned because of past associations with approaches now considered untenable. Instead, their utility for evolutionary archaeology should be judged by whether or not they improve our ability to understand the archaeological record as a product of selective retention of cultural variation. The example presented in the final section of this paper provides one piece of evidence pertinent to such a judgment: we believe our comparative observations establish the importance of environmental predictability as a source of selective pressures that shape patterns of local specialization and movement of materials through space; by so doing, they help establish the viability of comparative methods within evolutionary archaeology.

ACKNOWLEDGMENTS

We are extremely grateful to R. Lee Lyman and Michael J. O'Brien, both of whom read and commented extensively on an earlier draft of this paper. Their astute criticisms helped us clarify our thinking on several key issues. We hasten to add that we alone are responsible for any remaining blunders or misinterpretations. We also thank C. Michael Barton and Geoffrey A. Clark for encouraging us to contribute a paper to this volume.

END NOTES

[1] The Panglossian paradigm appears also to have been recognized and named earlier in anthropology than it was in evolutionary biology. Harris (1997: 414) has noted recently that his own use of "Panglossian" to characterize naïve functionalist arguments in anthropology dates back to 1967 (Harris 1967), which is long before it was used by Gould and Lewontin in reference to adaptationist research in biology.

[2] Experimentation can, however, play an important role in evolutionary investigation. For instance, as discussed later, engineering experiments can be used to identify the optimal design for organisms or artifacts (e.g., O'Brien & Holland 1992, O'Brien et al. 1994), so that features that were optimized by natural selection may be inferred. The point we are making here is that evolutionary archaeology and evolu-

tionary biology are not experimental sciences like physics or chemistry; while chemical and physical processes can be recreated over and over again through experimentation, recreating the historical process of differential persistence of cultural traits that generated a particular configuration of archaeological data is no more possible than recreating the evolutionary process that generated a horse or a redwood tree. In all these cases, knowledge about the historical process comes from observation and comparison of the historical phenomena themselves (perhaps supported in part by experimentally- grounded inferences).

[3] This is not to suggest that cultural transmission is something that should be ignored by archaeologists. Neff (1993) has argued that an evolutionary approach to the archaeological record demands attention to both transmission and selection. Further, as argued by Neff (1993) and later by Tschauner (1994), culture-historical typologies represent hypotheses regarding lines of cultural inheritance and common descent, and these hypotheses are amenable to testing with further analysis.

[4] On methodological grounds, Gans (1993: 403) prefers not to define "adaptation" in a way that requires inference of selection in the past. If adaptation is defined instead to refer simply to traits that currently improve the fitness of an organism, then the concept becomes readily testable. The emphasis in Gans definition is on adequacy with respect to currently observable environmental variation. We thank R. Lee Lyman for bringing Gans' article to our attention.

[5] Williams (1992: 39), a biologist, has a more positive view of optimal foraging theory accounts of human behavior. Later on, we point out what we believe to be a more serious problem with the use of the evolutionary ecology framework, namely that its key variables tend not to be observable in the archaeological record.

[6] As Neff (1993) has argued, "types" are empirical units (groups) that in most cases can be regarded as expressions of an archaeologist's hypotheses about relationships of common descent in some collection of pottery. Tschauner (1994) elaborates a closely similar line of reasoning with respect to the goals of archaeological classification (see footnote 3).

[7] The experimental studies on which O'Brien and Holland's discussions are based as well as the detailed arguments about how selection drove the shift to shell-tempering are reported by Dunnell and Feathers (1991) and Feathers (1990).

[8] Provenance determination affords a means for testing the hypothesis implied when a group of sherds is given a type name (Neff 1993, Neff et al. 1997). The hypothesis is that the formal/decorative attributes represented prominently in that type (group of sherds) are a reflection of decorative conventions shared by potters within a restricted region during a restricted time interval in the past. If the group of sherds is found to be compositionally uniform, then the hypothesis cannot be rejected.

REFERENCES CITED

Arnold, Dean E.

1975 Ceramic ecology of the Ayacucho Basin, Peru: implications for prehistory. *Current Anthropology* 16: 183-205.

1978 Ceramic variability, environment, and culture history among the Pokom in the Valley of Guatemala. In *Spatial Organisation of Culture* (I. Hodder, ed.), pp. 39 - 60. Pittsburgh: University of Pittsburgh.

Beardsley, Richard K., Preston Holder, Alex D. Krieger, Betty J. Meggers, John B. Rinaldo, & Paul Kutsche

1955 Functional and evolutionary implications of community patterning. In *Seminars in Archaeology: 1955*. Memoirs of the Society for American Archaeology, No. 11 (R. Wauchope, ed.), pp. 129-157. Salt Lake City.

Bettinger, Robert L.

1991 *Hunter-Gatherers: Archaeological and Evolutionary Theory*. New York: Plenum Press.

Bettinger, Robert L., Robert Boyd, & Peter J. Richerson

1996 Style, function, and cultural evolutionary processes. In *Darwinian Archaeologies* (H. D. G. Maschner, ed.), pp. 133-164. New York: Plenum Press.

Binford, Lewis R.

1962 Archaeology as anthropology. *American Antiquity* 28: 217-225.

1964 A consideration of archaeological research design. *American Antiquity* 28: 425-441.

1965 Archaeological systematics and the study of culture process. *American Antiquity* 31: 203-210.

Boyd, Richard & Peter J. Richerson

1985 *Culture and the Evolutionary Process*. Chicago: University of Chicago Press.

Braun, David P.

1983 Pots as tools. In *Archaeological Hammers and Theories* (J. Moore & A. Keene, eds.), pp. 107-134. New York: Academic Press.

1987 Coevolution of sedentism, pottery technology, and horticulture in the Central Midwest, 200 B.C. -A.D. 600. In *Emergent Horticultural Economies of the Eastern Woodlands*, Occasional Paper no. 7 (W. F. Keegan, ed.), pp. 153-181.. Carbondale: Center for Archaeological Investigations, Southern Illinois University at Carbondale.

Braun, David P. & Stephen Plog

1982 The evolution of "tribal" social networks: theory and prehistoric North American evidence. *American Antiquity* 47: 504-525.

Cackler, Paul & Hector Neff

n.d. Coevolution: explaining the course of Maya history. Ms. available from the authors.

Chimalpahin Quauhtlehuanitzin, Diego Francisco de San Antón Muñón

1965 *Relaciones Originales de Chalco Amequemeca Escritas por Don Francisco San Antón Muñón Chimalpahin Quauhtlehuanitzin*. Translated by Sylvia Rendón. Mexico City: Fondo de Cultura Económica.

Crow, James F. & Motoo Kimura

1970 *An Introduction to Population Genetics Theory*. New York: Harper & Row.

Dawkins, Richard

1976 *The Selfish Gene*. Oxford: Oxford University Press.

Deutchman, Haree L.

1980 Chemical evidence of ceramic exchange on Black Mesa. In *Models and Methods in Regional Exchange. SAA Papers No. 1* (R. E. Fry, ed.), pp. 119-134. Washington, D.C.: Society for American Archaeology.

Dunnell, Robert C.

1971 *Systematics in Prehistory*. New York: Free Press.

1978a Style and function: a fundamental dichotomy. *American Antiquity* 43: 192-202.

1978b Archaeological potential of anthropological and scientific models of function. In *Archaeological Essays in Honor of Irving B. Rouse* (R. C. Dunnell & E. S. Hall, Jr., eds.), pp. 41-73. Mouton: The Hague.

1980 Evolutionary theory and archaeology. In *Advances in Archaeological Method and Theory*, vol. 3 (M. B. Schiffer, ed.), pp. 35-99. New York: Academic Press.

1982 Science, social science, and common sense: the agonizing dilemma of modern archaeology. *Journal of Anthropological Research* 38: 1-26

1986 Methodological issues in Americanist artifact classification. In *Advances in Archaeological Method and Theory*, vol. 9 (M. B. Schiffer, ed.), pp. 149-207. New York: Academic Press

1987 Comment on "History, phylogeny, and evolution in Polynesia," by Patrick V. Kirch & Roger C. Green. *Current Anthropology* 28: 431-456.

1989 Aspects of the application of evolutionary theory in archaeology. In *Archaeological Thought in America* (C. C. Lamberg-Karlovsky, ed.), pp. 35-49. Cambridge: Cambridge University Press.

1992 Archaeology and evolutionary science. In *Quandaries and Quests: Visions of Archaeology's Future* (L. Wandsnider, ed.), pp. 209- 224. Occasional Paper no. 20. Carbondale: Center for Archaeological Investigations, Southern Illinois University at Carbondale.

1995 What is it that actually evolves? In *Evolutionary Archaeology: Methodological Issues* (P. Teltser, ed.), pp. 33 - 50. Tucson: University of Arizona Press.

Dunnell, Robert C. & James K. Feathers
1991 Late Woodland manifestations of the Malden Plain, Southeast Missouri. In *Stability, Transformation, and Variation: the Late Woodland Southeast* (M. S. Nassaney & C. R. Cobb, eds.), pp. 21-45. New York: Plenum Press.

Feathers, James K.
1989 Effects of temper on strength of ceramics: response to Bronitsky and Hamer. *American Antiquity* 54: 579-588.

1990 *Explaining the Evolution of Prehistoric Ceramics in Southeastern Missouri.* Unpublished Ph.D. dissertation, University of Washington, Seattle.

Gans, Carl
1993 On the merits of adequacy. *American Journal of Science*: 293-A: 391-406.

Glowacki, Donna M.
1995 *Patterns of Ceramic Production and Vessel Movement in the Mesa Verde Region.* Unpublished M.A. Thesis, University of Missouri, Columbia.

Glowacki, Donna M., Hector Neff, & Michael D. Glascock
1995 Characterization of Mesa Verde Black-on-White ceramics from Southwestern Colorado using NAA. *Journal of Radioanalytical and Nuclear Chemistry* 196: 215-222.

Gould, Stephen J.
1986 Evolution and the triumph of homology, or why history matters. *American Scientist* 74: 60-69.

Gould, Stephen J. & Richard Lewontin
1979 The spandrels of San Marco and the panglossian paradigm: a critique of the adaptationist programme. *Proceedings of the Royal Society of London* 205: 581-598.

Gould, Stephen J. & Elizabeth S. Vrba
1982 Exaptation–a missing term in the science of form. *Paleobiology* 8: 4-15.

Harris, Marvin
1967 Reply to John Bennett. *Current Anthropology* 8: 252-253.

1997 Comment on "Causation and the struggle for a science of culture" by Tim O'Meara. *Current Anthropology* 38: 410-415.

Harvey, Paul H. & Mark D. Pagel
1991 *The Comparative Method in Evolutionary Biology*. Oxford: Oxford University Press.

Hawkes, Kristin & James F. O'Connell
1992 On optimal foraging models and subsistence transitions. *Current Anthropology* 33: 63-65

Hodge, Mary G.
n.d. The cultural sequence at Mound 65 in Chalco. In *Place of Jade: Society and Economy in Ancient Chalco* (Mary G. Hodge). University of Pittsburgh, Department of Anthropology Monographs in Latin American Archaeology (submitted).

Hull, David L.
1980 Individuality and selection. *Annual Review of Ecology and Systematics* 11: 311-332.

1988 *Science as a Process: An Evolutionary Account of the Social and Conceptual Development of Science.* Chicago: The University of Chicago Press.

Kirch, Patrick V.
1980 The archaeological study of adaptation: theoretical and methodological issues. In *Advances in Archaeological Method and Theory*, vol. 3 (M. B. Schiffer, ed.), pp. 101-156. New York: Academic Press.

Kirch, Patrick V. & Roger C. Green
1987 History, phylogeny, and evolution in Polynesia. *Current Anthropology* 28: 431-456.

Kojo, Yasushi
1996 Production of prehistoric southwestern ceramics: a low-technology approach. *American Antiquity* 61: 325-339.

Larson, Daniel O.
1994 California climatic reconstructions. *Journal of Interdisciplinary History* 25: 225-253.

Larson, Daniel O. & Joel Michaelson
1990 Impacts of climatic variability and population growth on Virgin Branch Anasazi cultural developments. *American Antiquity* 55: 227-249.

Larson, Daniel O., Hector Neff, Donald A. Graybill, Joel Michaelson, & Elizabeth Ambrose
1996 Risk, climatic variability, and the study of southwestern prehistory: an evolutionary perspective. *American Antiquity* 61: 217-241.

Leonard, Robert D.
1989 Resource specialization, population growth, and agricultural production in the American Southwest. *American Antiquity* 54: 491-503.

Leonard, Robert D. & George T. Jones
1987 Elements of an inclusive evolutionary model for archaeology. *Journal of Anthropological Archaeology* 6: 199-219.

Leonard, Robert D. & Heidi E. Reed
1993 Population aggregation in the prehistoric American Southwest: a selectionist model. *American Antiquity* 58: 648-661.

Lewontin, Richard C.
1977 Sociobiology: a caricature of Darwinism. *Proceedings of the 1976 Biennial Meeting of the Philosophy of Science Association*, vol. 2. F. Suppé & P. D. Asquith, eds. pp. 22-31.

1979 Sociobiology as an adaptationist program. *Behavioral Science* 24: 5-14.

Lipo, Carl, Mark Madsen, & Tim Hunt
n.d. Artifact style dynamics: cultural transmission, population structure, and frequency seriation. *Journal of Anthropological Archaeology* (in press).

Lyman, R. Lee & Michael J. O'Brien
n.d. The Concept of evolution in early Twentieth-Century Americanist archaeology. (Chapter 2 in current volume)

Mace, Ruth & Mark D. Pagel
1994 The comparative method in anthropology *Current Anthropology* 35: 549-564.

Maynard Smith, John
1994 [1978] Optimization theory in evolution. In *Conceptual Issues in Evolutionary Biology*. E. Sober, ed. pp. 91-118. Cambridge, MA: the MIT Press (originally published in *Annual Review of Ecology and Systematics* 9: 31-56 [1978]).

Mayr, Ernst
1982 *The Growth of Biological Thought*. Cambridge, MA: Harvard University Press.

1983 How to carry out the adaptationist program? *The American Naturalist* 121: 324-334.

1988 *Toward a New Philosophy of Biology: Observations of an Evolutionist*. Cambridge: Harvard University Press.

Meggers, Betty J.
1983 Explaining the course of human events. In *How Humans Adapt: a Biocultural Odyssey* (D. J. Ortner, ed.), pp. 163-183. Washington, D.C.: Smithsonian Institution Press.

Meggers, Betty J., Clifford Evans, & E. Estrada
1965 *Early Formative Period of Coastal Ecuador: the Valdivia and Machalilla Phases*. Smithsonian Contributions to Anthropology, Vol. I. Washington, D.C.: Smithsonian Institution Press.

Mills, Susan K. & John H. Beatty
1979 The propensity interpretation of fitness. *Philosophy of Science* 46: 263-286.

Naroll, Raoul
1961 Two solutions to Galton's problem. *Philosophy of Science* 28:15-39.

Mosiño Alemán, Pedro A. & Enriqueta García
1974 The climate of Mexico. In *Climates of North America* (R. A. Bryson & F. Kenneth Hare, eds.), pp. 345 - 404. World Survey of Climatology Volume 11. New York: Elsevier Scientific Publishing Co.

Neff, Hector
1989 The effect of interregional distribution on Plumbate pottery production. In *Ancient Trade and Tribute: Economies of the Soconusco Region of Mesoamerica* (B. Voorhies, ed.), pp. 249-267. Provo: University of Utah Press.

1990 Culture contact and ceramic evolution: examples from Mesoamerica. *Ceramics and Civilization V: the Changing Roles of Ceramics in Society* (W. D. Kingery, ed.), pp. 159-182. Westerville, OH: American Ceramic Society, Inc.

1992 Ceramics and evolution. In *Archaeological Method and Theory*, vol. 4. M. B. Schiffer, ed. pp. 141-193. Tucson: University of Arizona Press.

1993 Theory, sampling, and technical analysis in archaeological pottery analysis. *American Antiquity* 58: 23-44

1995a A role for "sourcing" in evolutionary archaeology. In *Evolutionary Archaeology: Methodological Issues* (P. Teltser, ed.), pp. 69-112. Tucson: University of Arizona Press.

1995b The development of Plumbate ceramic ware in southern Mesoamerica. *JOM* 47(3): 52-56.

n.d. Units in chemistry-based provenance investigations of ceramics. In *Measuring Time, Space, and Material: Unit Issues in Archaeology* (A. F. Ramenofsky & A. Steffen). Salt Lake City: University of Utah Press (in press).

Neff, Hector & Barbara Arroyo
1996 The evolution of early formative ceramic traditions in Pacific coastal southern Mesoamerica. Presented in the symposium "Technology and Innovation in Social Context: the Development and Change of Pottery-Making Traditions," 1996 Chacmool Conference, November, Calgary.

Neff, Hector, Frederick J. Bove, & Barbara Arroyo
1995 Evolution of ceramic resource procurement on the Guatemalan central Pacific coast. *Applications of Evolutionary Theory in Archaeology* (R. C. Dunnell & R. D. Leonard). (in preparation; ms. available from the senior author.)

Neff, Hector, Frederick J. Bove, Eugenia Robinson, & Barbara Arroyo
1994 A ceramic compositional perspective on the formative to classic transition in southern Mesoamerica. *Latin American Antiquity* 5: 333-358.

Neff, Hector, & Mary G. Hodge
n.d. Ceramic production at Chalco, A.D. 500 - 1500: INAA of decorated serving dishes and raw clays. In *Place of Jade: Society and Economy in Ancient Chalco* (M. G. Hodge, ed.). University of Pittsburgh Memoirs in Latin American Archaeology (in preparation).

Neff, Hector & Daniel O. Larson
n.d. Exchange and other human activities that move ceramics through space: an evolutionary perspective. In *Doing Darwinian Archaeology* (P. T. McCutcheon & J. L Lanata, eds.). (in preparation; ms. available from the author).

Neff, Hector, Daniel O. Larson, & Michael D. Glascock
1997 Evolutionary implications of formal and chemical diversity in Anasazi ceramics from southern Utah. *Journal of Field Archaeology* (in press).

Neiman, Fraser
1995 Stylistic variation in evolutionary perspective: inferences from decorative diversity and interassemblage distance in Illinois Woodland ceramic assemblages. *American Antiquity* 60: 7-36.

O'Connell, James F.
1987 Alyawara site structure and its archaeological implications. *American Antiquity* 52: 74-108.

1995 Ethnoarchaeology needs a general theory of behavior. *Journal of Archaeological Research* 3. 205-255.

O'Brien, Michael J.
1996 The historical development of an evolutionary archaeology. In *Darwinian Archaeologies* (Herbert D. G. Maschner, ed.), pp. 17-32. New York: Plenum Press.

O'Brien, Michael J. & Robert J. Hoard,
1996 Ceramic vessels. In *Middle and Late Woodland Subsistence and Ceramic Technology in the Central Mississippi River Valley: Selected Studies from the Burkemper Site, Lincoln County, Missouri. Illinois State Museum, Reports of Investigations* No. 52. By Michael J. O'Brien, pp. 91-144.

O'Brien, Michael J., & Thomas D. Holland.
1990 Variation, selection, and the archaeological record. In *Archaeological Method and Theory*, vol. 2 (M. B. Schiffer, ed.), pp. 31-79. Tucson: University of Arizona Press.

1992 The role of adaptation in archaeological explanation. *American Antiquity* 57: 36-59.

1995a The nature and premise of selection-based archaeology. In *Evolutionary Archaeology: Methodological Issues* (P. A. Teltser, ed.), pp. 175-200. Tucson: University of Arizona Press.

1995b Behavioral archaeology and the extended phenotype. In *Expanding Archaeology* (J. M. Skibo, W. H. Walker, & A. E. Nielsen, eds.), pp. 143-161. Salt Lake City: University of Utah Press.

O'Brien, Michael J., Thomas D. Holland, Robert J. Hoard, & Gregory L. Fox
1994 Evolutionary implications of design and performance characteristics of prehistoric pottery. *Journal of Archaeological Method and Theory* 1: 259-304.

Plog, Fred T.
1983 Political and economic alliances on the Colorado Plateaus, A.D. 400 - 1450. *Advances in World Archaeology* 2. New York: Academic Press, 289-330.
1984 Exchange, tribes, and alliances: the northern Southwest. *American Archaeology* 4: 217-223.

Reid, J. Jefferson, Michael B. Schiffer, & William L. Rathje
1975 Behavioral archaeology: four strategies. *American Anthropologist* 77: 864-869.

Rindos, David
1980 Symbiosis, instability, and the origins and spread of agriculture: a new model. *Current Anthropology* 21: 751-772.
1984 *The Origins of Agriculture: An Evolutionary Approach*. New York: Academic Press.
1989 Undirected variation and the Darwinian explanation of cultural change. In *Archaeological Method and Theory*, vol. 1 (M. B. Schiffer, ed.), pp. 1-45. Tucson: University of Arizona Press.

Sahlins, Marshal D. & Elman R. Service (editors)
1960 *Evolution and Culture.* Ann Arbor: University of Michigan Press.

Schiffer, Michael B.
1996 Some relationships between behavioral and evolutionary archaeologies. *American Antiquity* 61: 643-662.

Schiffer, Michael B. & James M. Skibo
1987 Theory and experiment in the study of technological change. *Current Anthropology* 28: 595-622.
1997 The explanation of artifact variability. *American Antiquity* 62: 27-50.

Smith, Eric A.
1983 Anthropological applications of optimal foraging theory: a critical review. *Current Anthropology* 24: 625-651.

Smith, Eric A. & Bruce Winterhaler
1992 Natural selection and decision-making: some fundamental principles. In *Evolutionary Ecology and Human Behavior* (E. A. Smith & B. Winterhalder, eds.), pp. 25 - 60. New York: Aldine de Gruyter.

Sober, Elliot
1984 *The Nature of Selection: Evolutionary Theory in Philosophical Focus.* Cambridge: MIT Press.
1993 *Philosophy of Biology.* Boulder, CO: Westview Press.

Stark, Miriam
1991 Ceramic production and community specialization: a Kalinga ethnoarchaeological study. *World Archaeology* 23: 64-77.

Steward, Julian
1955 *Theory of Culture Change: the Methodology of Multilinear Evolution.* Urbana: University of Illinois Press.

Tolstoy, Paul
1975 Settlement and population trends in the Basin of Mexico (Ixtapaluca and Zacatenco phases). *Journal of Field Archaeology* 2: 331-349.

Triadan, Daniela
1994 *White Mountain Redware: Expensive Trade Goods or Local Commodity? A Study of the Production, Distribution and Function of White Mountain Redware During the 14th Century in the Grasshopper Region, East-Central Arizona.* Unpublished Ph.D. Dissertation, Frei Universität, Berlin.

Trigger, Bruce
1989 *A History of Archaeological Thought.* Cambridge: Cambridge University Press.

Tschauner, Hartmut
1994 Archaeological systematics and cultural evolution: retrieving the honour of culture history. *Man* (n.s.) 29: 77-93.

Vivó Escoto, Jorge A.
1964 Weather and climate of Mexico and Central America. *Handbook of Middle American Indians,* vol. 1. (R. C. West, ed.), pp. 187-215. Austin: University of Texas Press.

Walker, William H., James M. Skibo, & Axel E. Nielsen
1995 Introduction: expanding archaeology. In *Expanding Archaeology* (James M. Skibo, William H. Walker, & Axel E. Nielsen, eds.), pp. 1-12. Salt Lake City: University of Utah Press.

Weigand, Phil C., Garman Harbottle, & Edward V. Sayre
1977 Turquoise sources and source analysis: Mesoamerica and the southwestern U.S.A. In *Exchange Systems in Prehistory* (T. K. Earle & J. E. Ericson, eds.), pp. 15 - 34. New York: Academic Press.

White, Leslie A.
1949 *The Science of Culture: a Study of Man and Civilization.* New York: Grove Press.

Willey, Gordon R., Charles C. Di Peso, William A. Ritchie, Irving Rouse, John H. Rowe, & Donald W. Lathrap
1955 An archaeological classification of culture contact situations. In *Seminars in Archaeology: 1955.* Memoirs of the Society for American Archaeology, No. 11 (R. Wauchope, ed.), pp. 1-30. Salt Lake City.

Willey, Gordon R. & Philip Phillips
1958 *Method and Theory in American Archaeology.* Chicago: University of Chicago Press.

Williams, George C.
1992 *Natural Selection: Domains, Levels, and Challenges.* New York: Oxford University Press.

Winterhalder, Bruce
1990 Open field, common pot: harvest variability and risk avoidance in agricultural and foraging societies. In *Risk and Uncertainty in Tribal and Peasant Economies* (E. Cashdan, ed.), pp. 67-87. Boulder: Westview Press.

Zedeño, Maria Nieves
1994 *Sourcing Prehistoric Ceramics at Chodistaas Pueblo, Arizona: the Circulation of People and Pots in the Grasshopper Region.* Anthropological Papers of the University of Arizona, no. 58. Tucson: University of Arizona Press.

6

Content as Variability, Result as Selection: Toward a Behavioral Definition of Technology

Peter Bleed
University of Nebraska, Lincoln

ABSTRACT

Technology has long been of interest to anthropologists but anthropological study of the topic has lagged behind consideration of other human institutions. In part this is because technology has remained poorly understood and incompletely defined and because technology's material and non-material components have been either confused or treated in isolation. To clarify the relationship between its material and non-material aspects, this paper defines the contents and results of technology in behavioral terms. Material results are an intrinsic part of technology that offer immediate, clear feedback. For that reason, technological behaviors are easy to evaluate and adjust. This makes technology especially subject to selective pressure and evolutionary change.

Studies of primitive technology probably rank among the oldest and least interesting of all anthropological work (Hames and Hames 1976: 3).

INTRODUCTION

A common campus stereotype holds that anthropologists have interesting offices. This may not be the primary image our discipline would want, but as a stereotype, it is remarkably apt. Anthropologists regularly do decorate their offices with a nice basket, a hand woven fabric, an old bow, or some other bit of ethnographica. The reality of the stereotype carries some irony, however, because material culture and technology have become such minor concerns to the modern discipline that most anthropologists can ignore them, except when they outfit their work space! Concern for the things people make remains a coarse bit of anthropology's historic baggage, an heirloom in our disciplinary attic with little utility but too much heritage to simply jettison.

There are signs this situation may be changing. Over the past decade, at least a few anthropologists have presented optimistic assessments about the potential of anthropological study of technology (Appadurai 1986, Conkey 1989). There has even been a suggestion that anthropological study of technology "is poised for a comeback" (Pfaffenberger 1992: 491). And a new journal has recently appeared to "cater to the increasing interest in material culture studies" (see *Journal of Material Culture* aim and scope statement).

Many of these optimistic assessments of a renewed anthropological interest in technology respond to interests in material culture and technology that have developed outside of anthropology in fields that have traditionally emphasized social and intellectual institutions. A number of historians (Basalla 1988, Schlereth 1980, Quimby 1978, Nye 1994, Strasser 1982) and philosophers (Ellul 1964, Gendron 1977) as well as political scientists (Winner 1980, 1986), psychologists (Norman 1988), engineers (Petroski 1992), designers (Pye 1964, 1968), and architects (Deasy 1985) have turned their attention to either material culture or technology. This broad interest should encourage anthropologists to reconsider the topic since the search for regularities and inter-connections that marks much of the non-anthropological interest in technology, as well as much of the vocabulary it uses–evolution, selection, adaptation, and, indeed, material culture–are familiar to our discipline.

Beyond all of those developments, progress toward a modern anthropological interest in technology is also to be seen in a number of recent discussions by anthropological archeologists who have aimed at developing theoretically robust means of explaining variability and change in material systems. Some of these studies are identified with the "behavioral" research program, which builds on the work of Schiffer (e.g., 1976 *et seq.*) and his colleagues (Schiffer & Skibo 1996). Others use explicit evolutionary perspectives that were introduced to archeology by Dunnell (1978, 1980). Evolutionary approaches to the study of material culture see material culture as central to the human phenotype and, therefore, subject to pressure from either "selection" or "adaptation" (Braun 1990, Leonard & Jones 1987, Leonard & Reed 1993, O'Brien & Holland 1990, 1992, 1995). By arguing that "replicative success" of artifacts–of the sort documented by long-term tends in the archeological record–must mean that technological systems have undergone evolutionary selection, selectionist archeologists have focused renewed attention on the study of technology and material culture.

Although marked by theoretical differences (see O'Brien & Holland 1995, Schiffer 1996a), behavioral and evolutionary perspectives on technology have actively pursued inquiry into basic causes of artifact variability. They have both emphasized the behavioral nature of technology. Both have also advocated development of theoretical means of explaining material variability and change in terms of ideas that have proved compelling in biology and other realms. Not surprisingly, then, there is clear evidence that these archeological studies of material culture are coalescing (see Schiffer 1996; Skibo, Walker and Nielsen 1995) in a way that may give scientific apparatus to the study of human material systems.

This paper seeks to contribute to both evolutionary and behavioral studies of material culture by looking at some historical roots of anthropological interest in material culture and presenting a behavioral definition of technology. The primary goal of this definitional approach is to foster specificity–if not precision–to the study of technology and material culture. Evolutionary considerations of technology have tended to consider the topic from a coarse scale. Simple documentation of long-term trends that appear directional are not the same as pointing to specific technological variables that can be shown to have been selected by agents of the human environment (Abbott, Leonard & Jones 1995). The approach to technology offered here may clarify the variability that exists within technology and in that way show the potential for evolutionary change within material systems. Finally, this approach points to specific, substantive overlaps between the behavioral and adaptationist approaches to the study of material culture.

Descriptions of exotic artifacts and primitive tools were among the first contributions to the field that became anthropology. Thoughtful appreciations of material culture and its cultural setting further helped to crystallize the discipline and established a close link between it and the museums of the world. If the content of the major anthropology texts reflects the values of the discipline, familiarity with material culture remained a fundamental part of American anthropology up through the early 1960's (e.g., Beals & Hoijer 1953, Hoebel, 1958). Even before that time, however, the study of the social, ecological, and symbolic structures of human existence had moved squarely to the center of modern anthropology and theoretical study of the production and use of artifacts had languished.

The disappearance of technologically unacculturated societies and anthropology's migration from museums to universities certainly were factors in the shift away from the systematic study of human constructions. More importantly, studies by such early anthropological students of material culture as Mason (1895), Balfour (1893), Sayce (1933), and Pitt-Rivers (Lane Fox 1906) did not lead the discipline to more powerful investigations of humankind. Studies of social and intellectual institutions flourished, on the other hand, as frameworks were developed to show functional, evolutionary, ecological and cognitive interrelationships. Current anthropological theoreticians rarely ignore technology entirely, although, in practice for the past half-century or more, they have been able to ignore the hardware of human existence. To substantiate that assertion one need only consider the content of modern cultural anthropology texts. With one exception (Oswalt 1988), all of the current texts ignore material culture and treat technology, if at all, only in terms of subsistence patterns. Cotterall and Kamminga (1990: 6) summarize the situation with a bleak assessment, "Today there is no robust theory of evolution of material culture."

DEFINING TECHNOLOGY

In part, the problems modern anthropologists face in studying technology reflect basic difficulties the social sciences have had in addressing the material world (Oswalt 1976: 4-13). Early anthropological discussions (e.g., Nelson 1932, Wissler 1926) tended to treat technology as essentially synonymous with things, the study of which has invariably been considered less worthy than discussion of social and intellectual activities. Defining technology in this way had the double negative effects of offering only an imprecise assessment of the topic while stigmatizing it as "antiquarian."

Boas and his students were not particularly interested in technology, but they brought progress to the study of technology in two ways. First, their holistic interests led to the view that "the study of technology is essential for an understanding of culture" (Herskovits 1948: 41). The detailed ethnographic reports they produced remain an invaluable resource for the study of technological diversity. As a second contribution to the study of technology, Boas's followers recognized and asserted a clear distinction between *technology*–customs of manipulating the physical world–and *material culture*–the products resulting from those customs.

The distinction between technology and material culture has been retained in the popular anthropology texts published after the 1930's (Herskovits 1948, Kroeber 1948) so that it has become orthodox and, by now, even time-honored. Researchers outside of anthropology (Schlereth 1980) have also taken up the terminology although it has not necessarily clarified the relationships between material and non-material aspects of technology.

Americanist archeologists have embraced the Boasian distinction between technology and material culture although the single most commonly cited source of insight on the matter has been Leslie White (1949) who offered a cogent defense of the importance of technology that encouraged both anthropologists and archeologists interested in material issues. White's definitions and discussion of technology were, however, markedly abstract and tended to confuse the behavior of technology with its material products. To White, technology was "composed of the material, mechanical, physical, and chemical instruments, together with the techniques of their use, by means of which man, as an animal species, is articulated with the natural habitat" (White 1949: 364). Such statements made technology the "hero of the piece" (White 1949: 390) and allowed White to address the philosophical importance of dealing with the physical world and capturing energy. Aside from encouragement, though, White's definition offered little practical guidance to anthropologists wishing to study the operation and composition of technological systems. Perhaps for that reason, study of technology has only recently begun to experience conceptual refinements (e.g., Schiffer & Skibo 1987) comparable to those that let social anthropologists recognize within social structure, kinship organization with descent systems marked by distinctive qualities like cousin terminology.

Since technological change is central to the process of acculturation, a number of anthropologists interested in culture change have addressed technology (see Spicer 1952, Bernard & Pelto 1987). By showing the pivotal role technology has within cultural systems, these studies encouraged anthropologists to remain interested in technological issues. In general, however, since the hardware of the modern technologies has appeared to be outside the scope of anthropological attention, culture change studies have focused on the social implications of technological innovation. For that reason, they have not encouraged synthetic consideration of technology per se.

Americanist archeologists who might have addressed technology were drawn instead in the direction of rigorous description of artifacts, primarily through the definition of types and attributes. Initially, this work established descriptive comparability and communication between researchers and proved useful for the recognition of culture historical markers (Dunnell 1982). Once they had been established, creative investigations also made the descriptive categories interesting in other regards. Distribution studies showed that they could be interpreted as reflections of the kinds of patterns described by social anthropologists (e.g., Longacre 1968). Replication studies (e.g., Flenniken 1981) and microwear analyses (Keeley 1980) let type and attribute variations be interpreted in functional terms and linked to other kinds of economic and ecological patterns. This research was very stimulating to the field, but tended to draw archeologists in the direction of either ecological or social issues. It did not encourage conceptual study of technology itself.

A leader of recent American anthropological interest in technology and material culture has been Wendell Oswalt (Oswalt 1973, 1976; Lustig-Arecco 1975). His most basic assertion is that artifacts are the empirical "facts" of technology that can convey the otherwise immaterial ideas of their creators. Beyond that, Oswalt has proposed an approach to the analysis of material culture that is at least somewhat subjective under the best condition and certainly most suited to consideration of complete items. It is not easily applied to archaeological data. Thus, although Oswalt's approach as been generally stimulating to those interested in technology and useful for the consideration of some general issues of technology (Shott 1986, Torrence 1983; see also Bamforth & Bleed, Chapter 7), it has not inspired the general study of technology.

As a context for the description of some primitive technologies, Robert Spier (1970) offered a very concrete approach to the conceptualization of technology. Spier described technology in terms of its "inputs" and "outputs." In Spier's model, the inputs of technology are *knowledge, resources*, and *labor*. In comparably straightforward terms, he lists as the outputs of technology *material culture* and *environmental modifications*. Spier's goal was descriptive rather than theoretical and his treatment was cursory so he did not give concrete amplification to the variables he listed and did not address the relationship between technology's "inputs" and "outputs." The input-output model affords consideration of the specific variables that compose technology. This kind of model also invites consideration of the dynamic relations between various aspects of technology. Those are certainly the reasons why similar conceptualizations of technology were developed by researchers concerned with fostering technological innovation (NSF 1983).

TECHNOLOGY AS BEHAVIOR

Beyond being explicit and straightforward, the subdivisions Spier identifies within technology are useful because they can be linked to specific behavioral variables. In that regard, it parallels elements of behavioral archeology strategy. Schiffer (1976 *et seq.*) has repeatedly shown that giving concrete behavioral reality to otherwise diffuse topics is an effective way to begin understanding subjects that may be initially obscure. The behavioral approach has led to many analyses and understandings that have become central to modern archeology: site formation processes, artifact life histories, and use-phase analyses and more. In all of these cases, a major utility of the behavioral perspective is an attempt to make variables concrete and explicit while emphasizing their inter-relationships. As presented in Figure 6.1, the definition of technology presented here replaces Spier's inputs and outputs with behavioral *contents* and *results.*

Since technology is the immediate point of contact between people and the environment, consideration of technology must recognize the limiting and shaping roles of physical and social surroundings. These are, of course, large topics and for present purposes they can be no more than listed. The physical environment is certainly the primary frame within which technology must be analyzed since it shapes the schedule, occurrence, and abundance of natural resources. Beyond those factors, in most situations and for much of human history, technologies have existed within social environments that have influenced the course and rate of technological change. Finally, it is important to understand how the human body–with all of its anatomical characteristics and limitations–constitutes another of the environmental frames within which technology operates (see e.g., Pheasant 1986)

TECHNOLOGICAL RESULTS

Following Spier, the two most apparent outputs or 'results' of technology are *material culture* and *environmental modifications.* As discussed earlier, material culture has now universally come to refer to the *artifacts* or objects made and altered by a society. This category obviously includes a great range of materials. It is probably only worth noting that material culture includes residues of both "primary" customs–which have been consciously taught and learned–and "secondary" customs–or those which have been unconsciously acquired and practiced (Osgood 1951). Thus, both carefully crafted and highly interpreted objects as well as the incidental residues of life fall within this category. Environmental modifications encompass a comparably broad range of features. What is important about these material results is that they are what happens when people *do* technology. This makes them are part of technological behavior and, therefore, among the variables that comprise technology's contents.

Material culture is the most concrete result of technology, but many anthropologists and others have pointed out that there are as well other, less tangible but very important, technological outputs. Technological alternatives, for example, place demands on the nature and size of *work groups.* In the extreme, it has been suggested that different technologies embody alternative social structures and political arrangements (Dobres & Hoffman 1994, Winner 1980). A cultural materialist position familiar to most American archeologists strikes a less stridently deterministic tone by asserting that technology interacts with productive and family organization to give shape to cultural systems (Harris 1978).

However they are viewed, the social implications of technology must be considered 'proximate' results. Technological patterns reflect the facility with which people deal with their environment. Viewed in evolutionary terms, then, technology must also be seen as having long-term, or ultimate, implications for the *adaptive success* of both individuals and social groups (Leonard & Jones 1987, O'Brien & Holland 1990). That realization may be easier to make than it is to operationalize, and the currency in which technological fitness might be measured–energy capture, efficiency, time–remains uncertain.

To summarize this discussion of technological results, two things must be emphasized. First, the results of technology include both concrete and nonmaterial 'things', encompassing everything from the most trivial environmental modifications to a society's ultimate survival. Beyond that, since all of these results are objectively linkable to the behavior of technological activities, they are essentially behavioral variables. They can be approached as concrete realities that are inexorably linked to the variables that can be described as the content of technology.

TECHNOLOGICAL CONTENT

Material results are informed by behavior that constitutes the content of technology. These are the behaviors that people *do* when they make and use material culture. Conceptually this part of technology may be separated from material culture and other 'results.' Practically, it is hard to separate technological activities from the materials that result from them. The utility of sepa-

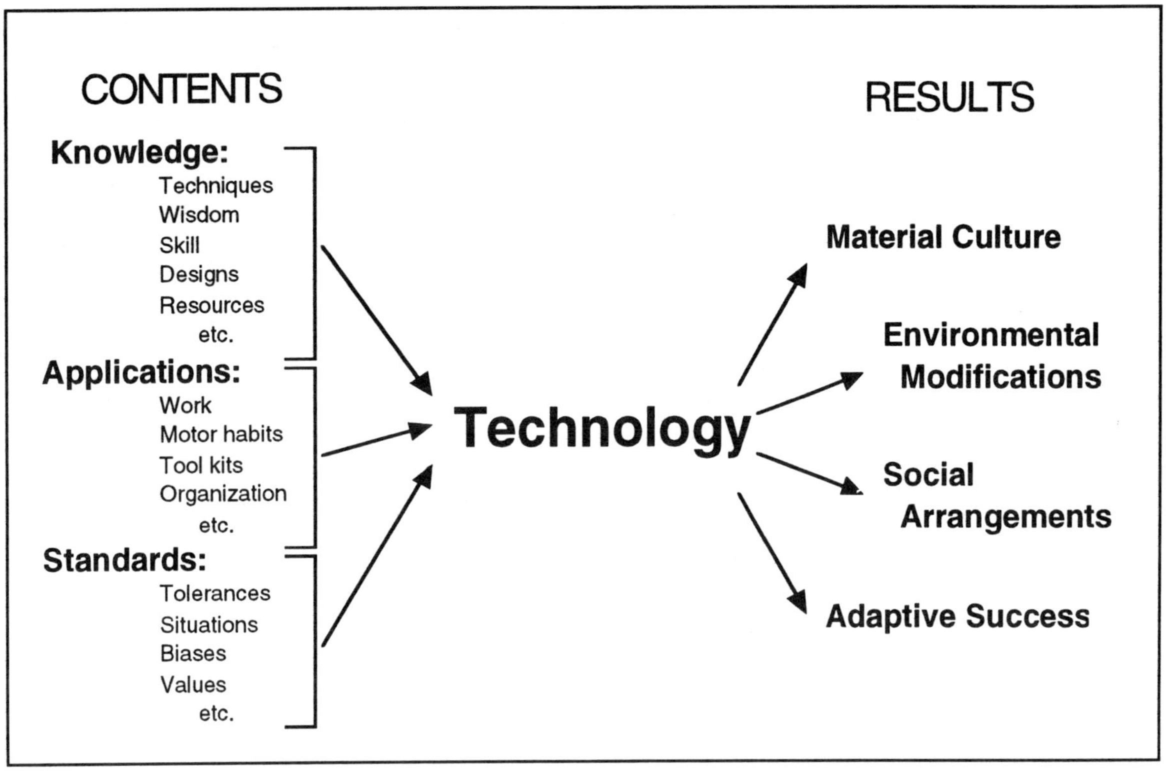

Figure 6.1 Physical, social and biological contexts: contents and results.

rately recognizing technological content is that it helps to assess and recognize the behavioral component of technological results. It lets us, indeed, treat the results of technology in behavioral terms.

As a way of addressing the variability that exists within technology, the content of technology can be conceived in terms of *knowledge, applications*, or *standards*. That is, the variables that compose the behavioral content of technology can be discussed in terms of what people know, how they apply their knowledge, and the standards that guide their technological behavior. As with all custom, content variables may have both etic and emic sides, but they certainly can be addressed in behavioral terms. They can be objectively described since they are observable in action, and, as importantly, in the material results of a given technology. It is also appropriate to point out that, like other aspects of culture, technological content has 'population-level properties' (Boyd & Richerson 1985: 6). The common human institutions of sharing, specialization, and exchange mean that every practitioner need not possess the entire technological content.

Knowledge

Knowledge of how to do things lies at the heart of technology, but clearly technical knowledge is of several different kinds (Schiffer & Skibo 1987: 596-8). At the base, every technology is associated with a vocabulary of *techniques*, or particular methods of manipulating the world. As White says, these may be mechanical, physical, or chemical. Technological knowledge no doubt exists on several levels and is of many different types. Schiffer and Skibo (1987: 597) have begun the process of teasing these levels apart by showing that in extant technologies techniques are informed by bodies of *wisdom* or *lore* and *techno-science* that organize and rationalize technical actions.

Furthermore, not all technical knowledge is discrete and nominal. *Skill* in the execution of technique can, for example, be viewed as a kind of knowledge that is differentially distributed between and within societies and, thus, the source of technological variation. At the same time, it must be viewed as a continuous variable of some complexity.

Tool designs can be considered a type of technological knowledge because workers operate within the constraints of a given assemblage of tools. They are not free to create a limitless range of tools, but instead must select from the array their technology presents them. Defining that array is a necessary step in describing a technology. It is possible because for any given technology, the array is finite. In the same sense, *resources* should be considered as yet another kind of technological knowledge. To be sure, resources are buffered by environment, but from a technological point of view, they reflect what people *know* how to obtain and use.

Applications

The term applications describes how technical knowledge is put into action. *Work* is the most obvious way of doing this, but this term, and indeed the concept, is notoriously hard to define (Erasmus 1955). Whether work is seen as a result or a content of technology, *scheduling* and *organization* of work are appropriate points of technological analysis since they are exactly the points of articulation between the technological and social sectors of a culture. They are not, however, the only technological applications that need to be considered.

Tool use forms a distinctive kind of technical application. Patterns of tool use exist at several levels, beginning with *motor habits*, or routine uses and postures of the body. These appear to be a highly regularized part of every technology (Spier 1970: 26-29) that interacts with functional requirements to produce distinctive tool designs (Bleed & Bleed 1987). The ways tools are assembled into *tool kits* (Kuhn 1994) constitute another level of tool use and another variable of technical application. Since tool applications are flexible and can vary quite independently from tool form, it is as important to monitor the ways tools are actually used as it is to consider their formal diversity.

Following Binford (1980), the ways artifacts are used in relation to environmental conditions has come to be called the *organization* of a technology (Nelson 1991). This is a highly synthetic term that encourages archeologists to consider the multiple interactions of such diverse conditions as resource distributions, seasonality, game abundance, and social patterns. At the base, however, it refers to variability in the domain of technology's applications.

Standards

Standards are the rules of appropriateness that govern technical activities. Cultural relativity has made many anthropologists unwilling to address these rules and there is not even a good vocabulary available to describe them. Part of the latter problem results from the fact that terms that might describe these aspects of an artisan's behavior have become theoretically loaded. Thus, Schiffer and Skibo (1996: 43) say that the notion of "stylistic" choice lacks behavioral referents because this term has become enmeshed in attempts to conceptually separate stylistic and functional variables (But see Barton, chapter 8). Still, these complexities should not keep us from addressing an undeniably important variable of technology. Archeologists associated with post-processual perspectives (Hodder 1986, Lemonnier 1986, 1992) as well as cognitive anthropologists (Keller & Keller 1996) have proposed ways of addressing cognitive rules reflected in technology and shown the potential utility of that kind of approach (see also Wiessner, Chapter 9). Lechtman (1977) has usefully shown that technological "styles" transcending specific techniques can develop to be distinctive features of entire technologies. This research has been stimulating and it is not necessary to adopt a mentalist approach to see that tool production and use are done within the context of intellectual patterns that influence the material results. Thus, people must decide when something is "done" or "good enough." However those decisions are informed, they are observable in behavioral terms both in technological activities and the resulting material expressions. Ethnographic anecdotes and common experience suggest that such *tolerances* are situational so that what is "right" in some cases is not in others. Understanding those *situations* and the range of variation that is encompassed within a technological system seems a worthwhile goal. Experience also tells us that *esthetic biases* and other *values* are a part of all technologies. Recognizing the existence of such factors and even diverse kinds of standards such as elegance–variability in the execution of a job–in no way suggests that these are independent variables. It merely allows them to be recognized as a source of technological variation so that their interaction with social and environmental contexts can be examined.

SELECTION, EVOLUTION, AND THE PROCESS OF TECHNOLOGY

If only because the world exists in a context of time that is continuous and unidirectional, technology is intrinsically processual. Technological activity must begin at some point, move through more or less discrete steps, and come to some kind of an end. Technological activities are hardly unique in this regard since all cultural behavior must be similarly processual in some terms. The processual operation of technology is, however, especially marked and especially important in determining the form and development of technological systems. It is an important starting point of the study of technology.

Viewing technology in processual terms is hardly a new idea. Many archeologists have recognized technology's processual nature and developed models to describe and analyze technological operations. Even simple constructs that differentiate between the *procurement* of raw materials, the *production* of usable tools, and their *use* (Collins 1975), have proved to be very useful in sorting out the diverse range of things that makes up the archeological record. More complex representations of technological processes have been used to understand how technological contents and results change and interact with one another as technological operations proceed (Bleed 1991, Bradley 1982, Lemonnier 1992, Sellet 1994, Schiffer 1976). These representations are useful because they show that technological operations are made up of *activities*, when something is done, and choice points, when *decisions* are made about what to do. These kinds of representations show that technological processes involve dynamic interactions of the variables that make up technology's content and results. Technological activities are most often easier to recognize because they involve directly observable actions. The decisions that link actions together can also be treated as concrete behaviors, observable both in action and in material results. They offer an objective basis for analysis since they are real things we can observe to study the operation of technology.

Beyond that, because technological decisions are bound to material results, they have several distinctive qualities that make them unlike the comparable components of other kinds of cultural behavior. Technological decisions, for example, may be clearer and more tangible than the activities that make up social or ideological processes simply because they involve material resources that present themselves in a limited number of ways. Furthermore, materials–unlike either human or supernatural beings–bring neither memory nor independent motives to their interactions with people.

Technological decisions are also *practical* in that they involve completion of tasks that are intrinsically more concrete than comparable social or ideological activities. This is the case because technological activities are circumscribed by their material results. This makes them markedly concrete and discrete. More importantly, it also means that technological behaviors have greater potential for immediate and direct feedback than do other kinds of cultural behaviors. The material results of technology give technological behaviors what Schiffer and Skibo (1987) refer to as "performance characteristics" that are intrinsically concrete. In most cases, the implications of technological decisions, and the success or failure of technological behaviors are readily apparent. At all levels–from the most specific technical action like cutting a twig or forming the wall of a clay vessel, to systemic level operations like selecting a subsistence strategy–material results quickly let people involved in technological activities know how they are doing. Because adjustment to unsatisfactory results can be made quickly, technology can respond easily to external conditions.

These two qualities–clear evidence of failure and relatively easy adjustment to negative outcomes–mean that technology is directly subject to selective pressures and, therefore, to evolutionary change. Variables in the domain of technology's content are subject to selective pressures created by the material results. Technology, thus, has a marked capacity to evolve in the strict Darwinian sense more easily and rapidly than other types of cultural behavior.

Technology's evolutionary potential has been discussed by many others and from several perspectives (e.g., White 1959, Dunnell 1980, O'Connell 1992, Basalla 1988, Dawkins 1982: 196-208; see also Boyd & Richerson 1985). Explicit recognition of the vital role material results play in setting up technology's evolutionary potential adds a significant step by indicating how material results contribute to technology's ability to change, move in particular directions, and maintain a high degree of patterning.

CONCLUDING REMARKS

Applying an evolutionary perspective carries two challenges. First, it is necessary to understand the sources of variation in the system under concern. Beyond that, it is necessary to identify the selection pressures that operate to favor some variations and not others. The content/results model of technology helps in both these regards. First, it identifies in concrete terms the variation available to the practitioner of specific technologies. Like the gene pool of a biological species, the behavioral content of technology is where technological variation exists. This content is observable in action and in material reflections so that it is concrete and knowable. Most importantly, the content of technology cannot be practically separated from technological results. The domain of results is where technological variability is readily and continually expressed and evaluated. Results demonstrate the effectiveness of content variations and, thus, operate like technology's phenotype.

The link between a technology's content and its results is certainly complex. Because of that complexity, and the fact that the two sides tend to require different technical expertise, most studies of technology have separated the two facets. In practice, however, the two are inexorably and dynamically linked. Neither is passive. Each reflects the other and they both are actively involved in shaping technological and human evolution.

ACKNOWLEDGEMENTS

A version of this paper was presented at the 1993 annual meeting of the Society for the History of Technology. In various forms it has been read by Michael Schiffer, Douglas Bamforth, and LuAnn Wandsnider. I gratefully acknowledge their comments and the insights of 25 years' worth of the University of Nebraska students who enrolled in "Primitive Technology." If I still haven't got it right, it certainly isn't their fault.

REFERENCES CITED

Abbott, A.R. Leonard & G. Jones
1995 Explaining the change from biface to flake technology: a selectionist application. In *Darwinian Archaeologies* (H. Maschner ed), pp. 33-42. New York: Plenum Press.

Appadurai, A. (editor)
1986 *The Social Life of Things: Commodities in Cultural Perspectives*. Cambridge: Cambridge University Presse.

Balfour, H.
1893 *The Evolution of Decorative Arts*. London: Rivington Percival.

Basalla, G.
1988 *The Evolution of Technology*. Cambridge: Cambridge University Presse.

Beals, R. & H. Hoijer
1953 *An Introduction to Anthropology*. New York: Macmillan.

Bernard, H. R. & P. Pelto
1987 *Technology and Social Change*. second edition. Prospect Heights: Waveland Press.

Binford, L.
1980 Willow smoke and dogs' tails: hunter-gatherer settlement systems and archaeological site formation. *American Antiquity*. 45: 4-20.

Bleed, P.
1991 Operations research and archaeology *American Antiquity* 56: 19-35.

Bleed P. & A. Bleed
1987 Energy efficiency and hand tool design: A performance comparison of push and pull stroke saws. *Journal of Anthropological Archeology* 6: 189-197

Boyd, R. & P. J. Richerson
1985 *Culture and the Evolutionary Process*. Chicago: University of Chicago Press.

Bradley, B.
1982 Flaked stone technology and typology. In *The Agate Basin Site: A Record of Paleoindian Occupation of the Northwest Great Plains* (G. Frison & D. Stanford, eds.), pp. 181-208. New York: Academic Press.

Braun, D.
1990 Selection, and evolution in non-hierarchical organization, in *The Evolution of Political Systems: Sociopolitics in Small-Scale Sedentary Societies* (S. Upham, ed.), pp. 62-86. Cambridge: Cambridge University Press.

Collins, M.
1975 Lithic technology as a means of processual inference. In *Lithic Technology: Making and Using Stone Tools* (E. Swanson, ed.), pp. 15-34. The Hague: Mouton.

Conkey, M.
1989 The place of material culture in contemporary anthropology. In *Perspectives on Anthropological Collections from the American Southwest* (A. Hedlund, ed.). Tempe: Arizona State University Anthropological Research Paper No. 40.

Cotterall, B. & J. Kamminga
1990 *Mechanics of Pre-Industrial Technology: An Introduction to the Mechanics of Ancient and Traditional Material Culture*. Cambridge: Cambridge University Press.

Dawkins, R.
1982 *The Extended Phenotype*. New York: Oxford University Press.

Deasy , C. M.
1985 *Designing Places for People: A Handbook on Human Behavior for Architects, Designers, and Facilities Managers*. New York: Whitney Library of Design.

Dobres, M. & C. Hoffman
1994 Social agency and the dynamics of prehistoric technology. *Journal of Archaeological Method and Theory* 1: 211-58.

Dunnell, R.
1978 Style and function: a fundamental dichotomy. *American Antiquity* 43: 192-202.

1980 Evolutionary theory and archeology. In M. B. Schiffer ed. *Advances in Archaeological Method and Theory* 3: 35-99. New York: Academic Press.

1982 Science, social science and common sense. *Journal of Anthropological Research* 38: 1-25.

Ellul, J.
1964 *Technological Society*. New York: Vintage.

Erasmus, C.
1955 Work patterns in a Mayo village. *American Anthropologist* 57: 322-334.

Flennikin, J. J.
1981 *Replication Systems Analysis: A Model Applied to the Vein Quartz Artifacts from the Hoko River Site*. Pullman: Washington State University Laboratory of Anthropology Reports of Investigation No 59.

Gendron, B.
1977 *Technology and the Human Condition*. New York: St. Martin's Press.

Hames, R. & I. Hames
1976 Ye'kwana basketry: Its cultural context *Anthropologia* 44: 3-58.

Harris, M.
1978 *Cultural Materialism*. New York: Random House.

Hodder, I.
1986 *Reading the Past, Current Approaches to Interpretation in Archaeology*. Cambridge: Cambridge University Press.

Hoebel, E. A.
1958 *Man in the Primitive World: An Introduction to Anthropology*. New York: McGraw-Hill.

Herskovits, M. J.
1948 *Man and His Works: The Science of Cultural Anthropology*. New York: Knopf.

Keeley, L. H.
1980 *Experimental Determination of Stone Tool Uses*. Chicago: University of Chicago Press.

Keller C. & J. Keller
1995 *Cognition and Tool Use: The Blacksmith at Work*. Cambridge: Cambridge University Press.

Kroeber, A. L.
1948 *Anthropology*, New York: Harcourt, Brace and World.

Kuhn, S.
1994 A formal approach to the design and assembly of mobile toolkits. *American Antiquity*. 59: 426-442.

Lane Fox, A.
1906 *The Evolution of Culture*. Oxford: Clarendon.

Lechtman, H.
1977 Style in technology-some early thoughts, In *Material Culture, Styles, Organization, and Dynamics of Technology* (H. Lechtman & R. Merrill, eds.), pp. 3-20. St. Paul: Westview.

Lemonnier, P.
1986 The study of material culture today: Toward an Anthropology of technical systems. *Journal of Anthropological Archaeology*. 5: 147-186.
1992 *Elements for an Anthropology of Technology*. Anthropological Papers of the Museum of Anthropology, University of Michigan No. 88. Ann Arbor.

Leonard, R. & G. Jones
1987 Elements of an inclusive evolutionary model for archaeology. *Journal of Anthropological Archaeology*. 6: 199-219.

Leonard R. & H. Reed
1993 Population aggregation in the prehistoric American Southwest: a selections model. *American Antiquity* 58: 648-661.

Longacre, W. A.
1968 Some aspects of prehistoric society in east-central Arizona, In *New Perspectives in Archeology* (S. R. Binford & L. Binford, eds.), pp. 89-102. New York: Aldine.

Lustig-Arecco, V.
1975 *Technology: Strategies for Survival*. New York: Holt, Rinehart, and Winston.

Mason, O.
1895 *The Origin of Invention*. London: Scott.

Mumford, L.
1933 *Technics and Civilization*. London: Routledge.

Nelson, M.
1991 The study of technological organization. In *Archaeological Method and Theory Vol. 3* (M. Schiffer, ed.), pp. 57-100. Tucson: University of Arizona Press.

Nelson, N. C.
1932 The origins and development of material culture. *Sigma Xi Quarterly* 20: 102-123.

Nye, D.
1994 *American Technological Sublime*. Cambridge: M.I.T. Press.

Norman, D.
1988 *The Psychology of Everyday Things*. New York: Basic Books.
The 'author' 1984 Productivity Improvement Research Section, Division of Industrial Science and Technological Innovation.
1983(May) *The Process of Technological Innovation: Review the Literature*. Washington: National Science Foundation.

O'Connell, J.
1992 The fine-tuning of the golden ear: high-end audio and the evolutionary model of technology. *Technology and Culture* 33: 1-37.

O'Brien, M. J. & T. D. Holland
1990 Variation, selection, and the archeological record. *Archaeological Method and Theory* 2: 31-79
1992 The role of adaptation in archaeological explanation. *American Antiquity* 57: 36-59.
1996 Behavioral archaeology and the extended Phenotype, In *Expanding Archaeology* (J. Skibo, W. Walker & A. Nielsen, eds.), pp. 143-61. Salt Lake City: University of Utah Press.

Osgood, C.
1951 Culture: its empirical and non-empirical character. *Southwestern Journal of Anthropology* 7: 202-214.

Oswalt, W. H.
1973 *Habitat and Technology, The Evolution of Hunting* New York: Holt, Rinehart, and Winston.
1976 *An Anthropological Analysis of Food-Getting Technology* New York: Wiley-Interscience.
1986 *Life Cycles and Lifeways: An Introduction to Cultural Anthropology*. Palo Alto: Mayfield.

Peasant, S.
1986 *Anthropometry, Ergonomics and Design*. London and Philadelphia: Taylor and Francis.

Pfaffenberger, B.
1992 Social anthropology of technology. *Annual Review of Anthropology* 21: 125-42.

Petroski, H.
1992 *The Evolution of Useful Things*. New York: Knopf.

Pye, D.
1964 *The Nature of Design*. New York: Reinhold.
1968 *The Nature and Art of Workmanship*. New York: Van Nostrand Reinhold.

Quimby, I. (editor)
1978 *Material Culture and the Study of American Culture*. New York: Norton.

Richter, M. N.
1982 *Technology and Social Complexity*. Albany: State University of New York Press.

Sayce, R. U.
1933 *Primitive Arts and Crafts*. Cambridge: Cambridge University Press.

Sellet, F.
1994 Chaîne opératoire: the concept and its application. *Lithic Technology*. 18: 106-112.

Schiffer, M. B.
1975 Behavioral chain analysis: Activities, organization, and the use of space. In Chapters in the Prehistory of Eastern Arizona, IV. *Fieldiana: Anthropology*. 65:103-119.
1976 *Behavioral Archeology*. New York: Academic Press.
1996 Some relations between behavioral and evolutionary archaeologies. *American Antiquity* 61: 642-662.

Schiffer, M. B. & J. Skibo
1987 Theory and Experiment in the Study of Technological Change. *Current Anthropology* 28: 595-622.
1997 The explanation of artifact variability. *American Antiquity* 62: 27-50.

Schlereth, T. J.
1980 *Artifacts and the American Past*. Nashville: American Association for State and Local History.

Shott, M.
1986 Settlement mobility and technological organization: An ethnographic examination. *Journal of Anthropological Research* 42: 15-51.

Skibo, J. W. Walker & A Nielsen (editors)
1995 *Expanding Archaeology.* Salt Lake City: University of Utah Press.

Spicer, E. (editor)
1952 *Human Problems in Technological Change.* New York: Wiley.

Spier, R. F. G.
1970 *From the Hand of Man.* Boston: Houghton Mifflin.

Strasser, S.
1982 *Never Done: A History of American Housework.* New York: Pantheon.

Torrence, R.
1983 Time budgeting and hunter-gatherer technology. In *Hunter-Gather Economy in Prehistory* (G. Bailey, ed.), pp. 11-22. Cambridge: Cambridge University Press.

White, L.
1949 *The Science of Culture.* New York: Farrar, Straus and Cudahy.
1959 *The Evolution of Culture.* New York: McGraw-Hill.

Wissler, L.
1926 *The Relations of Nature to Man in Aboriginal America.* New York: Oxford University Press.

Winner, L.
1980 Do artifacts have politics? *Daedalus* 109: 121-136.
1986 *The Whale and the Reactor: A Search for Limits in an Age of High Technology.* Chicago: University of Chicago Press.

Part III
Material Culture, Behavior & Middle Range Theory

Introduction

The five papers in this section turn on aspects of middle range theory, and on efforts to refine the logic of inference relating to pattern in particular classes of material culture residues. As such, they are sometimes less explicitly evolutionary than the essays in Part II. However, all proceed from the assumption that behaviour is part of the phenotype (upon which selection operates), that the differential persistence of some behaviours, at the expense of others, is due to the operation of natural and cultural selection, and that 'success' in the evolutionary long run can only be measured in terms of the reproductive fitness of individuals. Four essays (Bamforth & Bleed, Barton, Bettinger & Eerkens, Shott) focus on various aspects of stone artifact assemblages, and have important implications for the development of a more satisfactory middle range theory of lithic technology.

Following Torrence (1983), Douglas Bamforth and Peter Bleed (Chapter 7) argue that many critical aspects of flaked stone technology can be understood best from a 'risk management' perspective. Risk has two major components, they contend, and, while efforts to assess the *probability of technological failure* are central to most current approaches, efforts to take the *costs of technological failure* into account are practically non-existent. Using a literature search, they erect a conceptual framework that takes a closer look at 'risk', supporting it with ethnographic and archaeological case studies, and then consider problems arising from efforts to apply the model to archaeological data. They conclude that our limited ability to estimate failure probabilities and costs precludes rigorous testing of the model in archaeological contexts. However, ethnographic support for the approach tends to validate it, and its archaeological potential is demonstrated by examining the production of Folsom fluted projectile points and ditch and palisade defensive works along the Missouri River.

C. Michael Barton's essay (Chapter 8) focuses on 'style' in lithic artifact assemblages, widely regarded as providing the best information about social group affiliation. 'Style', however, is not a unified concept, and includes both passive variability resulting from stochastic processes and actively encoded social information, constrained by selection and manipulated by the makers and users of stone tools. He proposes a neo-Darwinian analytical framework in order to isolate identity-conscious social units and to trace their temporal and spatial boundaries. He argues that identifying prehistoric social units is roughly analogous to identifying biological taxa and constructing phylogenies from them. The approach allows him to generate expected patterns of variation in archaeological index types that should allow us to determine whether or not an identity-conscious social unit is making them. The approach is illustrated with an example from the epipaleolithic of the southern Levant.

Noting the universality of style as a means to communicate aspects of identity, Polly Wiessner (Chapter 9) takes an evolutionary approach to style that considers it a means of information exchange in the service of a specific, human cognitive process: the establishment of personal and social identity by a process of social comparison, and the projection of those identities onto surrounding 'others'. She argues that this approach can be distinguished from other approaches to style because evolutionary process does not operate directly on material culture per se, but rather on the human behavioural repertoire that produces it; that style does not select for efficiency in information transfer, as Wobst (1977) would have it; and that, following an evolutionary psychology perspective, the cultural transmission of stylistic information must be understood from the viewpoint of evolving human cognitive processes, and not from that of formal models constructed on the basis of logical possibilities. She explores some of the implications of identification by comparison with a tightly-focused diachronic study of !Kung San subsistence, settlement and lifestyle changes over a 35 year interval (1960-1995) during which the historical context and meanings associated with two prominent aspects of !Kung San stylistic behaviour - arrowheads and beadwork - changed in response to changes imposed on these erstwhile foragers by policies instituted by the national governments within which they find themselves imbedded.

Adopting the conceptual framework of Boyd and Richerson's cultural transmission theory (1985), Robert Bettinger and Jelmer Eerkens (Chapter 10) analyze a large sample of Great Basin projectile points to test aspects of the general CTT prediction that variation in socially-transmitted behaviour will vary inversely with its complexity, the complexity of the social and technical context in which it occurs, and the number of individuals involved in transmission. These CTT tenets allow them to predict that complex point shapes will be less variable than simple ones in the same series, that arrowheads will be less variable than dart points, and that temporally persistent dart points will be more variable than temporally sensitive dart points in the same regions, and more variable than their temporally sensitive equivalents in other regions. These expectations about pattern were only partly supported by a least squares regression on six metrical attributes originally compiled by Thomas (1981). The authors show that metrical variability in Great Basin projectile points is strongly scaled to size, and that the relationship is linear, but that most metrical variability is accounted for by the mean of the attribute in question, suggesting that this variation is independent of the mode of transmission and primarily due to relative, rather than absolute, error tolerances in production. More complex point shapes do in fact show less metrical variation than simple forms, and arrowheads show less variation than dart points, as predicted. However, many results run counter to conventional wisdom in projectile point studies, indicating the utility of the approach.

Michael Shott's essay (Chapter 11) is also grounded in cultural transmission theory, also uses biface metric data (dart and arrowpoints from the Range 'site' - an archaeological palimpsest in the Mississippi Valley of southern Illinois), and both complements and critiques the CTT applications of Bettinger and Eerkens. Shott uses robust EDA methods well-suited to the analysis of the 'dirty' data sets with which archaeologists most often contend. He focuses on measures of dispersion (by settlement type, by order in time), and gets results that lead him to conclude that CTT only weakly accounts for variation and selection through time in Range site metric point attributes. According to CTT, social transmission (frequency-dependent adoption, indirect bias) should become more important with increases in population size and social complexity. However, the Range site spans several cycles of occupation with these properties without corresponding variation in the central tendencies and dispersion of metric point values predicted by CTT, leading him to suggest that CTT, and other selectionist approaches, could benefit from the development of scale and context-sensitive site formation theories. Arguing that all archaeologists who seek to constructively engage the archaeological record are selectionists to some degree, and in an effort to avoid dogmatic commitment to a particular, narrowly-defined 'kind' of selectionism, he advocates an eclectic approach grounded in critical realist ontologies and modified objectivist epistemologies (Clark 1993).

REFERENCES

Boyd, R. & P. Richerson
1985 *Culture and the Evolutionary Process.* Chicago: University of Chicago Press.

Clark, G. A.
1993 Paradigms in science and archaeology. *Journal of Archaeological Research* 1: 203-234.

Thomas, D.
1981 How to classify the projectile points from Monitor Valley, Nevada. *Journal of California and Great Basin Archaeology* 3: 7-43.

Torrence, R.
1983 Time-budgeting and hunter-gatherer technology. In *Prehistoric Hunters and Gatherers in Europe* (G. Bailey, ed.), pp. 11-22. Cambridge: Cambridge University Press.

Wobst, H. M.
1977 Stylistic behavior and information exchange. In *Papers for the Director* (C. Cleland, ed.), pp. 317-342. Ann Arbor: University of Michigan Museum of Anthropology, Anthropological Papers No. 61.

7

Technology, Flaked Stone Technology, and Risk

Douglas B. Bamforth
University of Colorado

Peter Bleed
University of Nebraska

ABSTRACT

Recent theoretical studies of flaked stone technology have identified many factors that affect the ways in which human beings make and use tools. However, these studies lack a unified body of theory that might help to integrate their diverse perspectives. This paper expands recent anthropological discussions of risk as the basis for such a theory. We begin by defining risk, emphasizing two distinct components of this concept–the probability that some problem will occur and the cost of such an occurrence–and argue that technology can be seen as a means of reducing such probabilities in the face of unacceptably high costs. We support this argument using cross-cultural data on hunter-gatherer technology and archaeological and historic data on the construction of defensive works on the northern Great Plains. Next, we consider specific problems in applying this perspective archaeologically, concluding that existing limits on our ability to estimate failure probabilities and costs prevent us from testing the ideas outlined here in archaeological contexts. However, accepting this perspective as provisionally validated by ethnographic data allows us to see how it can illuminate archaeological cases, and we exemplify this by comparing the production of Araya microblade projectile points in the Japanese Paleolithic and Folsom fluted projectile points on the North American Great Plains.

This is a paper outlining a way to think about flaked stone tools, contained in a volume addressing the role of Darwinian theory in archaeological explanation. We think that this volume is an appropriate place for this paper to be, but we also think that the relation between Darwinian theory and stone tools requires some explication.

The version of Darwinian theory which is currently popular in archaeology is often referred to as "selectionist", following O'Brien and Holland (1990) and others. Selectionist archaeologists make two important claims, one regarding the mechanisms by which cultural change occurs and the other regarding the causal forces driving such change. First, selectionists argue that changes in human ways of life occur as the result of the differential persistence of one variant of a pattern of behavior or kind of artifact over another, with successful variants or artifact types "selected" by external, usually environmental, conditions; they often contrast such a pattern with transformational views of change. Second, selectionist scholars argue further that Darwinian forces drive this pattern of change, which means, or ought to mean, that one pattern of behavior or artifact type is selected over another on the basis of its effects on the reproductive fitness of individual human beings.

It is absolutely essential to distinguish these two arguments, because addressing them archaeologically poses very different problems. These problems are most severe in the case of attributing change to the causal force of reproductive fitness, which, no matter how we dance around it, is what Darwinian theory is about. Direct observation, archaeologically or otherwise, of the processes on which Darwinian evolution depends–the effect of differences in reproductive fitness among individuals within a breeding population–is exceedingly difficult, and we would argue that archaeologists who have attempted to look directly at reproductive fitness have achieved ambiguous results at best. For example, Leonard and Reed (1993; also see Larson et al. 1996) take regional population growth as a measure of fitness, despite the fact that Darwinian theory explicitly refers to the differential reproductive success of individual organisms, and that the spread of one phenotype at the expense of another therefore represents evolution by natural selection whether total population increases, decreases, or remains the same. Overall population levels may well measure the overall biological success of a regional population, but they absolutely do not automatically indicate Darwinian evolution in that population.

As Smith (1991: 34) notes, "Reproductive competition among individual members of a population may be such as to confound any generalization we might wish to make about the effect of selection on entire populations: selection might *result* in a decline in population size, an increase in mortality or strife, a decline in mean population fitness, or even extinction, even as it *favors* the individual traits that produce such effects" (emphasis in original). It seems to us that an emphasis on regional population growth comes perilously close to relying, implicitly or not, on arguments based on group rather than individual selection (compare, for example, Leonard & Reed [1993] with Kaplan & Hill [1985]). Group level processes may turn out to be particularly important in cultural, as opposed to biological, change, but such processes will not necessarily be compatible with, and may actually produce results which are quite divergent from, processes of Darwinian evolution (also see Smith 1988: 230-231).

Perhaps because of problems like these, the great majority of selectionist research focuses on the ways in which artifacts or behavioral patterns are likely to enhance fitness under specific conditions and then evaluates the effects of changes in those conditions, but does not address individual reproductive fitness directly. Arguments like these typically show that one artifactual or behavioral variant is particularly efficient or effective under some conditions and not under others, and that changing conditions select for it over other, less efficient or effective, variants; alternatively, some analysts attempt to isolate aspects of artifactual variation which are argued not to be subject to selective forces (i.e., Neff 1992, Neiman 1995).

The distinction we drew above between a selectionist pattern of change and the possibility that differential reproductive fitness drives such change is important here: without showing that fitness differences among individual humans are the ultimate cause of archaeological change, these less direct selectionist arguments use Darwinian theory basically as a heuristic metaphor, applying concepts developed to understand evolutionary changes in organisms as tools to clarify our thinking about temporal changes in material culture. To rephrase our earlier arguments, enhancing the survival of entire populations of humans is not Darwinian evolution, and, to the extent that selectionist arguments pertain to such population-level benefits, the causal processes to which they appeal differ little, if at all, from those of more traditional views which the selectionist program is intended to supplant (also see Schiffer 1996: 647).

Our discussion falls into this latter, heuristically "Darwinian", category of selectionist research, and we therefore proceed with some caution. Stone tools were essential to the ways in which humans coped with the conditions around them for over two million years. Understanding how these tools enhanced such coping is thus important to archaeological interpretation, and our goal here is to outline a framework for analyzing the links between different technological patterns and the context in which they were designed to operate; it will be clear that we view external environmental conditions as particularly important in shaping particular technological stances. However, we suspect that virtually all lithic datasets derive from the collective activities of groups of people and rarely tell us much about differences (reproductive or otherwise) among individual human beings. We are quite comfortable with arguments for selectionist patterns of change, and we believe that the framework we present here is particularly suited to such views. However, we also believe that archaeology is a long way from showing that truly Darwinian forces cause such change: we believe that a selectionist framework helps to make sense out of diachronic change and synchronic variation in lithic technology, but it is not clear to us that such technology tells us a lot about Darwinian process.

Our goal here is to outline a basis for building a body of theory relevant to important aspects of technology, particularly flaked stone technology, in order to address a general lack of synthesis and integration in the archaeological literature on this topic. A recent expansion of the attention given by archaeologists to the interpretive potential of flaked stone tools has clearly increased dramatically the range and sophistication of the archaeological problems that lithic data sets can help to solve. In the past two decades, archaeologists focused on flaked stone tools have derived substantial insights into the relations between technology and such factors as mobility patterns, complex economic organization, raw material availability, and cultural or ethnic preferences (Nelson [1991] reviews much of this literature; also see Andrefsky 1994, Bamforth 1991a; Barton 1991, Chapter 8; Bousman 1993, Carr 1994, Flenniken 1984; Hayden 1981, 1987; Jeske 1992, Kornfeld et al. 1990, Kuhn 1994, Ricklis & Cox 1993, Sackett 1982). Although many issues remain to be settled, there is no doubt that the field has made great strides. However, despite the range of topics lithic analysts have addressed and the progress made over the past decade, Jeske (1996) points out that this branch of archaeological thought seems stalled: studies of mobility and other patterns derived from Binford's (1977, 1979, 1980) work dominate the field, for example, and a single descriptive term–the "curation" concept (Binford 1973, 1977, 1979)–structures virtually every study in the published literature. Furthermore, theoretical lithic analysis is plagued by seemingly unnoticed mutually contradictory arguments, a wide range of distinct definitions for important terms (most notably "curation"), and little or no synthesis.

Our discussion begins by considering (briefly) what technology is, and then turns specifically to focus on the range of behavior involved in flaked stone technology. In this first section, we will argue that archaeologists (1) have been too quick to equate technology with tools and (2) have tended to focus on too narrow a range of behavior in their attempts to explain technological patterns. We then turn to build on Torrence's (1989) discussion of the importance of risk as a determinant of patterns of tool production and use as the basis for beginning to develop a more synthetic account of the determinants of technological behavior. This discussion emphasizes the problem of defining "risk" and translating the definition we use into the analytic framework we develop in the first two sections; our arguments here particularly rely on cost-benefit arguments, but point out that there are likely to be situations in which such arguments may not be important. We close by contrasting two specific archaeological examples in light of our arguments.

TECHNOLOGY AND FLAKED STONE TECHNOLOGY

We follow Bleed (Chapter 6) in defining "technology" as a society's "customary means of manipulating the physical environment". In this view, technology encompasses a wide range of complex behavior and knowledge regarding how to appropriately acquire and put tools to use, and is not simply equivalent to those tools themselves: technological options include knowledge both of what kinds of tools to make and of alternative ways to use those tools, alternatives that may involve such factors as work group organization or the geographic location or time of year in which a particular activity is carried out. Thus, for example, the technological knowledge required to make a living as a Late Prehistoric bison hunter on the Northwestern Great Plains included not only that pertaining to the manufacture and use of the bow and arrow, but also that pertaining to the conditions under which to hunt individually or communally. We therefore need to distinguish between what Bleed (Chapter 6) refers to as technological content (knowledge, standards regarding the proper way to carry out a particular activity, etc.) and technological results (material culture and modifications to the environment): as archaeologists, we study the second of these and hope to learn something about the first.

This view of technology has two important implications for our thought. First, emphasizing technological knowledge helps to remind us that all humans have more than one means of manipulating their environment; we know more than one way to accomplish any given goal, and we make different choices from our range of alternatives in different situations. Second, such an emphasis also reminds us that technological knowledge includes not only knowledge about how to use tools but also about how to obtain them, and that this knowledge can be manifest archaeologically (that is, it produces technological results) in the form of material residues generated at all steps along the continuum of behavior involved in tool production and use. In the specific context of flaked stone technology, many archaeologists have pointed out that this continuum includes the procurement of raw material, production of useable tools, use, optional maintenance and/or recycling of worn tools, and, finally, discard (Callahan 1978, Collins 1975, Crabtree 1968, Rick 1978). For present purposes, we follow Jeske (1989) in dividing this continuum into three fundamental domains of behavior: first, raw material from which tools can be made must be obtained; second, tools must be produced from this material; and, third, these tools must be applied to some task or tasks. In this scheme, tool maintenance and recycling fall into the domain of application.

As Nelson and Lippmeier (1993: 286-287) note, "the most fundamental condition influencing tool design is the task for which a tool is expected to be used; all tools must be minimally effective at the task for which they were produced." Recognizing that this is so, most archaeological attempts to theorize about the factors structuring technology have emphasized the domain of application, arguing, for example, that mobile groups are likely to make portable tools in order to minimize transport costs (Shott 1986) or that design differences between functionally identical implements result necessarily from ideational factors (Lemmonier 1986). However, we will argue here that, although it is obviously true that the utilitarian requirements of the task to be accomplished set fundamental limits on possible technological approaches to that task, technological strategies incorporate essential knowledge about the full continuum from procurement to application. Furthermore, the outcomes of choices made in the earlier points along this continuum strongly condition the options available at later points: successful application of a tool relies on successful production of that tool, which relies in turn on successful raw material procurement.

All flaked stone technologies must therefore solve the problems of obtaining sufficient material, producing adequate tools, and accomplishing necessary tasks, and any successful technological adaptation must therefore take advantage of opportunities and cope with constraints in all three of these areas. The problem in deriving general theory that helps to understand why technological behavior varies in time and space, then, is to understand how these opportunities and constraints are balanced against, or perhaps integrated with, each other in specific cases. The failure of most theoretically-oriented lithic analysis to systematically address the importance of technological decisions in the domains of procurement and production (with the major exception of Rick's [1978] discussion of heat alteration) is one of the important factors limiting our understanding of technological variability. Recognizing this, though, introduces the additional problem of finding a conceptual framework within which to analyze the balance (or integration) of technological decisions in the contexts of procurement, production, and application in specific cases. For the remainder of this paper, we attempt to build on Torrence's (1989) discussion of risk as an important factor structuring technological choices to lay the basis for such a framework.

RISK

All tools must meet certain baseline requirements of effectiveness and accessibility, each of which may be buffered by social factors and limits of knowledge. In general, it is likely that these buffers are minor determinants of technological decisions in most situations, particularly in the case of flaked stone technology: they are background and not the main arena of technological design (although there are presumably cases in which this is not so, as we discuss below). We therefore focus here on the practical or utilitarian factors affecting technology. Following Torrence (1989), we will argue that many critical aspects of human technological behavior, with flaked stone technology as a specific example, can be understood as a means of managing risk.

Although the notion of "risk" has a fairly clear intuitive meaning for most of us, it is not always apparent how it can be transformed into a construct that is precise enough for systematic analysis. As Winterhalder (1986a: 383) puts it, "defining risk is like juggling a porcupine: whatever way you throw it up, it comes down prickly". Anthropological interest in risk (see below) grows out of the recognition that the world can be an unpredictable and, sometimes, perilous place, and that an understanding of human affairs ought to take this into account. In common usage, we often refer to unpredictable and perilous situations as "risky", and this term has been carried over into anthropology and other disciplines. However, as is often true for the academic application of intuitively "obvious" concepts, different scholars have operationalized "risk" in a variety of ways that fit their different interests and perspectives: different analysts have applied the term "risk" to different (but related) concepts. As Cashdan (1990a; also see Christenson 1982: 422-423) notes, the existence of multiple operational definitions of the intuitive notion of "risk" is best taken as evidence that the world is unpredictable and perilous in more than one way and that different analytic tools are necessary to examine unpredictability and peril in different contexts. However, this implies that it is important to specify exactly what we mean when we use the word "risk" and why that meaning is relevant to the problem we are interested in solving. The following section therefore addresses this issue.

"Risk" in the Dictionary and Elsewhere

Webster's New Collegiate Dictionary presents a complex definition of "risk", the most general aspect of which defines the term as "Hazard; peril; exposure to loss or injury". A second, more precise, section of the definition derives from the insurance industry: "a. The chance of loss or the perils to the subject matter of insurance covered by the contract; also, the degree of probability of such loss. b. Short for amount at risk, that is, the amount which the company may lose." The first, more general, of these clearly refers to the common usage of the word "risk" to refer to the possibility that something bad may happen, while the second distinguishes clearly between two components of this usage: the probability of a bad thing occurring and an estimate of just how bad that bad thing is.

Systematic usage of the concept of risk in a variety of disciplines (including economics [Jones-Lee 1989], insurance [Hammond 1968], environmental hazards assessment [Sewall 1990, Suter 1993, Westman 1985], and others) generally focuses clearly on the importance of assessing both the probability and the magnitude of hazards in evaluating levels of risk. This literature has developed substantially as a basis for specifying decision rules on which to base financial and policy decisions, and, as a result, has developed a variety of highly quantitative approaches to operationalizing and analysing this interaction, approaches that are generally designed to help to identify the course of action that most effectively balances risks and the costs of avoiding them (Rescher [1983] presents an extended conceptual, as opposed to mathematical, discussion of many of the essential issues and debates that research in these areas has identified). Among the outcomes of work in this area is a widely (although not universally) applied distinction among risky situations, in which the possible outcomes of a choice (both negative and positive) and their respective probabilities are known, uncertain situations, in which the possible outcomes are known but their probabilities are not, and situations of incomplete knowledge, in which the range of possible outcomes is unknown (Mishan 1971).

Economic/business forecasting relies on assessments of investment risks which distinguish clearly between probabilities and costs, but often recasts the notion of cost in terms of "utility", or "the strength of our desire for something" (Bernstein 1996: 71). Although this branch of research has produced important insights into human responses to risk, it is worth noting that the equation of cost and utility blurs the distinction between the probability of obtaining some desired benefit and that of incurring some undesired harm. For example, Bernstein (1996: 71) quotes an anonymous seventeenth century author who wrote that "Fear of harm ought to be proportional not merely to the gravity of the harm, but also to the probability of the event", but goes on to argue that "We could turn this assertion around and state that a decision should involve the strength of our desire for a particular outcome as well as the degree of our belief about the probability of that outcome". Distinguishing clearly between costs and benefits, between losses and gains, is analytically particularly important because research into the psychology of risk-taking (see Bernstein 1996: 269-283 for a lucid summary) indicates that people make systematically different choices in relation to situations of loss than of gain, even when the outcomes of these situations are mathematically identical.

Evolutionary ecology has also attended to issues of risk. Limited early attention to the topic (i.e., Horn 1968) addressed the "predictability" of resource distributions, and more abundant recent work has explicitly applied the term "risk", resulting in important refinements of optimal foraging models (Real & Caraco 1986). However, although ecological applications of the risk concept generally begin from the commonsense observation that all organisms inhabit a world in which bad things may occur, and often note issues like the distinction between risk and uncertainty (i.e., Real & Caraco 1986: 373, Stephens 1990), ecological analyses tend to operationalize "risk" in a distinctive way. Such analyses deal almost exclusively with two kinds of risks: inadequate food intake and predation (and sometimes with both of these simultaneously–see, for example, Bednekoff & Houston 1994, Bulmer 1994: 112-116, Oksanen & Lunberg 1995). Not surprisingly, studies of predation generally take the magnitude of the hazard at issue–death–as fixed, and examine organisms' reponses to the varying probabilities of encounter with a predator; as we might expect, such studies (i.e., Bouskila 1995, Kennedy et al. 1994, Suhonen 1993) typically find that organisms alter their behavior to avoid contact with predators when predators are present.

In contrast, studies of dietary risk generally proceed by treating food availability as fluctuating, and considering the ways in which organisms' foraging decisions vary in response to these fluctuations; increases in the magnitude of resource fluctuations are seen as increases in risk. In general, theoretical analyses of "risk-sensitive foraging" suggest that an organism should act in ways that reduce the variance in foraging yields when resources are abundant relative to that organism's needs (that is, they should be "risk-averse") and should act in ways that increase variance when resources are scarce (that is, they should be "risk-prone"), and a number of studies have confirmed these predictions empirically (Bird, Chapter 16; Bulmer 1994: 111-112, Caraco et al. 1980, 1990; McNamara & Houston 1992, Real & Caraco 1986, Stephens & Charnov 1982). Alternatively, Stephens (1990) points out a second kind of variance that can be referred to as risk, variance in the degree to which return is delayed, and notes that experiments show that animals often select more variable over less variable delays. A few analyses (Caraco & Chasin 1984, Houston & McNamara 1986) have also examined the effects of the direction of skewing in the distribution of resource fluctuations. These studies complicate this issue somewhat, suggesting that positively skewed distributions should be favored when nutritional reserves are either low or high, while negatively skewed distributions should be favored in intermediate situations.

We discuss the difference between ecological usage of "risk" and the usage of this term in other fields below, in the context of a summary of anthropological usage of the term. However, it is worth noting that at least some of the analyses that simultaneously consider predation and dietary risks (see Bulmer's [1994: 112-116] discussion for an example) move quite close to the emphasis on both probabilities and costs in fields like insurance and hazards assessment

Risk in Anthropology and Archaeology

Both anthropology and archaeology have recognized the importance of addressing the role that risk plays in structuring human ways of life, although virtually all discussions of risk in our discipline have focused on subsistence risk (the risk of failing to obtain sufficient amounts of food) to the exclusion of other risks that humans may encounter (war, disease, accidents, etc.). In some cases (Gould 1991, Halstead & O'Shea 1989, Jochim 1981, Minnis 1996), anthropologists and archaeologists have used terms like "risk" and "uncertainty" without defining them, relying on our common intuitive understanding of these terms; in others, the substance of the risk concept-the possibility of encountering problems-enters into analyses under a different label (i.e., "predictability" or "reliability"; Bamforth 1988, Heffley 1981, Hayden 1981, Lee 1968, Tanaka 1976, Wilmsen 1973). However, other writers have variously drawn explicitly from usage in the literature in insurance/economics and evolutionary ecology.

Not surprisingly, definitions of risk in economic anthropology (Calavan 1984, Cancian 1980, Cashdan 1985; Weissner 1982a, 1982b; Wharton 1971) follow those in economics and business, emphasizing "risk" as the probability of loss (generally the loss of subsistence resources, particularly agricultural produce, due to environmental or other factors). In contrast, anthropological work grounded in evolutionary ecology often shows important definitional ambiguities. For example, Keene (1981: 179) defines risk as the probability of loss, but then uses the term in contexts that clearly emphasize only loss probabilities, with no attention to loss magnitudes (i.e., Keene 1981: 179, 180). Similarly, Winterhalder (1981: 94) refers to "species with low short-term risk of capture failure" and Smith (1981: 46) distinguishes between "low-risk, low-return resources . . . [and] relatively high-risk (erratic) resource types", clearly equating risk with loss probabilities in both cases; more recently, Baksh and Johnson (1990: 195) make the same equation in referring to estimates of the "risk" of rocket failure in the NASA space shuttle program during the 1980's that ranged from "1 in 100,000" to "1 in 70". Tainter (1996: 15) also seems to see risk primarily in terms of probabilities, distinguishing between "stress" ("the consequences that arise when the energy needs of a human population are not met") and "risk" ("the likelihood that energy needs will not be met), and Rautman (1996: 200) follows this usage in asserting that "Regional social networks extending beyond the local group reduce the risk of resource stress"; "stress" in this sense is essentially the same as the "loss" component of the economic definition of risk. However, many recent analyses draw explicitly on the specific approach taken by theoretical ecologists, and formal examinations increasingly equate risk explicitly with variance in resource availability (Bettinger 1991: 118-127, Hegmon 1996, Kelly 1995, Kohler & Van West 1996, Smith 1988; Winterhalder 1986a, 1986b; also see most of the authors in Cashdan 1990b). In one particularly clear example of this usage, Larson et al. (1996) measure subsistence risk in the precontact Southwest by examining temporal patterns in year to year variability in rainfall reconstructed from tree rings; however, these authors supplement this measure by also focusing specifically on years in which rainfall was below-average.

All of these studies address the means that humans have devised to prevent, or at least minimize, shortfalls in the food supply, and a common (although not universal) conclusion is that resource sharing within communities, and variability in the degree to which such sharing occurs, is related to the level and geographic homogeneity of subsistence risk in the environment; the emphasis on fairly abundant and spatially predictable plant foods among many hunter-gatherer groups is also regularly attributed to attempts to reduce subsistence risk (although this is often referred to as increasing subsistence reliability). Furthermore, the classic arguments that many hunter-gatherer groups maintain long-distance social ties in order to have access to distant resources during times of scarcity (i.e., Burch & Correll 1972, Gould 1991, Lee 1976, Smith & Boyd 1990), although not generally explicitly grounded in the literature on risk, clearly link important components of social relations to unpredictable subsistence problems. Archaeologists have also recognized the importance of unpredictable problems as factors structuring human adaptations (although they have not always used the word "risk" to describe these), but have tended to emphasize larger-scale, more archaeologically visible adaptations: for example, Smith (1995; also see Flannery 1986) attributes the rise of food production to hunter-gatherers' attempts to get through lean years, and a number of authors (Nicholson 1988; Rautman 1993, 1996; Spielmann 1991) emphasize interregional exchange as a means of evening out resource shortfalls over large areas .

Risk and Lithic Technology

As this discussion illustrates, and as Torrence (1989: 59) and Bousman (1993: 69) point out, anthropological discussions have tended to focus on social rather than technological responses to risk (such as sharing or exchange), although at least some authors (Cashdan 1985, Weissner 1982) have considered the conditions under which storage may help to reduce subsistence risk. However, there are excellent and obvious reasons for arguing that considerations of risk should be important in making technological choices. As Torrence notes (1989: 58-59), human beings use technology to manipulate the environment in such a way that they satisfy their needs. Because the environment varies in time and space and because this variation is only partially predictable, it is always possible that something may go wrong and either complicate or prevent human actions. Specific problems that do not create major difficulties do not necessarily require adaptive responses, but more significant problems do. That is, most human activities have some probability of failing and in many cases such a failure can have high costs, which is one way to say that some level of risk is built in to most human enterprises. Where risk is unacceptably high, it can be reduced either by lowering costs or by reducing failure probabilities, and, as we argue below, archaeologists seem generally to agree that that technology plays an important role in such reductions, particularly in the latter area.

However, relatively few archaeologists have specifically discussed the relation between risk (or "reliability"; Hayden 1981) and technology, including flaked stone technology. All of the studies that explicitly link risk and technology refer to Torrence's (1989) discussion, making her work an important starting point (note that we focus here on Torrence's definition of risk; we discuss her specific arguments regarding risk and technology in a later section). Torrence explicitly recognizes the importance of both failure probabilities and failure costs, defining risk in the context of subsistence activities as "the probability of failing to meet dietary requirements". She then argues from this definition that risk in this sense is not simply determined by the probability of failing to capture an animal, but by the cost of such a failure, and makes the essential point that this cost depends on the availability of alternative sources of food. However, although she stresses failure costs in portions of her discussion (i.e., Torrence 1989: 63), her cross-cultural analysis (see below) treats risk almost entirely as the probability of failure, as is indicated, for example, in assertions that "people do not attempt to minimize the expenditure of time for its own sake, but because the length of time a resource is available is an important measure of the probability of failing to capture it, i.e., of the degree of risk involved" and that tools used to procure aquatic resources are complex (have many parts) because of "the higher probability of loss for a resource captured in the water as opposed to on land" with no reference to the cost of such a loss (Torrence 1989: 60, 61).

Similarly mixed use of the term, as well as the equation of "risk" levels with estimates of probabilities, is evident in other studies of risk and technology. For example, Myers (1989) does not explicitly define risk, but he clearly uses it primarily with a similar emphasis on probabilities: for example, he argues that "the mobility of animals introduces risk through the capacity of the resource to avoid detection and capture. By the same token, plants are incapable of significant motion and cannot avoid capture once detected" (Myers 1989: 84). Myers focuses on a shift from Early to later British Mesolithic weapon technology from relatively simple, unstandardized stone projectile points and regular use of bone and antler points to complex, highly standardized stone barbs with almost no use of antler and bone points, explaining this as a response to a shift from an environment in which game animals migrated in large herds regularly between predictable locations to one in which these animals moved erratically in small groups. In this view, Myers sees the technological changes as enhancing weapon function, ease of repair, and accuracy, all of which became more important given the increased difficulty of locating mobile prey and thus the increased importance of ensuring a successful kill when such prey was sighted.

Similarly, Nelson (1996) cites an early version of the present paper in distinguishing between the probability of loss and the cost of that loss, and argues for important differences in technological strategies resulting from efforts to reduce subsistence risk by specializing on a small set of resources (which she argues should result in reliable [in Bleed's [1986] terms; see below], efficient, and specialized tools that require high investments of production time) and strategies resulting from efforts to reduce such risk by diversifying resources (which she argues should result in reliable, versatile, and portable tools that require low investments of production time). However, she follows Torrence in shifting from a recognition of the importance of both probabilities and costs to a primary focus on probabilities in assertions like "specialization and diversification are economic decisions intended to ameliorate stress or reduce the risk and cost of shortfall", that "the risk in production of cultigens is less than the risk in hunting", or that "while people engaged in hunting a wide variety of game during some time periods, this practice may not have been effective as a risk-reducing strategy in times of food shortage or anticipated risk of shortages" (Nelson 1996: 111, 134).

In contrast to these, Bousman (1993) explicitly refers to both the insurance/economic and ecological uses of the term as part of a more general discussion of hunter-gatherer mobility and technological change: "Risk is defined as the probability of economic loss, and perhaps it is best conceived in terms of unpredictable environmental variations that influence getting enough food to support a given population" (Bousman 1993: 64). However, his use of the term in reference to technological choices follows Weissner (1982a) and clearly adheres to the probability of loss definition: he argues that technology reduces risk primarily by preventing loss, and that increased variability in access to food should result in (1) increased discard of worn but still-usable tools, to reduce the chances that worn tools will fail in use, and (2) more intensive repair/maintenance of tools, for the same reason. Bousman's usage of "risk" then, links technology fairly specifically to attempts to control the "probability" component of the "probability of loss" definition.

Hiscock (1994: 276) makes this link somewhat more explicitly, equating risk with "a low probability of foraging success" which he notes can "incur large energy costs on any group but, in some instances, may endanger individuals or groups". Focusing on the mid-Holocene appearance in Australia of "highly retouched, finely shaped implements" (Hiscock 1994: 277), he argues that the technological patterns he discusses develop during times of rapid environmental change and when humans were in the process of colonizing new territory (see also Webb & Rindos, Chapter 13). Under these circumstances, Hiscock follows Kelly (1988; also see Parry & Kelly 1988) in arguing that humans know less about specific resource distributions than when conditions are stable or when they have a long acquaintance with a region, and therefore find it difficult to predict when and where they will move. In such a context, designing tools for easy transport, multiple uses, and relatively long use lives should help to minimize the possibility of running out of tools with no replacements available.

The probabilities that these authors implicitly or explicitly emphasize refer to the chances that tools either will not be available or will fail to perform as needed, resulting in failure of the activity in which they are used. We will refer to these as "failure probabilities". Despite a lack of overt references to "risk" in the great majority of the literature on flaked stone technology, a close analysis of most of the arguments that archaeologists have made in this literature indicates that they rely ultimately on some consideration of failure probabilities and thus, however incompletely or implicitly, on considerations of risk.

For example, maintaining access to raw material (Bamforth 1986) is essential to any viable technological strategy in order to ensure that it is possible to manufacture replacement tools, without which activities requiring those tools may fail. Similarly, Bleed (1986) discusses two aspects of tool design: maintainability, or the degree to which a design emphasizes features that make it easy and quick to return a broken tool to useful condition, and reliability, or the degree to which a design ensures that a tool will operate effectively under a range of conditions. As Bleed's discussion makes clear, emphasizing one or the other of these alternative (although not mutually exclusive) strategies amounts to selecting a way to ensure that useful tools will be available when they are needed, which is to say that these are different ways of reducing the probability of technological failure. Hayden (1981) argues along very similar lines to Nelson (1996), referring to "resource reliability" rather than risk; in particular, he predicts that specialization on a few highly productive resources to increase reliability should be accompanied by complex and specialized tools, while strategies that increase reliability by diversifying the diet should be accompanied by "complex and diversified" tools, although he sees somewhat different links between subsistence patterns and technology than Nelson (Hayden 1981: 526). Jeske (1989: 35) also links the complexity and diversity of tools used to procure food to the need to reduce "activity failure rates" (in our terms, failure probabilities), and, like Nelson (1996) and Myers (1989), points out that mobile resources (particularly aquatic resources) are likely to have substantially higher failure rates than immobile resources like plants.

In fact, the logic of virtually all of the theoretical literature on flaked stone technology (see Nelson [1991] for a useful review) rests on the assumption that the central problem that technological strategies have to solve is that of ensuring that the tools on which humans rely are available and in useful condition when they are needed, and links variability in these strategies to variability in the conditions under which different human groups live and have lived (Jochim 1981, Nelson 1991). There is, in effect, essentially universal implicit agreement among archaeologists that technology reduces risk (in Wiessner's [1982a] terms) by preventing loss, or at least reducing the probability that loss will occur (Bousman 1993: 68-69, Jochim 1981: 114, Torrence 1989: 64), implying that the risk concept may offer the basis for building a useful synthetic conceptual framework for work in this area. We therefore turn to discuss some of the implications of the varying definitions of risk we have outlined here, as a basis for assessing how risk might be useful in the study of technology.

Why Definitions Matter

It is clear that incorporating risk into anthropological and archaeological analyses has expanded our understanding of the determinants of human adaptations in important ways. However, varying definitions of risk in general and inconsistent use of chosen definitions limit the possibility of building on existing work and complicate synthesis. As we have noted, technological studies have rarely appealed to the "risk-as-variance" perspective, making the basic distinction in the non-archaeological literature between this view and the "risk-as-probability-of-loss" view less of a problem here, but it is still important to consider the differences between them.

Smith (1988: 231) refers to the risk-as-variance definition as "technical", and argues that it subsumes the "colloquial" probability of loss definition. We disagree. Using a measure of variance to define risk is certainly mathematically convenient, and thus may be technical (but see Larson et al. [1996] for an equally rigorous, and equally "technical", measure of risk based on frequencies of drought years in the pre-Contact American Southwest), and it does subsume losses, at least in the sense that the mathematical variance of a distribution encompasses both positive and negative deviations from the mean. Nevertheless, equating risk with variance includes within the operationalized definition of the term both unpredictable shortfalls and unpredictable windfalls, and few people would view the latter as contributing to risk under most circumstances. The "surprising" (Cashdan 1990, Stephens 1990) conclusion that organisms should be "risk-prone" under certain conditions derives from this, and this conclusion is much less surprising from the probability-of-loss view: "risk-prone" behavior is sometimes adaptive because, when expected returns are below the levels required for survival, seeking out variable returns introduces the possibility of obtaining unpredictable windfalls, thereby reducing the probability of suffering a loss. We might therefore clarify important issues somewhat by accepting the "probability of loss" definition and referring to "variance-prone" and "variance-averse", rather than "risk-prone" and "risk-averse", foragers.

Regardless, the work that is most important here clearly is most compatible with risk conceived as probability of loss. However, reliance on this definition requires explicit attention to its two distinct components: the probability that a loss will occur and the magnitude (or cost) of that loss are fundamentally different things, and risk in this sense results from the interaction between them. Rescher (1983) discusses this interaction at length, but it can be clarified by a simple example: the probability of falling from a tightrope is much the same whether the rope is one foot or 100 feet above the ground, but the cost of falling, and thus the risk entailed by walking the tightrope, is much higher in the latter case than in the former. Analysing risk and its relation to technological decision-making thus requires attention to both costs and the probabilities of bearing these costs and, although Torrence (1989) has made this point, archaeologists have yet to integrate it into our analyses. This is particularly important because, as Torrence (1989) points out, events with similar probabilities of occurrence may results in very different costs in different situations, and may therefore select for very different technological responses.

Focusing on definitional issues therefore highlights two particularly important points. First, it seems clear that at least one important component of risk, the probability of failure, is central to most current approaches to the study of flaked stone technology. Second, though, it is also clear that archaeologists have rarely explicitly attended to the importance of the second component of risk, the costs of such failure. Following Torrence (1989), we will therefore argue for the remainder of this paper that explicitly attending to failure costs as well as to failure probabilities will greatly help to expand our current understanding of technological variability. In addition, we will also argue, (1) that a second kind of costs, the costs of making particular technological choices (which we will refer to as "technological costs"), must also be built into our discussions (also see Bleed 1986), and (2) that it is necessary to assess failure costs, failure probabilities, and technological costs not only in the domain of application but throughout the continuum of tool production and use defined above.

RISK, COST, AND TECHNOLOGY: A CONCEPTUAL FRAMEWORK

> Only the pathologically risk-averse make choices based on the consequences without regard to the probability involved . . . only the foolhardy make choices based only on the probability of an outcome without regard to its consequences (Bernstein 1996: 100).

At the most basic level, common sense obviously suggests that high failure costs ought to select for technological responses to reduce failure probabilities: if an event would have truly awful results, it is to our benefit to ensure that it does not occur, and designing technology to this end is one effective way to do this. Again, this is, in effect, the logic most archaeologists have followed in discussing technology, but it is important to note that this assertion also implies that low-cost failures do not have this selective force and that, as Torrence (1989) notes, the same activity may have different associated costs when it is carried out under different circumstances. In this view, Jeske (1989), Myers (1989), and Nelson (1996) are incorrect in arguing that hunting unpredictably mobile animals is necessarily riskier than gathering immobile plants or relying on cultivated plants: hunting is likely to be more failure-prone than gathering and, in at least some circumstances, than cultivation, but the cost of hunting failure, and thus the risk of hunting, depends not on the chances of killing an animal but on the likelihood that failing to kill an animal will cause the hunter to starve (although, as Christenson [1982: 422] points out, hunting is often riskier than plant-gathering or farming in the sense that wounded or threatened animals may injure or kill the hunter). If there are abundant alternative resources available, this cost is low; if there are no alternatives, this cost is high (Torrence 1989: 64).

To generalize beyond the domain of tools used to obtain food, the degree to which it is important for technology to reduce failure probabilities should vary with the costs of those failures. However, humans live in a world with limited time and resources, and we therefore have a limited ability to "invest" in technology. Making tools takes time and energy in procurement and production, and choosing to spend large amounts of time or energy producing tools, organizing labor, etc., makes it impossible to spend that time or energy doing something else–there are "opportunity costs" (Rescher 1983, Winterhalder 1987) entailed by investing in technological solutions to problems. Furthermore, much of the recent attention to risk management in environmental hazards work (i.e., Jones-Lee 1989, Rescher 1983, Sewall 1990, Westman 1985) has observed that greater reductions in the probability of encountering such hazards often require greater investments of resources, and that, while it may be technologically feasible to reduce such probabilities to zero, it is often not economically feasible. We should thus expect humans to vary in the degree to which they can afford to bear technological costs, and situations may exist that preclude a reliance on high-cost technology even when such technology would reduce the probability of high-cost failures.

In brief, then, technological strategies should be linked to risk by their ability to reduce failure probabilities in the face of high failure costs, but a group's ability to bear the costs of relying on such strategies is likely to limit potential technological options. Arguments that search for a balance between costs and benefits, as ours obviously does, follow the logic of optimization reasoning. A number of recent authors (i.e., Bettinger 1991, Bousman 1993, Kelly 1995, Nelson 1996, Torrence 1989, Winterhalder 1987) have discussed this logic at some length, and it is unnecessary to reiterate these discussions here. However, it is important to make one point in this context. Optimization models rely on idealized views of likely adaptive responses to limited sets of variables, and all optimization reasoning incorporates the unspoken caution that optimal predictions refer to a situation in which the behavior being studied is not influenced by any important factors that are not included in the optimization model at issue. The great value of optimization models is their ability to clarify our thoughts and direct our attention to specific aspects of the world; they offer a clear baseline against which to assess the world, which is almost never completely consistent in detail with those models. As Winterhalder (1987: 313, 315) notes, "optimal foraging models are valuable. . .not as lawful statements about reality but as structured forms of inquiry, more interesting to stalk than to live by. . . We seek, then, models that are good to think".

Cultural anthropologists, faced with the spectacular complexity of day to day human activities, are generally very clear that optimization predictions are guides to help us investigate the real world, and focus explicitly on both the fit between models and reality and the likely factors that produce differences between models and reality. However, archaeologists are often not clear in this sense, and sometimes treat untested optimization predictions as facts; such predictions are sometimes then taken as rules that allow us to interpret patterns in archaeological data as evidence for specific kinds of adaptive responses. This is a mistake: theoretical arguments like those presented here, no matter how plausible, require empirical support. Furthermore, it is extremely difficult to test such arguments adequately against purely archaeological data. Most tests of optimization predictions, even those that ultimately focus on archaeological situations, therefore examine those predictions in an ethnographic context (i.e., Bamforth 1988; Bird, Chapter 16; Jochim 1976). We follow this procedure by examining cross-cultural data on recent hunter-gatherer technology in light of our discussion so far. However, some archaeological data are easier to deal with than others, and we also provide a case study of technological change that relies on reasonably clear archaeological data before turning to deal specifically with flaked stone technology.

RISK AND TECHNICAL OPTIONS: CROSS-CULTURAL AND ARCHAEOLOGICAL DATA

As have a number of other studies of technology (Bousman 1993, Shott 1986; Torrence 1983, 1989), we turn to a set of data on hunter-gatherer subsistence technology compiled by Oswalt (1976) to illustrate the general relations we have proposed between risk and technology; our additional analysis here builds primarily on Torrence's work. These data, which include information on the number of tools used to acquire food (technological diversity) and number of parts in each of these tools (technological complexity) for a series of hunter-gatherer groups, provide at least some basis for examining the relations we have suggested above and thus documenting empirically that there is indeed an interesting relation between technology and risk. Our analysis has three parts. First, we consider Torrence's specific use of these data to examine the link between technology and risk and show that sampling problems in the data make this link at once both more complex and somewhat more problematic than she asserts; particularly, her arguments depend on equating risk with failure probabilities, and this is incorrect. Second, we show that considering the "cost" component of risk helps to eliminate some of these problems and, third, we argue that that integrating information on technological costs helps to eliminate them as well.

Table 7.1 Technological and Dietary Data on 20 Hunter-Gatherer Groups (from Torrence [1983, Table 3.1] and Murdock [1967]).

In columns 2 through 6, the first number refers to the total number of tools in the category and the second refers to the total number of parts that make up those tools; the average number of parts per tool can be obtained by dividing the second number by the first.

Group	lat.	instr.		weap.		tended facil.		untended facil.		total		% aqu. res.
Tiw	12	3	6	6	6	2	2	0	0	11	14	10
Andamanese	12	4	8	4	31	3	12	0	0	11	51	40
Ingura	14	3	3	6	19	3	8	1	2	13	32	20
Chenchu	16	7	13	7	26	6	16	0	0	20	55	5
Naron	19	2	5	5	19	3	3	2	11	12	40	0
Aranda	24	4	7	4	21	7	10	1	4	16	42	0
Owens Valley Paiute	37	4	9	9	44	10	30	5	24	28	107	10
Surprise Valley Paiute	42	7	15	9	27	19	41	4	14	39	97	30
Tasmanians	42	3	3	3	3	4	8	1	1	11	15	20
Klamath	43	9	18	7	35	22	70	5	28	43	151	50
Twana	48	4	7	12	70	19	96	13	64	48	237	60
Tlingit	58	4	7	8	25	8	34	8	55	28	121	60
Tanaina	60	7	13	16	83	3	17	14	111	40	224	50
Ingalik	62	6	14	13	64	15	61	21	157	55	296	50
Nabesna	63	1	1	8	36	8	23	8	45	25	105	20
Caribou	63	3	12	10	39	13	37	8	30	34	118	40
Angmaksalik	66	4	18	18	151	9	20	2	13	33	202	80
Iglulik	69	3	8	20	142	8	27	11	48	42	225	50
Copper	70	4	16	8	53	11	36	4	17	27	122	60
Tareumiut	71	1	3	18	133	10	41	6	28	35	205	70

Latitude, Diet, and Technology

Torrence (1983) initially showed that several characteristics of the individual tools and assemblages of tools used by hunter/gatherers to procure subsistence resources vary on a global scale with latitude. Following Oswalt (1976), Torrence recognizes four distinct classes of tools: instruments (used mainly to procure immobile resources, particularly plants), weapons (used to procure mobile resources, including aquatic and terrestrial resources), tended facilities (such as fish weirs or hunting blinds, that are used to take animals but that function only when humans are present), and untended facilities (such as dead-fall traps, that take animals whether humans or present or not).

Torrence finds several important world-wide patterns in her data. First, instruments dominate the tool kits of low-latitude groups, being replaced, first, by weapons and then, successively, by tended facilities and untended facilities as latitude increases. Second, both the total number of tools used by a group (technological diversity) and the number of component parts per tool (technological complexity) increase with latitude (Pearson's r between latitude and diversity is 0.69 [df = 18, .001 > p] and between latitude and complexity is 0.75 [df = 18, .001 > p]). In addition, untended facilities tend to be the most complex class of tools, followed by tended facilities, weapons, and implements. Finally, tools used to procure aquatic resources are more complex than tools used on land. In a later publication, Torrence (1989) explains these patterns by arguing that, for hunter/gatherers, subsistence risk increases with latitude, and that the trends she identifies are adaptations to this increase.

In this view, risk increases with latitude because increasing seasonality reduces the period of the year during which resources are available, because potential resources tend to be fewer and less abundant towards the poles, and because high latitude hunter/gatherer groups tend to rely on mobile (animal) resources whose ability to escape potential predators makes them inherently riskier resources than immobile resources such as plants. Under these conditions, Torrence argues that it is optimal to use tools that are carefully designed to acquire specific resources and that therefore reduce the probability that an attempt to obtain a resource will fail. As risk increases, this should lead to a reliance on a diverse set of specialized tools, each of which is intended for a specific purpose and is therefore poorly designed for other purposes. Similarly, she argues that complex tools, or tools with many parts, increase the speed with which tasks can be accomplished and, possibly, the ease with which broken tools can be repaired; such tools should also be used in riskier contexts. Finally, the complexity of tools used to procure aquatic resources reflects the risk inherent in a reliance on such resources: fish and sea mammals are more difficult to attack than terrestrial fauna, and cannot be tracked as a wounded land mammal can be; fishing tackle and other gear used to take aquatic resources are often specialized (i.e, are useful for a limited range of tasks) and complex (i.e., have many parts) because retrieval mechanisms are built into them (Torrence 1983: 20-21). Thus, the global trends in Oswalt's data can be explained as technological means of reducing risk.

These data seem to provide relatively strong support for the argument that risk strongly conditions technology. However, Torrence's analysis nowhere explicitly considers the cost of technological failure for the societies in her sample, and this is a problem because of a serious sampling bias in Oswalt's data: these data include a highly selected, non-random sample of hunter/gatherer societies (Eskimo groups, for example, are dramatically overrepresented relative to other groups), and there is a serious bias in them when they are considered in terms of latitude. This bias derives from the relationship in the sample between dependence on aquatic resources and latitude: virtually all of the high latitude groups in Torrence's sample rely heavily on aquatic resources, and there are very strong correlations between the percent of such resources in the diets of the groups in her sample (Table 7.1, data from Murdock 1967) and technological diversity ($r=0.61$) and complexity ($r=0.76$; also see Bousman 1993: 73). As might be expected, there is also a very strong correlation between latitude and the percent of aquatic resources in the diet ($r=0.72$).

This is very important, because Torrence explicitly distinguishes betwen risk resulting from increasing seasonality at high latitudes and failure probabilities inherent in the pursuit of aquatic resources. As just noted, Torrence follows Oswalt (1976) in noting the relationship between dependence on aquatic resources and complex tools, and this fact is essential to interpreting the patterning in these data. That is, it is possible to argue that tools in Oswalt's data set that are used to take aquatic resources are complex not because they are used under riskier conditions but because capturing fish and sea mammals requires complex tools. Because the available data confound dependence on aquatic resources and the many aspects of environmental variation measured by latitude, Torrence's analyses do not allow us to to determine whether differences in technological diversity and complexity reflect differences in risk due to (1) such global factors as seasonality, (2) the requirements of the specific tasks that the tools at issue were designed to accomplish, or (3) some combination of the two.

This problem can be addressed quantitatively by computing the first-order partial correlations between latitude, percent of aquatic resources in the diet, and the two technological variables. First-order partial correlations measure the degree of relationship between two variables when the effects of a third variable have been removed; statistically, they represent the correlation between the residuals of these two variables when each has been regressed against the third. They therefore make it possible to identify spurious correlations resulting when two variables are related not because one determines the other but because they are both determined by some third variable, and they thus make it possible to control for the bias in the dataset and distinguish between the effects on technology of reliance on aquatic resources and of seasonality, at least as it is monitored by latitude.

Table 7.2 summarizes the results of this analysis both for the measures of overall toolkit diversity and complexity and for each of the four specific classes of tool defined above. Torrence measures complexity as the total number of parts in the overall set of tools used by a society, a measure that clearly increases as toolkit diversity increases whether the tools being used have many parts or not; the average number of parts per tool measures complexity more directly than total number of parts, and we use this measure for the remainder of this section.

Blalock (1960: 337-343) discusses the interpretation of partial correlations in some detail. Following his discussion, the results in Table 7.2 reduce to five patterns of interaction. First (and simplest), the diversity of instruments is uncorrelated with both latitude and depen-

Table 7.2 Simple and First Order Partial Correlations Between Latitude, Percent Dependence on Aquatic Resources, and Technological Diversity and Complexity.

Variable	1	2	3	4
All Tools				
diversity	0.69	0.60	0.47	0.19
complexity	0.67	0.76	0.28	0.53
Instruments				
diversity	-0.09	0.03	-0.17	0.15
complexity	0.48	0.52	0.18	0.28
Weapons				
diversity	0.72	0.68	0.45	0.33
complexity	0.42	0.59	-0.01	0.46
Tended Facilities				
diversity	0.41	0.35	0.24	0.09
complexity	0.48	0.62	0.07	0.45
Untended Facilities				
diversity	0.63	0.45	0.49	-0.01
complexity	0.41	0.43	0.14	0.19

1 simple correlation with latitude
2 simple correlation with percent dependence on aquatic resources
3 first-order partial correlation with latitude, controlling for percent dependence on aquatic resources
4 first-order partial correlation with percent dependence on aquatic resources, controlling for latitude

dence on aquatic resources whether we compute simple or partial correlations. Second, in three cases (overall complexity and the complexity of weapons and of tended facilities), partialling greatly reduces the correlation between latitude and technology but leaves a substantial correlation between dependence on aquatic resources and technology, indicating that latitude is essentially irrelevant in these cases once diet is taken into account. In three cases (overall diversity and the diversity of tended and untended facilities), the situation is reversed, with latitude important and diet irrelevant.

The remainder of the results (the complexity of instruments and untended facilities and the diversity of weapons) have a more complex interpretation. In these cases, partialling reduces both of the correlations, a pattern that is generally expected when the relationship being examined involves an independent variable affecting a dependent variable through an intervening variable. Controlling the independent variable in this case should reduce the correlation between the intervening and dependent variables (Blalock 1960: 342). In the present case, the reduction in both partial correlations suggests that there is a complex interaction between the many environmental characteristics measured by latitude and the subsistence patterns measured by percent dependence on aquatic resources, with the effects of each mediated by the other and, probably, by other variables as well. Because of the high correlation between latitude and diet in the available data, it is probably not possible to sort this interaction out using these data.

These results reveal a complex pattern of interactions that varies with the specific category of tool being considered and that suggests a somewhat different set of conclusions than was indicated by Torrence's initial analysis. Overall, these data suggest that increasing latitude seems generally to lead human beings to produce more kinds of tools but not necessarily more complex tools. In contrast, increasing dependence on aquatic resources seems generally to lead human beings to produce more complex tools as well as more different kinds of tools, although there appear to be important but presently uninterpretable interactions between these variables that affect some aspects of both toolkit diversity and complexity.

Integrating Failure Costs

As we note above, latitude monitors many aspects of environmental variation, making it difficult to attribute the increased diversity of tools at higher latitudes to any one of these. However, the seeming absence of a connection between technological complexity and latitude merits a closer evaluation. Taken at face value, this result would seem to refute Torrence's argument regarding the importance of complex technology in reducing risk, and, given the higher technological costs of more complex tools, would also seem to pose problems for our aguments here. However, Torrence's arguments linking both technological diversity and technological complexity to risk treat this concept primarily as probability of failure, and our arguments explicitly require attention to failure costs, which the analysis above does not consider. One important reason for the great dependence of high latitude groups, particularly arctic groups, on aquatic resources is the relative paucity of alternative resources at high latitudes. To demonstrate that the patterns in the available data relate to the risk associated with exploitation of aquatic resources, instead of simply to the mechanical problems of retrieving such resources, it is necessary to consider failure costs. As Torrence points out, in the context of subsistence, the risk entailed by dependence on aquatic resources derives not only from the the practical problems of capturing them but also from the abundance of alternative sources of food. Considering this issue puts the data in Tables 7.1 and 7.2 in a somewhat different light.

The narrow range of resources available to most high-latitude hunter-gatherers provides one key to doing this. This observation suggests that risk increases towards the poles not because of increases in reliance on mobile prey, which increases the probability of procurement failure, but, rather, because of the absence of alternative sources of food, which raises the costs of failure when attacking mobile prey. More diverse toolkits at high latitudes, then, can be seen as attempts to reduce higher failure probabilities because of high costs. In support of this, we can measure dependence on mobile prey as the proportion in the diet made up by hunting and fishing. This measure correlates almost perfectly with latitude ($r = 0.96$), obviously indicating that the abundance of plants in the diet drops directly with latitude (Lee 1968). The simple correlation between the number of weapons in a toolkit and dependence on mobile resources is 0.73, suggesting that reliance on mobile resources requires a diverse set of tools. However, holding latitude constant by computing a partial correlation reduces this value to 0.29, implying that reliance on mobile prey–that is, reliance on resources that are inherently difficult to capture–does not by itself predict technology very accurately. This suggests that it is also necessary to consider the effects of the decreasing availability of alternative foods, and therefore the increasing costs of procurement failure, with increasing latitude.

Oswalt's data do not offer any obvious means of doing this, but we can examine this issue further by considering the complexity of a specific class of weapons and comparing some of these data to information on a group not included in his sample. One means of considering the effects on technology of failure costs is to examine the characteristics of tools used in the same activity when it is reasonable to infer that the costs of that activity differ. Although it is difficult to ensure that activities carried out by different societies are indeed "the same", it is reasonable to infer that they are nearly so in at least some cases. The case that we consider here is hunting sea mammals from a boat in open water. The most complex weapons in Oswalt's sample are harpoons used for this purpose by northern Eskimo groups such as the Iglulik, Tareumiut, and Angmaksalik, that average 20 to 25 parts and range as high as 33 parts (Oswalt 1976: 99-101, 281-294). The complexity of these tools derives in large part from the addition to them of lines, drags, and floats designed to tire a wounded animal and to assist in retrieving it.

Like the Eskimo, the Chumash, a complex maritime hunting and gathering group on the California coast, used harpoons to take sea mammals from boats in open water. However, the Chumash did so under very different conditions than did the Eskimo. The Chumash relied heavily on a very wide range of relatively abundant aquatic resources, including shellfish, and also exploited a wide range of terrestrial foods (Landberg 1965). Given a diverse and productive subsistence base, the cost of losing any one resource, and therefore the risk entailed in attempting to acquire it, must have been lower for the Chumash than for high latitude coastal groups such as the Eskimo. Given weaker selection for success in any one subsistence pursuit, then, the arguments we have made above suggest that Chumash harpoons should be simpler than those used by the Eskimo.

The excellent descriptions of Chumash material culture compiled by Hudson and Blackburn (1982) indicate that, as predicted, Chumash harpoons used in open water are much less complex than those used by the Eskimo, having only seven or eight parts. These descriptions do not suggest, however, that the Chumash relied on a less diverse range of tools than did most high-latitude Eskimo groups, although we have not attempted to quantify this. Similarly, the maximum complexity of any kind of harpoon used by Athapascan groups such as the Ingalik and Tanaina is 11 parts (Oswalt 1976), and these groups inhabited environments that are richer and more diverse in both terrestrial and aquatic resources than those exploited by the Eskimo groups producing more complex tools (although it is difficult to view the Ingalik and Tanaina environments as rich or productive in any absolute sense). Although data on such a small number of societies provide only a partial test of any hypothesis, these examples are clearly consistent with our arguments, suggesting that failure probabilities and failure costs must be considered separately in assessing risk and that technological decisions integrate considerations of both of these.

To briefly summarize, we have made two important points in this section. First, a reanalysis of Oswalt's (1976) data on hunter/gatherer technology provides only partial support for Torrence's (1983, 1989) interpretation of these data. In particular, increasing latitude, which Torrence takes as a proxy measure of increasing subsistence risk due to greater and greater seasonality, seems to relate primarily to toolkit diversity and not to toolkit complexity. Degree of dependence on aquatic resources, which is highly correlated with latitude in Oswalt's data, appears instead to be the major conditioner of technological complexity in this dataset, and also plays an important role in conditioning some aspects of diversity as well. It is therefore not possible to appeal to the correlation between complexity and latitude as evidence supporting the importance to tool design of overall, seasonally-based "risk"; this correlation is largely the spurious result of systematic sampling bias in the available data.

The links between aquatic resources, latitude, and technology might suggest that tool design responds to at least one component of risk, the probability of failure. However, we further pointed out that such a conclusion neglects the other component of risk, failure costs. Focusing on this problem, we were able to show that the available ethnographic data support the importance of failure costs as conditioners of technology. We therefore conclude that Oswalt's data suggest that at least some important aspects of technological design respond to failure costs rather than failure probabilities, as our arguments above predict.

Integrating Technological Costs

However, we noted earlier that a second domain of costs, opportunity costs entailed by choosing to invest in a particular technological strategy, are also likely to be important conditioners of technological behavior: both the full continuum of behavior involved in the production and use of stone tools and the connections between this continuum and other domains of human activity are essential to understanding the overall character of any technological adaptation. Torrence's arguments regarding the data we have just considered deal almost exclusively with the domain of application, and do not consider either procurement or production (as is evident, for example, in such arguments as "when the risk of failure to procure food is high, hunter-gatherers respond by increasing their overall investment in technology and in the diversity of tools" [Torrence 1989: 61]). Our analysis of these data to this point similarly focuses only on the domain of application.

However, an exclusive emphasis on the domain of application requires the assumption that there are no constraints operating in the domains of procurement or production, an assumption that can seriously limit our understanding of the forces that structure technology. Ebert (1992: 35) offers a particularly clear example of such limited understanding, asserting that "Bamforth (1986) argues that it is not the subsistence strategies of a group that determine whether its technology is curated or expedient, asserting that curation is instead a response to raw material shortages. This is tantamount to thinking that someone will not use a pocket knife for cutting if there is a source of steel nearby with which an expedient blade could be fashioned". Unfortunately, the rhetorical panache of this argument masks a failure to grasp some of the most important differences between stone and metal tools, differences that lie particularly in the dramatically higher production costs associated with most metal technology. Many flaked stone tools can be manufactured in an instant, and even the most complex flaked stone tools (for example, fluted points) take a skilled stoneworker no more than an hour to produce; flaked stone tool production also requires very simple tools, most often no more than a hammerstone, an antler billet, and a pressure flaker. It is likely that most stone tool users made most, if not all, of their tools themselves. Manufacturing metal tools, in contrast, virtually always requires substantial investments in production technology, labor specialization, etc., investments that make it effectively impossible for the great majority of metal tool users also to be metal tool producers. Direct comparisons of stone and metal technology that, like Ebert's, fail to take this crucial distinction into account are inevitably misleading (Barton [1991] points out in more detail a range of other important differences between stone and metal tools).

Aspects of Oswalt's data indicate that constraints in the domain of production are extremely important determinants of technology, and considering this issue highlights one specific relationship that is likely to be important in many situations. It is possible to invest substantial amounts of time and/or energy in tool design and construction only when it is possible to take time and/or energy away from other activities without compromising those activities. That is, technological costs as well as failure costs may be high or low. There appear to be two major contexts within which time-intensive (or high cost) technological adaptations are feasible. First, substantial amounts of time may be available to all members of a society when they rely primarily on stored food for some portion of the year, or, at least, when they have some predictable, extended period of "down-time" available. Second, substantial amounts of time may be available to some members of a society if the labor of others is sufficient to generate a surplus of needed goods to support them. Neither of these alternatives is feasible in all circumstances. To illustrate these points, we turn to the technological contrasts evident in Table 1 between the Tanaina and Ingalik, Athapascan groups in western Alaska, and the Caribou Eskimo of the region just west of Hudson's Bay. These groups occupy territories at almost identical latitudes and incorporate similar proportions of aquatic resources in their diets[1], yet the Tanaina and Ingalik show a far more diverse and complex toolkit than do the Caribou Eskimo.

Like any other general technological pattern, this difference probably reflects more than one factor. In part, for example, it is likely that differences in the density and specific character of the aquatic resources on which these groups relied affected the ways in which those resources might be taken. However, it is likely that the much greater diversity and complexity of tools among the Athapascan than the Eskimo groups may also reflect different constraints operating in the domain of production.

Both the Tanaina and the Ingalik lived in permanent settlements and supported themselves on stored food for substantial portions of the year (Osgood 1937, 1940, 1958); these groups thus had substantial and predictable periods of down-time during which tools could be made and repaired. In contrast, the Caribou Eskimo followed a relatively unpredictable pattern of seasonal movement, and hunted year-round; although they laid in winter stores, these often were not enough to support family groups until spring (Birket-Smith 1929). Under these conditions, the opportunity to devote large amounts of time and effort to the manufacture of diverse and complex tools is not available, whether or not it might be useful to have such tools, because this time and effort must be taken away from subsistence pursuits.

A few specific examples highlight this pattern. Oswalt (1976) catalogs only two parts for the single type of fish weir constructed by the Caribou Eskimo, but four to eight parts for two different kinds of weirs among the Ingalik. The Caribou Eskimo also produced a single type of fish trap with six parts for use at weirs, while the Ingalik used four different kinds of traps consisting of eight to 12 parts each (mean = 11.0). In procuring land mammals, the Caribou Eskimo relied on four types of untended facilities (pitfalls and cages) made from two to five parts (mean = 3.5), while the Tanaina used six, and the Ingalik eight, such facilities (in both cases, these are snares and deadfalls) made from five to 12 parts (mean = 9.0).

The significance of this pattern is emphasized further by the likelihood that the costs of failure in any given procurement attempt were notably higher for the Caribou Eskimo than for either the Ingalik or the Tanaina. Periodic starvation was relatively common among this group, making all subsistence resources and all efforts to procure these resources important. In contrast, as noted above, both the Ingalik and the Tanaina exploited relatively more diverse and productive environments. That is, these patterns suggest that the Caribou Eskimo inhabited an environment in which high failure costs gave a selective advantage to more complex and diverse toolkits, but which at the same time limited the ability of a human group to bear the high technological costs entailed by producing such toolkits.

There are, of course, other possible expanations for these patterns. For example, we have argued that the higher and more unpredictable mobility of the Caribou Eskimo made it difficult to devote time to the construction of complex tools, but we might also follow Shott (1986) and argue that the simple act of moving from place to place frequently should impact technology by selecting for tools that are compact and easy to transport. However, it is not likely that this explanation accounts for the patterns we have just noted, because the tools that we examined are not used in mobile subsistence activities: fish weirs, traps used at weirs, and untended facilities like deadfall traps are all designed and constructed for use at specific points on the landscape; indeed, in many places, fish weirs are permanent facilities to which human groups return year after year. Better ethnographic data on very specific aspects of tool production, transport, and use would certainly help to sort factors like this out, but the currently available information supports our arguments here.

This example is particularly important because it emphasizes the importance of technological costs as conditioners of technological choices. We have stressed the central importance of failure costs to technology, and a logic much like the one we have followed in arguing this importance underlies most or all archaeological analyses linking technology to overall adaptive patterns. However, the constraints on production evident among the

Caribou Eskimo highlight the potential problems with this logic. Human beings inhabit a world in which demands on time and other resources vary in time and space. Where such demands are limited, humans can choose from a wide range of technological alternatives; where they are not, high-cost alternatives may be impractical regardless of the pressures that might select for them (cf. Maynard Smith [1978] and Pyke [1984] on analogous limits on optimization models in general). We also emphasize the Caribou Eskimo example because it seems likely that the interaction between advantages in the domain of application and constraints on the domain of production is important in many situations, particularly among highly and unpredictably mobile hunter-gatherers and among horticultural groups who rely on plants or animals that require substantial, year-round labor investments and do not produce large surpluses. Whether this is correct or not, though, the Eskimo example highlights the importance of considering risk in more than one domain of behavior: technological organization is not simply determined by the domain of application (also see Bleed & Bleed 1987: 196).

An Archaeological Case: Ditch and Palisade Defenses

The cross-cultural data just discussed thus fit fairly well with the general relations we have outlined here: important aspects of technology can be seen as attempts to reduce failure probabilities in response to high failure costs, with the ability to bear technological costs limiting possible technological options for doing this. A specific archaeological and historical example also provides support for this argument, and emphasizes as well that technology responds to risks in addition to those resulting from uncertainties in the food supply. This example is the construction of ditch and palisade defenses around settled communities along the Missouri River in North and South Dakota from the 1300's through the 1800's.

Warfare among settled horticultural groups along the Missouri River in North and South Dakota (and between these groups and neighboring hunter-gatherers) during the Post- Contact period is well-documented by both archaeology and ethnohistory, and this warfare often resulted in massacres and destruction of entire towns (Lehmer 1971, Owsley et al. 1977, Thwaites 1906). Furthermore, indisputable archaeological evidence indicates that essentially identical patterns of conflict, which sometimes resulted in massacres of the populations of entire towns, existed in the region well before contact, at least as far back as the early 1300's (Bamforth 1994, Willey 1990).

One of the principal technological means of dealing with the risk entailed by this conflict was the construction of extensive defensive works, including the excavation of ditches up to 10 feet deep and 20 feet across which were backed by continuous wooden palisades; many of these defenses included bastions, presumably to allow defenders to set up a cross-fire when necessary. Although towns were often situated on narrow ridges overlooking the Missouri River and defenses were sometimes constructed only across the neck of these ridges, in many cases they completely encircled entire communities. For example, the Arzberger site (Spaulding 1956) was defended by a continuous bastioned defensive perimeter 1.5 miles long, and bastions in a similar perimeter at the Huff site were apparently strengthened by a secondary row of sharpened posts (Wood 1967). The importance of these defenses is indicated at the Crow Creek site, a town whose inhabitants were massacred early in the 1300's: excavations of the fortifications suggest that the site was overwhelmed when its inhabitants had partially dismantled an existing palisade in the course of building a new one (Kivett & Jensen 1971, Willey 1990).

The Post-Contact period saw particularly intense conflict between the settled farming groups (the Mandan, Hidatsa, and Arikara) and the nomadic Lakota, and the frequency of Lakota attacks on farming towns seems to have increased from the 1700's into the 1800's. In the framework outlined here, this implies that the probability of attack (and therefore, presumably, of suffering the consequences of military defeat) was either constant or increasing during this time. Furthermore, contemporary observers and archaeological data show clearly that the costs of defeat in these conflicts (including annihilation of all of the occupants of a community and destruction of their property) were about as high as they could possibly be. However, technological investment in defensive works declined during the Contact period. The elaborate bastioned defenses just noted predate White contact; bastions essentially disappear during the 1700's, and sites dating to the 1800's were often defended only by a palisade without a supporting ditch. The decreasing sophistication of defensive technology in spite of an increasing probability of attack along the Missouri River almost certainly reflects the other major Post-Contact factor that disrupted Indian ways of life there: depopulation resulting from the introduction of epidemic diseases like smallpox. By the 1800's, many settled communities apparently could not field the labor (that is to say, they could not bear the technological cost) required to build elaborate fortifications, and in some cases fortifications were built for Indian groups by Whites (Lehmer [1971: 131-179] discusses the Post-Contact archaeology of the Missouri River valley in detail).

Construction of ditch and palisade defenses along the Missouri River thus appears to fit well with our arguments here: these defenses clearly represented technological mechanisms for reducing risk by minimizing failure probabilities, but constructing them even in situations where they were strongly selected for apparently depended on the ability to bear substantial technological costs. This example also highlights the important point that subsistence is not the only domain of human activity in which we face risks and that technology can be important in other domains as well.

RISK AND FLAKED STONE TECHNOLOGY

Our goal for the remainder of this paper is to sketch an outline of how risk as we have defined it can be considered in the specific context of flaked stone technology. At the outset of this discussion, it is important to reiterate two points. The first is that the risk entailed by any overall technological adaptation results from interactions among the probabilities and external costs of failure and the technological costs entailed by activities in the domains of procurement, manufacture, and application. Any technological adaptation represents a balance of all of these, with the interaction between the domains of production and application perhaps being particularly important. Second, the technological options available to human beings do not all involve tool design and production: as we noted in the first section of this paper, technology includes knowledge about how to apply tools, and a single toolkit can often be used in many different ways to perform a single activity (also see Schiffer & Skibo 1987). Bamforth (1988) has argued, for example, that variation in the ways in which human beings organized themselves on the Great Plains for communal bison hunts was an adaptive response to the environmental conditions under which those hunts were conducted, but follows Fawcett (1986) in noting that essentially the same kinds of tools appear to have been used in this activity throughout the Plains for at least 10,000 years. Not all technological options, then, involve choices about what kinds of tools to make or how to make them, and archaeology needs to develop analytic means of recognizing such options.

Costs of Failure in Procurement, Production, and Application

It is possible to derive a simple heuristic model of the failure costs to which technology responds by conceiving them as dichotomous variables operating within each of the three domains of technological behavior: failure costs can be high or low in raw material procurement, in tool production, or in tool application, producing a total of eight possible combinations (see Figure 7.1). Reality is obviously more complex than the model depicted in Figure 7.1–costs should ideally be measured on a continuous rather than a dichotomous scale, for example (Winterhalder 1986: 385-386) and Figure 7.1 does not incorporate technological opportunity costs–and we present this illustration only as a tool to help think about the issues we have raised here.

If technology helps to minimize risk, we can predict that high cost situations should select for solutions that minimize the probability of technological failure. Predicting technological responses to low-cost situations, though, is somewhat less straightforward. Durham (1976) has pointed out that, although medium- and high-cost behavior is very likely to be constrained by adaptive considerations, low-cost behavior is not, and may vary considerably for idiosyncratic ideological or other reasons. That is, if technological failure just does not create any substantial problems, we may not be able to predict many aspects of technological choices very accurately. Considered in the framework outlined here, this problem may also arise if failure probabilities are extremely low. Durham further points out that the cost of any behavior depends on the "budget" of the organism(s) engaged in that behavior, and this may vary from society to society or from individual to individual or season to season within a single society: the same pattern of behavior may thus be costly in one context and not in another.

This ambiguity can perhaps be reduced, although not eliminated, by considering the costs of investing in a given technological option. As discussed earlier, when an option's costs are unacceptably high, that option should be avoided whether it would reduce risk or not. This suggests that the circumstances that are most likely to be ambiguous from the perspective of the general conceptual framework we have presented here are those in which some aspect of a particular way of life makes it possible to invest heavily in technology even though such investments are not required to reduce failure probabilities in response to high failure costs: for example, group that spend a substantial portion of the year in one place and live off of food stores that can be acquired easily and reliably with simple tools may be able to produce elaborate tools even though they are not "necessary" in the utilitarian sense we emphasize here. Such production might instead grow out of regional patterns of economic specialization or internal social and political processes. However, as failure costs and probabilities rise, risk should play an increasingly important role in structuring technological choices, within the limits set by a society's ability to bear technological costs.

Bearing in mind these probable limits on the conditions under which risk is likely to be a key determi-

		Procurement Costs: high		Procurement Costs: low	
		Production costs		Production Costs	
		high	low	high	low
Application Costs	high	A proc high prod high appl high	B proc high prod low appl high	C proc low prod high appl high	D proc low prod low appl high
	low	E proc high prod high appl low	F prod high prod low appl low	G proc low prod high appl low	H proc low prod low appl low

Figure 7.1 A heuristic model of failure costs in the domains of procurement, production, and application.

nant of technological strategies, it is useful to consider how failures in flaked stone tool procurement, manufacture, and application inflict costs on human beings. The most obvious cost of failure in procurement derives from the absence of the material needed to produce tools. If suitable stone is not available when needed, scheduled production may be impossible, special trips may be required to procure it, and tools may not be available when they are needed, resulting in failure in some other activity. The degree to which a failure to have stone available is risky, of course, depends on the costs of the activities disrupted and of making special efforts to obtain additional material. This latter problem clearly depends on the abundance and distribution of material within a region relative to the locations of human settlement and on a society's ability to take time from other activities and devote it to procuring raw material. Archaeologists studying hunters and gatherers have often minimized this problem by appealing to Binford's (1979) observation that the Nunamiut Eskimo "embed" lithic procurement in other activities, a habit that is often asserted to eliminate or, at least, minimize procurement costs while simultaneously ensuring access to raw material. However, the modern Nunamiut virtually never actually use the stone that they procure this way: the Nunamiut depend on metal tools and use stone tools, on those rare occasions when they use them at all, as an emergency backup. Whether a stone-dependent group could supply itself without periodically making special efforts to obtain raw material is at least open to question, and Gould (1978) has discussed the substantial efforts that stone tool-dependent Australian groups made to obtain raw material.

If tools fail during production, additional time must be invested in producing more, and, if failure rates are too high, tools may not be available when they are needed. High failure rates also require access to substantial amounts of raw material, often requiring greater efforts in procurement. Reliance on tools that are particularly difficult to make may also require at least part-time support for specialist producers and substantial investments of time in training these individuals. Less directly, time spent manufacturing tools is time that cannot be spent in other activities, and spending substantial amounts of time in tool production may impose severe limits on the other things people can do, as we note above. In addition, under conditions where tools are needed continuously or unpredictably, or where long periods of time cannot be set aside for single purposes, it may be difficult to allocate sufficient time to produce some kinds of tools. Finally, failure in application, which may result from tool breakage, poor tool design, or the lack of the needed tools (as well as from other factors, including deficits in the skill or knowledge of the tool user), can mean that the opportunity to obtain a resource or to carry out some other activity that cannot be scheduled at will is lost, or may require additional investments of time and effort attempting the application again; when failure results in the loss of essential resources like food, human groups may face starvation.

Problems like these arise in different ways and to differing degrees in the varying contexts in which humans need and use technology, and technological solutions to these problems are also affected by the varying opportunities and constraints such contexts offer. Recognizing this, archaeologists have identified a wide array of such solutions which appear to be effective under different conditions of (for example) raw material availability, movement across the landscape, etc. Table 7.3 summarizes many of these strategies (also see Nelson 1991), grouped into those that potentially reduce the probability of failure in each of the three domains of technology we recognize here. These options represent our interpretation of the topics that have been discussed in the archaeological literature, as well as our own assessment of possible courses of behavior. Although we do not claim that Table 7.3 presents every course of action that might possibly reduce the risks associated with a reliance on stone tools, it does present the major options that archaeologists have recognized and have not rejected. Ideally, it would be possible to bring together the selective conditions presented in Figure 7.1 with the technological options summarized in Table 7.3 to derive a series of reasonably specific predictions regarding the ways in which stone tool users should have adapted their behavior. In the real world, though, there are serious difficulties in doing this.

These difficulties result most obviously from the paucity of more than impressionistic data regarding the failure probabilities and technological costs of the options in Table 7.3. Data on failures reflected in the archaeological record are rarely presented in systematic form, in part because there have been few methods for organizing and analyzing such information (Bleed 1991). Furthermore, the implications of failure are hard to estimate because there has been little experimental or ethnoarchaeological investigation of this topic. Certainly, estimating failure costs is not a simple matter. Broken preforms or nearly finished points with series of deep step fractures can be interpreted as production failures, and high frequencies of such items may sometimes allow us to infer high failure costs. However, the concept of production failure in our sense here refers not to failure rates for individual items, but to the inability to produce adequate numbers of tools. Even high rates of preform breakage, then, may not result in high failure probabilities if knappers have enough time and raw material to reach their production goals. Analogous points can be made for the domains of procurement and application.

Table 7.3 Potential technological options for reducing failure probabilities in the domains of procurement, production, and application.

Procurement:
- storage of stone
- strategic caching of stone
- economizing raw material use:
  - designing long use-life tools*
  - designing multi-function tools*
  - tool recycling
  - tool resharpening
- exploit many different raw materials sources
- use of durable stone
- permanent residence near a quarry or cache
- transport of large amounts of stone

Production:
- stockpile/cache partially finished tools
- advance production of tools/production during "down" time
- production at a quarry
- production by specialists
- reliance on technologically simple tools*
- use of easily workable stone
- heat alteration

Application:
- careful design for specific purposes*
- overdesign to prevent breakage*
- design for easy repair*
- transport of extra tools
- tool use by specialists/experts
- cooperative group organization of tool use

* option involving artifact design

In addition, it is essential to recognize the distinction between "tools" as archaeologists use the term and "tools" as actual functioning implements in the past. We reject the common equation drawn between "tools" and "retouched pieces", on the grounds that systematic microwear analyses universally show that unmodified flakes were a central component of every prehistoric technological adaptation known (also see Barton 1991, Gero 1990). In this context, we also note that archaeologists are often not very good at identifying used flakes, a serious problem in any lithic analysis (Young & Bamforth 1990). In addition, the presence of retouch on a piece of flakeable stone in no way guarantees that that piece was actually used to carry out a task.

However, a deeper problem in dealing with failure or technological costs is the equation of the flaked stone objects that archaeologists recover with the tools that prehistoric people used. In the case of unhafted, handheld implements, such an equation is reasonable, but it is not in the case of hafted tools. In general, the stone blade of a hafted tool is the "cheapest" part of that tool; manufacturing the haft often takes much more time than manufacturing the disposable blades that are set into it (Keeley 1982). Although the investment of time and energy represented by the haft is made only once over the useful life of the tool as a whole, using a blade in a haft alters the conditions under which a worn or broken edge can be replaced, because replacing a well-hafted blade takes time and, often, such special facilities and materials as fire, sinew or other material for wrapping, and/or some form of glue. The common association of hearth areas in residential sites with spent stone blades (such as projectile point bases) from hafted tools (Keeley 1982) illustrates this clearly. The notion of "production costs" for hafted tools therefore refers not only to the costs of making the stone portions of those tools, the component of such implements with which archaeologists are most familiar, but also to the the costs of manufacturing hafts and of replacing worn or broken blades in those hafts.

In at least some cases, techniques like "event tree analysis " (Bleed 1991) can be used to reconstruct failure probabilities for at least some categories of stone tools. From such calculations and adequate data on the behavioral contexts within which technological failure occurred, acceptable assessments, perhaps even quantitative assessments, of failure and technological costs for flaked stone technology might be approachable; the final section of this paper presents an attempt to do this. However, consideration of the ideas presented here need not wait for these developments. There are serious limits on our ability to rigorously test the ideas we have presented here against most archaeological data. However, if we take the cross-cultural and archaeological data discussed above as provisional evidence that these ideas are useful, it is possible to consider how they illuminate archaeological cases and guide analysis towards interesting questions; to paraphrase Winterhalder (1987: 313, 315), we can treat these ideas as the basis for a form of "structured inquiry" and see whether or not they are "good to think" in specific archaeological contexts. We thus turn to summarize and compare two examples to show that it is possible for archaeologists to address variables like the probabilities and costs of failure in stone age technologies and to suggest that understanding thse variables illuminates interesting aspects of the ways of life that relied on those technologies.

TWO PATTERNS OF PROJECTILE POINT PRODUCTION

The cases we consider are production of microblades during the Late Paleolithic at the Araya site in Niigata, Japan (Bleed 1996, Sutoh et al 1990) and of fluted points during the Folsom period on the North American Great Plains (Akerman & Fagan 1986, Crabtree 1968, Flenniken 1978, Frison & Bradley 1981, Sollberger 1985, Tunnell 1977, Winfrey 1990). We begin by contrasting rates of production failure in these two technologies and then turn to consider the implications of such failure rates.

As reconstructed in several studies (Kobayashi 1970, Yoshizaki 1959), production of microblades in Japan and elsewhere (Andrefsky 1987, Morlan 1970) entailed a number of routinized steps, carried out in precise sequence, with great skill. Based on artifacts found in their hafts in Siberian sites (Kajiwara 1989), it is inferred that microblades were used as replaceable barbs and edges of hunting projectile points. It is also assumed that the small blades were important to their makers, because they are very common in the sites that yield them. However, beyond that, little is known about the operation of the cultures that made the microblades.

To address failure and cost rates in the Araya microblade technology, this example uses "event tree analysis," an operations research technique developed by engineers to model the operation of technical systems (Bleed 1991, 1996; Winfrey 1990). Event trees describe activities in terms of "events," or steps that contribute to completion of the system, making it easy for them to accommodate failed and rejected production residues which are, of course, direct and accurate reflections of the failures associated with a technology.

Essentially all of the flaked tools at Araya were made from a fine-grained hard shale that was brought to the site in the form of bifaces that were used both as cores and as chopping tools. Bifaces were also the starting point for production of microblade cores. Six distinct events were involved in this process. As shown in a graphic model (Figure 7.2), each event is reflected by distinctive residues that can be considered here in two ways. First, they can be counted to calculate the total number of cores that were worked at the site. Secondly, since failed residues are direct reflections of production failures, they can be used to determine failure rates for individual events and for the sequence of events as a whole (Bleed [1996] discusses these data in detail).

Taking refitted pieces into account, archaeological evidence thus indicates that 37 separate cores were made or modified at the Araya site. If we consider the failed pieces shown on the event tree to be concrete evidence of

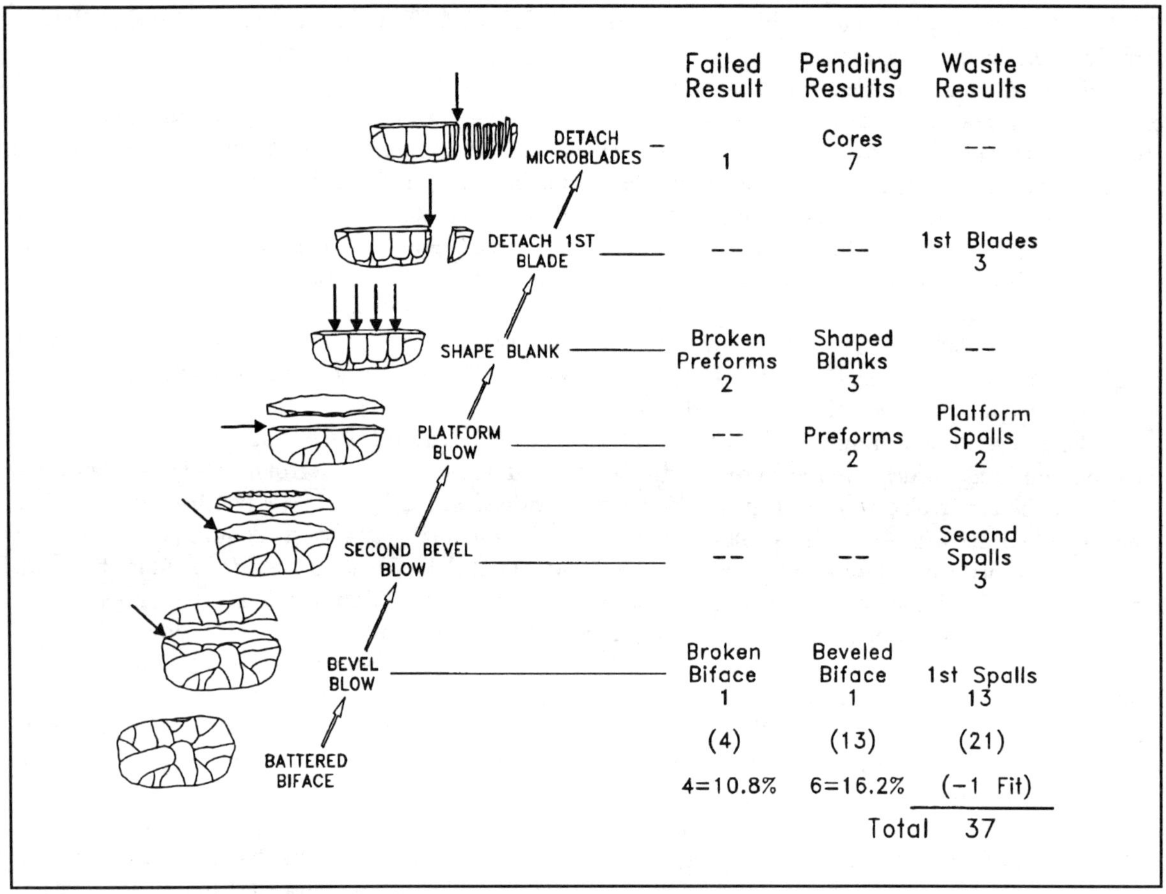

Figure 7.2 Event tree model of Araya microblade production (from Bleed 1996).

the actual failure rates involved in microblade production, the four broken cores in the assemblage suggest that the overall production failure rate for Araya microblade cores was only 10.8 percent. It is hard to interpret that rate, since there is little comparative information available on failure rates in other stone tool technologies. As the next example shows, however, reconstruction of Folsom point failure rates is much higher, which suggests that the 10 percent production failure rate reconstructed for the Araya cores is quite low. Furthermore, the point in reduction where failure occurred is interesting: cores left at Araya failed in all steps *except* during microblade removal. Clearly, then, failure added little to the production cost of the microblades. Some thick and irregular microblades in the assemblage indicate that microblade removal may not always have made perfectly "efficient" use of the core, but it appears never to have caused total failure of the cores. The ability to detach microblades so predictably must have entailed great skill. This degree of skill suggests that microblade production may have been an activity carried out by practiced experts, if not by "specialists." It also seems likely that microblades were detached in "batches" and that this further reduced failure probabilities. Batch processing reduces "set-up" costs and makes it possible to schedule production during times that do not conflict with other opportunities.

As is true for Araya microblade production, fluted point manufacture clearly involves a complex and technically difficult sequence of stages. Although a variety of forms can provide an initial blank (see Callahan 1979: 41-66), the sequence of reduction from this blank to a finished point involved bifacial flaking to shape the overall preform, more refined flaking to carefully contour the surface to be fluted, preparation of the platform from which to drive the flute, and removal of the flute itself; once one face of the preform was fluted successfully, this sequence was repeated on the other face. Finally, the fluted preform was pressure-flaked into its finished form and its edges were ground for hafting. Taking this sequence of steps as outlined by Frison and Bradley (1981) and arraying frequencies at the Hanson and Agate Basin sites of production failures at each step in an event tree, Winfrey (1990) estimates an overall failure rate in Folsom point production of approximately 50 percent. Unsurprisingly, Winfrey's analysis indicates that most

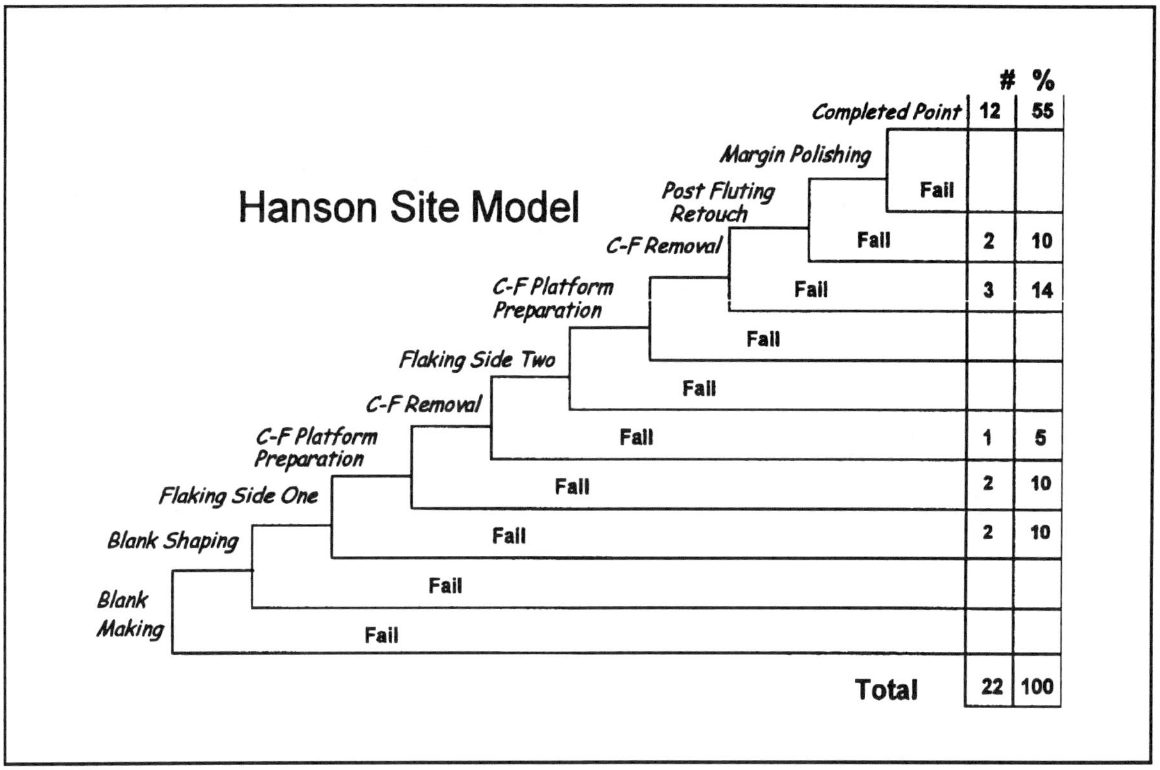

Figure 7.3 Event tree model of Folsom point production at the Hanson site (from Winfrey 1990). "C-F" is an abbreviation for "channel flake".

failures–roughly half of the total–occurred during removal of the flutes, particularly the second flute (Figures 7.3 and 7.4).

The substantial difference between failure rates in the Araya and Folsom examples is particularly interesting because both patterns of production appear to have required high levels of knapping skill: as we have noted, consistent removal of microblades is difficult, and it is clear that it takes many years of practice to flute points consistently. To return to an earlier issue, though, consistent reliance on these patterns of production over relatively long periods of time and large areas implies that both patterns consistently produced adequate supplies of tools despite the need for highly skilled artisans in both cases, and particularly despite the great waste of raw material and production time in the Folsom case. Considering Araya and Folsom production in terms of our general sequence of procurement, production, and use helps to see how this was accomplished.

The procurement cost of the cores used for Araya microblade production was very low since they all appear to have be made out of bifaces that had already had long use histories: reduction to a microblade core appears to have been a step taken only after a biface had been much used and become too small to be useful as a core or a beveled chopping tool. Placing microblade production at the end of the recycling system kept procurement costs low and meant that other operations of the stone tool technology resulted in raw material suitable for use as microblade cores. There was, then, a low probability of failing to have raw material suitable for microblades.

Archaeologists have made similar arguments in the Folsom case, reconstructing a pattern in which large bifacial cores were transported, initially used as sources of flake blanks for other tools, and finally transformed into fluted point preforms (Hofman 1992, Ingbar 1992, Kelly & Todd 1988); Hofman (1992) further suggests that Folsom knappers produced unfluted points (which make up a substantial proportion of the total points dated to the Folsom period [Bamforth 1991b]) when replacement material was scarce. However, such a pattern clearly imposes higher procurement costs than the Araya case. First, a 50 percent failure rate implies that Folsom groups would have had to transport two preforms for every fluted point they expected to produce, which adds weight and bulk to a mobile toolkit rather rapidly. Second, the importance of carefully contouring the surface to be fluted places important limits on a knapper's ability to strike large flakes from a bifacial core. Relative to the Araya case, then, Folsom point production imposed high procurement and raw material transport costs.

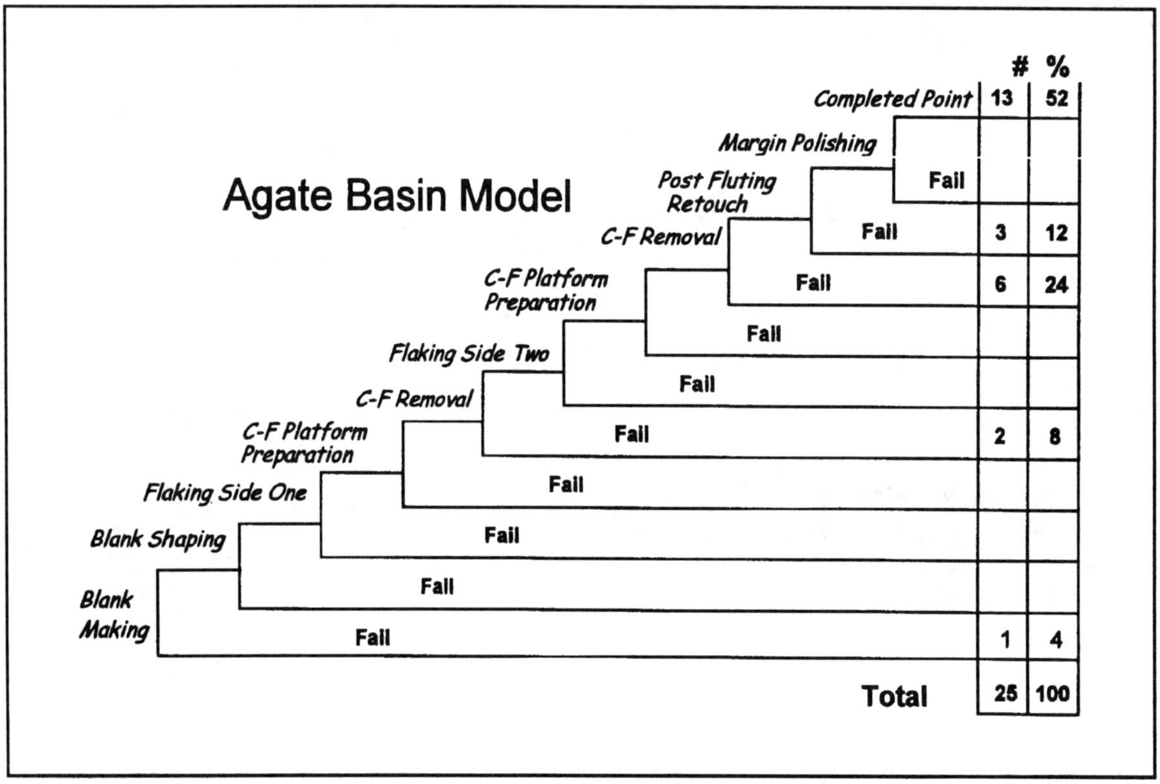

Figure 7.4 Event tree model of Folsom point production at the Agate Basin site (from Winfrey 1990). "C-F" is an abbreviation for "channel flake".

Araya and Folsom production share some important costs in the context of production, particularly in their apparent reliance on high levels of knapping skill. Skillful execution like that observed in the removal of microblades or Folsom flutes is easiest to achieve when repetition results in a positive "practice effect"; in combination with the likelihood that a reliance on batch processing helps to reduce failure rates, these examples must have required Araya and Folsom society to support lengthy periods of training and predictable periods of uninterrupted work time. It is likely that it takes longer to produce a single fluted point than to produce a series of microblades, but we note again that making a stone blade for a hafted tool is only part of the cost of producing that tool: projectile points are useless without shafts, and fixing more than one stone blade in the haft to produce a single useful projectile certainly increased Araya production time over that needed for at least some alternative weapons. Lacking estimates of production time, we infer tentatively that there is no reason to reconstruct substantial differences in overall production costs for the Folsom and Araya cases, although it is clear that Folsom and Araya toolmakers incurred these costs in different ways.

Finally, because microblades appear to have been used as barbs and edge blades in composite projectiles and because Folsom points were clearly used to tip projectiles, failure in application could carry serious costs: without stone blades, these projectiles would not have been functional. It is likely, however, that use of microblades as tips and barbs in compound projectile helped to minimize the probability of such failure. First, multi-barb projectiles are "over-designed" weapons in which individual blades served as redundant elements (Bleed 1986). Failure to have a microblade, therefore, or inadequately fastening or breaking one of several microblades in its haft, would not cause the projectile to fail completely. Second, production of microblades in batches meant that microblades could be produced in groups that could cover periods of anticipated need. Thus, batch processing would contribute to a low probability of no blades whatsoever being available at times of need.

Interestingly, it is difficult to make comparable arguments for the practical benefits of Folsom points in the context of use. Although several archaeologists have suggested that the fluting process played an important ceremonial or other non-utilitarian role in Folsom society (Frison & Bradley 1982), there is no evidence that fluting increased the killing power or accuracy of the weapons to which fluted points were attached; certainly,

Plains Indian hunters (including, as we noted earlier, at least some Folsom period hunters) slaughtered bison with great success for millenia without using fluted points and it is difficult to see how the details of the stone tip of a projectile would greatly influence the effectiveness of that projectile, so long as that tip is sufficiently sharp to penetrate its target (but see Nelson 1996). For example, Parks (ms.) shows that points of the size used on both arrows and darts in prehistoric North America have no influence on the flight of either type of projectile. Alternatively, M. Peterson (personal communication) has suggested that fluted Folsom points may be less breakable in use than unfluted points because the material of the haft, which presumably extended along the flutes on both face, may have provided more support for the stone blade on impact than hafts that are confined the base of a point. However, lower rates of use-damaged points probably have more significance in reducing replacement rates (that is, production costs) than in reducing rates of application failure. Folsom points were sometimes resharpened (Hofman 1991), thereby extending their use-lives, but there is similarly no evidence that fluting made this easier to do: later Paleoindian groups on the Plains also resharpened their non-fluted points (i.e., Wheat 1977). We thus see nothing specific to fluted points which reduces failure probabilities in the context of use.

In our terms, the Araya case thus seems to represent an investment in the domain of production which helps to minimize failure probabilities in the domains of procurement and application. In contrast, Folsom point production has relatively high procurement costs (although these may have been somewhat reduced by using bifaces as both cores and preforms), and also imposes high production costs, but seems not to compensate for either of these by reducing failure in application. Following our arguments above, we might interpret the Araya pattern as evidence for a way of life in which (1) access to raw material was not assured and raw material conservation was extremely important, (2) it was important to maximize the chances of hunting success, and (3) it was possible to invest time and effort in production to deal with these problems. The Araya case might then fall into cell B in Figure 7.1, although the paucity of supporting evidence on Araya lifeways leaves no way to test these possibilities at present.

In contrast, there appears to be no reason to suppose that the increased procurement and production costs associated with fluted points were compensated by practical benefits at other steps of the Folsom technological continuum. Appealing again to our earlier discussion, we suspect that Folsom hunters may have found themselves either in an adaptive context in which both technological (opportunity) costs and failure costs were low or one in which technological costs were low and failure probabilities were reduced by means other than tool design. Archaeological data suggest that the costs of failure in application were probably high in Folsom times: fluted and unfluted points were used to hunt bison and probably other prey on which Folsom hunters depended for their survival, and fluted points appear to be disproportionately common in large bison kills (Bamforth 1991b). Instead of designing a tool which reduced failure in application through its design, we suspect that Folsom groups reduced failure probabilities in this activity through the organization of hunting behavior, probably by relying (perhaps seasonally) on relatively large-scale cooperative bison hunting.

At minimum, it seems clear that something about the Folsom way of life ensured that Folsom stoneworkers would have substantial amounts of time to devote to their craft as well as adequate supplies of raw material. In this regard, it is particularly interesting that failed preforms are discarded so regularly in Folsom sites (i.e., Ingbar 1992, Wilmsen & Roberts 1979), often apparently without being reused for other purposes. This implies, arguably, that raw material was not at a great premium in the contexts in which most Folsom points were manufactured. Such a pattern of behavior is difficult to reconcile with a way of life in which residences are always moved unpredictably or with great frequency, which has obvious implications for reconstructions of Folsom mobility patterns. Depending on our view of Folsom raw material procurement, the Folsom case might thus fall into either cell H or cell F in Figure 7.1.

Investigating these, or other, possibilities for either the Araya or the Folsom case is beyond the scope of this paper, and is almost certainly probably beyond the present capacity of the Araya data base. Since the purpose of these illustrations is to show that archaeologists can address risk in various technological domains, exploring all of the implications of these data is beyond the scope of this discussion. However, these cases do il lustrate ways in which our perspective helps to predict and explore linkages between technology and the adaptive context in which it operated.

SUMMARY AND CONCLUSIONS

To summarize, we have argued, following Torrence (1989), that risk is an important determinant of technological adaptations. However, we believe that it is essential to recognize two distinct components of the risk concept-probability and costs of failure-and we view technology as attempts to reduce failure probabilities in the face of high failure costs. Our analysis of ethnographic data on hunter-gatherer technology supports this assertion, but also indicates that choosing any technological option also imposes costs that may themselves be unacceptable in certain contexts: the fact that a particular option may be advantageous in a given situation does not ensure that it will develop–to be chosen, options must feasible as well as effective. We further considered the special problems of examining these issues in the context of flaked stone technology, and concluded that the absence of reasonable estimates of failure probabilities and failure and technological costs in prehistoric contexts seriously limits our ability to test these ideas archaeologically. Despite these limits, this conceptual framework has the potential to provide important insights for archaeologists, and we demonstrated this potential by examining construction of ditch and palisade defenses along the Missouri River and comparing the production of microblades during the Japanese Paleolithic and of Folsom fluted projectile points on the North American Great Plains.

Our arguments here clearly derive from the optimization perspective that underlies much of the recent theoretical research in lithic analysis (i.e., Bousman 1993; Nelson 1991, 1996; Torrence 1989). However, the emphasis here on the full continuum of technological decision making from raw material acquisition to tool discard, the recognition that the ability to bear high technological costs may constrain technological choices, and the possibility that the combination of low failure and low technological costs may produce technological choices that reflect idiosyncratic cultural rather than utilitarian factors all combine to lead our arguments towards new conclusions. In particular, this perspective allows us to begin to analyze the diversity of technological responses to superfically similar needs, as exemplified in the contrasts between Eskimo and Chumash harpoons and between the complexity of Caribou Eskimo and Athabascan tools and facilities discussed earlier. Such diversity is difficult to accomodate within most existing arguments, which often rely on relatively vaguely defined descriptions of technology, particularly Binford's (1979) notions of curated and expedient technological strategies, and empirically unsupported assertions about the links between these descriptions and selected aspects of hunter-gatherer mobility patterns (see Bamforth 1986, 1990). Design differences among functionally similar items have thus often been attributed solely to such factors as ethnic preferences (i.e., Lemmonier 1986, Sackett 1982), and the framework we have presented here suggests that this is a serious oversimplification.

Finally, by helping to identify more clearly the forces which structure technological choices, our perspective helps to make technological change understandable within a selectionist perspective, although, again, it is not clear that it helps to identify truly Darwinian processes of change. The example of Folsom points above shows this particularly clearly. As we noted, despite a common archaeological emphasis on fluted Folsom points, many points dated to the Folsom period are unfluted. Fluted points disappear from North America after Folsom times, and later Paleoindian points, although still manifesting great stoneworking skill, are finished by pressure flaking, a shift which can easily be seen as selection for unfluted points at the expense of fluted points. Our discussion above suggests that the basis for this selection is likely to be the high procurement and production costs of fluting. For example, the size of large bison kills appears to increase after Folsom times (Judge n.d.), and it may be that slaughtering larger herds required so many projectile points that a 50 percent failure rate simply required more stone than Paleoindian groups could transport. Alternatively, changing social and/or environmental patterns may have made it more difficult to rely on only the most skilled stoneworkers to produce important tools (Bamforth 1988: 188, 1991).

Assessing possibilities like these requires data and arguments which go well beyond the scope of this paper. However, we think that the ability of our perspective to help to raise them is a measure of its usefulness. Although the ideas we have presented here are useful to archaeologists in their current form, there is no doubt that they will require refinement and elaboration. One one hand, a number of authors have identified aspects of "risk" beyond those we have discussed here (for example, its frequency and seasonal pattern–Cashdan 1992, Rescher 1983), and different kinds of risks will certainly entail different kinds of responses; the existing literature, as well as most of our discussion here, emphasizes subsistence risks, and other domains of risk may entail very different kinds of analyses. It is also likely that many of the changes that will be needed to "fine tune" this perspective will result from deeper insights into the ways in which non-technological factors affect technological choices and from a better understanding of the range of technological choices that do not involve tool design. Like all other aspects of human ways of life, technology is embedded in, and conditioned by, many other aspects of culture, and the interactions among these various conditioners are likely to be complex. However, the basic framework we have presented here provides, at least, a way of structuring our thought about these issues and thereby helping to integrate a wide-ranging and exciting, but theoretically dispersed and sometimes contradictory, body of literature.

ACKNOWLEDGEMENTS

We have discussed the issues in this paper with many people over the inordinately long time it has taken to write it, and the ideas we present here owe a great debt to many other people who have inspired or challenged us over the past few years. In particular, Cathy Cameron, Linda Cordell, Ray Hames, Mike Jochim, Larry Keeley, Mary Lou Larson, Steve Lekson, Peggy Nelson, Payson Sheets, Andrew Stewart, Robin Torrence, and Barbara Voorhies have provided references, useful critiques, and general intellectual stimulation. Stan Parks produced the Araya event tree.

NOTES

1. Murdock's estimate of the proportion of fish in the Caribou Eskimo diet derives from the work of Birket-Smith (1929), the same work which provided Oswalt's (1976) data on tool form and construction. Recent research by Burch (1986, 1988) indicates that Birket-Smith's travels among the Caribou Eskimo occurred towards the end of an extended period of extreme deprivation, during which caribou were scarce and human population was reduced by as much as half by starvation. Murdock's figure is therefore probably not an accurate estimate of the role of fish in the Caribou Eskimo diet over the long term. However, this observation has few or no implications for our analysis. First, the conditions under which Birket-Smith observed the Caribou Eskimo had existed for some time and, to survive under these conditions, this group must have had to alter many aspects of its behavior, including its diet and the frequency of use of the tools needed to support that diet. The tools in use during Birket-Smith's visit must therefore have been those required to survive under the conditions which Birket-Smith observed. If key resources, particularly caribou, were scarce at this time, and normally less important resources, particularly fish, were therefore essential to suvival, we might expect that conditions should have favored the construction of the tools required to maximise success in procuring these other resources. The persistent differences between Caribou Eskimo and Ingalik/Tanaina technology under these conditions emphasize rather than damage our arguments.

REFERENCES CITED

Akerman, K. & J. Fagan
1986 Fluting the Lindenmeier Folsom: a simple and economical solution to the problem, and its implications for other fluted point technologies. *Lithic Technology* 15: 1-7.

Andrefsky, W.
1987 Diffusion and innovation from the perspective of wedge-shaped cores in Alaska and Japan. In *The Organization of Core Technology* (J Johnson and C. Morrow, eds.), pp. 13-44. Boulder, CO: Westview.
1994 Raw material availability and the organization of technology. *American Antiquity* 59: 21-35.

Baksh, M. & A. Johnson
1990 Insurance policies among the Machiguenga: an ethnographic analysis of risk management in a non-Western society. In *Risk and Uncertainty in Tribal and Peasant Economies* (E. Cashdan, ed.), pp. 193-228. Boulder, CO: Westview.

Bamforth, D.
1986 Technological efficiency and tool curation. *American Antiquity* 51: 38-50.
1988 *Ecology and Human Organization on the Great Plains*. New York: Plenum Press.
1990 Mobility, raw material, and lithic procurement in the central Mojave desert. *Journal of Anthropological Archaeology* 9:70-104.
1991a Technological organization and hunter-gatherer mobility: a California example. *American Antiquity* 56: 216-235.
1991b Flintknapping skill, communal hunting, and paleoindian projectile point typology. *Plains Anthropologist* 36: 309-322.
1994 Indigenous people, iIndigenous violence: precontact warfare on the North American great plains. *Man* 29: 95-115.

Barton, C. M.
1991 Retouched tools, fact or fiction? Paradigms for interpreting the paleolithic chipped stone. In *Perspectives on the Past* (G. Clark, ed.), pp. 143-163. Philadelphia: University of Pennsylvania Press.

Bednekoff, P. & A. Houston
1994 Dymanic models of mass-dependent predation, risk-sensitive foraging, and premigratory fattening in birds. *Ecology* 75: 1131-1140.

Bernstein, P.
1996 *Against the Gods: The Remarkable Story of Risk*. New York: Wiley and Sons.

Bettinger, R.
1991 *Hunter-Gatherers: Archaeological and Evolutionary Theory*. New York: Plenum Press.

Binford, L.
1973 Interassemblage variability–the Mousterian and the "functional" argument. In *The Explanation of Culture Change* (C. Renfrew, ed.), pp. 227-254. London: Duckworth.
1977 Forty-seven trips. In *Stone Tools as Cultural Markers*. R. V. S. Wright, ed. pp. 24-36. Canberra: Australian Institute of Aboriginal Studies.
1979 Organization and Formation Processes: Looking at Curated Technologies. *Journal of Anthropological Research* 35: 255-273.
1980 Willow smoke and dog's tails: hunter-gatherer settlement systems and archaeological site formation. *American Antiquity* 45: 4-20.

Birket-Smith, K.
1929 The Caribou Eskimo. *Report of the Fifth Thule Expedition*, v. 5., pts. 1 and 2 Washington: Bureau of American Ethnology.

Blalock, H.
1960 *Social Statistics*. New York: McGraw-Hill.

Bleed, P.
1986 The optimal design of hunting weapons: maintainability or reliability? *American Antiquity* 51: 737-747.
1991 Operations research and archaeology. *American Antiquity* 56: 19-35.
1996 Risk and cost in Japanese microblade technology. *Lithic Technology* 21: 95-107.

Bleed, P. & A. Bleed
1987 Energetic efficiency and hand tool design: a performance comparison of push and pull stroke saws. *Journal of Anthropological Archaeology* 6: 189-197.

Bouskila, A.
1995 Interactions between predation risk and compeition: a field study of kangaroo rats and snakes. *Ecology* 76: 165-178.

Bousman, C.
1993 Hunter-gatherer adaptations, economic risk, and tool design. *Lithic Technology* 18: 59-86.

Bulmer, M.
1994 *Theoretical Evolutionary Ecology*. Sunderland, MA.: Sinauer Associates.

Burch, E.
1986 The Caribou Inuit. In *Native Peoples: The Canadian Experience* (R. Morrison & C. Wilson, eds.), pp. 106-133. Toronto: McClelland and Stewart.

1988 Knud Rasmussen and the "original" inland Eskimos of southern Keewatin. *Inuit Studies* 12: 81-100.

Burch, E. & T. Correll
1972 Alliance and conflict: inter-regional relations in North Alaska. In *Alliance in Eskimo Society* (L. Guemple, ed.), pp. 17-39. Proceedings of the American Ethnological Society, Supplement. Seattle: University of Washington Press.

Calavan, M.
1984 Prospects for a probabilistic reinterpetation of Chayanovian theory: an exploratory discussion. In *Chayanov, Peasants, and Economic Anthropology* (E. Durrenberger, ed.), pp. 51-69. New York: Academic Press.

Callahan, E.
1979 The basics of biface manufacture in the Eastern fluted point tradition. *Archaeology of Eastern North America* 7: 1-180.

Cancian, F.
1980 Risk and uncertainty in agricultural decision-making. In *Agricultural Decision-Making: Anthropological Contributions to Rural Development* (Peggy Barlett, ed.), pp. 171-176. New York: Academic Press.

Caraco, T., S. Martindale, & T. Whittam
1980 An empirical demonstration of risk-sensitive foraging preferences. *Animal Behavior* 28: 820-830.

Caraco, T., W. Blankenhorn, G. Gregory, J. Newman, G. Recer, & S. Zwicker
1990 Risk-Sensitivity: ambient temperature affects foraging choice. *Animal Behavior* 39: 338-345.

Caraco, T. & M. Chasin
1984 Foraging preferences: response to reward skew. *Animal Behavior* 32: 76-85.

Carr, P.
1994 *The Organization of North American Prehistoric Chipped Stone Technologies*. Ann Arbor: International Monographs in Prehistory, Archaeological Series No. 7.

Cashdan, E.
1985 Coping with risk: reciprocity among the Basarwa of Northern Botswana. *Man* 20: 454-474.

1990a Introduction. In *Risk and Uncertainty in Tribal and Peasant Economies* (E. Cashdan, ed.), pp. 1-15. Boulder, CO: Westview.

1990b *Risk and Uncertainty in Tribal and Peasant Economies*. Boulder, CO: Westview.

Christenson, A.
1982 Maximizing clarity in economic terminology. *American Antiquity* 47: 419-426.

Collins, M.
1975 Lithic technology as a means of processual inference. In *Lithic Technology: Making and Using Stone Tools* (E. Swanson, ed.), pp. 15-34. The Hague: Mouton.

Crabtree, D.
1968 A stoneworker's approach to analyzing and replicating the Lindenmeier Folsom. *Tebiwa* 9: 3-39.

Durham, W.
1976 The adaptive significance of human behavior. *Human Ecology* 4: 89-121.

Fawcett, W.
1986 Variation in the lithic assemblages from bison kills and associated camp and processing areas. *Wyoming Archaeologist* 29: 9-28.

Ebert, J.
1992 *Distributional Archaeology*. Albuquerque: University of New Mexico Press.

Flannery, K.
1986 *Guila Naquitz*. New York: Academic Press.

Flenniken, J.
1978 Reevaluation of the Lindenmeier Folsom: a replication eExperiment in lithic technology. *American Antiquity* 43: 473-480.

1984 The past, present, and future of flintknapping: an anthropological perspective. *Annual Review of Anthropology* 13: 187-203.

Frison, G. & B. Bradley
1982 Fluting Folsom points. In *The Agate Basin Site* (G. Frison & D. Stanford, eds.), pp. 209-212. New York: Academic Press.

Frison, G. & D. Stanford
1982 *The Agate Basin Site: A Record of Paleoindian Occupation of the Northwestern Plains*. New York: Academic Press.

Gero, J.
1990 Genderlithics: women's roles in stone tool production. In *Engendering Archaeology* (J. Gero & M. Conkey, eds.), pp. 163-193. Cambridge, MA: Blackwell.

Gould, R.
1978 The Anthropology of human residues. *American Anthropologist* 80: 815-835.

1991 Arid-land foraging as seen from Australia: adaptive models and behavioral realities. *Oceania* 62: 12-33.

Halstead, P. & J. O'Shea
1989 *Bad Year Economics: Cultural Responses to Risk and Uncertainty*. Cambridge: Cambridge University Press.

Hammond, J.
1968 *Essays in the Theory of Risk and Insurance*. Glenview, IL: Scott, Foresman.

Hayden, B.
1981 Reserch and development in the Stone Age: technological transitions among hunter-gatherers. *Current Anthropology* 22: 519-548.

1987 From chopper to Celt: the evolution of resharpening techniques. *Lithic Technology* 16: 33-43.

Heffley, S.
1981 The relationship between Northern Athapaskan settlement patterns and resource distribution: an application of Horn's model. In *Hunter-Gatherer Foraging Strategies* (B. Winterhalder & E. Smith, eds.), pp. 126-147. Chicago: University of Chicago Press.

Hegman, M
1989 Risk reduction and variation in agricultural economies: a computer simulation of Hopi agriculture. In *Research in Economic Anthropology* (B. Isaac, ed.), pp. 89-121. Greenwich: JAI Press.

1996 Variability in food production, strategies of storage and sharing, and the pithouse-to-pueblo transition in the northern Southwest. In *Evolving Complexity and Environmental Risk in the Prehistoric Southwest* (J. Tainter & B. Tainter, eds.), pp. 223-250. New York: Addison-Wesley.

Hiscock, P.
1994 Technological responses to risk in Holocene Australia. *Journal of World Prehistory* 8: 267-292.

Hofman, J.
1991 Folsom land use: projectile point variability as a key to mobility. In raw material economies among prehistoric hunter-gatherers (A. Montet-White & S. Holen, eds.), pp. 335-356. *Publications in Anthropology* No. 19. Lawrence: University of Kansas.

1992 Recognition and Interpretation of Folsom technological variability on the southern plains. In *Ice Age Hunters of the Rockies* (D. Stanford & J. Day, eds.), pp. 193-224. Niwot: University Press of Colorado.

Horn, H.
1968 The adaptive significance of conlonial nesting in the Brewer's blackbird. *Ecology* 49: 682-694.

Houston, A. & J. McNamara
1986 Evaluating the selection pressure on foraging decisions. In *Relevance of Models and theoreis in Ethology* (R. Campan & R. Zayan, eds.), pp. 61-75. Toulouse: Privat.

Hudson, T. & T. Blackburn
1982 *Material Culture of the Chumash Interaction Sphere, Vol 1: Food Procurement and Transportation.* Los Altos, CA: Ballena.

Ingbar, E.
1992 The Hanson site and Folsom on the northwestern plains. In *Ice Age Hunters of the Rockies* (D. Stanford & J. Day, eds.), pp. 169-192. Denver: Denver Museum of Natural History.

Jeske, R.
1989 Economies in raw material use by prehistoric hunter-hatherers. In *Time, Energy, and Stone Tools* (R. Torrence, ed.), pp. 34-45. New York: Cambridge University Press.

1992 Energetic efficiency and lithic technology: an upper Mississippian example. *American Antiquity* 57: 467-481.

1996 Review of *The Organization of North American Prehistoric Chipped Stone Technologies*, by P. Carr. *American Antiquity* 61:175-176.

Jochim, M
1976 *Hunter-Gatherer Subsistence and Settlement.* New York: Academic Press.

1981 *Strategies for Survival.* New York: Academic Press.

Jones-Lee, M.
1989 *The Economics of Safety and Physical Risk.* New York: Basil Blackwell.

Judge, J.
n.d. Paleoindian: Plains and Southwest. In *Handbook of North American Indians* Vol. 3 (W.C. Sturtevant, ed.). Washington: Smithsonian Institution Press.

Kajiwara, H.
1989 Shiberia, Kyukuto, Arasuka no saisekijin bunka. *Kokogaku Jyanaru* 306: 2-5.

Kaplan, H. & K. Hill
1985 Hunting ability and reproductive success among male Ache foragers: preliminary results. *Current Anthropology* 26: 131-133.

Keeley, L.
1982 Hafting and retooling: effects on the archaeological record. *American Antiquity* 47: 798-809.

Keene, A.
1981 Optimal foraging in a nonmarginal environment: a model of prehistoric subsistence strategies in Michigan. In *Hunter-Gatherer Foraging Strategies* (B. Winterhalder & E. Smith, eds.), pp. 171-193. Chicago: University of Chicago Press.

Kelly, R.
1995 *The Foraging Spectrum.* Washington: Smithsonian Press.

Kelly, R. & L. Todd
1988 Coming into the country: early paleoindian hunting and mobility. *American Antiquity* 53: 231-244.

Kennedy, M., C. Shave, & H. Spencer
1994 Quantifying the effects of predation risk on foraging bullies: no need to assume an IFD. *Ecology* 75: 2220-2226.

Kivett, M. & R. Jensen
1976 *Archaeological Investigations at the Crow Creek Site (38BF11).* Nebraska State Historical Society Publications in Anthropology No. 7.

Kobayashi, T.
1970 Microblade industries in the Japanese Archipelago. *Arctic Anthropology* 7: 38-58.

Kohler, T. & C. Van West
1996 The calculus of self-interest in the development of cooperation: sociopolitical development and risk among the northern Anasazi. In *Evolving Complexity and Environmental Risk in the Prehistoric Southwest* (J. Tainter & B. Tainter, eds.), pp. 169-196. New York: Addison-Wesley.

Kornfeld, M., K. Akoshima, & G. Frison
1990 Stone tool caching on the North American plains: implications of the McKean site tool kit. *Journal of Field Archeology* 17: 301-323.

Kuhn, S.
1994 A formal approach to the design and assembly of mobile toolkits. *American Antiquity* 59: 426-442.

Landberg, L.
1965 The Chumash Indians of southern California. *Southwest Museum Papers* No. 19. Los Angeles: Southwest Museum.

Larson, D., H. Neff, D. Graybill, J Michaelson, & E. Ambos
1996 Risk, climatic variability, and the study of southwestern prehistory: an evolutionary perspective. *American Antiquity* 61: 217-242.

Lee, R.
1968 What hunters do for a living, or, how to make out on scarce resources. In *Man the Hunter* (R. Lee & I. Devore, eds.), pp. 30-48. Chicago: Aldine.

1976 !Kung spatial organization: an ecological and historical perspective. In *Kalahari Hunter-Gatherers: Studies of the !Kung San and Their Neighbors* (R. Lee & I. Devore, eds.), pp. 73-97. Cambridge: Harvard University Press.

Lehmer, D.
1971 *Middle Missouri Archaeology.* Washington, D.C.: National Park Service.

Lemmonier, P.
1986 The study of material culture today: toward an anthropology of material systems. *Journal of Anthropological Archaeology* 5: 147-186.

Leonard, R. & H. Reed
1993 Population aggregation in the prehistoric American Southwest: a selectionist model. *American Antiquity* 58: 648-661.

Maynard Smith, J.
1978 Optimization theory in evolution. *Annual Review of Ecology and Systematics* 9: 31-56.

McNamara, J. & A. Houston
1992 Risk-sensitive foraging: a review of the theory. *Bulletin of Mathematical Biology* 54: 355-378.

Minnis, P.
1996 Notes on economic uncertainty and human behavior in the prehistoric North American Southwest. In *Evolving Complexity and Environmental Risk in the Prehistoric Southwest* (J. Tainter & B. Tainter, eds.), pp. 57-78. New York: Addison-Wesley.

Mishan, E.
1971 *Cost-Benefit Analysis*. London: Allen and Unwyn.

Morlan, R.
1970 Wedge-shaped core technology in Northern America. *Arctic Anthropology* 7: 17-37.

Murdock, G.
1967 Ethnographic atlas. *Ethnology* 6: 109-236.

Myers, A.
1989 Reliable and maintainable technological strategies in the Mesolithic of mainland Britain. In *Time, Energy, and Stone Tools* (R. Torrence, ed.), pp. 78-91. New York: Cambridge University Press.

Neff, H.
1992 Ceramics and evolution. *Archaeological Method and Theory* 4: 141-193.

Neiman, F.
1995 Stylistic variation in evolutionary perspective: inferences from decorative diversity and interassemblage distance in Illinois Woodland ceramic assemblages. *American Antiquity* 60: 7-36.

Nelson, M.
1991 The study of technological organization. *Archaeological Method and Theory* 3: 57-100.

1996 Technological strategies responsive to subsistence stress. In *Evolving Complexity and Environmental Risk in the Prehistoric Southwest* (J. Tainter & B. Tainter, eds.), pp. 107-144. New York: Addison-Wesley.

Nelson, M. & H. Lippmeier
1993 Grinding stone design as conditioned by land-use pattern. *American Antiquity* 58: 286-305.

Nicholson. B.
1988 Modeling subsistence strategies in the forest/grassland transition zone of Western Manitoba during the Late Prehistoric and Early Historic periods. *Plains Anthropologist* 33: 351-368.

O'Brien, M. & T. Holland
1990 Variation, selection, and the archaeological record. *Archaeological Method and Theory* 2: 31-79.

Oksanen, L. & P. Lunberg
1995 Optimization of reproductive effort and foraging time in mammals: the influence of rResource level and predation risk. *Evolutionary Ecology* 9: 45-56.

Osgood, C.
1937 Ethnography of the Tanaina. *Yale University Publications in Anthropology* No. 16. New Haven: Yale University Press.

1940 Ingalik material culture. *Yale University Publications in Anthropology* No. 22. New Haven: Yale University Press.

1958 Ingalik social culture. *Yale University Publications in Anthropology* No. 53. New Haven: Yale University Press.

Oswalt, W.
1976 *An Anthropological Analysis of Food-Getting Technology*. New York: Wiley and Sons.

Owsley, D. H. Berryman, & W. Bass
1977 Demographic and osteological evidence for warfare at the Larson site, South Dakota. In Trends in Middle Missouri Prehistory (W. Wood, ed.), pp. 119-131. *Plains Anthropologist Memoir* No. 13.

Parks, S.
1990 Getting to the point: the physics and morphology of projectiles. Unpublished Manuscript in author's possession.

Parry, W. & R. Kelly
1987 Expedient core technology and sedentism. In *The Organization of Core Technology* (J Johnson & C. Morrow, eds.), pp. 285-308. Boulder, CO: Westview.

Pyke, G.
1984 Optimal foraging theory: a critical review. *Annual Review of Ecology and Systematics* 15: 523-575.

Rautman, A.
1993 Resource variability, risk, and the structure of social networks: an example from the orehistoric Southwest. *American Antiquity* 58: 403-424.

1996 Risk, reciprocity, and the operation of social networks. In *Evolving Complexity and Environmental Risk in the Prehistoric Southwest* (J. Tainter & B. Tainter, eds.), pp. 197-222. New York: Addison-Wesley.

Real, L. & T. Caraco
1986 Risk and foraging in stochastic environments. *Annual Review of Ecology and Systematics* 17: 371-390.

Rescher, N.
1983 *Risk*. Washington: University Press of America.

Rick, J.
1978 Heat-altered cherts of the Lower Illinois Valley: an experimental study in prehistoric technology. *Northwestern University Archaeological Program Prehistoric Records* No. 2. Evanston.

Ricklis, R. & K. Cox
1993 Examining lithic technological organization as a dynamic cultural subsystem: the advantages of an explicitly spatial approach. *American Antiquity* 58: 444-461.

Sackett, J.
1982 Approaches to style in lithic archaeology. *Journal of Anthropological Archaeology* 1: 59-112.

Schiffer, M.
1996 Some relationships between behavioral and evolutionary archaeologies. *American Antiquity* 61: 643-662.

Schiffer, M. & J. Skibo
1987 Theory and experiment in the study of Technological Change. *Current Anthropology* 28: 595-622.

Sewall, K.
1990 *The Tradeoff Between Cost and Risk in Hazardous Waste Management*. New York: Garland.

Shott, M.
1986 Technological organization and settlement mobility: an ethnographic examination. *Journal of Anthropological Research* 42: 15-51.

Smith, B.
1995 *The Emergence of Agriculture.* New York: Scientific American Library.

Smith, E.
1981 The application of optimal foraging theory to the analysis of hunter-gatherer group size. In *Hunter-Gatherer Foraging Strategies* (B. Winterhalder & E. Smith, eds.), pp. 36-65. Chicago: University of Chicago Press.

1988 Risk and uncertainty in the "original affluent society": evolutionary ecology of resource sharing and land tenure. In *Hunters and Gatherers 1: History, Evolution, and Social Change* (T. Ingold, D. Riches, & J. Woodburn, eds.), pp. 222-251. New York: St. Martin's.

1991 *Inujjuamiut Foraging Strategies: Evolutionary Ecology of an Arctic Hunting Economy.* New York: Aldine de Gruyter.

Smith, E. & R. Boyd
1990 Risk and reciprocity: hunter-gatherer sociecology and the problem of collective action. In *Risk and Uncertainty in Tribal and Peasant Economies* (E. Cashdan, ed.), pp. 167-192. Boulder, CO: Westview.

Sollberger, J.
1985 A technique for fluting folsom. *Lithic Technology* 14: 41-50.

Spaulding, A.
1956 The Arzberger site, Hughes County, South Dakota. *Occasional Contributions of the University of Michigan Museum of Anthropology* No. 16.

Spielmann, K. (editor)
1991 *Farmers, Hunters, and Colonists: Interaction between the Southwest and Southern Plains.* Tucson: University of Arizona Press.

Stephens, D.
1990 Risk and incomplete information in behavioral ecology. In *Risk and Uncertainty in Tribal and Peasant Economies* (E. Cashdan, ed.), pp. 19-46. Boulder, CO: Westview.

Stephens, D. & E. Charnov
1982 Optimal foraging: some simple Stochastic models. *Behavioral Ecology and Sociobiology* 10: 251-263.

Suhonen, J.
1993 Risk of predation and foraging sites of individuals in mixed-species Tit flocks. *Animal Behavior* 45: 1193-1198.

Suter, G.
1993 *Ecological Risk Assessment.* Ann Arbor: Lewis.

Sutoh, T. et al.
1990 *The Araya Site: Results of the Second and Third Term Excavations.* Sendai: Department of Archaeology, Tohoku University. (Japanese with English title and abstract).

Tainter, J.
1996 Introduction: prehistoric societies as evolving complex systems. In *Evolving Complexity and Environmental Risk in the Prehistoric Southwest* (J. Tainter & B. Tainter, eds.), pp. 1-24. New York: Addison-Wesley.

Tanaka, J.
1976 Subsistence ecology of Central Kalahari San. In *Kalahari Hunter-Gatherers: Studies of the !Kung San and Their Neighbors* (R. Lee & I. Devore, eds.), pp. 98-119. Cambridge: Harvard University Press.

Thwaites, R.
1906 *Early Western Travels.* Cleveland: Arthur H. Clark.

Torrence, R.
1983 Time-budgeting and hunter-gatherer technology. In *Prehistoric Hunters and Gatherers in Europe* (G. Bailey, ed.), pp. 11-22. New York: Cambridge University Press.

1989 Re-tooling: towards a behavioral theory of stone tools. In *Time, Energy, and Stone Tools* (R. Torrence, ed.), pp. 57-66. New York: Cambridge University Press.

Tunnell, C.
1977 Fluted point production as revealed by lithic specimens from the Adair-Steadman site in northwest Texas. *Museum Journal* 17: 140-168.

Weissner
1982a Risk, reciprocity, and social influences on !Kung San economics. In *Politics and History in Band Society* (E. Leacock & R. Lee, eds.), pp. 61-84. New York: Cambridge University Press.

1982b Beyond willow smoke and dogs' tails: a comment on Binford's analysis of hunter-gatherer settlement systems. *American Antiquity* 47: 171-178.

Westman, W.
1985 *Ecology, Impact Assessment, and Environmental Planning.* New York: Wiley and Sons.

Wharton, C.
1971 Risk, uncertainty, and the subsistence farmer: technological innovation and resistance to change in the context of survival. In *Studies in Economic Anthropology* (G. Dalton, ed.), pp. 151-178. Washington: American Anthropological Association.

Wheat, J.
1977 Technology, typology, and use patterns at the Jurgens site. In E. Johnson (ed.), Paleoindian Lifeways. *The Museum Journal,* 17: 126-139.

Willey, P.
1990 *Prehistoric Warfare on the Great Plains.* New York: Garland.

Wilmsen, E.
1973 Interaction, spacing behavior, and the organization of hunting bands. *Journal of Anthropological Research* 29: 1-31.

Wilmsen, E. & F. Roberts
1978 *Lindenmeier, 1934-1974: Concluding Report on Investigations.* Washington: Smithsonian Contributions to Anthropology No. 24.

Winfrey, J.
1990 An event tree analysis of Folsom point failure. *Plains Anthropologist* 35: 263-272.

Winterhalder, B.
1981 Foraging strategies in the Boreal forest. In *Hunter-Gatherer Foraging Strategies* (B. Winterhalder & E. Smith, eds.), pp. 66-98. Chicago: University of Chicago Press.

1986a Diet choice, risk, and food sharing in a stochastic environment. *Journal of Anthropological Archaeology* 5: 369-392.

1986b Optimal foraging: simulation studies of diet choice in a stochastic environment. *Journal of Ethnobiology* 6: 205-223.

1987 The analysis of hunter-gatherer diets: stalking an optimal foraging model. In *Food and Evolution* (M. Harris & E. Ross, eds.), pp. 311-339. Philadelphia: Temple University Press.

Wood, W. R.
1967 *An Interpretation of Mandan Culture History.* Washington: Bureau of American Ethnology, River Basin Surveys Bulletin 198, paper No. 39.

Yoshizaki, M.
1961 Shirataki Iseki to Hokkaido no Mudoki Bunka. *Minzokugaku* 26: 13-23.

Young, D. & D. Bamforth
1990 On the macroscopic identification of used flakes. *American Antiquity* 55: 403-409.

8

Stone Tools, Style, and Social Identity: an Evolutionary Perspective on the Archaeological Record

C. Michael Barton
Arizona State University

ABSTRACT

Because of their prevalence in the archaeological record, chipped stone assemblages have long been used for the identification of social entities and the tracing of cultural relationships through space and time. To do this, archaeologists have focused primarily on variations in lithic morphology. Although the forms of stone artifacts are determined by a combination their utilitarian function, 'style', and the physical constraints of knapping different cryptocrystalline rocks, there is widespread belief that style provides the best information about social group membership. Style, however, is not a unified concept, including both passive variability resulting from stochastic processes and actively encoded social information, constrained by selection and manipulated by the makers and users of artifacts.

A neo-Darwinian framework is used to evaluate differing concepts of style and their applicability to the lithic archaeological record. Identifying prehistoric social entities and tracing cultural relationships is loosely analogous, methodologically and theoretically, to identifying taxa and tracing ancestor/descendant relationships in biology. Lithic technology is also examined from the point of view of neo-Darwinian evolutionary theory to identity sources of morphological variability most likely to mark group social identity, and suggest methodologies best able to identify and differentiate prehistoric social groups.

INTRODUCTION

Since the discovery of the first recognized human ancestor in the Neander Valley of Germany nearly 150 years ago, archaeologists have been concerned with the identity of prehistoric individuals and the social groups in which they lived. For the preagricultural economies that represent most of the human past, stone artifacts–primarily spatial and temporal variation in the macroscopic form of these artifacts–comprise the predominant form of behavioral residue used to reconstruct prehistoric social identity.

In recent decades, the ability of morphological variability in stone artifacts to permit reliable inference of social identity has been increasingly questioned, leading to a series of debates in the archaeological literature. The best known of these debates was between Lewis Binford and François Bordes (Binford 1973, Binford & Binford 1966, Bordes 1973, Bordes & de Sonneville Bordes 1970; see also Dibble 1995) over the meaning of variability in Middle Paleolithic assemblages. Similar debates have focused on North American projectile point forms (Bettinger, O'Connell, & Thomas 1991; Thomas 1986, Flenniken & Wilke 1989, Flenniken & Raymond 1986, Wilke & Flenniken 1991, Rondeau 1996, Hoffman 1985) and Eurasian microliths (Neeley & Barton 1994, Barton & Neeley 1996, Kaufman 1995, Goring-Morris 1996, Henry 1996, Clark 1996). In general these discussions have focused on the causes and interpretation of variability in stone artifact forms. Also crucial to these debates, however, is the notion of group-level social identity (sometimes referred to as 'ethnicity'), and to what extent it can be recognized in the archaeological record.

Neo-Darwinian evolutionary theory provides a means to explore the inference of social identity with reference to stone technology in particular and material culture in general. While the focus here is on the hunter/gatherer societies that comprise most of the human past and the lithic assemblages that are their most common behavioral residues, this discussion is pertinent at a more general level to more complex societies and other forms of material culture.

SOCIAL IDENTITY AND STYLE

The primary objective of this chapter is to assess the potential for chipped stone artifacts to provide archaeologists with information about extra-individual social identity and social organization from an evolutionary perspective. However, studies of prehistoric social identity are generally closely interwoven with concepts of style in archaeology (Conkey 1990, Hegmon 1992, Shennan 1989). Hence, it is useful first to examine style, as it relates to social identity and stone tools, in an evolutionary context. This is not intended to be an exhaustive treatment (see Carr & Neitzel 1996, Conkey & Hastorf 1990, Hegmon 1992 for more wide ranging discussions), but to briefly review current concepts of style in an evolutionary framework.

Types of Style

Archaeologists generally agree that style can serve to demarcate boundaries between social units but disagree as to what constitutes and causes style, and the nature of the social units defined (see Conkey 1990, Hegmon 1992, Shennan 1989). To oversimplify somewhat for the sake of brevity, archaeological approaches to style fall into two groups. The approach with the longest history sees style as shared forms created by common social learning within social groups. Often called the 'social interaction' school of style, the degree of similarity between stylistic forms here reflects the social distance between the makers of artifacts in question (e.g., Plog 1990, Neiman 1995). Generally implicit in this view is that variability in stylistically relevant characteristics is not directly affected by variations in artifact manufacturing processes or artifact use (see Jelinek 1976, Sackett 1990, Neiman 1995 for explicit discussions of this criterion). In Sackett's words, style comprises culturally prescribed choices among the "spectrum of equivalent alternatives, of equally viable options, for attaining any given end in manufacturing and/or using material items." (1990: 33). This approach has been characterized as viewing style as 'passive' and reflecting a normative perspective on human cultural systems (e.g., Clark 1989, Conkey 1990, Sackett 1990).

Using neo-Darwinian concepts, Dunnell (1978a, 1978b) suggested that style is best considered as stochastic variability in material culture. That is, style is represented by those artifact characteristics that are under so little selective pressure that their variability in time and space can be described by transmission effects (i.e., social learning patterns) and chance (see also Neiman 1995). Such traits are effectively neutral with respect to selection. Dunnell contrasts these stylistic characteristics of material culture with functional ones that are affected by selection (see also Jones, Leonard, & Abbott 1995; Teltser 1995; O'Brien & Holland 1990, 1995). This is similar in effect to the social interaction approach to style, especially as described by Sackett (1990). Neiman (1995) has used simulation modeling to show quantitatively how stochastic processes can produce the type of stylistic variability proposed by Dunnell and social interaction theorists. Such stochastic variation is still 'passive', in the sense that it *results from* social interaction rather than *mediates* social interaction, but it not tied to a normative view of culture.

An alternative approach to style, popularized by Wobst (1977), views style as those characteristics of material culture that convey information (overtly or covertly) about the social roles of the individuals who make and use it (Conkey 1990; Wiessner 1983, 1985). In this 'information exchange' approach, style is considered an important aspect of artifact function–one that communicates social information. While style may vary according to the social distance between its creators, it is primarily maintained at the group level by selection related to the maintenance (or minimization) of social boundaries and the nature of social networks (see also Barton, Clark, & Cohen 1994).

With respect to the nature of the social units reflected by style, archaeologists have generally remained rather vague (Conkey 1990, Shennan 1989). It is not that archaeologists are naive about the complexities of social identity or are uninterested in the topic (e.g., Wiessner 1983, Shennan 1989). But the nature of the archaeological record–especially its coarse temporal resolution (usually incapable of distinguishing a human generation) and the tendency for artifact assemblages that are archaeologically visible as 'sites' to be palimpsests of repeated use of a given piece of the landscape–often makes it difficult to relate archaeological social units, such as 'cultures', to social units noted in living societies (Shennan 1989). An added difficulty especially relevant to the discussion here is that ethnographic and ethnoarchaeological studies that have investigated relationships between style and social identity have generally focused on behavioral residues other than stone artifacts (Nelson 1991; Wiessner 1983, 1985; but see Sinopoli [1991] for a study that does include stone artifacts).

Style and Evolution

In many ways, inferring social identity at the group level parallels the identification of descent groups in biology (Jones, Leonard, & Abbott 1995; Teltser 1995). Archaeologists seek to define artifact assemblages whose similarities derive from the common social descent of their makers rather than similar (especially techno/economic) uses. In evolutionary terms, archaeologists interested in group social identity are seeking shared, derived characteristics (i.e., synapomorphies) in the archaeological record.

Similarity of form alone is not sufficient to demonstrate common descent, of course. Similarity can also result from convergent evolution (i.e., homoplasy). That is, common selective pressures can produce similar forms. Projectile tips throughout the world tend to be pointed, not because their makers shared a common social descent, but because pointed tips make more effective weapons than non-pointed ones and are, hence, more likely to be replicated in contexts where successful hunting contributes positively to fitness. Focusing on artifact forms whose variability is largely a result of stochastic processes (i.e., drift and transmission effects), rather than selection, helps avoid confusing similarities due to common descent with those resulting from convergence (Neiman 1995).

This does not mean, of course, that artifact properties useful for reconstructing social group identity are non-functional, simply that relevant variation among these assemblages is due primarily to social descent (including social distance) rather than common selective pressures. Nor does it mean that variability due to selection is useless for differentiating social groups. Those employing an information exchange approach to style can argue that selection for the maintenance of social group boundaries is one of the causes of stylistic variation (e.g., Barton, Clark, & Cohen 1994; Clark, Barton, & Cohen 1996; Hegmon 1992, Wobst 1977, Wiessner 1983). This is somewhat analogous to divergent selection associated with niche differentiation. Under such circumstances, stylistic variation in material culture can mark distinct social groups (e.g., Conkey 1980). Not all such 'active' style (*sensu* Clark 1989, Conkey 1990) is equally useful for identifying social groups, however. Those forms, primarily those termed 'emblemic style' by Wiessner (1983) and 'iconic style' by Sackett (1985, 1990), that have other social groups as their targets can best serve to reliably mark group boundaries. As proposed by Wobst and subsequently elaborated by others, those forms that are most effective in boundary definition should be easily visible and recognizable by members of other social groups, have a relatively long use-life, employ a sufficiently complex design and/or manufacture to encode unambiguous social information, and be used in contexts where they are likely to be encountered by members of other social groups (see also Gero 1989; Wiessner 1983; Clark 1989; Clark, Barton, and Cohen 1996).

In sum, from an evolutionary perspective, those forms in the archaeological record that should serve most reliably as markers of social descent, social group identity, and social distance should be:

- those whose heritable variation is primarily due to stochastic processes, and
- those whose variation is due to selection for group boundary maintenance.

As a shorthand to these concepts I will use the terms *passive style* and *active style* respectively. Stochastic and selective style might be more accurate, but I shrink from further muddying terminological waters already filled with labels like 'assertive', 'iconological', 'emblemic', and 'isochrestic'. These different forms are not mutually exclusive. Variation originally due to stochastic processes may later come under the control of selection with changes in the social landscape (Sackett 1985, Shennan 1989). Given these observations, the potential of chipped stone artifact assemblages to provide reliable information about group-level social identity can be evaluated in a neo-Darwinian evolutionary framework.

LITHICS, ACTIVE STYLE, AND SOCIAL IDENTITY

Lithics and Active Style

As mentioned above, Wobst and others have proposed a number of artifact properties that would allow these artifacts to effectively carry information about group boundaries and membership. This makes it a relatively straightforward task to evaluate the ability of chipped stone artifacts to carry such messages. Wobst (1977) suggests that forms most likely to convey information about social group affiliation are those that are most visible, most often play a role in social interactions, and are encountered by the most individuals.

Lithics, generally, are small objects made, used, and discarded by individuals in the performance of economic tasks. This means that an individual must be physically so close to a lithic artifact–and its user–in order to be able to note morphological features that might convey social information, that any such messaging would be of minimal value (Wobst 1977). Furthermore, most lithic artifacts have such short use-lives as to be relatively poor carriers of social group information from the point of view of active style (Clark 1989, Gero 1989).

There are exceptions, of course. Cores and some retouched artifacts–especially hafted ones–probably enjoyed longer use-lives. Given the regular presence of raw materials from sources quite distant from a site, cores may have traveled long distances and/or may have been traded among different social groups. Such exchange can obviate the need to be in close proximity to the artifact maker. Hafting can improve the effectiveness of lithic artifacts and extend their uselife by allowing them to be resharpened to a smaller size than would be possible if they were simply hand-held. This makes it more likely that they would be seen by more individuals. Wiessner (1983), for example, documents the exchange of arrows among San groups. Other hafted artifacts also could have been exchanged, increasing their potential visibility and distribution, and giving them a greater potential for transmitting information about social identity from one group to another. Whether or not this information can inform about group-level social identity is another matter, however.

Information about group membership must be encoded in lithic artifacts in such a way that it can be decoded by prehistoric recipients *and* by archaeologists. Cores, for example, achieve the form in which archaeologists recover them via the process of flake or blade removal. Making the reasonable assumption that exhausted cores would not be exchanged among groups, this means that the morphology of exchanged cores in archaeological contexts would be primarily determined by the flake/blade producing activities of the recipients, not the producers. Any information about group affiliation encoded in core morphology by the producers would be obliterated by the recipients during core reduction. The forms of very many retouched tools–even hafted ones–would also tend to follow the same pattern. The morphology of endscrapers, sidescrapers, burins, and many other retouched artifacts reflects a history of use, resharpening, and changes in function before they eventually become useless and are discarded (Barton 1991, Barton, Olszewski, & Coinman 1996, Dibble 1987, 1995, Frison 1968, Jelinek 1976). Even bifaces, including projectile points, do not seem immune from such morphological dynamism (Frison 1968, Flenniken & Wilke 1989, Hoffman 1985, Kelly 1988). While some artifacts may break and be discarded before such processes alter their form dramatically, others may not. Again, the morphology of the artifact recovered by archaeologists more likely reflects the activities of the recipient than the social identity of the manufacturer. An archaeological assemblage could contain, therefore, examples of forms manufactured by the users and forms manufactured by others, in various states of alteration from their manufactured forms.

It should also be remembered that even *hafted* lithic artifacts are typically the shortest lived and most easily replaced components of compound tools. If there is a selective advantage to signaling social identity in these forms of material culture, it would likely favor encoding in the larger (hence, more easily visible) and longer-lived haft than the lithic portion, providing a better chance of unambiguous decipherment both by prehistoric recipients and archaeologists. In a study of Great Basin arrows, Sinopoli (1991) found significant associations between aspects of shaft form and social group membership, but no such relationship with the morphology of the stone projectile tips.

There are, however, a few lithic forms that have a better chance of embodying information about social identity. Good examples are the intricately flaked 'eccentrics' from Mesoamerica (Shele & Miller 1986, Berrin & Pasztory 1993) and western North America (Hayden & Schulting 1997). Sometimes flaked into the profile of humanoid figures (possibly deities) they show no evidence of use in the types of economic activities in which most lithic artifacts are employed. Furthermore, they show no evidence of maintenance by resharpening and, hence, retain their original forms throughout their possibly extended use-lives. They display a form that was clearly the intent of the manufacturer, and a potentially high visibility that makes them potential carriers of active style. Large, thin bifaces like those found in some North American Early Woodland (e.g., Adena), and European Upper Paleolithic (e.g., Solutrean) contexts *may* also be candidates for markers of social identity. However, care should be taken about automatically assuming that large bifaces reveal social identity. They also serve as efficient cores, especially under conditions of high mobility (Parry & Kelly 1987, Kelly 1988). In this latter case, forms would be more strongly controlled by selection related to economic concerns rather than social group boundary maintenance.

Lithics and Passive Style

While selection to differentiate social groups may affect only a very restricted class of lithic artifacts, stochastic variability has the *potential* to affect a much wider range of lithic variability. Sackett (e.g., 1990) is very clear on the potential for any form of artifact to exhibit passive style (i.e., isochrestic variability). This potential is the underlying basis for most claims for the identification of social groups from the analysis of chipped stone artifact assemblages (e.g., Bordes 1973, Close 1989, Kaufman 1995, Henry 1996, Rick 1996, Sackett 1990). However, beyond addressing the potential for passive style in any kind of artifact, there has been little in the way of concrete predictions about the form it might take or the contexts in which it might appear. This makes it considerably more difficult to evaluate the capacity for lithics to convey information about social identity in this form.

The technological decisions and motor habits responsible for the production and use of lithic artifacts are generally learned in a social context and, hence, potentially transmittable. In order to address the potential for lithics to passively carry information about social identity via stochastically produced variation, it is necessary to first examine the effects of selection on lithic assemblages. Chipped stone technology has been a significant aspect of the human phenotype for at least two and a half million years. Its presence among all known examples of the genus *Homo* and its world-wide persistence until the last three millennia argue strongly for its maintenance by selection.

Throughout this enormous time span, the extremely conservative nature of lithic technology is notable. The basic technique of removing flakes from an amorphous core may predate the appearance of our genus and was still practiced during the Bronze and Iron Ages (e.g., Rosen 1996)–and until this century in some areas. Bifacial technology appeared somewhere around a million years ago and also continued until the recent extinction of lithic technology in the last century. Blade technology initially appeared at least 400,000 years ago and also continued through the early metal ages in the Old World and until historic times in the New World. The overall effect of these few major changes in lithic technology was to increase the efficiency of stone use. That is, they allowed humans to obtain an increasing amount of cutting edge per unit of stone (Hayden 1987, Parry & Kelly 1989). Again, this argues for rather tight and persistent selective control over the forms produced over the lifespan of lithic technology.

This is unremarkable for a number of reasons. First and foremost, chipped stone technology has functioned primarily to acquire and modify a wide variety of resources needed for human survival. In this sense, failure of the lithic technological system could have meant the failure of human society. However, increasing lithic efficiency involves increasing energy investment in lithic technology (Parry & Kelly 1989)–in the acquisition of better quality stone, increased time spent in core preparation, and more investment in knapping tools and training. On the other hand, lithics themselves are not *directly* necessary for human survival, and increasing investment in lithic technology means less can be invested in other pursuits that are more immediately relevant for survival.

Furthermore, the mechanics of knapping impose rather strict constraints on the ways in which lithic artifacts can be produced, including various aspects of core morphology, the angle and force of removal blows, and the resilience of the hammer (Cotterell & Kaminga 1987, Dibble & Whittaker 1981). Although the works of Michelangelo, Easter Island figures, and the temple of Abu Simbel make it clear that there *can* be a wide variety of ways to shape stone, there are a very limited number of methods available for knapping a usable edge. The fundamental principles of knapping were worked out by Pliocene hominids and have been elaborated only a modicum since. In the words of Geof Clark , "there are only a few ways to back a bladelet, all of them immediately apparent to even an exceptionally stupid rock-knocker." (1996: 139) The same is true of lithic technology in general. Schiffer (1996) has compared the commonalties of lithic use-life trajectories that appear widely throughout space and time to ontological development. While not an entirely an apt analogy, given the greater potential for variation in the length and path of such lithic trajectories, it must be remembered that ontology, too, is governed by selection.

The overall result of these constraints on lithic technology has been the evolution–at least as early as the appearance of *H. erectus*–of a flexible system that responded to processing and extraction needs on one hand, and the effective availability of stone on the other. Increased stone availability and/or decreased need for stone in resource acquisition and processing permitted reduced investment in flake production efficiency (e.g., a shift from bifacial to amorphous cores). Decreased stone availability (due to increased residential mobility or the spatial disjunction of stone and other resources such as food and water, as well as limited stone sources) and/or increased need for stone (due to increases in extraction and processing needs or increased population) was balanced by a shift to more efficient core technologies (bifaces or blades) and use-life extension strategies such as retouch and hafting. In neo-Darwinian terms, there is an underlying body of heritable information (analogous to genotype) responsible for technological behavior that can be variably expressed, depending on the environmental context, in phenotypic form as different kinds of lithic assemblages.

There are also selection driven shifts in the technological information that is transmitted. As has been pointed out by design theorists, the design and manufacture of stone tools responds to spatial and temporal variation in resource distribution and levels of uncertainty in these distributions (Bleed 1986, Torrence 1989, Nelson 1991). Myer (1989), for example, suggests that changes in compound projectiles employing microliths were tied to changing hunting techniques and prey selection in Mesolithic Britain. Similarly, Shott (1996) links temporal change in projectile point forms from the American Bottoms to hafting techniques and projectile shaft dimensions. These, in turn, are constrained by projectile delivery systems, prey selection, and hunting methods. Neeley and Barton (1994) focus on the relationships between compound tool manufacture and maintenance, mobility, and microlith forms. Nelson (1991) presents yet other analogous examples. The point here (pun intended) is that lithic morphology has little freedom to vary independently of the larger techno-economic system in which it plays a fundamental role.

Even if there were more permissible morphological variation in chipped stone artifacts, other factors would limit its potential for expression. There is sufficient uncertainty in the knapping process that complex morphologies, such as the eccentrics mentioned above, can be produced through knapping only with a considerable investment in time, skill, and high quality raw material. A usable edge can be produced in seconds, however. Related to this, a skilled knapper can produce a very serviceable biface in less than an hour, but it could easily take considerably more time to create an *exact* duplicate of the first. Furthermore, morphological embellishment has the potential to reduce the effectiveness of lithic tool. Given the extremely short use-life of many lithic artifacts, it is likely that there are selective constraints *against* the potential for a wide range stochastic variability.

Identifying Passive Style in Archaeological Assemblages

It is clear that while lithic morphology may vary, most expressed variability is the product of selective constraints. This means that there are rather stringent limitations on the expression of passive style in lithic assemblages. The extent to which variation due to stochastic processes, operating within these limits and indicating group social identity, can be recognized in archaeological assemblages is a further consideration. Especially relevant here is the fact that the lithics encountered by archaeologists are generally discard assemblages. That is, archaeological assemblages are comprised of artifacts that are no longer in usable condition and, in fact, may differ to a considerable extent from their original morphology (Barton 1991, Frison 1968, Jelinek 1976). Morphological features arising from stochastic processes during manufacture are likely to be altered or obliterated during the use-life of many lithic artifacts. This is the case not only for simple flake tools, but also for more complex forms such as bifacial projectile points (Flenniken & Wilke 1989, Hoffman 1985). Hence, the final artifact forms encountered by lithic analysts are the result of use-life alterations overlaid on the original design.

Many attribute analyses and typologies currently in use conflate features that most likely vary according to intended use or in the course of maintenance during use (Barton 1991, Barton & Neeley 1996; Barton, Olszewski & Coinman 1996, Dibble 1987, 1995, Flenniken & Wilke 1989, 1995, Thomas 1981). However, Whittaker (1987) examines individual-level variations in biface morphology, using projectile points from mortuary contexts, that might give rise to group-wide patterns through stochastic processes. Additional studies of such detailed attributes (e.g., flake-scar orientation angles) may be of value for distinguishing social groups (see also Sackett [1989] for examples with non-bifacial artifacts). Whittaker's work also raises questions as to whether variation at the individual level (either isochrestic or assertive [*sensu* Wiessner 1983] in nature) might be great enough to obscure stochastic variation at the group level.

Following the lead of methodologies successfully employed with ceramics, many archaeologists have used assemblage-level frequency seriations of chipped stone artifacts to infer the boundaries of social groups. That is, variation among assemblages in the relative frequencies of artifact types is interpreted to represent differences in social descent among the makers of the assemblages. Perhaps the best known example of such an approach is the interpretation of assemblages-level variability in the Middle Paleolithic by François Bordes (1973, Bordes & de Sonneville-Bordes 1970, see also Jelinek 1988). The underlying implication is that such variation is stochastic in nature (e.g., Bordes 1973, Close 1989). When frequency seriation produces an accurate chronological ordering, the case for stochastic processes and common social descent is felt to be even stronger (e.g., Teltser 1995). For several reasons, however, frequency seriation applied to lithic assemblages is unlikely to provide information about social group membership in the context of the small scale societies that characterize mobile foragers and even simple agricultural communities.

Primary among these is the relationship between stochastic variability and population size. In brief, the amount of stochastically maintained variability present in a population at any one time is inversely proportional to the size of the population. In stochastically controlled variability, one (isochrestic) variant of a form will eventually become fixed in a population and others will disappear due to chance and transmission processes. The rate at which this process occurs is inversely proportional to population size. This has long been known to be true for biological populations and has recently been modeled for behaviors transmitted via social learning (Neiman 1995). In large populations the rate at which a variant becomes fixed is slower and the rate at which new variations are introduced (a function of population size) is higher. Hence, a number of selectively neutral variants will be present at any given time. In small populations, however, variants go to fixity much more rapidly and new variation is introduced more slowly than in large populations, meaning that fewer neutral variants will be present at any one time. For hunter-gatherer groups of

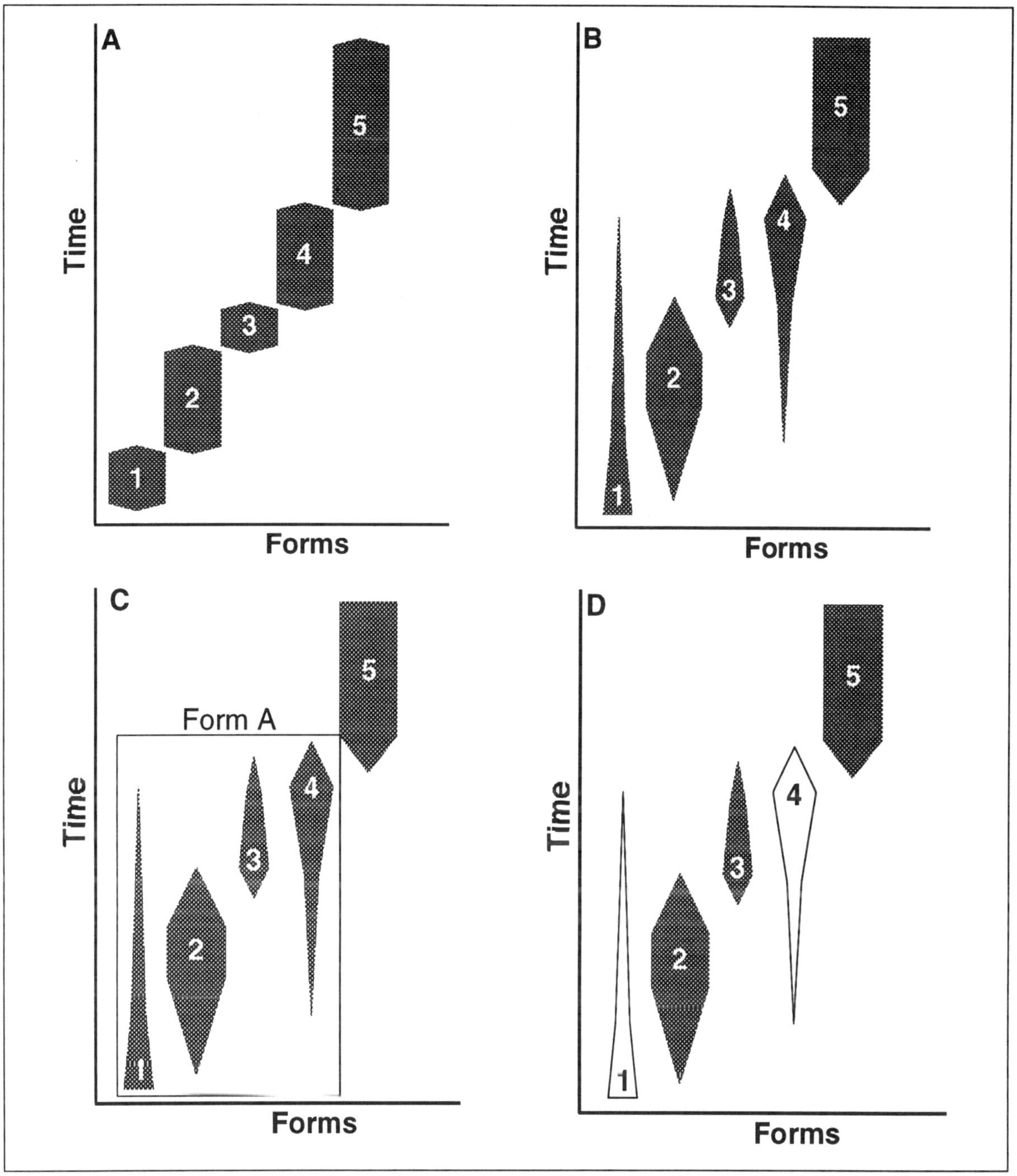

Figure 8.1 Stochastic processes and artifact form. Graphs A through D represent hypothetical frequency distributions over time of a particular artifact class, such as projectile points, within the assemblages from a stratified archaeological site. Each frequency polygon (i.e., battleship curve) represents the change in frequency of a particular morphological form (a.k.a. 'style') over time. Graph A shows the expected distribution over time of forms controled by stochastic processes in small social groups. Note that only one form is represented at any given time, except during rapid transitions between forms. The distribution of forms shown in graph B can either result from selection maintaining multiple (i.e., functionally distinct) forms or stochastic processes in large social groups. Graphs C and D show alternative explanations for the distribution in graph B, especially relevant for small groups. In graph C, forms 1-4 are simply arbitrary divisions of continuous variability in discard morpholgy in a single form ('form A'). Form 5 is morphologically distinct (either functionally or stylistically) and replaces form A. In graph D, forms 2, 3, and 5 represent stochastic change in artifact morphology within a single social group. Forms 1 and 4 derive from different social group; their presence at the site could represent exchange or territorial group boundary fluctuations.

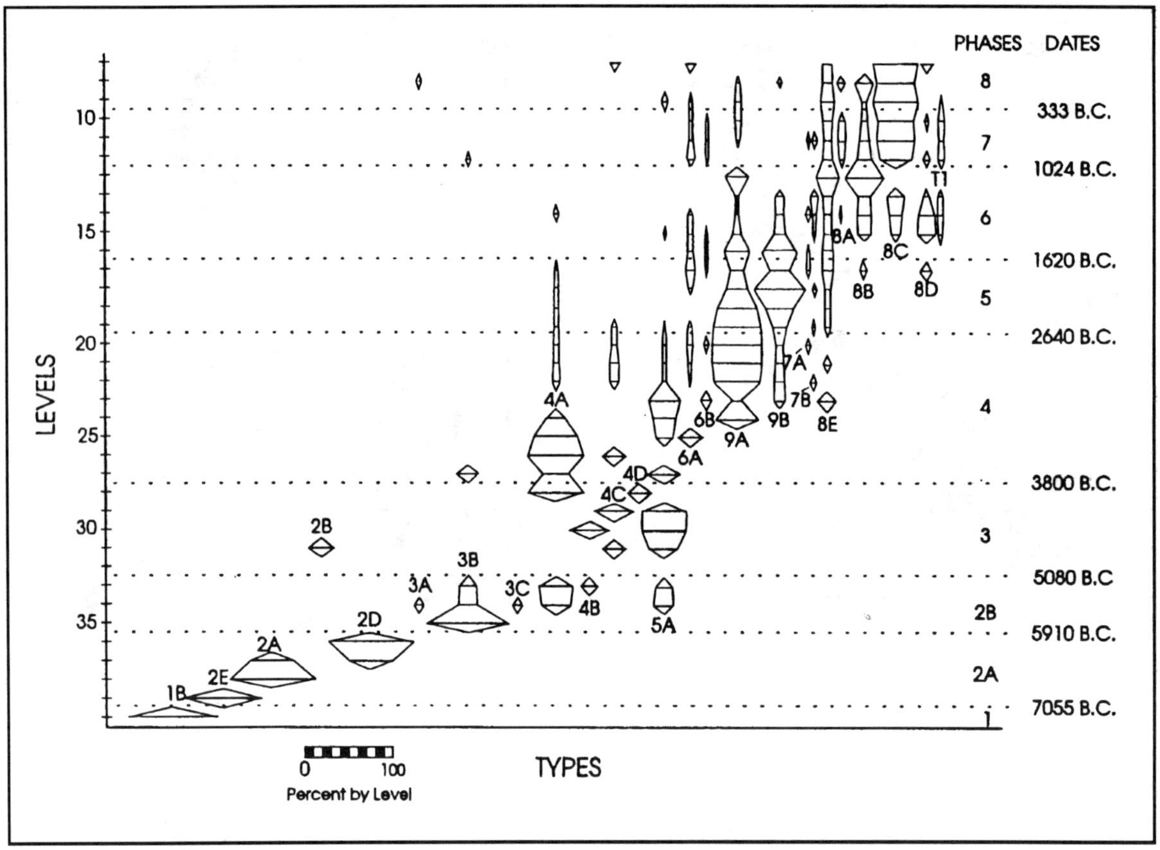

Figure 8.2 Projectile point relative frequencies from Panaulauca Cave, Peru (from Rick 1996: Figure 2).

ca. 25 individuals, only a few of whom would be actively making stone tools, this means that it is highly likely that *only one* stochastically controlled variant of an artifact form would be present at any given time. Any new variant introduced would either rapidly disappear or completely replace the pre-existing variant. Hence, stochastic variability in chipped stone artifact assemblages produced by a given social group should be manifest at archaeological resolutions as a succession of single forms through time rather than changing frequencies of a set of forms (Figure 8.1a).

To cast this situation in terms more familiar to many anthropologists, the stone tools discarded by a band of foragers were made by some subset of the total band. These few individuals were likely closely related, shared in the same socialization processes, and usually worked together. Each individual probably made stone artifacts in a slightly different way (Whittaker 1987, Wiessner 1983). However, at a group level, it is very difficult to imagine a realistic situation in which this small group of knappers made several different styles (in the isochrestic sense) of, say, projectile points and highly unlikely that these styles were maintained as distinct, unchanging forms over many generations of their descendants.

For situations where several apparent variations of a single artifact class are maintained through time, it is highly unlikely that the different forms are simply different 'styles' (in the sense of passive style) of a particular type. A particularly clear case in point is Rick's (1996) study of the Archaic Period lithic assemblages from Peru. In some of the assemblages from Panaulauca rockshelter, as many as ten different 'styles' are thought to have been produced in a single occupation (Figure 8.2). Some of these styles (e.g., type 9A) are thought to have been continuously produced for over 2,000 years. Rick maintains that the different point forms are stylistic alternatives within a single functional class (i.e., isochrestic style [1996: 251]).

In the light of the preceding discussion, this is extremely unlikely. This is not to say that Rick has inaccurately described the projectile point morphology and its variability. However, given that Panaulauca was occupied by bands of Archaic foragers, the number of contemporaneous styles proposed by Rick probably often exceeded the number of active knappers in residence there at a given time. Several other alternatives are more likely (Figure 8.1b-c). One is that this sort of formal variation is maintained by selection, analogous to the common genetic case of balanced polymorphisms (Figure 8.1b). That is, selection, not stochastic processes, probably maintains morphological diversity in point forms in this and many similar contexts. An alternative explanation (Figure 8.1c) is that the apparent discrete artifact types are arbitrary divisions of a range of continuous morphological variability due to variation in the life histories of artifacts with respect to wear, breakage, and maintenance (e.g., Barton 1990, 1991; Dibble 1995, Flenniken & Wilke 1989, Wilke & Flenniken 1991). As discussed above, both the range of such variability and the uselife trajectories that particular artifact forms may take are ultimately under selective control. On the basis of his illustrations (1996: Figure 4), this may account for many of Rick's types. A third alternative explanation (Figure 8.1c) is the movement (i.e., diffusion) of forms among different social groups due to exchange of artifacts or people (as is the case with San arrows documented by Wiessner [1983]). This latter explanation is equivalent to gene flow in biology and will be addressed in more detail below.

These considerations do not necessarily mean that such seriations are useless for constructing chronologies, a point that Thomas emphasizes (1981, 1986). In fact, given that stochastic variants are likely to go to fixity very rapidly and differ among widely spaced, mobile forager bands, stochastic processes may not produce the same kind of temporally patterned variation that they do with larger, more sedentary social groups. Changes in lithic assemblages due to directional selection, cutting across social group boundaries and encompassing larger geographic regions, may be more useful for the construction of chronological frameworks (e.g., Shott 1996).

LITHICS AND GROUP IDENTITY

It is clear that variability in most features of lithic macromorphology–including overall size, shape, and retouch configuration that form the basis for widely used typologies are likely to be under selective control. Hence, similarity in artifact form and assemblages composition are more likely to be a result of selection favoring one form over others or evolutionary convergence in similar contexts, than indicators of common social descent. Examples include the widespread use of backed microblades in compound tools from the late Pleistocene through the mid-Holocene over much of temperate Eurasia and the spread of small projectile points over much of North America about 2,000 years ago. The development of specialized blade industries in the complex societies of southwest Asia (Rosen 1983), the Indus valley of India (Biagi & Cremaschi 1991), and Mesoamerica (J. Clark 1987) seems best explained by convergence, as do the presence of chopping tools in the Spanish Mesolithic (Clark 1983) and the Oldowan and the appearance of sidescrapers in European Mousterian and Arizona Hohokam assemblages.

In comparing the effects of stochastic processes and selection on lithic form, it should be clear that the recognition of passive style in lithic artifacts is difficult at best. Nevertheless, this is not an impossible task. In fact, the identification of lithic residues of social groups is relatively easy at one level. Given the sparse nature of the *known* archaeological record compared to the amount of material culture originally created and used by prehistoric people, we can say without fear of contradiction that each lithic assemblage from a single occupation (to the extent this can be ascertained given the variable temporal resolution of the archaeological record) minimally represents the residue of a single identity conscious social group. Of course, many (perhaps most) such assemblages probably represent the combined residues of more than one group that left their artifacts at a particular location, of course; only very rarely do we find more than one assemblage deposited by what was demonstrably the *same* group of people. However, this is not what most archaeologists want to know about social identity.

As Teltser (1995) notes, we want to construct 'true phylogenies' (see also Jones, Leonard, & Abbott 1995). In diachronic studies, we need to identify social descent lines in which change takes place. Across space, we want to map the social distance between groups that left contemporaneous residues. In more general terms, archaeologists often need to trace networks of information transmission (including the direction and intensity of 'social interaction' or 'exchange'), and changes in such networks through time (e.g., Barton, Clark, & Cohen 1994, Clark, Barton, & Cohen 1996). Although determining social distance and group membership are not essential to the research questions posed in many studies, there are many more where such information is critical.

My own research on the eastern Spanish Mesolithic and Neolithic is a case in point. Within the region of Valencia there are apparently contemporaneous sites in which only the remains of wild plants and animals are found and sites from which remains of domestic cereals and ovicaprids have been recovered. Did the same group of people (or groups of people) leave the assemblages in both types of sites or are they the residues of two different groups employing different subsistence behaviors? The answer to this question structures the explanations proposed for the evolution of food production in the region.

Given the preceding discussion, several suggestions can be offered for guiding a research program to identify group-level social identity in the archaeological lithic record of prehistoric foragers. An important principal to both structure such research and empirically evaluate its results is the relationship between self-conscious social groups and geography. That is, for foragers–and quite possibly most small-scale societies prior to the evolution of social complexity–social groups are coterminous with a piece of the landscape (Shennan 1989) and there is at least some evidence that such geographical relationships can be stable well into the range of temporal resolution for archaeological data (Hayden, Bakewell, & Gargett 1996). This should come as no surprise, as it is characteristic of the vast majority of mammalian species.

Selection and Social Identity

Evidence for group-level social identity can potentially be found in artifacts whose forms are the result of selection for social boundary maintenance. Logically, such selective pressures are likely to be more clearly discernible in contexts of sedentism, land ownership, and social complexity than among mobile foragers for whom the potential for flexibility in group membership can reduce risk (see Shennan 1989, Wiessner 1983). Where it does occur, in most cases selection will probably favor non-lithic artifacts as more effective media for communicating group membership (Barton, Clark, & Cohen 1994, Clark, Barton, & Cohen 1996, Clark 1989, Conkey 1980, Gero 1989). However, a few lithic forms, such as the previously mentioned eccentrics, may also serve this function. To operate in this role, lithic artifacts should show no evidence of usewear or maintenance, indicating that their morphologies were stable over the course of their uselives. They should have long uselives, though this may be difficult to demonstrate. They should have rather complex morphologies that vary minimally within the context of a single social group in order to unambiguously communicate information. Because communication comprises an important (perhaps primary) function, their discard patterns should be distinct from lithics used in more pedestrian direct economic activities like resource extraction and processing. At a regional level, they should be found within a contiguous geographic region and their distribution should be dichotomous or steeply clinal. This latter criterion, along with discard patterns, should help to distinguish lithic forms that communicate information about group membership from those that are linked to individual roles (Gero 1989, Wiessner 1983).

Geography and Social Identity

The relationship between social groups and geography can provide important clues to social identity. Most useful are resources with fixed, identifiable provenances, that are preserved and can be identified in the archaeological record. Sourcing studies have become increasingly valuable in ceramic studies for identifying social groups and their interrelationships (e.g., Neff 1992, 1993; Abbott 1994; Abbott & Walsh-Anduze 1995). Although there are complications, compared to ceramic sourcing, in identifying the source localities of many of the rocks used in lithic technology (Bush & Sieveking 1987), many of these problems are potentially solvable at least at a local scale (Hayden, Bakewell, & Gargett 1996; Hoffman, Todd, & Collins 1991). Sourcing studies for European Middle Paleolithic (Féblot-Augustins 1993), North American Paleoindian (Seeman 1994), and North American Archaic (Shackley 1986) assemblages exemplify the value of this approach for demarcating social groupings and group-level interaction at a regional scale. In a provocative recent study, Hayden and colleagues (Hayden, Bakewell, & Gargett 1996) present evidence suggesting long-term (i.e., more than a millennium) control over specific lithic resources by local-scale social groups and their descent lines. If such work can be replicated elsewhere, it could provide archaeologists with information about social identity at a resolution much finer than has previously been considered possible for prehistoric foragers.

Stochastic Variation and Social Identity

Finally, expectations about patterning for stochastic variability can be used to evaluate the potential for specific artifact forms to provide information about group-level social identity. Within a given social learning context, stochastic processes should result in a single neutral variant rapidly becoming fixed–that is, only a single variant should be produced–in small forager groups. Thus, individual social groups should be characterized by particular variants of a lithic artifact class, reminiscent of the artifact index types (*fossiles directeurs*) commonly used by archaeologists in the first half of this century. Because of the potential for fluidity in group boundaries (both geographic and social), for exchange, and for individual knappers to occasionally change group membership (see, e.g., Wiessner 1983, 1985) the artifact forms specific to one social group might be discarded with the material residues from other groups in varying frequencies. The frequency at which this is likely to occur should be proportional to the combined social and geographic distances between groups.

In a landscape populated by several forager social groups, the spatial distribution of contemporaneous stochastically controlled variants should appear as a series of overlapping dichotomous to clinal distributions (depending on the social distance between the groups). The central part of each distribution, representing more or less the home range of each group, should have variant frequencies approaching unity. Variant frequencies should

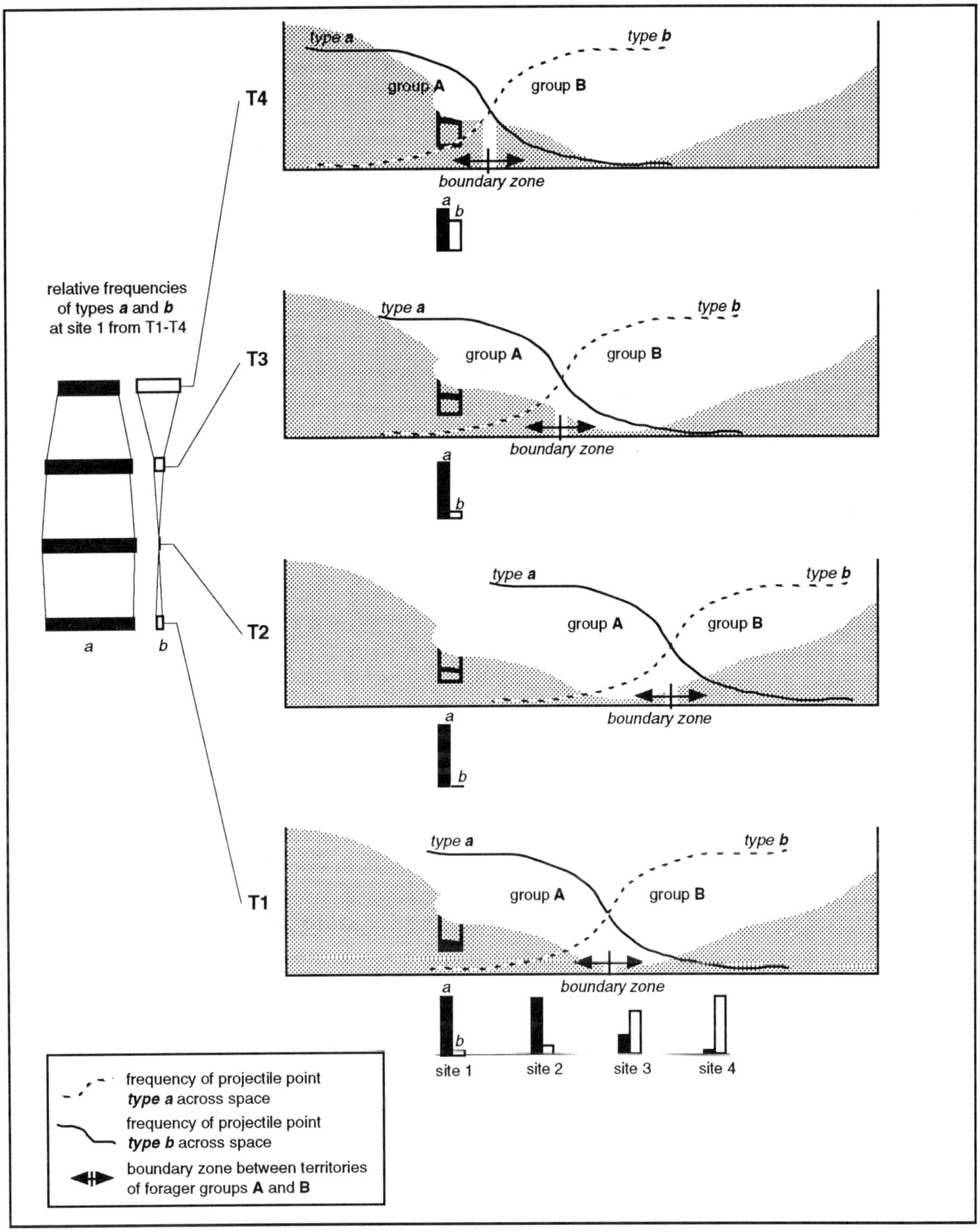

Figure 8.3 Spatial and temporal frequency distributions predicted for two projectile point (or other artifact) types (types *a* and *b*) whose forms are the result of stochastic processes within two groups of mobile forager (A and B). Patterning in the relative frequencies of the two types across space is shown for four time periods (T1-T4). These frequencies display clinal distributions across space, as seen in the relative frequencies of the point types at 4 settlements (site 1-site 4) occupied during time period T1. Note the shift in the spatial (i.e., horizontal) position of the boundary between groups A and B from time periods T1 through T4. The changing relative frequencies of types *a* and *b* at site 1 through time reflect the concommitent changes through time in the distance of site from the boundary zone between the forager groups. These diachronic frequency changes are re-expressed as the "battleship" curves shown at the left side of the figure.

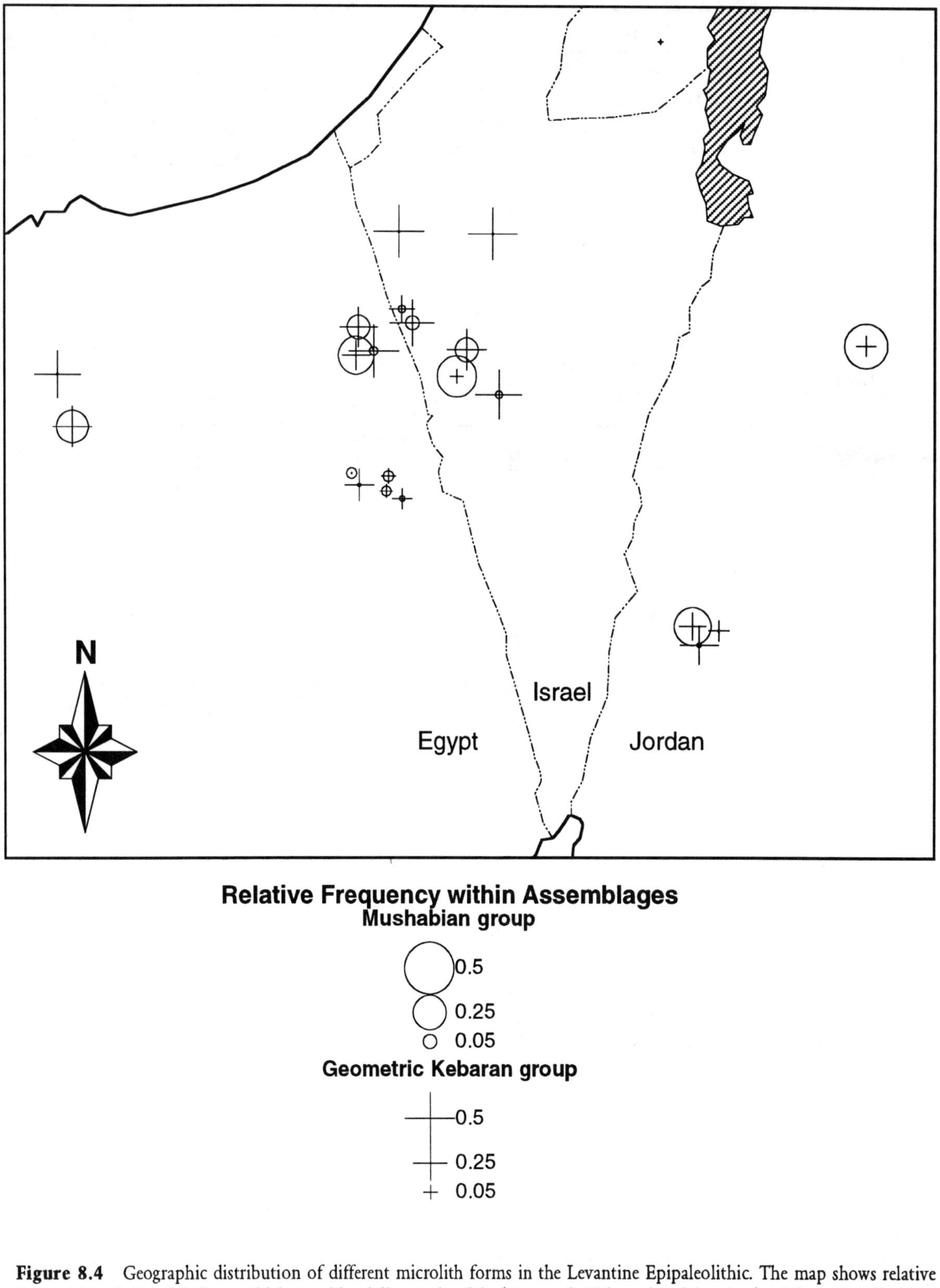

Figure 8.4 Geographic distribution of different microlith forms in the Levantine Epipaleolithic. The map shows relative frequencies of microlith forms widely considered diagnostic of the 'Geometric Kebaran' social entity (straight backed bladelets and trapeze/rectangles), and microlith forms considered diagnostic of the Mushabian' social entity (arched backed and scalene bladelets, and La Mouillah points) in assemblages from sites in the southern Levant. Symbol size is proportional to relative frequency within assemblages (after Barton and Neeley 1996: Figure 1).

decline across boundaries at different rates as indicated above (Figure 8.3). In a recent study of late Epipaleolithic assemblages from Southwest Asia, Neeley and I used this approach to evaluate claims that particular microlith classes were indicative of social group membership (Barton & Neeley 1996, Neeley & Barton 1994). Similar methodologies using ceramic data are discussed in Neiman (1995) and Plog (1990).

Neeley and I found that the artifact classes in question did not behave in the way predicted here, suggesting that an alternative explanation for associated morphological variability is warranted. As can be seen in Figure 8.4 the distributions of the artifact classes in question-Geometric Kebaran type microliths, and Mushabian type microliths-are erratic across space. They are neither differentiated spatially nor do they show a clinal geographic distribution. We suggested alternatively that the distribution of these artifacts was the result of differential discard of compound tool manufacturing and maintenance residues by foragers for whom we cannot yet distinguish distinct social groupings within the region. Nevertheless, there may be other lithic forms in these same assemblages that do distribute as expected for stochastically controlled variants and that might serve to differentiate group membership and boundaries. Analyzing spatial patterning in the discard frequencies of different artifact classes, as described here, would be more informative in this respect than the common practice of classifying lithic assemblages according to the relative frequencies of standard artifact types. Future research in this direction is clearly warranted given the importance of Southwest Asia to understanding the evolution of domestication economies.

CONCLUSIONS

In both non-Darwinian and Darwinian approaches to human behavior, strong cases have been made for partitioning the archaeological record according to social descent groups. As Teltser (1995) and Dunnell (1995) point out, this is where the effects of selection play out. For most of the human past, lithic artifacts are the only phenotypic residues available for inferring prehistoric social identity. The theoretical literature has become increasingly sophisticated during the past two decades at linking artifactual variability with social identity via concepts of style (e.g., Hegmon 1992, Conkey & Hastorf 1990). However, this work has primarily focused on nonlithic residues of human behavior. Furthermore, profound differences as to what constitutes style and its expression have made it even more difficult to apply the theoretical concepts to real archaeological assemblages.

One result for lithic studies is that style is relegated to a garbage category after other explanations have failed to account for variability, and social identity is a fallback explanation when others do not seem to suffice (e.g., Close 1989, Henry 1996). What I advocate here is that archaeologists take a much more proactive approach to identifying social identity that has generally been the case in the past. Neo-Darwinian evolutionary theory provides a framework to reconcile the apparently conflicting concepts of style that are currently debated in the archaeological literature and their relationships both with social identity and artifact assemblages. More importantly it permits us to make specific empirical predictions about the nature of the archaeological record as it pertains to group-level social identity. This, in turn, permits the evaluation of claims of prehistoric social organization based on variability in assemblages of chipped stone artifacts, and the development and refinement of new models of social organization. As several evolutionary theorists have recently pointed (Jones, Leonard, & Abbott 1995; O'Brien & Holland 1995) out partitioning the archaeological record according to social descent is an important part of proposing robust, scientific explanations in a Darwinian framework. I would hope that the suggestions made here can help further this goal.

REFERENCES CITED

Abbott, David R.
1994 Ceramics among the Hohokam: modeling social organization and exchange. In *Archaeometry of Pre-Columbian Sites and Artifacts* (D.A. Scott & P. Meyers, eds.), pp. 248-258. Los Angeles: Getty Conservation Institute.

Abbott, David R. & Mary-Ellen Walsh-Anduze
1995 Temporal patterns without temporal variation: the paradox of Hohokam Red Ware ceramics. In *The Organization of Ceramic Production in the American Southwest* (B.J. Mills & P.L. Crown, eds.),pp. 88-114. Tucson: University of Arizona Press.

Barton, C. Michael
1990 Beyond style and function: a view from the Middle Paleolithic. *American Anthropologist* 92: 57-72.
1991 Retouched tools: fact or fiction? Paradigms for interpreting chipped stone. *Perspectives on the Past: Paradigmatic Biases in Circum-Mediterranean Hunter-Gatherer Research* (G.A. Clark, ed.), pp. 143-163. Philadelphia: University of Pennsylvania Press.

Barton, C. Michael, G. A. Clark, & Allison E. Cohen
1994 Art as information: explaining Upper Paleolithic art in western Europe. *World Archaeology* 26: 185-207.

Barton, C. Michael & Michael P. Neeley
1996 Phantom cultures of the Levantine Epipaleolithic. *Antiquity* 70: 139-147.

Barton, C. Michael, Deborah I. Olszewski, & Nancy R. Coinman
1995 Beyond the graver: reconsidering burin function. *Journal of Field Archaeology* 23: 111-125.

Berrin, Kathleen & Esther Pasztory
1993 *Teotihuacan: Art from the City of the Gods.* New York: Thames and Hudson.

Bettinger, Robert L., James F. O'Connell, & David H. Thomas
1991 Projectile points as time markers in the Great Basin. *American Anthropologist* 93: 166-171.

Biagi, P. & M. Cremaschi
1991 The Harappan flint quarries of the Rohri Hills. *Antiquity* 65: 97-101.

Binford, Lewis R.
1973 Interassemblage variability-the Mousterian and the 'functional' argument. In *The Explanation of Culture Change: Models in Prehistory* (C. Renfrew, ed.), pp. 227-254. London: Duckworth.

Binford, Lewis R. & Sally R. Binford
1966 A preliminary analysis of functional variability in the Mousterian of Levallois facies. *American Anthropologist* 68: 238-295.

Bleed, Peter
1986 The optimal design of hunting weapons: maintainability or reliability. *American Antiquity* 51: 737-747.

Bordes, F.
1973 On the chronology and contemporaneity of different paleolithic cultures in France. *The Explanation of Culture Change: Models in Prehistory* (C. Renfrew, ed.), pp. 217-226. London: Duckworth.

Bordes, F. & D. de Sonneville Bordes
1970 The significance of variability in paleolithic assemblages. *World Archaeology* 2: 61-73

Bush, P.R. & G. de G. Sieveking
1987 Geochemistry and the provenance of flint axes. *The Scientific Study of Fint and Chert* (G. de G. Sieveking & M.B. Hart) pp. 133-140. Cambridge: Cambridge University Press.

Carr, C. & J. Neitzel (editors)
1996 *Style, Society, and Person*. New York: Plenum

Clark, G.A.
1983 *The Asturian of Cantabria: Early Holocene Hunter-Gatherers in Northern Spain.* Tucson: Anthropological Papers of the University of Arizona No. 41.

Clark, G.A.
1989 Romancing the stones: biases, style, and lithics at La Riera. *Alternative Approaches to Lithic Analysis* (D.O. Henry & G.H. Odell) pp. 27-50. Washington: Archeological Papers of the American Anthropological Association No. 1.

1996 Plus français que les Français. *Antiquity* 70: 130-147.

Clark, G.A., C.M.Barton & A.E.Cohen
1996 Explaining art in the Franco-Cantabrian refugium: an information exchange model. In *Debating Complexity* (D. Meyer et al., eds.), pp. 241-253. Calgary: Archaeological Association of the University of Calgary.

Clark, John E.
1987 Politics, prismatic blades, and Mesoamerican civilization. *The Organization of Core Technology* (J.K. Johnson & C.A. Morrow, eds.), pp. 259-284. Boulder, CO: Westview Press.

Close, Angela E.
1989 Identifying style in stone artefacts: a case study from the Nile Valley. *Alternative Approaches to Lithic Analysis* (D.O. Henry & G.H. Odell, eds.), pp. 3-26. Washington: Archeological Papers of the American Anthropological Association Number No. 1.

Conkey, Margaret W.
1980 The identification of prehistoric hunter-gatherer aggregation sites: the case for Altamira. *Current Anthropology* 21: 609-630.

1990 Experimenting with style in archaeology: some historical and theoretical issues. *The Uses of Style in Archaeology* (M.W. Conkey & C.A. Hastorf, eds.), pp. 5-17. Cambridge: Cambridge University Press.

Conkey, Margaret W. & Christine A. Hastorf (editors)
1990 *The Uses of Style in Archaeology.* Cambridge: Cambridge University Press.

Cotterell, B. & J. Kamminga
1987 The formation of flakes. *American Antiquity* 52: 675-708.

Dibble, Harold L.
1987 The interpretation of Middle Paleolithic scraper morphology. *American Antiquity* 52: 109-117.

1995 Middle Paleolithic scraper reduction: background, clarification, and review of the evidence to date. *Journal of Archaeological Method and Theory* 2: 299-368.

Dibble, Harold L. & John C. Whittaker
1981 New experimental evidence on the relation between percussion flaking and flake variation. *Journal of Archaeological Science* 8: 283-296.

Dunnell, Robert C.
1978 Archaeological potential of anthropological and scientific models of function. *Archaeological Essays in Honor of Irving B. Rouse* (R.C. Dunnell & E.S. Hall, eds.), pp. 41-73. The Hague: Mouton.

1978 Style and function: a fundamental dichotomy. *American Antiquity* 43: 192-202.

1995 What is is that actually evolves? *Evolutionary Archaeology: Methodological Issues* (P.A. Teltser, ed.), pp. 33-50. Tucson: University of Arizona Press.

Féblot-Augustins, J.
1993 Mobility strategies in the late Middle Paleolithic of Central Europe and Western Europe: elements of stability and variability. *Journal of Anthropological Archaeology* 12: 211-265.

Flenniken, J. J. & W. Raymond
1986 Morphological projectile point typology: replication, experimentation, and technological analysis. *American Antiquity* 51: 603-614.

Flenniken, J. J. & P.J. Wilke
1989 Typology, technology, and chronology of Great Basin dart points. *American Anthropologist* 91: 149-158.

Frison, George C.
1968 A functional analysis of certain chipped stone tools. *American Antiquity* 33: 149-155.

Gero, Joan M.
1989 Assessing social information in material objects: how well do lithics measure up?. *Time Energy and Stone Tools* (R. Torrence, ed.), pp. 92-105. Cambridge: Cambridge University Press.

Goring-Morris, Nigel A.
1996 Square pegs into round holes: a critique of Neeley and Barton. *Antiquity* 70: 130-135.

Hayden, Brian
1987 From chopper to celt: the evolution of resharpening techniques. *Lithic Technology* 16: 33-43.

Hayden, Brian, Edward Bakewell, & Rob Gargett
1996 The world's longest-lived corporate group: lithic analysis reveals prehistoric social organization near Lillooet, British Columbia. *American Antiquity* 61: 341-356.

Hayden, Brian & Rick Schulting
1997 The Plateau Interaction Sphere and late prehistoric cultural complexity. *American Antiquity* 62: 51-85.

Hegmon, Michelle
1992 Archaeological research on style. *Annual Review of Anthropology* 21: 517-536.

Henry, Donald O.
1996 Functional minimalism versus ethnicity in explaining lithic patterns in the Levantine Epipaleolithic. *Antiquity* 70: 135-136.

Hoffman, C. M.
1985 Projectile point maintenance and typology: assessment with factor analysis and canonical correlation. *For Concordance in Archaeological Analysis* (C. Carr, ed.), pp. 566-612. Prospect Heights, IL: Waveland Press.

Hofman, J.L., L.C. Todd, & M.B. Collins
1991 Identification of central Texas Edwards chert at the Folsom and Lindenmeier sites. *Plains Anthropologist* 36: 297-308.

Jelinek, Arthur J.
1976 Form, function, and style in lithic analysis. *Cultural Change and Continuity* (C.E. Cleland, ed.) pp. 19-35. New York: Academic Press.

1988 Technology, typology, and culture in the Middle Paleolithic. *Upper Pleistocene Prehistory of Western Eurasia* (H.L. Dibble & A. Montet-White, eds.), pp. 199-212. Philadelphia: University of Pennsylvania Museum.

Jones, George T., Robert D. Leonard, & Alysia L. Abbot
1995 The structure of selectionist explanations in archaeology. *Evolutionary Archaeology: Methological Issues* (P.A. Teltser, ed.), pp. 13-32. Tucson: University of Arizona Press.

Kaufman, Daniel
1995 Microburins and microliths of the Levantine Epipaleolithic: a comment on the paper by Neeley & Barton. *Antiquity* 69: 375-381.

Kelly, Robert L.
1988 The three sides of a biface. *American Antiquity* 53: 717-734.

Myers, Andrew
1989 Reliable and maintainable technological strategies in the Mesolithic of mainland Britain. *Time, Energy and Stone Tools* (R. Torrence, ed.), pp. 78-91. Cambridge: Cambridge University Press.

Neeley, Michael P. & C. M. Barton
1994 A new approach to interpreting late Pleistocene microlith industries in southwest Asia. *Antiquity* 68: 275-288.

Neff, Hector
1992 Ceramics and evolution. *Archaeological Method and Theory* 4: 141-193.

1993 Theory, sampling, and analytical technique in the archaeological study of prehistoric ceramics. *American Antiquity* 58: 23-44.

Neiman, Fraser D.
1995 Stylistic variation in evolutionary perspective: inferences from decorative diversity and interassemblage distance in Illinois Woodland ceramic assemblages. *American Antiquity* 60: 7-36.

Nelson, Margaret C.
1991 The study of technological organization. *Archaeological Method and Theory* 3: 57-100.

O'Brien, Michael J. & Thomas D. Holland
1990 Variation, selection, and the archaeological record. *Archaeological Method and Theory* 2: 31-79.

1995 The nature and premise of a selection-based archaeology. *Evolutionary Archaeology: Methological Issues* (P.A. Teltser, ed.), pp. 175-200. Tucson: University of Arizona Press.

Parry, W.J. & R.L. Kelly
1987 Expedient core technology and sedentism. *The Organization of Core Technology* (J.K. Johnson & C.A. Marrow, eds.), pp. 284-304. Boulder: Westview Press.

Plog, Stephen
1990 Sociopolitical implications of stylistic variation in the American Southwest. *The Uses of Style in Archaeology* (M.W. Conkey & C.A. Hastorf, eds.), pp. 61-72. Cambridge: Cambridge University Press.

Rick, John W.
1996 Projectile points, style, and social process in the Preceramic of central Peru. *Stone Tools: Theoretical Insights into Human Prehistory* (G.H. Odell, ed.), pp. 245-278. New York: Plenum Press.

Rondeau, Michael F.
1996 When is an Elko? *Stone Tools: Theoretical Insights into Human Prehistory* (G.H. Odell, ed.), pp. 229-243. New York: Plenum Press.

Rosen, Steven A.
1983 The Canaanean blade and the Early Bronze Age. *Israel Exploration Journal* 33: 15-28.

1996 The decline and fall of flint. *Stone Tools: Theoretical Insights into Human Prehistory* (G.H. Odell, ed.), pp. 129-158. New York: Plenum Press.

Sackett, James R.
1985 Style and ethnicity in the Kalahari: a reply to Wiessner. *American Antiquity* 50: 154-159.

1989 Statistics, attributes, and the dynamics of burin typology. *Alternative Approaches to Lithic Analysis* (D.O. Henry & G.H. Odell, eds.), pp. 51-82. Washington: Archeological Papers of the American Anthropological Association No. 1.

1990 Style and ethnicity in archaeology: the case for isochrestism. *The Uses of Style in Archaeology* (M.W. Conkey & C.A. Hastorf, eds.), pp. 32-43. Cambridge: Cambridge University Press.

Schele, Linda & Mary E. Milla
1986 *The Blood of Kings: Dynasty and Ritual in Maya Art.* New York: George Braziller, Inc.

Schiffer, Michael B.
1996 Some relationships between behavioral and evolutionary archaeologies. *American Antiquity* 61: 643-662.

Seeman, Mark F.
1994 Intercluster lithic patterning at Nobles Pond: a case for 'disembedded' procurement among early Paleoindian societies. *American Antiquity* 59: 273-278.

Shackley, M. S.
1986 Obsidian geochemistry and lithic technology: inferences for archaic hunter-gatherer procurement ranges. *Prehistoric Hunter-Gatherers of South Central Arizona: the Picacho Reservoir Archaic Project* (F.E. Bayham, D.H. Morris, & M.S. Shackley, eds.), pp. 171-190. Tempe: Arizona State University Anthropological Field Studies No. 13.

Shennan, Stephen J.
1989 Introduction: archaeological approaches to cultural identity. *Archaeological Approaches to Cultural Identity* (S.J. Shennan, ed.), pp. 1-32. London: Unwin Hyman.

Shott, Michael J.
1996 Innovation and selection in prehistory: a case study from the American Bottom. *Stone Tools: Theoretical Insights into Human Prehistory* (G.H. Odell, ed.), pp. 279-309. New York: Plenum Press.

Sinopoli, Carla M.
1991 Style in arrow: a study of an ethnographic collection from the western United States. *Foragers in Context* (P.T. Miracle, L.E. Fisher, & J. Brown, eds.), pp. 63-87. Ann Arbor: University of Michigan Discussions in Anthropology No. 10.

Teltser, Patrice A.
1995 Culture history, evolutionary theory, and frequency seriation. *Evolutionary Archaeology: Methological Issues* (P.A. Teltser, ed.), pp. 51-68. Tucson: University of Arizona Press.

Thomas, David H.
1981 How to classify the projectile points from the Monitor Valley, Nevada. *Journal California and Great Basin Anthropology* 3: 7-43.

1986 Points on points: a reply to Flenniken and Raymond. *American Antiquity* 51: 619-627.

Torrence, Robin
1989 Retooling: towards a behavioral theory of stone tools. *Time, Energy and Stone Tools* (R. Torrence, ed.), pp. 57-66. Cambridge: Cambridge University Press.

Whittaker, John C.
1987 Individual variation as an approach to economic organization: projectile points at Grasshopper Pueblo, Arizona. *Journal of Field Archaeology* 14: 465-479.

Wiessner, P.
1983 Style and social information in Kalahari San Projectile Points. *American Antiquity* 48: 253-276.

Wiessner, Polly
1985 Style or isochrestic variation? A reply to Sackett. *American Antiquity* 50: 160-165.

Wilke, Philip J. & J. J. Flenniken
1991 Missing the point: rebuttal to Bettinger, O'Connell, and Thomas. *American Anthropologist* 93: 172-173.

Wobst, H. M.
1977 Stylistic behavior and information exchange. *For the Director: Research Essays in Honor of James B. Griffin* (C.E. Cleland, ed.), pp. 317-344. Ann Arbor: Anthropological Papers of the Museum of Anthropology No. 61.

9

Seeking Guidelines through an Evolutionary Approach: Style Revisited among the !Kung San (Ju/'hoansi) of the 1990s

Polly Wiessner
Research Group for Human Ethology
in the Max Planck Society
Andechs, Germany

ABSTRACT

This paper will seek to situate style as a means of communication in a fundamental human cognitive process identified in social psychology: the establishment of positive personal and social identities via social comparison and the projection of these to others. First, identification by comparison and its link to stylistic communication will be discussed. Second, how this cognitive process came to be in our evolutionary history, and the evolutionary developments through which style might have arisen will be identified. Third, the implications of this process for formulating stylistic questions and guiding analysis will be developed. The focus will be on diachronic rather than synchronic stylistic analysis, namely, on analysis of the changing relations between the individual and society and how history is drawn on selectively in creating or transforming identities. Finally, the utility of the approach will be tested in a follow-up case study of stylistic change in projectile points and beadwork among the !Kung San (Ju/ 'hoansi) of Botswana and Namibia.

INTRODUCTION

Virtually all known human societies produce variation in material culture that falls with in the general concept of style as 'a way of doing or expressing things', though definitions used in archaeological studies differ. The reciprocal also holds true–not only do most humans engage in stylistic behavior, but observers seek to interpret it. For example, a visitor to a foreign culture who is ignorant of the language may look for stylistic clues to determine how people identify and classify themselves, whether this be by age, sex, or social group. Because of the close association of material culture with the bearer, often as an extension of the body, style tends to be given meaning and become emotionally loaded–it can incite feelings of joy, worth, brother/sisterhood in the bearer and attract, repel, appease, or provoke the observer.

The universality of style in material culture as a means to communicate aspects of identity, the tendency of the receiver to consider that it has such meaning, and its emotional tagging all indicate that style is grounded in human cognition. In the approach taken here, I will seek to situate style as information exchange in the service of a fundamental cognitive process: the establishment of positive personal and social identities via social comparison and the projection of these to others (Wiessner 1984: 191-195). This process, identified by social psychologists, is universally found in human societies and appears to be fundamental to psychological wellbeing (Lemaine 1974; Tajfel 1974, 1978, 1982a, 1982b; Turner 1975). I will depart from other evolutionary approaches in a number of respects. First, natural selection will not be seen as operating directly on material culture, the 'fitness of artifacts', but on the humans and their behaviors that produce style. It is people–not artifacts–who reproduce. Second, although style will be treated as a form of non-verbal communication, it will not be argued that style maximizes efficiency of information transfer, as in Wobst's formulation (1977). Style is chosen for effectiveness within a given cultural context, sometimes involving costly display. Third, style will not be analyzed within the functional framework of maximizing the reproductive success of its bearers in any single given instance. Rather, following the schools of thought from human ethology (Eibl-Eibesfeldt 1989) and evolutionary psychology (Tooby & Cosmides 1989), cognitive dispositions will be seen as having conferred reproductive advantage when they evolved, and thereafter more often than not in our evolutionary history.

The goal of this paper is to explore the cognitive processes underlying the cultural transmission of style, their role in structuring variation in material culture, and thereby identify relevant questions to guide stylistic analysis. Though there is much of interest and value in the work that has been done on cultural transmission to date (Boyd & Richardson 1985, Neiman 1995), it will be argued that cultural transmission must be understood from the viewpoint of human cognitive processes and not formal models constructed on the basis of 'logical possibilities.' Mathematical models for predicting variation in material culture that require no interpretation may be ideal for a scientific archaeology, as Neiman (1995: 30-31) argues. However, if cognitive processes underlying stylistic behavior involve considerable interpretation and intentionality on the part of the actors, then one can hardly construct viable models of cultural transmission without taking these factors into account.

To look at style within the framework of identification via comparison is not new (see Wiessner 1983, 1984). Here, however, I would like to develop facets of this approach that I have not developed or emphasized in other publications. The first is to make some suggestions as to how and why the cognitive process of identification via comparison came to be in our evolutionary history, and thereby try to identify the evolutionary developments with which style might be most closely linked. The second is to examine the process of identification via comparison in order to derive guidelines for the kinds of questions that can be asked of style in a diachronic perspective. Finally I will try to test the utility of this approach in a follow-up case study of the !Kung San (Ju/'hoansi) 23 years later, after major changes had taken place in subsistence, settlement, and lifestyle. While, in my 1970s research, I could only explore the potential of synchronic stylistic analysis of !Kung artifacts for yielding information on boundaries and interaction, the diachronic perspective that can be taken after a longer term study makes it possible to work with an entirely different range of information contained in style: insights into the changing relations between the individual and society and how history is drawn on selectively in creating or transforming identities.

IDENTIFICATION VIA SOCIAL AND STYLISTIC COMPARISON

A vast literature in social psychology has documented the need of humans to establish self-images or identities with both personal and social components. Personal identity usually denotes specific attributes of the individual such as such as psychological characteristics, bodily attributes, feelings of competence, ways of relating to others, intellectual concerns, personal tastes, and so on (Gergen 1971). Material culture of the bearer may be an extension of this image. Social identity, according to Tajfel, is "the part of individuals' self concepts which derive from their knowledge of their membership of a social group(s) together with the value and emotional significance attached to that membership" (1982a: 2). Both social and personal components of identity are crucial to the formation of self images: persons are unable to form identities in the absence of group membership, and conversely, individuals feel uncomfortable in situations of extreme conformity and seek to differentiate themselves on a personal basis (Fromkin 1972, Lemaine et al. 1978). A sense of personal or social identity may come to the fore in certain situations; social identity is the cognitive mechanism that makes group behavior possible (Turner 1982). Finally, humans not only have a need to establish personal and social identities, but they seek to project these to others in a positive light.

Much evidence has accumulated to demonstrate that the process of identity formation is social comparison, that is, the tendency of humans to evaluate their abilities, opinions, and attributes against those of surrounding others (Festinger 1954). So well established is the identification via comparison that it is assumed by many authors to be a basic cognitive process in humans (Lemaine 1974; Tajfel 1978, 1982a; Turner 1975). That the contents of identities are to a great extent culturally and historically determined is well-recognized (Wetherell 1982). Identities are created selectively from historical and current considerations, and chosen markers are used to form, maintain or transform corresponding boundaries (Barth 1969).

Style in material culture is one of several important channels by which personal and social self-images can be presented to others to receive recognition. Stylistic choices are made in a process parallel to that of identity formation, through comparison with surrounding others followed by statement of position: for instance, expression of similarity, differentiation, opposition, or complementary. Style is particularly appropriate for transmitting information about identity because: (1) as an extension of the body it can enhance certain attributes; (2) it can present several aspects of identity at one time, which might influence or modify one another; (3) it can make use of aesthetic experience to reinforce identity representations as well as bonding those who respond to it with pleasure (Eibl-Eibesfeldt 1989); (4) style can be tactful–it communicates without demanding response, leaving others time to formulate their reactions; and (5) style is produced in artifacts that are often durable and replicated from generation to generation. It is thus suitable for conveying information regarding historical associations that bind people in dispersed social networks.

Thoughts on the Evolution of Identification via Comparison and Style

When, where, and how did this cognitive disposition come to be? Since the evolution of identification via comparison does not 'fossilize', I can only present some reflections on the processes that might have led humans to 'fix' stylistic activities in their behavioural repertoires. It is not difficult to envision selection pressures for the disposition to establish a sense of personal identity via comparison and to project this to others, for instance in mate selection, dominance struggles or, as Crook (1981) has proposed, to attract others in relationships of reciprocal altruism (Trivers 1971). To account for the evolution of social identity is more complex. Need for cooperation in hunting or defense are candidates, although I would propose that a much stronger impetus probably came with the evolution of socially defined kinship systems, that is, the extension of familial relationships to affinal relatives, distant kin, or even non-kin. Spheres of kinship would no longer be delimited in terms of immediate family membership or physical cohabitation, but by networks held 'in the mind and heart' composed of kin of varying degrees of relatedness in different locations. Socially defined kinship relations, together with their most important rights and privileges of delayed reciprocal exchange and access to the resources of others, made possible the construction of networks to reduce risk. These were crucial to the expansion and survival of human populations, for they permitted people to redistribute themselves over the social and natural resources of a region according to need. Barton, Clark and Cohen (1994; Clark, Barton & Cohen 1996) have pointed out that, prior to the last glaciation, the middle latitudes of the Old World were largely abandoned by human populations, while during the last glacial maximum they were able to remain at densities higher than ever before. Perhaps this can be taken as an indicator of the interval during which socially defined kinship systems might have developed; certainly the flourish in style and exchange suggests the same (Gamble 1982, 1986). Historical associations may have been chosen as important components of social identity with the evolution of social kinship networks because bonds between members of such dispersed networks are based on periodic association over an extended stretch of time rather than spatial contiguity.

Boehm (1993) has recently called attention to another development that could have greatly facilitated the evolution of 'pro-social' or altruistic behaviors critical to maintaining socially defined kinship systems and accompanying obligations for reciprocity, that is, egalitarianism. Unlike most great ape societies, human foraging societies and many simple horticultural societies are strongly egalitatian (Knauft 1991, Wrangham 1987). It is thus likely that egalitarian structures evolved out of hierarchical ones,[1] perhaps out of primate coalitions to rise in dominance hierarchies (De Waal 1982). With egalitarianism, inter-individual competition is greatly reduced and together with it the cost of 'altruistic behavior'. This is because recipients of altruistic behavior are barred by social rules and sanctions from using goods or assistance received to develop a position from which they can appropriate the resources of others or gain marked advantage in differential reproduction. Certainly in foraging societies today the dynamics of sharing and delayed reciprocal exchange hinge on the fact that what is given cannot be used to dominate or incur debt.

Style would have been an important means of communication to support the expansion of kinship networks and egalitatian relations because it could:

(a) express information about identity and its positive correlates, such competence, strength, or worth, without resorting to the overt, aggressive, and confrontational displays that are so characteristic of great ape societies with dominance hierarchies;
(b) provide indicators of similarity that modify physical differences;
(c) help establish homogeneity between individuals in different groups and hence facilitate the expansion of kinship networks;
(d) attract mates from other groups to forge the ties necessary for expanding socially defined kinship networks.

It appears that socially defined kinship systems were first constructed to elicit cooperation within a broader regional population. Only later, in the face of demographic and economic pressure were systems of social kinship elaborated into more pronounced corporate groups such as clans, tribes, and political hierarchies that form the infrastructure for most known pre-state societies. In this process, historical associations that bind people to the land and to each other would have become a much stronger component of social identity and the role of style broadened to support new kinship structures and maintain boundaries between these.

Identification via Comparison and Stylistic Information

If style is situated in process of identification via comparison, what guidelines can knowledge of this cognitive disposition give us for stylistic analysis? First, because the parameters for social and stylistic comparison are culturally defined, before stylistic analysis is undertaken it is essential to take cultural meaning into consideration (see also Hodder 1982, 1986; Shanks & Tilley 1987). This can only be done in the most general terms, for it is difficult to predict how humans will attach meaning to form, and there are but a few signals with biological roots whose range of meanings are similar in most societies (e.g., bared teeth, tooth patterns, or phallic display [Eibl-Eibesfeldt & Sütterlin 1992; Uher 1991]). An important first step is to ascertain whether or not a certain artifact or attribute bears culturally significant meaning. That is, once some items of material culture or attributes are given meaning, all have the potential to transmit information or be interpreted as such. Nonetheless, some are loaded and others hold little, a fact that must be reckoned with in analysis of variation in material culture.

For simplicity, I will refer to variation in material culture that is assigned little meaning as 'isochrestic' (Sackett 1982) and that attributed meaning as 'style'. Although there is not a distinct line between the two, (and style can lose meaning and isochrestic variation be given meaning), it is possible to use certain criteria to sort artifacts and attributes that are likely to contain social meaning from those that are not. A more obvious one is whether an artifact bears decoration or whether an attribute is decorative. Others include centrality of an artifact in subsistence or social systems making it suitable for stylistic communication; whether variation can be readily perceived with the naked eye; and whether this variation is the product of careful craftsmanship or casual production. Patterns of change through time can also give clues. For example, if social, economic, or political conditions are altered and style in material culture employed in associated activities remains constant, then it is less likely that these bear personal or social messages. A preliminary sorting of artifacts or attributes into those that appear to have much or little meaning, though not possible in every case, is an important first step, because it determines whether models such as those based on 'drift' should be applied (Neiman 1995) or those centering on social aspects of transmission.

If an attribute or set of attributes is assigned cultural meaning and can be counted as style, then it becomes an means of agency, as individuals and groups manipulate meanings in social strategies. Here I would like to sketch out two sources of information in style that are generated during the process of identification via comparison which can be analyzed to derive information about social agency in a diachronic perspective: (1) the changing relations between the individual and society and (2) the role of historical associations in constructing identities. Although these two points are rarely the goal of stylistic analysis, they hold great potential.

Let us turn to the first. Because people have a need to both form and project positive images of personal and social identity, style can be expected to carry information about the two (expression of individual identity is essentially what Macdonald [1990] has called 'panache', and social identity, 'protocol'). As personal identity is not constructed in a vacuum, but in relation to surrounding others, it expresses not only individual attributes but associations, usually those stemming from close personal relations. As a result, personal stylistic expression can be used as a measure of affiliative interaction. Social identity, by contrast, places the individual within larger social groups, and if these groups are contiguous in space, stylistic expression of social identity may yield information on groups and boundaries.[2]

The use of style to delimit boundaries and gauge rates of interaction or affiliation are the most conventional uses of style in archaeology (see Barton, Chapter 8). However, if the personal and social expressions of identity are analyzed in a diachronic perspective and in relation to one another, then style potentially holds information on the changing relations between the individual and society. From such relations it is sometimes possible to make inferences concerning units of social agency which initiated developments. Changes in the proportion of individual and group expression can be derived from the increase or decrease in the amount of stylistic variability through time in any given attribute or set of attributes within an area. That is, if certain styles exhibit innovation and increased variability over time, one might deduce accentuated personal differentiation; conversely, the appearance of homogeneity or standard emblems within a bounded area might suggest pronounced feelings of group solidarity. Furthermore, because certain situations predictably "switch on" expressions of group and personal identity respectively, it is not only possible to show that changes occurred in the proportion of individual to group expression, but to propose potential underlying causes. For example, inter-group competition, aggression, or other conditions that require or impose cooperation draw out expressions of group identity. Opportunities for personal gain, increased inter-individual competition, abundance, or break down in the social order foster individual expression. Such developments can then be further interpreted by other sources of information on politics, economy, or lifestyle.

A second and integral feature of identification via comparison is the role of history in the process of constituting identity. Though identification via comparison is a process found in all humans, the contents of identity are historically and culturally constituted. That is, when people compare themselves to other, whether as individuals or groups, and chose to express affiliation, differentiation, or opposition, track record and other historical associations play a crucial role. Likewise, when identity is communicated stylistically, choice of stylistic decisions are guided by historical meanings. For personal identity, many facets of a self-image may be formed within the person's life time; for group identity, however, historical associations are fundamental, for it is history and genealogy that bind people to land and either to link them to similar others or separate them from the same. Nonetheless, history does not determine identity. Rather, historical associations and their material expressions are called upon selectively in the process of creating, altering, or transforming identities. Examining which historical markers are evoked in the process of identity formation, by whom, and how they are combined with new elements can thus provide valuable information on the source, nature, and agency of change.

In summary, stylistic transmission operates through the interpretive processes of social and stylistic comparison, taking relationships and meaning from both the present and past to shape identity relations of the future. Models of cultural transmission that do not consider this process are likely to miss the mark. Understanding the cognitive dispositions underlying style gives a theoretical basis for broadening the use of stylistic data examine: (1) the changing relations between the individual and society, and (2) agency in cultural change as reflected in how history is selectively used for constructing identities.

A CASE STUDY AMONG THE JU/'HOANSI (!KUNG SAN)

The Ju/'hoansi of the Kalahari or !Kung San of northwest Botswana and northeast Namibia need no introduction to archaeologists or anthropologists. Their ethnography has been well-documented in the work of Lee (1979, 1986, 1993); Lee and DeVore (1976); Marshall (1976); Howell (1979); Biesele (1990, 1993); Shostak (1981); Wiessner (1977, 1982) and Yellen (1977) amongst many others. Subsequently their history and position in the larger political economy of southern Africa has been elucidated by Wilmsen (1989) and Gordon (1992). Here I will briefly review the ethnographic situation in the 1970s and go on to describe economic and political developments that have occurred in the meantime. Then, drawing on the above evolutionary framework for guidelines, I will turn to style.

Background

In the mid-1970s when I completed my fieldwork in northwestern Botswana, the Ju/'hoansi of the Dobe-/Kae/kae area were primarily foragers (Figure 9.1). Subsistence was obtained from gathering, hunting, or working for neighboring agro-pastoralists, in that order. Settlement patterns within a band's area of land rights *(n!ore)* consisted of dispersal into small scattered camps during the wet season and aggregation at large camps during the dry season. Both meat and vegetable foods were widely shared. While food was plentiful in the mid- and late wet season, Ju/'hoansi suffered shortages in the drier months, particularly as overgrazing by the herds of neighboring agro-pastoralists destroyed bush foods. Far-reaching networks of socially defined kinship,[3] activated by partnerships of delayed, reciprocal exchange called *xaro* (*hxaro*) (Wiessner 1977, 1982, 1986, 1994), gave Ju/'hoansi access to the resources of people in other *n!ore* within a radius of approximately 150 km, the boundaries of the central !Kung dialect group (excluding Ju/'hoansi who were settled on farms of other ethnic groups). Regular visiting on the basis of *xaro* relationships redistributed people over available resources in times of environmental or social hardship.

Strong egalitarian ethics prevailed in Ju/'hoan society, and those who sought to accumulate wealth or status were quickly leveled. Nonetheless, there were slight inequalities in status and unequal access to resources, for instance, older people had some priority over the young, and *n!ore kxao*, 'owner(s) of the place', over those who moved in from other areas. Gender differences between men and women existed, but these were not linked to unequal status or access to resources. Egalitarianism provided the social matrix for sharing, *xaro* exchange, and all accompanying forms of mutual assistance, for help was given with the knowledge that it could not be used to build positions from which one could dominate or coerce another and that it must be reciprocated when conditions of 'have' and 'have not' were reversed. It is important to note, however, that egalitarianism amongst the Ju/'hoansi was and is not grounded in conformity but rather in difference. The Ju/'hoansi not only tolerate but appreciate individual difference (Biesele 1993); ethics of equality and accompanying sanctions then level disparities in access to material or social resources that can result from such differences.

Site structure was circular or semi- circular and open, expressing both the obligation to share and facilitating control of sharing by making the resources of any

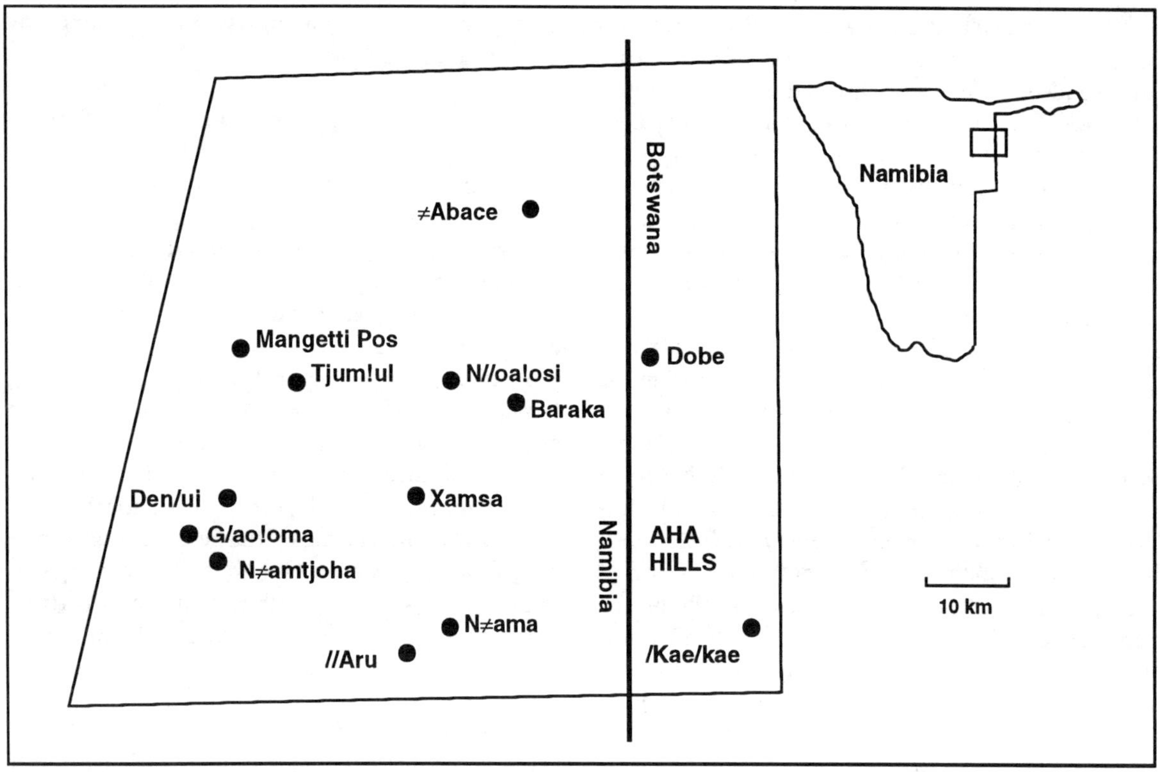

Figure 9.1 Map showing villages in Namibia and Botswana where beadwork and projectile points were recorded. Dashed lines delimit approximate area containg villages of the Nyae Nyae Farmer's Cooperative.

household in the camp fully visible to others. Spacing between huts increased with kinship distance and corresponding obligations to share. Houses were of mixed styles, including traditional grass huts as well as more substantial mud and thatch huts, a building style adopted from surrounding agro-pastoralists.

Near Sehitwa, approximately 150 km to the east of Dobe and /Kae/kae, where Ju/'hoansi from the same population were settled on farms of Tswana and Herero, the situation was quite different. Here Ju/'hoansi were integrated into the lifestyle of the agro-pastoralists for whom they worked. The greatest proportion of food obtained was domestic, that is, milk, meat, and maize obtained from employers. Huts were more permanent structures, plastered, thatched, and often fenced. *Xaro* exchange was maintained with a few relatives in distant locations, but else was on the decline. Guns were used for hunting and few Ju/honasi possessed bows and arrows. Most Ju/'hoan material culture was replaced with store bought items of that of agro-pastoralists, and few Ju/'hoan women continued to make beadwork. Systematic stylistic comparison of arrows and beadwork between this area and Dobe-/Kae/kae was thus not possible.

To the west of Dobe and /Kae/kae in the Nyae Nyae area, rapid change was initiated from the outside in 1959. A South African administrator was sent to Nyae Nyae to assemble people at Tjum!kui and establish a center offering a store, school, clinic, agricultural programs, permanent water, and wage labor. In 1970 an official Bushman homeland was proclaimed giving the southern part of traditional Ju/'hoan territory to the Herero and the northern part to a game reserve. In contrast to communities in Botswana, change in the Nyae Nyae area was imposed from the outside and not the product of integration with other ethnic groups. It generated problems within that had to be solved from within, for Ju/'hoansi were artificially kept in a world of other Ju/'hoansi by the borders of their homeland. By 1973 and 1975, as many as 900 Ju/'hoansi, largely from the Nyae Nyae area, were settled at Tjum!kui either in government-built 'location' houses, concrete bungalows, or in more traditional 'villages' surrounding Tjum!kui. The better part of subsistence came from rations provided by the government, supplemented by bush foods available in the vicinity of Tjum!kui. There were, in addition, ample opportunities for economic gain from wage labor or sale of crafts.

With such dense settlement and new economic opportunities, inter-individual competition and social inequalities began to develop. The pendulum at Tjum!kui swung between a tendency to show off new found wealth, in contravention to social norms, and attempts to restore harmony and equal distribution of wealth through giving. In the process, Tjum!kui Ju/'hoansi constructed complex *xaro* networks that wound their way through the villages of Tjum!kui, activating a wide range of kinship ties and redistributing wealth to appease. But such means of redistributing wealth was only a partial solution, for the employed were able to maintain two to four times as many *xaro* partners as the unemployed, brewing jealousy. Tension mounted and many turned to alcohol, greatly exasperating these problems. A new era of frequent physical violence began which took the lives of many Ju/'hoansi in the 1980s and early 1990s. Tjum!kui became both a magnet for its wealth and action and the dreaded 'place of death' marked by conflict, alcohol, and sickness. When problems at home could not be solved, Tjum!kui residents fled to live with *xaro* partners in Botswana for months at a time.

In the mid to late 1970s developments were also taking place in the Dobe-/Kae/kae area that are worth mentioning for the contrast they provide with the Nyae Nyae area of the 1990s. In the face of the destruction of traditional veldt resources by the herds of neighboring agro-pastoralists, Ju/'hoansi of different camps chose different options. Some had long ago attached themselves to the households of agro-pastoralist and engaged in casual labor in exchange for domestic foods. Others just gleaned what they could from any source. The more enterprising in certain camps of Dobe and /Kae/kae followed the agro-pastoralist model and began to plant gardens and build small herds on their own.[4] This was a laborious process—to earn the cash to purchase a single cow required months of wage labor or crafts production. During a narrow window of time, items that were acquired by hard work, for instance livestock, began to move into the category of private possession, though the meat and milk were shared. A certain number of blankets and other utilitarian items became accepted as essential household possessions which need not be given away at request of others. With an increase in livestock and material possessions, the structure of some camps began to change from an open circular one to a linear one with houses more widely spaced and fenced (Yellen 1984). Possessions were locked in trunks for the first time. Such developments were later curbed by the introduction of food relief programs in Botswana which reduced anxiety about future subsistence.

Style in the 1970s

I have discussed style in Ju/'hoan projectile points and beadwork of the 1970s in depth elsewhere (Wiessner 1983, 1984); here I will give a brief overview as background to understanding developments of the 1990s.

Projectile Points

In the mid-1970s, Ju/'hoan projectile points were of great significance for their functional attributes, social ones (the exchange of projectile points specifies obligations for meat sharing), and because an arrow fired in a moment of anger can kill. Certain attributes, namely length and width of projectile points, delimited the broader Ju/'hoan population that pooled risk, essentially the Ju/'hoan dialect group. These attributes indicate an implicit understanding that all within this pool held similar values with respect to land rights, sharing, and aggression avoidance, thereby underwriting networks of social kinship and reciprocal access to resources.

Ju/'hoan projectile points could be immediately distinguished from those of other linguistic groups such as the Nharo, G/wi, and !Xo on the basis of length and width. Although the points from all three linguistic groups followed the same functional principles (sufficient penetration for poison to be effective) and were used on a similar range of animals, Ju/'hoan points were about half the size of G/wi and !Xo points. Ju/'hoansi expressed some fear and suspicion about people who made their arrows so differently.

Within these limits of size, distal point shape, body shape, and base shape varied greatly, expressing personal identity. Individuals did not produce regular styles that would serve as marks of ownership. However hunters sought regularity in every batch of points, so that those given or loaned to others could be recognized and obligations for meat sharing carried out accordingly.

Glass beadwork

Sewn glass beadwork consisting of headbands, necklaces, aprons, bags, and belts were the most desired items for adornment and cherished gifts for *xaro* exchange. When circulated in *xaro*, they cemented relationships between giver and receiver. Though they might travel far in the *xaro* network, the name of the maker was generally retained. Headbands had particular significance for they were worn on important social and ceremonial occasions. Variation in headband design played upon of a repetoire of nine different major designs and five different background designs (see Wiessner 1984: Figs. 2, 3). Three of the former were considered to be headband designs, and the remaining were designs usually found on belts or aprons but occasionally used on headbands. Five different background designs were regularly employed. The combination of patterns used, size of beads, design structure, and kinds of items made expressed Bushman identity and separated desert Bushmen from peoples other linguistic groups, though no outstanding internal divisions were found between the Ju/'hoansi, G/wi, Nharo, and !Xo of Botswana.[5] Particularly important was design structure which consisted of a central design that wove its way through one or more background designs, usually not more than two (Fig. 9.2, a-d). This structure described was by some as evoking the image of a person 'walking softly', in other words exhibiting modesty, reserve, and consideration.

Within the standard repetoire of beadwork designs and structure, there was much room for personal expression in the use of color and combination of major and background designs. Ju/'hoan beadworkers played with these possibilities most creatively. The variation exhibited in Ju/'hoan beadwork encapsulated the relation of the individual to society. The individual, her diligence, skills, and creativity had room for expression (Fig. 9.2, c-d), but this was contained by design structure that symbolized values of modesty and reserved behavior held by the society as a whole. Personal markers denoting ownership were absent, for ultimately the greatest value of beadwork (and other items of material culture as well) was its social worth when given in *xaro* exchange.

Social Change and Style at Tjum!kui in the 1970s

At Tjum!kui, beadwork boomed with the construction of internal *xaro* paths to mediate emergent social inequalities and other problems of the densely settled community and unequal opportunities. New styles appeared as Ju/'hoan women, in comparing themselves with others and their past identities with current ones, innovated and differentiated. They drew on the traditional design repetoire, reproducing traditional items and design elements, but ruptured the former design structure which carried implications of modesty and reserve. For headbands, several major designs were combined on one piece, or more frequently, background designs were accentuated to fill the entire piece (Fig. 9.2, e-g). The structure of a central design meandering through a more discreet background, 'walking softly' was replaced by what was often a more ostentatious, although strikingly beautiful combination of designs. In short, with economic opportunities and dissolution of the social order, personal expression flourished, expanding the limits of design structure and altering the former balance of individual and group expression.

Beadworkers in Botswana had an interesting reaction to these new styles. On the one hand, they admired these pieces for their striking beauty and the more affluent lifestyle that they represented. On the other hand, they were said to be to be busy "like the many people talking (quarreling) at once at Tjum!kui" and not representative of the soft-spokeness and discretion so prized in Ju/'hoan society. Tjum!kui residents, also aware that these styles represented altered relations and values, tended to choose more traditional designs when making or selecting *xaro* gifts to give to relatives in Botswana, presumably as a token of respect for their lifestyle (Wiessner 1984).

The new designs, in recombining elements of the traditional repetoire, expressed developments initiated from within: the reworking of existing relations and values in response to economic change imposed from the outside. However, imports in beadwork styles occured simultaneously, reflecting new contact with other populations. One was a new form of beadwork, the European tie made with glass or ostrich eggshell beads to be worn to official meetings. It combined Ju/'hoan and western attire with a good touch of humor. Others were imported unaltered by those who traveled to other areas and not integrated into traditional Ju/'hoan styles. These included collars of 'lace beadwork' and necklaces of beads sewn into a 'rope' which are found in several other ethnic groups of southern Africa, but are not a part of the traditional Ju/'hoan repertoire.

Change in the 1990s

In 1978 and 1979, Ju/'hoansi of the Nyae Nyae area[6] (Tjum!kui) began joining the South African Defense force when bases were established in west Bushmanland, a formerly waterless region. Only about 10% of the population was employed by the defense force, largely males of suitable age and physical condition, but others flocked to the vicinity of the bases to reap benefits–cash from the employed, schooling, medicine, well-stocked stores, electronic entertainment, and so on. Pay was relatively high by Ju/'hoan standards, and previously unknown amounts of money poured into the area to be spent on alcohol, clothing, household goods, radios, phonographs, bicycles, and some livestock. The employed were permitted by their peers to keep more than others, provided that the better portion of their income was circulated within the broader population. Though many Ju/'hoansi look at this as a golden age of wealth, virtually all were repelled by the idea of involvement in the business of killing, even though few soldiers from the Nyae Nyae area actually saw combat. This situation was to last for the next ten years.[7]

In the late 1970s and early 1980s, the Ju/'hoansi of Nyae Nyae faced two major threats. One was the proposal of the South African Government to make a game reserve out of their former lands–since most Ju/'hoansi were settled at Tjum!kui or in police/army camps, the land was temporarily empty. A second source of pressure came from Herero living to the south who had their eyes on the rich and largely empty grassland of their Ju/'hoan

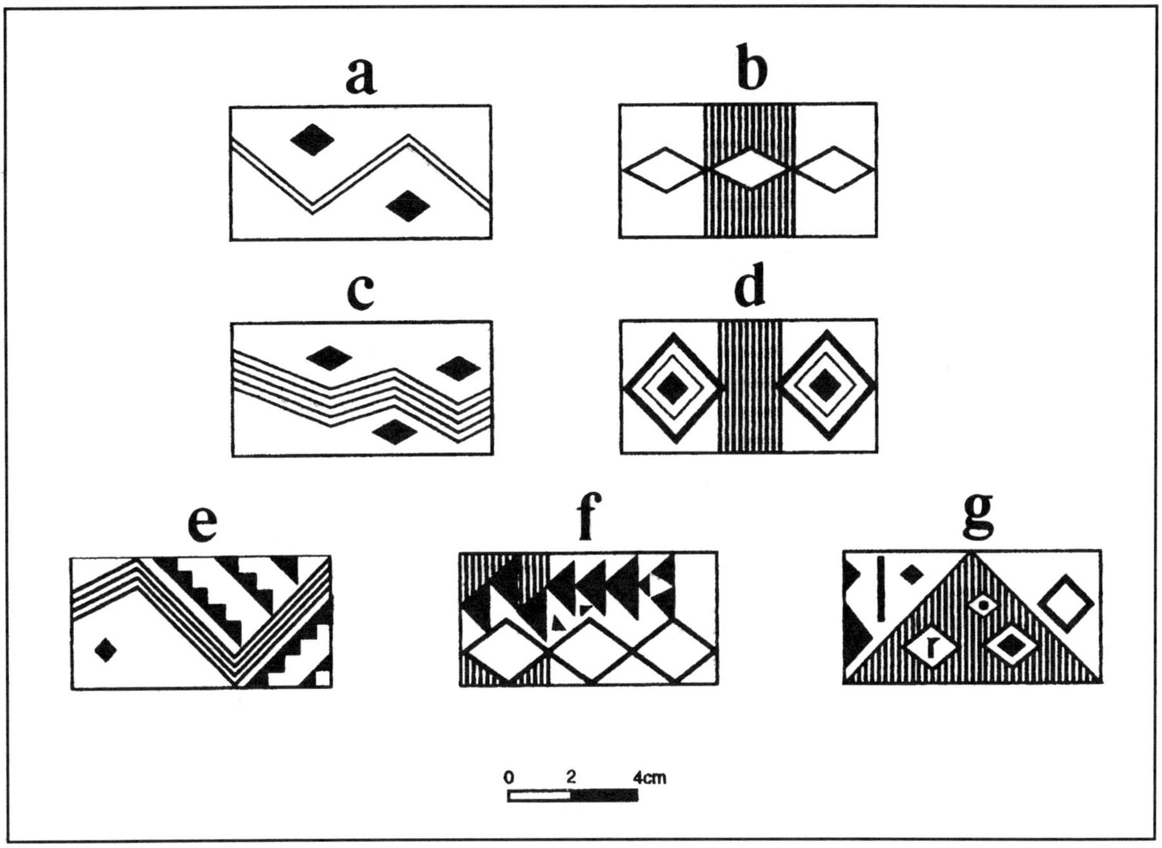

Figure 9.2 Design structure of Ju/'hoansi beaded headbands. A and B–basic design structure; C and D–developed design structure; and E, F, and G–complex design structure.

neighbors. In the early 1980s, small groups of Ju/'hoansi, repelled by the conditions at Tjum!kui, began to move back to their traditional lands. These initiatives received strong support from the anthropologists and film makers John Marshall and Claire Ritchie who together with the Ju/'hoansi founded the Nyae Nyae Development Foundation of Namibia. By 1992 over thirty groups had moved back to their traditional areas.

The ensuing events are both heartening and unique amongst the San of southern Africa. Here I can give but a brief outline; excellent descriptions can be found in Biesele (1990), Lee (1993), Marshall and Ritchie (1984), amongst others. The Ju/'hoansi formed the Nyae Nyae Farmer's Cooperative in 1989, and drawing on funds raised by the foundation, were able to have boreholes drilled, provide their members with small herds of cattle at highly subsidized prices (Ritchie 1989) , initiate agricultural programs, promote Ju/'hoan literacy in their own language, and organize a crafts marketing program. Wage labor, upon which many Ju/'hoan families had been dependent in the 1970s and 1980s, virtually ceased (outside of a few government jobs at Tjum!kui) and programs were aimed at self-sufficiency through a mixed economy of agriculture, hunting and gathering, and crafts production. An administrative and training center was established for the cooperative at Baraka. Decisions within the cooperative are made by a body of elected leaders representing each of the villages.

The Nyae Nyae Farmer's Cooperative is making great strides in securing and protecting the land of the Ju/'hoansi and providing them with a voice in the government of Namibia. In 1992, a further step to give Ju/'hoan rights over their resources came from a government and World Wildlife initiative to establish conservancies which would allow minority groups to take control over their own resources. Over a five year period, US $1,500, 000 is to be invested in the area for establishing the conservancy and planing utilization of resources (Marshall 1996). The proceeds are to be returned to communities. While all this was happening, the Namibian government of the 1990s began to invest in the area. During the better part of the period since independence, the government has provided communities with monthly 'relief' rations of maize meal, oil , and beans via an effective distribution system. These rations are sufficient to support households throughout the year, although the

Ju/'hoansi consider the diet tedious. Ju/ 'hoansi over 60 years of age receive monthly pensions from mobile ATM machines that spit out cash upon recognition of finger prints.

What does this mean for the life of the average Ju/ 'hoan? First and foremost, virtually all serious environmental risks caused by limited water and pronounced seasonal and yearly climatic fluctuations have been eliminated. The bulk of the diet comes from store bought food and relief rations. Hunting and gathering, small herds, and planting provide dietary supplements–though at least in recent years, harvests have yielded little and many cattle have been sold or slaughtered. Permanent settlements have been established at boreholes. Finally, with legislation backing efforts to keep agro-pastoralists from moving into Ju/'hoan land, Ju/'hoansi live in world composed primarily of Ju/'hoansi, and not together with agro-pastoralists as in Botswana. The only exception is at the administrative center of Tjum!kui where Ju/'hoansi reside with Herero, Okavango, Damara, Europeans, and others who have come there for work, business, or schooling. While at Baraka, there is a strong sense of Ju/'hoan identity forming amongst those in the Nyae Nyae Farmer's Cooperative who deal with external relations, this is not the case in most villages where people are concerned with daily needs and internal relations.

Upon returning to the Ju/'hoansi after a twenty year absence, I had a number of expectations based on knowledge of the altered economy, secure food and water, trends that were appearing in the 1970s, and the assumption that experiences during the defense force years would alter goals, values, relationships, and the meaning of material culture. These included: (1) decreased mobility, narrowing *xaro* networks, and less sharing on a daily basis; (2) more material possessions, minor capital accumulation, and accompanying changes in site structure toward wider hut spacing and a less open site plan; and (3) a larger proportion of goods purchased than received in exchange, increased wealth differentials between households, and a continuation of individual innovation in beadwork styles.

Fieldwork was to yield some surprises. The data presented here are preliminary results from ongoing research in three villages that are well separated in space: Xamsa, N≠ama and Mangetti Pos (Figure 9.1).[8] First subsistence, sharing, exchange and settlement patterns will be examined to document that changes that have occurred, and in this context, stylistic developments analyzed.

Mobility

Permanent settlement at boreholes, cash income, availability of foodstuffs at stores, and relief food have led to decreased mobility in two respects. The first is in seasonal rounds. Today Ju/'hoansi remain year-round in permanent villages, hunting and gathering within a radius of about 5-7 km. Some people, though few, occasionally go further afield on hunting-gathering trips of 1-5 days. The second is the decline in extended visiting trips to the camps of *xaro* partners 20-150 km away that so characterized Ju/'hoan life in the past. For instance, data from 1968-9 collected by Richard Lee and that for 1974 collected by myself for the Dobe-/Kae/Kae Ju/ 'hoansi indicate that the average Ju/'hoan spent approximately three months a year living in camps of *xaro* partners and hunting and gathering resources of those visited (Wiessner 1981, 1982). Day visits to nearby camps were extremely common. By contrast, extrapolating from visiting data collected for two weeks out of every month between August 1996 through February 1997 at Xamsa and Mangetti Pos, the average Mangetti resident now spends approximately 36 days a year away from home and the average Xamsa resident 23 days.[9] The most common destination (70% of all days spent away) is the administrative center of Tjum!kui, where people from widely dispersed villages meet, socialize, exchange gifts and news, shop, seek medical care, visit children at school, tend to official matters, and frequent drinking establishments run by people from other ethnic groups. Overnight visits to other camps comprise only 30% of days away, though day visits to villages within a radius of 5-15 km are common. Thus, with steady supplies of food and water in the villages, extended visiting is declining markedly except to Tjum!kui, the only true 'town' in Bushmanland, where people from all parts of Bushmanland converge to obtain non-traditional resources.

Subsistence and Food Sharing

The subsistence income of all families in the three villages was measured daily for two weeks out of every month from August 1996-February 1997, as well as food consumed and where it was obtained. Preliminary results show a 90% dependence on government rations and trade store foods purchased with crafts or pension money from the period of August to November for Xamsa and N≠ama. For Mangetti Pos, where mongongo nuts were available, there was roughly a 70% dependence on rations. At Xamsa a large animal was killed approximately every six weeks, at N≠ama no large animal was killed during this period, and at Mangetti Pos three medium-sized animals and a number of small animals were killed. Meat was widely shared as in the past. Collection of data on wet season subsistence is currently in progress - women go on gathering trips 2-3 times a week to collect wild vegetable foods that make up between 30-50% of the wet season diet. Bush foods are a welcome supplement to the dull diet of maize meal. In February, 1997, rations of maize meal

Table 9.1. Comparison of mean number of *xaro* partners by kinship relationship and location for 1974 and 1997. (In 1974, 59 individuals were interviewed at /Kae/kae, Dobe, and Nyae Nyae (Tjum!kui), and in 1996-7, 60 individuals from three villages in the Nyae Nyae area.)

	xaro partners mean no.	kinship relation[a] close	distant	affines	own camp	location[b] 2-25km	26-75km	75km+
1974	16.0	7.4	7.2	1.4	2.6	3.5	6.7	1.5
1997	6.9	3.8	2.0	1.1	2.9	1.8	2.1	0.1

[a] For location of partners in 1974, the sample consists of partners of 35 /Kae/kae Ju/'hoansi only, for at that time most Nyae Nyae Ju/'hoansi were resettled at Tjum!kui.
[b] Close kin include parents, children, siblings, grandparents parents' siblings, and parent's siblings' children.

were discontinued for two months, and in this year of bountiful veldt foods, Ju/'hoansi could support themselves by hunting, gathering, and purchasing food from the store or mobile peddlers. Since April, rations have been provided once again, and hunting and gathering is decreasing radically while major bush foods rot in the veldt.

Very interesting in view of the fact that households receive similar rations is that sharing of all types of food remains central to Ju/'hoan life. For instance, of 297 meals recorded for eight families at Xamsa from July, 1996, until January, 1997, 198 or 66% of meals were either consumed at the hearth of others or included food contributions from others. Food sharing in the other two camps followed a similar pattern. Widespread food sharing takes place for the pleasure of sociality, because reduced firewood supply at permanent settlements makes it more efficient to cook in one pot, and very importantly, as a way to uphold the practice of equal access to the resources of others. Were sharing to break down on a daily basis, it would be hard to activate sharing obligations for scarcer and more desired foods - meat, milk and store bought foods.

Xaro Exchange

Data on *xaro* partners of all adults in the three villages were collected and subsequently all possessions of the household recorded together with information on how and where they were obtained. As is clear in Table 9.1, the scope of *xaro* has narrowed greatly in the 1990s. The average Ju/'hoan of the Nyae Nyae area today has 6.9 *xaro* partners as compared to 16.0 for /Kae/kae, Dobe, and Tjum!kui (Nyae Nyae) Ju/'hoansi in the 1970s. Some individuals interviewed are in both the 1974 and 1997 samples.

In terms of kin relations, the greatest decline in *xaro* exchange is occurring with distant relatives, though *xaro* with close relatives is also waning. In space, average numbers of partners in a person's camp remains approximately the same, while substantial decreases are occurring in partnerships with Ju/'hoansi in other camps, a predictable result of declining mobility. Short *xaro* chains of 3-5 people still link families related by marriage, but only a few older Ju/'hoansi maintain chains that bring in gifts from afar.

While in the past, explanations of *xaro* had been consistent, today people of different ages groups have very different views, reflecting changes in the system. Older people see *xaro* partnerships as semi-formalized relationships which open access to the resources of partners in nearby and distant places, as one man put it "*xaro* is about unreliable food and water, or at least it was in the past". They are also aware that *xaro* gifts used to travel along chains of partners, as material goods were imported and exported to different areas (Wiessner 1994). Some lament the decline of networks as old *xaro* partners pass away. By contrast, most young people have no clear concept of *xaro* and no knowledge of former *xaro* chains. They only know that one engages in xaro exchange with particularly beloved relatives or friends for whom one feels strong obligations for mutual support.

Despite these trends, Ju/'hoansi still attach great importance to *xaro* exchange as evidenced by the fact that about 50% of the average Ju/'hoan's possessions are still received in *xaro* (see below). *Xaro* continues to designate kindred members to whom individuals have strong obligations and affections, underwrites affinal ties, redistributes material goods, and is one of several ways in which ties with other villages are maintained. People catch rides on trucks to visit *xaro* partners and bring or request gifts from them, enjoying the diversion and chance to socialize. During such visits a number of things transpire. If people of one village have received some things that another has not, part of these are shared with visitors. Moreover, much information about what is hap-

Table 9.2 Comparison of number of possessions of Xamsa households in 1997 with those of /Kae/kae and Dobe households in 1974 (see also Wiessner 1982).

	average number of possessions per household /Kae/kae 1974	Tjum!kui 1974	Xamsa 1997
	n=16	n=9	n=8
beadwork	14.5	22.1	6.0
blankets	3.8	5.3	4.1
clothing	11.0	16.4	12.5
kitchenware	10.1	10.4	9.4
hunting & gathering equipment	3.8	2.7	3.1
livestock	1.8	0.1	0.4

Tjum!kui residents were interviewed when visiting Botswana. Had they been interviewed at their homes in Tjum!kui, their number of possessions would have been substantially higher.

pening in other villages is gleaned during such visits: whether relief foods and pensions have reached other villages, where trucks from the Nyae Nyae Farmer's Co-operative have gone to buy crafts, which villages have been visited by tourists, and so on. Through maintaining such a networks of exchange, Ju/'hoansi keep a finger on the pulse of developments occurring within the broader region and thereby are able to take action to insure continuing access. And so, *xaro* of the 1990s is no longer about fluctuations in water and food as it was in the past, nor does it serve to mediate tensions and developing social inequalities as at Tjum!kui during the 1970s. Today it is hovering in a state of transition, keeping many doors open, and ready to be applied to the possibilities and problems that the future might bring.

Material Possessions

Comparison of material possessions between the mid-1970s for /Kae/kae and Tjum!kui Ju/'hoansi and Ju/'hoansi of Xamsa village in 1977 indicate that Ju/'hoansi of the 1990s have, on the average, no more possessions than did Ju/'hoansi surveyed in the 1970s (Table 9.2). The data for the two other villages, Mangetti Pos and N≠ama, have not yet been compiled; however, preliminary counts indicate that they do not differ substantially from Xamsa. The masses of material goods purchased during the South African Defense Force years have vanished on *xaro* networks, have been exchanged for beer, or lie broken in the sand for future archaeologists.[10]

Cash incomes of the 1980s and 1990s are much higher than they were in the 1970s for Botswana residents, store bought goods much more easily accessible, and there are no of barriers to accumulation caused by mobility. And so the fact that Ju/'hoansi possess no more today than in the 1970s is the result of the meaning attached to things, not access to possessions nor facilities to store them. For most Ju/'hoansi, accumulation of possessions beyond a basic repetoire to meet daily needs is pointless in view of the social value of material goods when used in exchange. Giving yields credit for generosity and strengthens social bonds; accumulation draws little admiration and may incite criticism if done conspicuously. The rapid turnover of goods which results does little to encourage conservation or repair, and with the exception of beadwork and a few other items, material goods have a short life. Finally, many material goods today are given to members of other ethnic groups for beer, together with the intensive social activity that drinking evokes.

That the principle value of material possessions, beyond a basic inventory necessary for household needs, is for exchange rather then accumulation is further confirmed by analysis of sources of possessions. Though number of hxaro partners has declined to less than half of what it was in the 1970s, the reduction percentage of household possessions received in *xaro* is small in proportion. In the 1970s, for Dobe-!Kae/kae and Tjum!kui Ju/'hoansi, 69% of all possessions were received in *xaro*, 27% bought or made and 4% were gifts from non-Ju/'hoansi in comparison with to 50% received in *xaro* at Xamsa in 1996-7, 31% bought or made, 5% received from non-Ju/'hoansi and 14% received in casual giving between friends. The latter is a product of less formally defined exchange relationships in the younger generation.

Table 9.3 Comparison of hut spacing and site structure at Dobe villages in 1963-4, 1968-9, 1975-6, and 1980 with villages in the Nyae Nyae area in 1997.

village	no. of huts	no. of hearths	average distance between huts
Dobe 1963-4 circular	7	7	7.0m
Dobe 1968-9 circular	15	14	7.1m
Dobe 1975-6 segmented: semi-circular segments	11	10	10.8m
Dobe 1980 linear	6	6	11.6m
Nyae Nyae 1997:			
Xamsa semi-circular	11	5	5.2m
Mangetti Pos segmented, semi-circular, Linear portions	13	6	9.9m (6.6m within segments)
N≠ama (1) circular	8	3	8.5m
Nama (2) semi-circular	6	3	7.0m
Nama (3) semi-circular	4	2	9.0m

Notes: Dobe figures are measured from Yellen's (1984) illustrations. These are base camps at permanent water. Spacing of huts in some small temporary bush camps is narrower (Yellen 1977).

For semi-circular camps, nearest neighbor was measured with the space between the first and last hut in the semi-circle excluded.

Village Structure

As John Yellen has so clearly shown for the Dobe Ju/'hoansi, the structure of camps in the 1960-1980s was altered as Ju/'hoan families began to cultivate their own gardens, establish their own herds, and accumulate more material possessions (Gelburd 1978). In the 1960s, when Dobe Ju/'hoansi owned no livestock, camp structure was roughly circular with the average distance between huts at the 1963-4 Dobe camp approximately 7 m (Table 9.3). An open plan made the possessions and foodstuffs of each household readily visible to others and subject to requests for sharing. In 1968-9 when the first goats had been acquired, the goat kraal was placed outside the hut circle at close proximity, but site structure and hut spacing were not altered. By 1975-6 with further increase in herd size, planting, and private possessions, the circular pattern was disrupted by the integration of the kraal into the hut circle, dividing the camp into two semi-circular clusters of huts that face the kraal. Measurements made from Yellen's illustrations indicate an increase in hut spacing to an average of 10.8 m between huts. The most interesting change came in the 1980 camp with an increase in the herd of livestock, western material possessions, and private property. The base camp was then constructed in a linear structure with houses facing the kraal, not each other. Spacing between huts widened to an average of 11.6 m. Residents of this camp possessed trunks in which private property could be concealed from others and locked away. And so, with an increase in food production as a result of individual family enterprise, accumulation of property, and reduction of sharing, site structure was altered from an circular open one with closely spaced huts to a linear one in which household food and belongings were not open to public view. Grass huts were replaced with more permanent plaster huts.

Are these lines of development continuing with permanent settlement and farming programs in the Nyae Nyae area today? It is difficult to generalize, for there is great variation in site structure of different villages in the Nyae Nyae area of the 1990s as well as experimentation each time a village is constructed anew. For instance, a few communities have built villages on a linear plan feeling that this is modern. However, site plans in these cases does not alter use of space, for people continue to live in these as if they were circular, concentrating all major activities in one or two communal areas adjacent to the line of huts. Most, villages, however, resemble the Dobe camp of the mid-1970s in that the traditional circular pattern has been disrupted by the integration of new communal facilities on one side - for instance the tap that provides water for the village, kraals, chicken coops, or small kitchen gardens with irrigation hoses. As vehicles deliver important resources, a parking area in full view of all households is integrated into many villages.

Most houses are larger than in the past and constructed from plaster and thatch, though traditional grass huts are still built by some. Most village plans remain open, usually in a semi-circular arrangement so that households can easily observe the activities of others. Huts are rarely fenced and only one locking trunk was recorded in the villages studied. Rather than the traditional one hearth per hut, the five villages mapped had an average of 0.4 huts per hearth, with two or more family units sharing one hearth, perhaps a response to shortage of firewood within close proximity. Hut spacing has not increased significantly compared to that of Dobe in the 1960s (Table 9.3). Thus, in villages of the Nyae Nyae area, the needs of permanent settlement have been accommodated by more substantial housing and village plans have been altered to incorporate new economic activities within or adjacent to the hut circle. However, site plans remain open and hut spacing relatively narrow, discouraging private accumulation.

Summary

The variety of economic strategies to make a living practiced over the past twenty years show a great openness on the part of the Ju/'hoansi to take advantage of new economic opportunities. Historical studies suggest that this has been the case for hundreds of years (Gordon 1992, Wilmsen 1989). Nonetheless, food sharing, *xaro* exchange, sources and numbers of material possessions, and site structure indicate that there is persistence in the primary value of material goods as a currency of social exchange and in the structure and terms of kinship relations and accompanying exchange. Outstanding among these is the strong current of equality and respect between Ju/'hoansi. And it is these terms that keep access to the resources of others in the region open, for people can welcome others without fear of exploitation, something which is not true of relations with non-Ju/'hoansi. As long as equality is preserved and obligations of kinship maintained, *xaro* exchange can be restructured to give Ju/'hoansi access to the resources of others in the broader region.

Conversations with Ju/'hoansi indicate that the desire to maintain such relations is not a remnant of the past, but rather strategy in view of the events of recent history. For the last thirty years the major economic opportunities that have come to the Nyae Nyae area have come from the outside in the form of government programs or development aid.[11] Accordingly Ju/'hoansi are optimistic about the future and have great expectations of the wealth that will flow from trophy hunters, ostrich farms, or a booming tourist business when their conservancy is approved. They are also well aware that if some get a foot in the door of inequality, these families will remain in a position of advantage. With the anticipation of wealth to come, most Ju/'hoansi seek to preserve equal access to resources of the region and thereby maintain an equal right to what the future will bring. This comes in contrast to the Ju/'hoansi in the Dobe-/Kae/kae area during the 1970s whose negative perception of future economic opportunities urged a change on the part of some towards the models of agro-pastoralism and with it the development of minor social inequalities and capital accumulation described by Gelburd (1978), Yellen (1984), Wiessner (1982) and Wilmsen (1989).

STYLISTIC CHANGE

Returning to the issues raised in the first part of this paper, I will address three questions in the following analysis of stylistic change: (1) Do beadwork and projectile points continue to have significant enough social meaning in Ju/'hoan society to be used to communicate information about identity? If not, is this possible to determine from formal variation? (2) Is it possible to glean information from stylistic analysis about the changing relations between the individual and society that occurred in the Nyae Nyae area over the last twenty-five years? (3) How are stylistic markers and their meanings from the past called up selectively in the process of re-constructing social identity in view of current social, economic, and political concerns?

Table 9.4 Comparison of design structure in Ju/'hoan beaded headbands for the 1970s and 1990s.

Developed major designs embellish simple designs, but remain within the traditional design structure. (Fig. 9.1, c-d). Complex designs break with traditional structure, combining several major designs on one headband or accentuating background designs to dominate central ones (Fig. 9.1, e-g).

	Basic		Developed		Complex		Total	
1974								
Dobe	11	(38%)	13	(45%)	5	(17%)	29	(100%)
/Kae/kae	51	(63%)	22	(27%)	8	(10%)	81	(100%)
Tjum!kui	4	(18%)	6	(27%)	12	(55%)	22	(100%)
1997								
Nyae Nyae	27	(46%)	28	(47%)	4	(7%)	59	(100%)

Beadwork

In answer to the first question, sewn glass beadworking is an art that is flourishing in the 1990s and its products remain the most valued items of self-adornment and most prized gifts in *xaro* exchange. Beadwork continues to be exchanged to underwrite networks of social kinship and their corresponding relationships of equality and respect.[12] By contrast, virtually all other forms of traditional material culture have been replaced with store bought ones due to the superior functional qualities of the latter and desire on the part of Ju/'hoansi to keep up with changing times.

In answer to the second question–yes, both the increase in personal expression that came with new economic opportunities and emerging social inequalities in the 1970s, as well as the subsequent swing back in the opposite direction during the 1990s, are indeed expressed in beadwork styles. As can be seen in Table 9.4, virtually gone are the complex headband designs developed at Tjum!kui in the 1970s which evoked a scene of 'many people taking at once' (Fig.9.2, e-g). In the mid-1990s, simple and developed designs predominate, demonstrating the competence of beadworkers, but keeping within the traditional design structure, 'to walk softly', associated with the modesty and respect of others that comes with equality. Gone too are styles imported in the 1970s from other ethnic groups–beaded necklaces sewn into lace or rope patterns. They are still produced for the crafts shops of Windhoek, not for Ju/'hoan women–only three such pieces were found amongst all the sewn beadwork recorded.

The range of designs has also narrowed. The three traditional headband-specific designs were used on 89% of pieces recorded in the mid 1990s by contrast to only 65% of Tjum!kui headbands in the 1970s. The simplest design (Wiessner 1984: Fig. 2c) which permits almost no embellishment with background designs increased in frequency from 4% in the 1970s to 24% in the 1990s. No designs from other ethnic groups were incorporated into the repetoire. It should be noted that parallel trends have taken place in the design of women's beaded pubic aprons and beaded bags, though analysis of such is beyond the scope of this paper.

In answer to the third question, the reconstruction of identity through drawing selectively on past styles and their associated meanings is indeed expressed in Ju/'hoan material culture. First, the continuing and enthusiastic production of beadwork, the principal currency used in support of social kinship networks, in the face of the decline in other forms of traditional material culture, indicates that it is traditional social networks that the Ju/'hoansi are actively maintaining despite rapid economic change. Recall for contrast that amongst Ju/'hoansi living on farms in the 1970s, social networks declined as people became attached to the households of agro-pastoralists, and correspondingly, the art of beadwork was maintained by only a few. Second, the fact that styles from the Tjum!kui era are neglected and that styles from an earlier period are selected represents a rejection of marked assertion of individual identity via style that occurred when Ju/'hoansi were settled at Tjum!kui and the reassertion of some aspects of social identity prevalent in pre-Tjum!kui times. Third, the fact that styles of beadwork are no longer being imported from surrounding populations locates the impetus for these developments to agency coming from within. Moreover, while the distribution of styles in space during the 1970s pointed to a localized center for innovation at Tjum!kui, the reversion to styles of earlier decades is occurring in all vil-

Table 9.5 Comparison of Ju/'hoansi projectile points recorded in 1974 and 1997.

	1974 n=55	1997 n=60
Length		
mean	9.5	9.8
sd.	1.4	1.3
Width		
mean	7.6	7.3
sd.	1.3	1.0
Width/Length		
Mean	.74	.75
sd.	.13	.09
Distal point shape		
sharp	19	22
rounded	28	29
very rounded	8	9
Body Shape*		
convex	19	20
convex/straight	22	26
straight	12	9
convex/concave	2	5
Base shape		
indented	30	32
straight	16	12
convex	9	16

The 1974 sample is from a wide range of hunters in N.W. Botswana who regularly cross the border to make extended visits to the Nyae Nyae area. The 1997 sample is composed of the arrows from the quivers eight hunters who are from widely dispersed villages in the Nyae Nyae area. All of these hunters also visit relatives in Botswana.

* Convex/straight means that one side is convex and the other straight.

lages and indicates that a broad base of the population is participating.

It should be added that some innovation in beadwork is occurring today, largely on the part of women residing Baraka, the administrative center of the Nyae Nyae Farmer's Cooperative. These women generally have higher cash income that other Ju/'hoan women and travel more widely to other parts of Namibia to attend meetings and workshops. Unlike most women in the villages, they are concerned with Ju/'hoan identity in relation to the broader population of Namibia. All innovations that I have seen so far on the part of these women involve taking traditional Ju/'hoan symbols and sewing them into (or onto) more modern forms. A good example can be seen in the trousers of the elected traditional leader, Bobo Tsamkxao. Under Namibia's newly formed traditional leaders council, Bobo represents the Ju/'hoansi of the Nyae Nyae area. As an older man who grew up with no formal education, he must seek to understand current trends and integrate them with Ju/'hoan culture. In keeping with this role, his wife has taken a pair of western trousers and embroidered them with circular patterns of glass beads called */uisa /aasi* (owls eyes) which are one of the most characteristic decorative elements in the traditional Ju/'hoan repertoire of embroidered beadwork. As Nyae Nyae Farmers Cooperative develops, new styles will be designed hand in hand with the restructuring of Ju/'hoan identity relations from generation to generation.

Style in Projectile Points

Today Ju/'hoan projectile points and other hunting gear remain a part of the material repetoire largely because of enforced legislation stipulating that Ju/'hoansi may only hunt by traditional means and because large numbers of bows and arrows are sold to tourists. New illegal forms of hunting–hunting on horseback with spears or guns and using snares for large animals–are practiced occasionally by younger people throughout Botswana and Namibia, but when caught the hunters are prosecuted. Because traditional hunting is strenuous and requires specialized knowledge, many Ju/'hoan men, particularly men under forty, do not hunt with bow and arrow today.

There are three categories of men who produce arrows for hunting. The first are the good hunters who learned to hunt from their fathers and have continued to hunt over the last thirty years. These are few (except in some villages of the southern or northernmost parts of the Nyae Nyae area), but they are highly respected for their skills. When a large animal is killed, meat sharing conforms to past practices. The second are the occasional hunters, men who learned to hunt during childhood, but ceased bow and arrow hunting with employment at Tjum!kui or by the army. Many of these men are competent in hunting on horseback, with guns, or in using snares. Such men possess quivers which they carry just in case they might come across opportunities or trouble when traveling in the bush. They rarely kill anything, and snickers are heard when these men say that they hunt. The third is comprised of men of all ages who make bows, quivers, and arrows to sell for the tourist trade. Due to stringent requirements on the part of buyers, many of these can only be distinguished from true hunting gear by the fact that the arrows are not poisoned.

As a result of these developments, most men do not possess functional quivers, the exchange of arrows is very limited, and few hunters hunt outside a 5-10 km radius of their own villages. The social significance of arrows has thus greatly declined except among a small inner circle of good hunters. It is probably fair to say that variability in arrows is slowly moving from being 'style', through which aspects of identity are communicated, to isochrestic variation, in which arrows are reproduced following 'tradition'. As can be seen in Table 9.5, hunters continue to produce arrows conforming to styles they acquired from their fathers, and so arrow styles of the 1990s do not differ significantly from those of the 1970s. However, 'occasional hunters' (not included in Table 9.5) tend to have arrows that are poorly made, in poor condition, and display a wide range of size and shape in one set. Quality thus decreases with loss of social meaning.

Because of the declining role of projectile points in Ju/'hoan villages today and their association with the crafts industry, projectile points are not used to transmit information in the process of re-structuring identity, by stark contrast to beadwork. It should be noted, however, that this was not the case in the past, for virtually all older Ju/honasi interviewed said that in the period from about 1910 to 1950, when metal began to be used and Ju/'hoansi began to have more regular contact with non-Bushmen, there was great experimentation, innovation, and change in styles of projectile points. Samples of projectile points resembling those produced in the past that were made at my request by older Ju/'hoansi demonstrate and great variability before the styles recorded in the 1970s became the mode.

CONCLUSION

Stylistic studies are complex due to the many possible sources of variation; it is not easy to get satisfying results. Perhaps for this reason a number of recent studies have begun by addressing the formidable problem of accounting for all sources of variation in artifacts. Most impressive efforts in this respect have been made in behavioral archaeology (Schiffer & Skibo 1997). For style in particular, Carr (1995) has made an exhaustive attempt to understand all the factors that can affect stylistic variation in his 'unified theory of artifact design'. Here I have taken quite a different tack because, important those these efforts are, if style is regarded as communication, it must be possible to narrow the field of questions asked. Human systems of non-verbal communication do not proceed following eighty-one pages of deliberations. Style may not maximize efficiency of information transfer, but the road is not long before inefficiency kills effectiveness. Stylistic communication is effective precisely because it follows certain lines that are comprehensible to others, comprehensible because they are grounded in human cognition. For this reason, I have tried to situate style in a fundamental cognitive process–personal and social identification via comparison–and through understanding this process, provide priority questions to be asked in stylistic analysis. Only then can the difficult methodological issues be addressed.

For the Ju/'hoan case, an evolutionary approach yields some promising results. Three questions were derived from the framework of identification via comparison as focal points for diachronic analysis of style. The first was whether sufficient cultural meaning was attached to an artifact to make it likely to be used as a means of communication. As shown by the lack of stylistic change in Ju/'hoan arrows in the face of substantial socio-economic change, when artifacts lose their meaning, they are no longer used to negotiate current identities. Second, a finding taken from social psychology, that identities in all humans have two components, a personal one and a social one, has given a theoretical basis to direct stylistic analysis toward the question of the changing relations between the individual and society through time. Results show that Ju/'hoan beadwork is indeed a sensitive indicator of such changes, even over a short time span of 20-30 years. Third, the understanding that even though identification via comparison is a fundamental human cognitive process, the components of identity are culturally and historically constituted, has raised the importance of history in identity formation. Because during the social and stylistic comparison process individuals draw on both current and historical associations and state their positions relative to these, stylistic studies can productively yield information on how identity markers are selectively used in the process of restructuring identities through time. The results can in turn provide insights into the nature and agents of change, by examining what is preserved, what is discarded, what is imported from outside, on what are innovations built, and whether innovations are highly localized or broadly based. In the Ju/'hoan case, economic change in the 1990s coming from outside is seen in subsistence remains and settlement patterns. The repertoire of material culture also indicates economic integration into the broader world economy, as traditional tools and clothing are replaced with western ones. However, style in beadwork presents another side of the story. It expresses the efforts of Ju/'hoansi to draw selectively on their past and maintain the time-tried relations of equality, respect, and mutual reciprocity between people that provide the foundation for broad based networks of social support. Through such efforts, Ju/

'hoansi compose their own response and identities in a rapidly changing world.

This paper is but a first attempt. However, the results do suggest that evolutionary approaches may help guide us through the many sources of variation in material remains and help us identify what is important, what we should seek in archaeological analyses, and what course might be followed.

Note on the Orthography

Patrick Dickens has published a standard orthography for the Ju/'hoan language in his *English-Ju/'hoan, Ju/'hoan-English Dictionary* (Rüdiger Köppe Verlag, Cologne, 1994). This has been accepted by the Nyae Nyae Farmer's Cooperative. Conforming to Dickens' orthography requires changing spellings so that they deviate from those used my previous publications. Noteworthy changes are Tjum!kui for Tsumkwe, /Kae/kae for /Xai/xai, and *xaro* for hxaro. !Kung has been replaced by Ju/'hoansi which is the dialect group of !Kung speakers to which the central !Kung belong. 'Ju/'hoan' is both the singular and adjectival form.

ENDNOTES

1. To my limited knowledge, no concerted attempts have been made in archaeological analyses to see at what point in human evolution egalitarian societies first appeared, though evidence might be found in a change from sites with poorly differentiated living floors to sites with distinct, structurally similar household units, amongst other things.

2. It should be cautioned, however, that the referents of styles during comparison can range from vague associations to distinct meanings, what I have called 'emblemic' and 'assertive style' (Wiessner 1983). Specificity of referent, in turn, has a pronounced impact on the kind of information one can derive from style, and is a second level of meaning that must be considered. Distinct referents for personal expression are used to mark ownership or status positions, amongst other things. For social groups they usually delimit internal or external boundaries, and so give no measure of interaction even when intensive interaction exists. Stylistic elements with non-distinct referents, by contrast , indicate less specific feelings of affiliation, similarity, or association. As a result, they move more easily over boundaries and can be used as measures of influence or interaction. Emblemic style for both personal and group expression is rare in foraging societies where people seek to construct far flung and seamless networks of reciprocity. It increases in frequency with social complexity, either with more intricate relations within a geographical region or hierarchical relations within a society.

3. The Ju/'hoan kinship system is complex. First, the kinship universe is defined through conventional kin terms. It is further expanded and manipulated through the name-sharing relationship (Marshall 1976). Finally the *'wi'* relationship (Lee 1986) comes into play, allowing elders to chose how to classify juniors from the many possibilities that the name relationship offers. *Xaro* exchange partnerships then activate the obligations of some kin relations and leave others dormant.

4. Additional impetus for this also came from anthropologists working in the area.

5. This may be because in the 1970s only a few Nharo, G/wi and !Xo women made headbands. By contrast, preliminary results from Namibia indicate that inter-linguistic group stylistic differences between Ju/'hoan and Nharo beadwork do exist.

6. This case study will be limited to the Nyae Nyae area of Namibia. Due to the current ban on anthropological research in Botswana, it has not been possible to follow developments at Dobe and /Kae/kae.

7. Scars remain from this time in the form of alcohol abuse and alienation owing to the fact that the enemy whom Ju/'hoansi were hired to fight would soon become their builders of their new nation.

8. It should be noted that no Ju/'hoan village today can be considered to be representative, for different villages have taken somewhat different paths. For example, a very few villages, usually those with immigrants from farming areas, have been successful in horticulture; a few have successfully retained cattle distributed by the Nyae Nyae foundation; in some villages, older men regularly hunt with bow and arrow, and two villages have started small projects to attract tourists.

9. Not including men who left for employment and single young men who travel widely to seek employment or visit girl friends.

10. Data on sources of possessions in all three villages show that very few material goods purchased during the defense force years are still in the hands of the buyer.

11. The decision to give Ju/'hoansi welfare comes from perception of government and development agencies that their lifestyle was inadequate for the modern world and their abilities to cope with change insufficient. In fact, the Ju/'hoansi are extremely resilient and adjust rapidly to new conditions and opportunities. Where real poverty exists, it is largely where the Ju/'hoansi have been dispossessed of their land.

12. Ostrich eggshell beads have become an important export for the Namibian tourist trade, and because of their role in the cash economy, they no longer circulate in *xaro* exchange and are not regularly used for personal adornment. Cash obtained from their sale is largely used to buy sugar, tea, tobacco, food, alcohol, and glass beads, all of which are either shared and consumed or, in the case of beads, circulated on exchange networks.

REFERENCES CITED

Barth, F.
1969 *Ethnic Groups and Boundaries*. Boston: Little, Brown.

Barton, C.M., G. Clark, G. & A. Cohen
1994 Art as information: explaining Upper Palaeolithic art in western Europe. *World Archaeology* 26:185-207.

Biesele, M.,
1990 *Shaken Roots: The Bushmen of Namibia*. Marshalltown: EDA Publications.

1993 Women like Meat. *The Folklore and Foraging Ideology of the Kalahari Ju/'hoan.* Bloomington: Indiana University Press.

Boehm, C.
1993 Egalitarian society and reverse dominance hierarchy. *Current Anthropolog.* 34: 227-54.

1996 Emergency decisions, cultural-selection mechanics, and group selection. *Current Anthropology* 37: 738-763.

Boyd, R. & P. Richardson
1985 *Culture and the Evolutionary Process.* Chicago: University of Chicago Press.

Carr, C.
1995 A unified middle-range theory of artifact design. In *Style, Society, and Person* (C. Carr and J. Neitzel, eds.), pp. 171-258. New York: Plenum.

Carr, C. & J. Neitzel (eds.)
1995 *Style, Society, and Person.* New York: Plenum.

Clark, G. A., C. Barton & A. Cohen
1996 Explaining art in the Franco-Cantabrian refugium: an information exchange model. In *Debating Complexity* (D. Meyer, P. Dawson & D. Hanna, eds.), pp. 241-253. Calgary: Archaeological Association of the University of Calgary.

Crook, J. H.
1981 The evolutionary ethology of social processes in man. In *Group Cohesion* (H. Kellerman, eds.), pp. 86-108. New York: Grune and Stratton.

De Waal, F.
1982 *Chimpanzee Politics*: Power and Sex among Apes. New York: Harper and Row.
1989 *Peacemaking among Primates.* Cambridge: Harvard University Press.

Eibl-Eibesfeldt, I.
1989 *Human Ethology.* New York: Aldine.

Eibl-Eibesfeldt, I. & C. Sutterlin
1992 *Im Banne der Angst.* München: Piper.

Festinger, I.
1954 A theory of social comparison processes. *Human Relations* 7: 117-140.

Fromkin, H.
1972 Feelings of interpersonal undistinctiveness: an unpleasant affective state. *Journal of Experimental Research in Personality* 6:178-185.

Gamble, C.
1982 Interaction and alliance in palaeolithic society. *Man* 17:92-107.
1986 *The Palaeolithic Settlement of Europe.* Cambridge: Cambridge University Press.

Gelburd, D.
1978 Indicators of culture change among the Dobe !Kung San. *Botswana Notes and Records* 10:27-36.

Gergen, K.
1971 *The Concept of Self.* New York: Holt, Rinehart, and Winston.

Gordon, R.
1992 *The Bushman Myth: The Making of a Namibian Underclass.* Boulder, CO: Westview.

Hodder, I.
1982 *Symbols in Action.* Cambridge: Cambridge University Press.
1986 *Reading the Past.* Cambridge: Cambridge University Press.

Howell, N.
1979 *Demography of the Dobe !Kung.* New York: Academic Press.

Knauft, B.
1991 Violence and sociality in human evolution. *Current Anthropology* 32: 391-428.

Lee, R. B. & I. DeVore (eds.)
1976 *Kalahari Hunter-Gatherers.* Cambridge: Harvard University Press.

Lee, R. B.
1979 *The !Kung San: Men, Women and Work in a Foraging Society.* Cambridge: Cambridge University Press.
1986 !Kung kin terms, the name relationship and the process of discovery. In *The Past and Future of !Kung Ethnography: Critical Reflections and Symbolic Perspectives. Essays in Honour of Lorna Marshall* (M. Biesele, R. Gordon & R. Lee, eds.), pp. 77-102. Hamburg: Helmut Buske Verlag.
1993 *The Dobe Ju/'hoansi.* New York: Harcourt Brace.

Lemaine, G., J. Kastersztein& B. Personaz
1978 Social Differentiation. In *Differentiation between Social Groups* (H. Tajfel, ed.), pp. 269-99. New York: Academic Press.

Lemaine, G.
1974 Social differentiation and social originality. *European Journal of Social Psychology* 4:17-52.

Macdonald, W.
1990 Investigating style: an exploratory analysis of some Plains burials. In *The Uses of Style in Archaeology* (M. Conkey & C. Hastorf, eds.), pp. 52-60. Cambridge: Cambridge University Press.

Marshall, L.
1976 *The !Kung of Nyae Nyae.* Cambridge: Harvard University Press.

Marshall, J.
1996 The need to be informed. *Anthropology Newsletter* 37: 4,5.

Marshall, J. & C. Ritchie
1984 *Where are the Bushmen of Nyae Nyae? Changes in a bushmen Society, 1958-1981.* Cape Town: University Centre for African Studies.

Neiman, F.
1995 Stylistic variation in an evolutionary perspective: inferences from decorative diversity and inter assemblage distance in Illinois Woodland ceramic assemblages. *American Antiquity,* 60: 7-36.

Ritchie, C
1989 *The Political Economy of Resource Tenure in the Kalahari.* Unpublished Masters thesis, Boston University.

Sackett, J.
1982 Approaches to style in lithic archaeology. *Journal of Anthropological Archaeology,* 1:59-112.

Schiffer, M. & J. Skibo
1997 The explanation of artifact variability. *American Antiquity,* 62: 27-50.

Shanks, M. & C. Tilley
1987 *Re-Constructing Archaeology: Theory and Practice.* London: Routledge.

Shostak, M.
1981 *Nisa: the Life and Words of a !Kung Woman.* Cambridge: Harvard University Press.

Tajfel, H.
1974 Social identity and intergroup behaviour. *Social Science Information* 13: 65-93.

Tajfel, H. (ed.)
1978 *Differentiation between Social Groups.* New York: Academic Press.
1982a Introduction. In *Social Identity and Intergroup Relations,* pp. 1-11. New York: Academic Press.
1982b *Social Identity and Intergroup Relations.* Cambridge: Cambridge University Press.

Tooby, J. & L. Cosmides
1989 Evolutionary psychology and the generation of culture, part 1. *Ethology and Sociobiology,* 10: 29-49.

Trivers, R.
1971 The evolution of reciprocal altruism. In *Readings in Sociobiology* (T. Clutton-Brock & P. Harvey, eds.), pp. 198-226. San Francisco: Freeman.

Turner, J.
1975 Social comparison and social identity: some prospects for intergroup behaviour. *European Journal of Social Psychology* 5: 5-34.

1982 Towards a redefinition of the social group. In *Social Identity and Intergroup Relations* (H. Tajfel, ed.), pp. 15-39. Cambridge: Cambridge University Press.

Uher, J.
1991 On zigzag designs: three levels of meaning. *Current Anthropology* 32: 437-8.

Wetherell, N.
1982 Cross cultural studies of minimal groups: implications for the social identity theory of intergroup relations. In *Social Identity and Intergroup Relations* (H. Tajfel, eds.), pp. 207-204. Cambridge: Cambridge University Press.

Wiessner, P.
1977 *Hxaro: A regional System of Reciprocity for Reducing Risk among the !Kung San*. PhD. Dissertation, University of Michigan. Ann Arbor: University Microfilms.

1981 Measuring the impact of social ties on nutritional status among the !Kung San. *Social Science Information* 20: 641-78

1982 Risk, reciprocity and social influences on !Kung San economics. In *Politics and History in Band Societies* (E. Leacock & R. B. Lee, eds.), pp. 61-84. Cambridge: Cambridge University Press.

1983 Style and social information in Kalahari San projectile points. *American Antiquity* 48: 253-76.

1984 Reconsidering the behavioral basis for style: a case study among the Kalahari San. *Journal of Anthropological Archaeology* 3: 190-234.

1986 !Kung San networks in a generational perspective. *In The Past and Future of !Kung Ethnography: Critical Reflections and Symbolic Perspectives. Essays in Honour of Lorna Marshall* (M Biesele, R. Gordon, & R. Lee, eds.), pp. 103-36. Hamburg: Helmut Buske Verlag.

1988 Style and the changing relations between the individual and society. In *The Meaning of Things: Material Culture and Symbolic Expression*. (I. Hodder, ed.), pp. 56-63. London: Unwin Hyman.

1994 The pathways of the past: !Kung San hxaro exchange and history. In *Überlebenstrategien in Africa* (M. Bollig & F. Klees, eds.), pp. 101-123. Köln: Heinrich-Barth-Institut.

Wilmsen, E.
1989 *Land Filled with Flies*. Chicago: University of Chicago Press.

Wobst, M.
1977 Stylistic Behavior and information exchange. In *Papers for the Director* (C. Cleland, ed.). Ann Arbor: University of Michigan, Museum of Anthropology Anthropological Papers No. 61.

Wrangham, R.
1987 The significance of African apes for reconstructing human social evolution. In *The Evolution of Human Behavior: Primate Models* (W. Kinzey, ed.), pp. 51-71. Albany: State University of New York Press.

Yellen, J.
1977 *Archaeological Approaches to the Present: Models for Reconstructing the Past*. New York: Academic Press.

1984 The integration of herding into prehistoric hunting and gathering economies. In *Frontiers: Southern African Archaeology Today* (M. Hall, G. Avery, D. Avery, M. Wilson & A. Humphreys, eds.), pp. 53-64. Cambridge: Monographs in African Archaeology No. 10, BAR International Series No. 207.

10

Evolutionary Implications of Metrical Variation in Great Basin Projectile Points

Robert L. Bettinger
University of California, Davis

Jelmer Eerkens
University of California, Santa Barbara

ABSTRACT

We used a sample of 5285 Great Basin projectile points to test several implications of the general prediction from Neo-Darwinist culture transmission theory that variation in a socially-transmitted behavior will vary inversely with: 1) its complexity; 2) the complexity of the social and technical context in which it occurs; and 3) the number of individuals that contribute to its transmission. Statistical analysis of these points demonstrates that 80% of the variation in metrical attributes (e.g., length, width) is accounted for by the mean of the attribute, suggesting that this variation is independent of the transmission process and primarily due to production in which error tolerances are relative, not absolute. The remaining variation, specifically, metrical variability that is substantially greater or less than normally occurs as a consequence of these relative production tolerances, is only partly consistent with transmission theory expectations. More complex point shapes show less metrical variation than simple forms, and arrow points show less metrical variation than dart points, as predicted. There is substantial variation by section, point type, and measurement, however, that remains unexplained by the transmission hypothesis.

INTRODUCTION

We note an unfortunate parallel between evolutionary archaeology and the weather: everyone *talks* about it but no one *does* anything about it. We think this is the fruit of research strategies charted some time ago in response to a debate among processualists about what was wrong with the New Archaeology and what should be done about it. In that case, the middle-range theorists (e.g., Binford 1977a,b, 1978a,b, 1979; Thomas 1986, Grayson 1986) held that the New Archaeology suffered from faulty arguments about the archaeological record itself, and what was needed was a set of tools that would permit direct engagement between theory and the archaeological record. A second group, mostly composed of individuals enamored with neo-Darwinian evolutionary theory (e.g., Dunnell 1980), held that the problem with the New Archaeology was more fundamental and had to do with flaws in its governing general theory - Whitean culture evolutionary theory (White 1959), for which some version of neo-Darwinian evolutionary theory should be substituted.

It seems quite clear to us, nearly two decades down the road, that these two programs cannot possibly succeed independently: middle range research requires uniformitarian assumptions and principles directly guaranteed by general theory (Bettinger 1987). And general theory, especially neo-Darwinian general theory, is far too complex and theoretically ambiguous to be articulated in the absence of specific case studies that show how particular components of the theory are to be interpreted. To us, this means it is pointless to debate the philosophical complexities of evolutionary theory in the abstract. The winning version will not necessarily be the most conceptually elegant. Nor will it necessarily be the one that seems most compatible with what we now think we know about the course of human culture history. It will be the one that can be applied to real cases so that we know more following the application than we did before. We intend here to provide such an example of an evolutionary analysis using a common subject matter (projectile points), common metrical data (length, width, etc.), common statistics (mean and standard deviation), and a middle range measure that relates patterning in the subject matter to theory.

BACKGROUND

It is hardly a secret that archaeologists are fascinated by projectile points. Much of this has to do with the fact that we think we know how and why projectile points were made and find enough of them so that we can regularly use what we know to make inferences about the past. Archaeologists, however, are probably more captivated by projectile point morphology. Projectile points seem to come in distinct shapes and sizes, varying outwardly in ways that are easy to measure. This diversity has been attributed to a wide range of factors of fundamental interest to the archaeologist, including time (Lanning 1963), function (Fenenga 1953), technology (Ericson 1982:144), and ethnicity/culture (True 1970:54). That it is difficult to decide which of these dimensions is mostly responsible for the variation observed in a given collection of points does not preclude the use of point morphology to index these dimensions. It is quite possible to devise a perfectly workable projectile point chronology without knowing why it works (e.g., Hester 1973). Nor is one required to argue that chronology is all that is being reflected in a projectile point typology that happens to tell time (e.g., Clewlow 1967). Regardless of root process, the same point classification can provide useful information about several archaeological dimensions. This makes it possible for one archaeologist to see in it function and another ethnicity, neither being sure why – but sure just the same that it is there because it seems to work. This is not something one wants to advertise. Closely inspected it amounts to an admission that we cannot distinguish the archaeological consequences of processes that theory tells us are separate and important. We think that some of the reasons for this have to do with the way archaeologists have developed, used, and thought about their classifications.

Traditional archaeological classification has understandably emphasized the definition of categories and means for identifying them. Here, what one wants to know most about the classification is whether it reliably indexes important dimensions: "good" ones do, bad ones don't. Most of the effort expended in describing different categories attends to attribute central location (mean, mode, median), which, in well-behaved attributes, is often linked to a material or symbolic function. For example, differences in the sizes of types of projectile points are often attributed to intended use (big vs. small game) or weapon type (atlatl vs. bow; e.g., Lanning 1963).

Much can be learned from this approach, but we think the emerging interest in evolutionary models and explanations justifies reversing the traditional logic of classification to ask what makes items in the same taxonomic category unlike. To do this we employ a measure of attribute dispersion akin to the coefficient of variation (i.e., standard deviation divided by the mean; Simpson, Lewontin & Roe 1960), to gauge the differences that exist between items within a category rather than those that exist between categories. We then use this to make inferences about cultural transmission in the populations these categories represent.

The contrasts between this approach and the traditional one, which emphasizes attribute central location, are not earth-shaking but they are important. Each measure of central location can be thought of as telling us about a commonality shared by items in a category, to which they are being driven by all the forces that make the category a category. Characterizing a category on this basis is problematic in two respects. Firstly, it is unclear whether all attributes are equally affected by the forces that "make" the category, for example length vs. thickness. Measures of central location provide no clues in this regard. They describe the commonalties uniting items in the category without telling us whether all those commonalties are equally important. It is implicitly assumed, of course, that most categories comprise a few salient, or key, attributes whose behavior is stable relative to that of the many more non-salient attributes, for example fluting in Folsom points. Secondly, it is much easier to statistically generalize or describe attribute central location, than it is to explain it. As just noted, most explanations of attribute central location are functional and heir to all the problems that beset such arguments (Orans 1975; Bettinger 1980, 1991).

Looking at archaeological categories from the perspective of attribute dispersion brings us a little closer to solving these problems. Dispersion can be thought of as measuring the extent to which an attribute is subject to stabilizing or centripetal forces, which account for its mean, and random or centrifugal forces, which account for its variance. This obviously has important implications for how we interpret and characterize our categories. Perhaps more importantly with regard to explanation, dispersion makes us think about populations rather than archetypal forms and lends itself to interpretation in a way that central location does not. The evidence from biology suggests that, when scaled relative to the mean, attribute dispersion takes on a narrow, and fairly stable, range of values across disparate populations and variables (Simpson, Roe & Lewontin 1960). Because of this it is possible to compare attribute dispersion in individual populations to an expected value derived from what is essentially a biological constant. Significant departures from expected values can frequently be explained in terms of simple evolutionary or ecological principles. If this applies in the case of human material culture, and our data suggest that in the case of Great Basin projectile points it does, we should be able to develop a statistic that permits characterization of attribute dispersion to a standardized scale. The Neo-Darwinian culture transmission theory (Boyd & Richerson 1985, 1987) provides a coherent framework for interpreting significant departures from this standard.

VARIABILITY AND CULTURE TRANSMISSION THEORY

At the core of culture transmission theory are four basic models of the way humans acquire cultural behaviors, each having different population-level effects on behavioral variation. In two of these, one termed *guided variation* (Boyd & Richerson 1985: 83-98), the other *direct bias* (Boyd & Richerson 1985: 137-146), traits are socially acquired, then tested by experiment. Specifically, in *guided variation* individuals acquire behaviors socially by "averaging" the behavior of one or more models (cultural parents) and then attempt to improve this acquired average behavior by independent trial and error (Bettinger 1991: 186-188). The social phase of this process levels differences between models and reduces variation at the population level. Subsequent experimentation, on the other hand, generates new behaviors, increasing population-level variability. *Directly biased* social transmission (Bettinger 1991: 188-190) is similar except that *two* alternative behaviors are socially acquired and tested, and the variant judged superior is retained unmodified. It is easy to see that, because only extant variants are tested and the winning variant is unchanged, directly biased transmission reduces population variation.[1] In short, the experimental component of guided variation increases variation, direct bias and the social component of guided variation reduce it.

The two remaining models of cultural transmission, one termed *frequency-dependent bias* (Boyd & Richerson 1985: 206-213), the other *indirect bias* (Boyd & Richerson 1985: 247-259), are both variation reducing. In each case, behaviors are acquired socially but not tested by field trial as in guided variation and direct bias. In the conformist version of *frequency-dependent transmission* (Bettinger 1991: 194-196), for example, individuals survey the local model pool, then adopt the most common behavioral variants in preference to less common alternatives. It is again easy to see that local populations will become homogeneous for attributes acquired this way, and variation will be reduced. *Indirectly biased transmission* is likewise exclusively social. In it, individuals evaluate a social model pool using preferences for a certain model profile previously acquired by social transmission (Bettinger 1991: 196-198). Models matching the profile are then copied for profile behaviors and additional useful behavioral traits functionally unrelated to the profile. For instance, individuals with socially-acquired preferences for models with high status may copy, in addition to high-status profile behaviors (e.g., clothing, mannerisms, etc.), any number of purely functional traits - house-building, for example. The social transmission of house-building behavior is then said to be *indirectly biased* by the preferred model profile, which uses status-related behavior as an indirect measure of house-building competence. This entrains a correlation between model profiles (i.e., preferences), profile behaviors, and indirectly biased traits for which the profile behavior serves as a proxy measure of competence. Intra-population homogeneity increases as a result.[2] Because indirect bias requires variation, its force will be strongest where exotic social models are routinely encountered, e.g., at boundaries between strongly contrasting adaptive strategies (Boyd & Richerson 1987). This gives indirect bias the power not only to decrease within population variation (as with frequency dependent bias), but to create, maintain, and exaggerate the behavioral differences between adjacent, but differently structured, adaptive poses.

Because guided variation and direct bias use field trial as a basis for decision-making, they are most effective in relatively simple technological settings, where alternative behaviors are susceptible to direct experiment. Using these modes of transmission each individual personally assesses the causal linkage between behavior and outcome by actual exposure and decides what works and what doesn't. However, as experimentation becomes more costly, individuals will be forced to commit to behaviors without understanding all or even most the causal connections between components or having any reasonable chance to figure them out independently. In general, one expects cultural transmission systems to respond to these circumstances by adjusting the weight of the social transmission and experimentation. A workable rule of thumb is that emphasis on social transmission will be directly proportional to the cost of trial and error, and inversely proportional to the fitness differences between alternative behaviors and social transmission costs (Pruett-Jones 1992, Boyd & Richerson 1985: 94-98).[3]

Many relevant costs of both kinds are likely correlated with cultural/technical complexity and population density. Experimental costs obviously increase as organization and technology become more complex and socially coordinated (Sugden 1986), especially where population density decreases latitude for individual behavioral flexibility. Conversely, since costly sampling errors associated with model pool averaging decrease with increasing model pool size, the costs of social transmission should decrease with population increase, provided that increase is reflected in model pool size.

One then expects cultural transmission systems to emphasize guided variation and direct bias, in that order, and the experimental component of guided variation, when population densities and techno-organizational complexity are low. As population grows and/or technology grows more complex, transmission systems should increasingly emphasize direct bias, which experiments only with extant, i.e., socially pre-tested, behaviors, at the expense of guided variation, the social component of which should increase. The importance of direct bias relative to guided variation, and the social component within guided variation, should both increase with subsequent increases in population density and social complexity, but the collective weight of both modes will gradually decrease in relation to the much cheaper, purely social modes of frequency-dependent and indirectly biased transmission. The implications of this for behavioral variation at the population level are clear. Variation should decrease as cultural/technical complexity and population density cause transmission systems to shift emphasis from experiment to model averaging social acquisition within guided variation, and from guided variation to direct bias, and subsequently from guided variation and direct bias to frequency dependent and indirectly biased transmission.

Social transmission and trial and error learning were surely both significant in prehistoric projectile point manufacture, depending on local circumstances, including complexity of intended form, tolerance of intended use and susceptibility to retouch, and population density. Point-makers and users likely experimented with morphology during production and resharpening in direct relation to expected cost and benefit. Social transmission may have been limited to simple model averaging, as in ordinary guided variation. However, as the weight of social transmission increases, alert individuals can usually do better than simple model-averagers by using relevant collateral information to rank potential models. Thus, Great Basin point-makers and users may have made choices about behavioral alternatives that involved observations about model prestige and hunting success (indirect bias) and variant frequency (frequency-dependent bias).[4]

In arguing that these predictions can be applied to understand variability in Great Basin projectile points, one need not contend that other ecological and functional processes are unimportant. Raw material type and availability, for instance, certainly constrain knapper ability to produce and resharpen idealized forms, amplifying or dampening metrical variability in projectile points. However, we are prepared to argue that, at the trans-sectional scale of this study, these effects are sufficiently "averaged out" to detect the effects of cultural transmission. The data presented below at least partly justify this assumption.

GREAT BASIN PROJECTILE POINT TYPOLOGY

The projectile point typology in most general use in the Great Basin today was developed as a chronological device in the 1960s by Robert Heizer, Martin Baumhoff, and their students at the University of California, Berkeley, on the basis of collections obtained in the course of excavations at several western and central Great Basin sites during the 1950s and 1960s (Baumhoff & Byrne 1959, Clewlow 1967, Heizer & Baumhoff 1961, Heizer, Baumhoff & Clewlow 1968, Heizer and Clewlow 1968, Heizer and Hester 1978, Lanning 1963, O'Connell 1967). They held that Great Basin points naturally segregated into distinct combinations of size and shape, that could be assigned to distinct periods of time, one after the other, that began and ended synchronously across the whole of the Great Basin. Although the nomenclature has changed, nearly all of the original series and their constituent types are still recognized (Holmer 1978, 1986; Thomas 1981): Humboldt, Large Side-notched (e.g., Northern Side-notched and Bitterroot Side-notched types), Gatecliff (Gatecliff Contracting Stem and Gatecliff Split Stem types), Elko (Elko Corner-notched and Elko Eared types), Rosegate, and Desert (Cottonwood Triangular and Desert Side-notched types).

The original scheme proved unacceptably simplistic (e.g., Aikens 1970) and has been frequently revised as new stratigraphic and radiometric data have become available from a much larger sample of sites (see Thomas 1981, Beck 1984, Wilde 1985, Holmer 1986, O'Connell & Inoway 1994 for reviews and syntheses). Of the original types, only the two major arrow point series, Rosegate and Desert, are still widely accepted as defining consecutively discrete (i.e., non-overlapping) periods that began and ended more or less simultaneously wherever found across the whole of the Great Basin (Rosegate series: A.D. 600 - A.D. 1300, Desert series: A.D. 1300 - Euroamerican contact).

The situation is much different for dart points. Humboldt, generally conceded to be the least time-sensitive of all Great Basin point types, was evidently made throughout the Great Basin during the entire period in which the atlatl was in use, ending around A.D. 600 (Thomas 1981). Large Side-notched dart points are not well defined or dated but nevertheless widely regarded as reliable time-markers in the northern, eastern, and some parts of the western Great Basin (Thomas 1981, O'Connell 1975). Most of the difficulty with Great Basin dart points centers on variants of the most important common series, Gatecliff and Elko, which seem to be time-sensitive in degrees that vary with location. In the western and central Great Basin, it appears that Gatecliff series and Elko series points do in fact characterize mutually exclusive and consecutive periods of time (Gatecliff series: 2500 B.C. - 1200 B.C.; Elko series: 1200 B.C. - A.D. 600) in much the same way that Rosegate and Desert series arrow points do throughout the whole of the Great Basin later in time (Thomas 1981, Bettinger 1989). In the eastern and northern Great Basin, however, Gatecliff and Elko display very different historical trajectories. In these areas the Elko Corner-notched type in particular displays unusual longevity and is broadly overlapping in time with Gatecliff, limiting its utility as a time marker (Holmer 1986).

The question one wants to ask is why these outwardly similar dart points display such disparate historical trajectories and why dart points in general seem more susceptible to this than arrow points. We think culture transmission theory offers the potential for explaining these differences and devote the rest of this discussion to a test of this by measuring differences in metrical variation in large collections of projectile points from different parts of the Great Basin. In general we are prepared to argue that points are time-sensitive as a consequence of the relative weighting of the social and experimental components of cultural transmission. Time sensitivity results when, for any number of reasons, the social component of cultural transmission contributing to point shape is especially large. This, of course, requires that we explain why the weight of the social component of cultural transmission should differ between different kinds of projectile points in different parts of the Great Basin.

Expectations

In general, culture transmission theory leads us to expect that the weighting of social transmission in projectile point form should vary directly with population size and technical complexity. It follows that the metric attributes of complex, high-risk, or error-sensitive point shapes should vary less than those of simpler, more robust, and easily generated shapes. It is relatively simple to assess relative complexity of form within a point series in these terms: as illustrated in Fig. 1, Elko Eared points are more complex than Elko Corner-notched points because they are essentially Elko Corner-notched points modified by basal notching or indentation, i.e., they involve one additional step. Similarly, Gatecliff Split Stem is the Gatecliff Contracting Stem form with basal notching added, and Desert Side-notched is the Cottonwood Triangular form with side, and generally basal, notching added. Large Side Notch and Humboldt points cannot be related in this way but Humboldt is a simple, shoulderless form obviously comparable to Cottonwood Triangular. Similarly, the Large Side-notched form is obviously a complex form comparable to Desert Side-notched. Culture transmission theory makes the counter-intuitive prediction that within these series pairs, the more complex form will be metrically less variable than its simpler counterpart. This explanation is consistent with the observation that Elko Corner-notched – a simple form predicted here to be more prone to experiment, hence less time-sensitive, that contributes most of the length to the so-called "long chronology," i.e., is the least time-sensitive (Holmer 1986: Fig. 6, 12, 23).

It is also reasonable to think that technical differences relating to the use of the atlatl and bow were potentially large enough to affect mode and weighting of cultural transmission and hence metric variability in dart points relative to arrow points. That arrow points are as broadly distributed in space, yet more discretely distributed in time, than dart points, suggests greater importance of social transmission in the former and lends some credence to this (Holmer 1986: Fig. 16). The obvious possibility is that bow and arrow technology is more technically complex than atlatl technology, the bow involving more and moving parts. This might discourage tinkering with arrow point shape and favor greater emphasis on the social component in cultural transmission, causing arrow points to be less variable than atlatl points. That late-prehistoric, bow-using, populations seem to have been more numerous than their atlatl-using predecessors, similarly predicts greater late prehistoric emphasis on social transmission, hence less variability within arrow points.

Differences in the temporal persistence of certain dart point forms (e.g., the Elko series), on the other hand, suggest that the details of their transmission differed by section (Holmer 1986: Fig. 23). We hazard the hypothesis that the neat, replacive succession of dart points in the central and western Great Basin is the result of cultural transmission in which the social component was heavily weighted, perhaps because population densities were relatively high. These factors combined to produce homogenous populations favoring first one form, then another. The accompanying prediction is that where these neat patterns of dart point replacement are lacking, as in the eastern and northern Great Basin, cultural transmission was more heavily weighted to individual learning, which, perhaps because densities were lower and behaviors less socially-coordinated, maintained a large range of general forms over long intervals.

In summary, we expect: (1) complex point shapes (Desert Side-notched, Elko Eared, Gatecliff Split-stem, and Large Side-notched) to be less variable than simple ones in the same series (Cottonwood Triangular, Elko Corner-notched, Gatecliff Contracting Stem); (2) arrow points (Cottonwood Triangular, Desert Side-notched, and Rosegate) to be less variable than dart points (Elko series, Gatecliff series, Humboldt and Large Side-notched); (3) temporally persistent dart shapes (Elko Corner-notched in the eastern and northern Great Basin) to be more variable than more temporally sensitive dart shapes (Gatecliff Split-Stem, Gatecliff Contracting Stem) in the same sections, and substantially more variable than their temporally-sensitive formal equivalents in the central and western Great Basin.

The Humboldt Concave-base form furnishes a baseline measure against which these expectations can be compared. This simplest of dart point forms is nowhere regarded as time-sensitive and by the line of reasoning developed above would represent an extreme instance of the more general case of a point form maintained by cultural transmission dominated by individual learning. In this view, Humboldt was a generally-useful, easily-imitated, robust shape that could be cobbled up on the spot to suit the special needs of individual hunters. This would minimize the contribution of social transmission and cause the Humboldt form to vary widely and independently of model group size.

APPLICATION AND RESULTS

To apply this transmission-based model, we turned to data compiled in an exhaustive survey of Great Basin projectile points by David Hurst Thomas in the early 1970's (Thomas 1981). The study examined and classified roughly 5900 pieces drawn from 38 well-documented collections representing all corners of the Great Basin (Fig. 10.2, Table 10.1). The sample used here comprises 5285 specimens from 37 collections in eight areas, termed here sections: North Lake in the northern Great Basin; South Lake, Lahontan and Reno in the western Great Basin; West Central and East Central in the central Great Basin; and Tonopah and Bonneville in the eastern Great Basin.[5] Thomas measured 9 attributes on each projectile point and assigned each one to a traditionally recognized Great Basin type using criteria initially established to classify projectile points from the Monitor Valley in central Nevada. Our observations here are confined to analyses of summary metrical data from six variables measured in units of length (i.e., mm): maximum length, axial length (i.e., centerline length), maximum width, basal width, neck width, and thickness.

Talking about variability in projectile points with respect to these different measures is easier than quantifying it. One cannot simply compare variances between different point types or measures without understanding how that variance is scaled. It would be unfair, for example, to compare variances in height of corn stalks against variance in the height of redwood trees. Redwood trees, being taller, simply provide more room for variability. Likewise, projectile points come in different sizes, and larger ones will naturally vary more than smaller ones. Does this mean they are "more variable" in the terms we have discussed above? The answer, of course is, "No." In order to talk about metrical variability in projectile points, one must scale that variation to a common denominator. In the case of Great Basin projectile points, the scale that relates mean and standard deviation is linear and surprisingly strong (Fig. 10.3).

This line is noteworthy in several respects. It shows, firstly, that, in Great Basin projectile points, as the mean of any basic metrical attribute (e.g., maximum length, maximum width, etc.) increases, so does its standard deviation by a factor of 0.24, which is the slope of the line relating mean and standard deviation (i.e., $y = .24x + 0.30$, where y is standard deviation, x is mean of any given metric attribute). Least-Squares regression shows that for all the major measured dimensions of all major Great Basin projectile points from all sections of the Great Basin, variation in measurement mean accounts for 81 percent of the variation in measurement variability ($r = .899$). This suggests that the production/rejuvenation/discard tolerances are scaled by size.

A relationship was to be expected, of course, but several alternatives are clearly possible. Variation or tolerance in the cultural world, for instance, is frequently independent of scale. Thus, a carpenter asked to cut a board to length will generally produce a board with fixed variation, independent of overall board length (say, plus or minus one-quarter inch). It doesn't matter how big or how long the board wants to be, the carpenter's ability to read a tape measure and cut the board is the same, resulting in a fixed error independent of size. Hupa craftsmen calculated finished canoe height in just this way, using an accepted standard the length of which was tattooed on the leg for easy reference (Goddard 1903-1904: 50). In Great Basin projectile points, by contrast, the variation in metrical attributes is relative to scale. The regression line demonstrates that variation in all measured at-

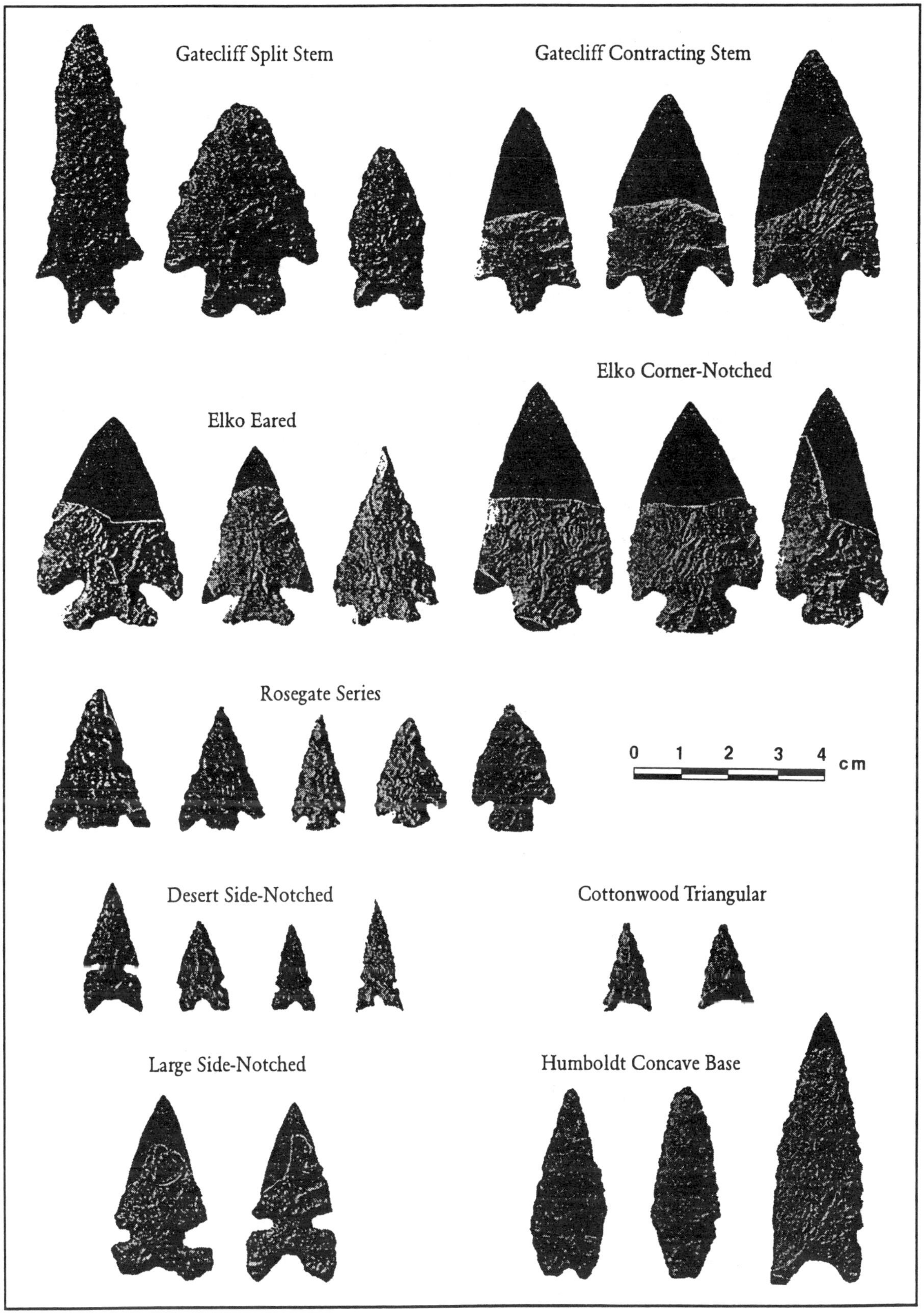

Figure 10.1 Great Basin Projectile Point Types.

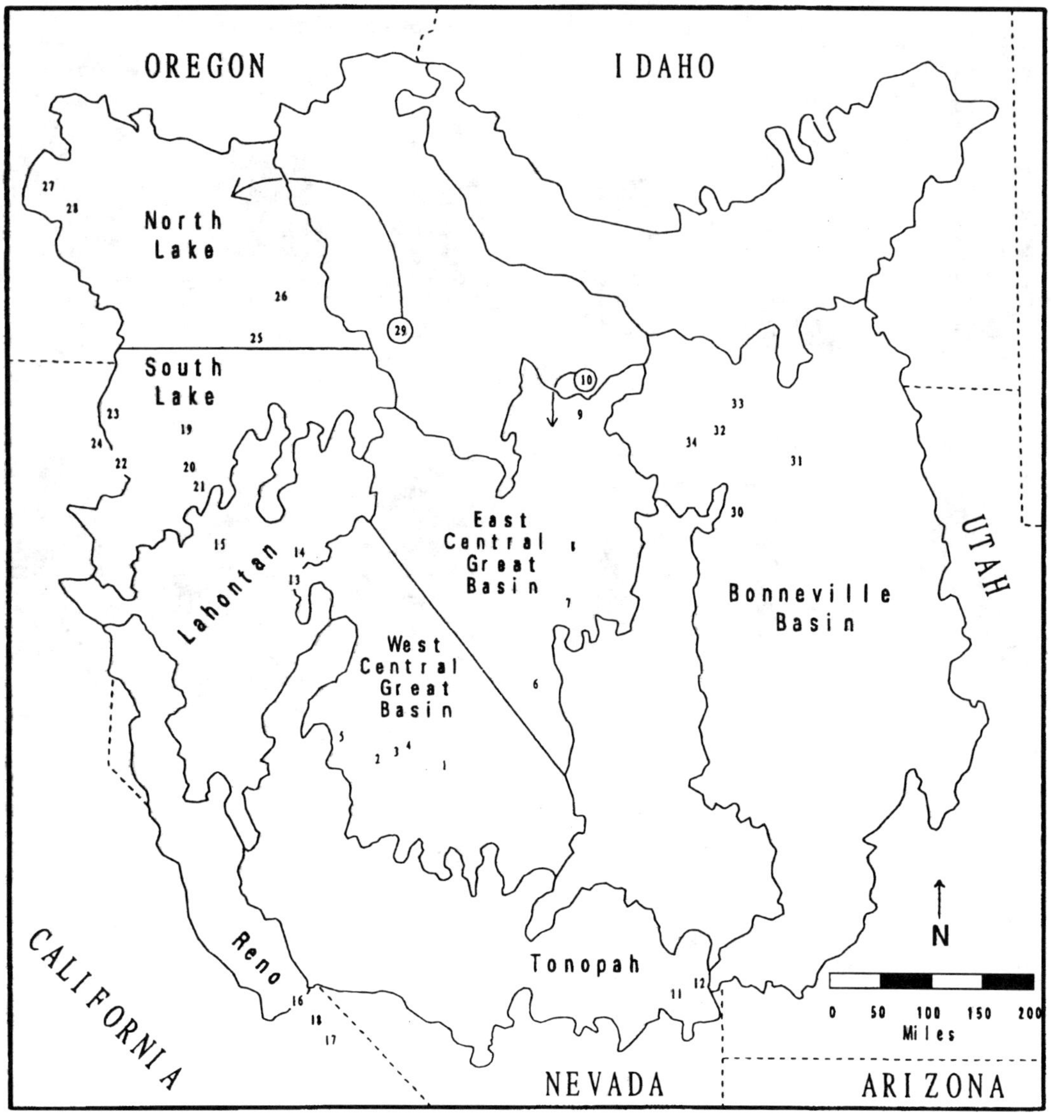

Figure 10.2 Great Basin sections and locations of projectile point collections used in the analysis. 1. Monitor Valley: Gatecliff Shelter, Triple T Shelter, Monitor Valley survey, Other Monitor sites; 2. Reese River I survey; 3. Reese River II survey; 4. Mateo's Ridge; 5. Wagon Jack Shelter; 6. Newark Cave; 7. Ruby Cave; 8. South Fork Shelter; 9. Deer Creek Cave; 10. Freightor's Defeat; 11. Conaway Shelter; 12. O'Malley Shelter; 13. Humboldt Lakebed site; 14. Hesterlee site; 15. Black Rock Desert survey; 16. Owens Valley survey; 17. Rose Spring site; 18. Stahl Site; 19. Hanging Rock Shelter; 20. Smokey Creek Cave; 21. Silent Snake Springs; 22. Rodriquez site; 23. Menlo Baths; 24. Bare Creek Cave; 25. Catlow Cave No. 1; 26. Roaring Springs Cave; 27. Fort Rock Cave; 28. Connely Caves; 29. Dirty Shame Rockshelter; 30. Danger Cave; 31. Hogup Cave; 32. Swallow Shelter; 33. Beatty Springs; 34. No Name Valley.

tributes conforms to the same basic scalar relationship. Variation in axial length, for example, is basically the same scale as variation in thickness, maximum width, and so on. This runs counter to much received wisdom, intuiting that certain of these metrical attributes are qualitatively more stable than others (e.g., basal width, Thomas 1981:15). Our data show that there are differences in the variability of different metrical attributes but that they are quantitative and generally situational, varying from place to place and form to form (see below). Thirdly, the line says that the slope of the regression is much steeper than the 0.04 - 0.10 slopes normally associated with biological traits under simple genetic control (Simpson, Lewontin & Roe 1960). This suggests projec-

Table 10.1 Distribution of Projectile Points Used in the Analysis.

Region		Desert Side-notched	Cotton-wood Tri-	Rose-gate	Elko Corner-notched	Elko Eared	Gate-cliff Contract-ing Stem	Gate-cliff Split Stem	Hum-boldt	Large Side-notched	TO-TAL
West Central	collections[1]	8	8	8	8	8	8	7	8	5	8
Great Basin	specimens[2]	128	72	281	535	244	143	105	190	45	1743
East Central	collections	5	3	5	5	4	5	5	5	5	5
Great Basin	specimens	44	7	91	135	14	30	29	66	66	482
Tonopah	collections	2	2	2	2	1	2	1	1	1	2
	specimens	17	27	89	89	10	104	9	48	15	408
Lahontan	collections	3	2	2	2	2	1	2	2	2	3
Basin	specimens	22	13	28	38	22	33	30	22	14	222
Reno	collections	3	3	3	3	3	2	3	3	0	3
	specimens	39	37	61	78	53	15	50	63	0	396
Southern	collections	3	2	6	6	5	3	6	6	5	6
Lake	specimens	23	2	61	95	91	10	60	79	47	468
Northern	collections	5	2	4	4	4	4	2	4	4	5
Lake	specimens	56	25	173	140	76	12	11	46	46	585
Bonneville	collections	4	3	4	5	5	5	5	5	5	5
	specimens	20	29	120	362	73	62	98	90	127	981
TOTAL	regions	8	8	8	8	8	8	8	8	7	8
	collections	33	25	34	35	32	30	31	34	27	37
	specimens	349	212	904	1472	583	409	392	604	360	5285

1. Number of collections in which a form is represented.
2. Thomas does not report actual numbers by collection or region. This is the sum of minimum values for each type and collection given by the number of measured specimens for the variable represented by the greatest number of measurements (observations) of that form in that collection.

tile points were not scaled with reference to anatomical features (e.g., digit length) or to other projectile components (e.g., shaftments or foreshaftments) produced with reference to such scales (e.g. Kroeber 1976: 197). This is not the place to explore the implications of this potentially important observation except to note that supports the notion of Boyd & Richerson (1985: 150) that cultural traits provide sufficient variation to make relatively weak bias forces effective in producing evolutionary change. The fourth point of interest is that the Y-intercept of the regression at -0.3 is non-zero, and is statistically less than zero at the 95% confidence interval. Since there are neither data points with negative standard deviation nor ones with means of zero, this likely implies that the relationship between mean and standard deviation is at least slightly curvilinear, being initially relatively flat, then increasing in slope to a value close to the one observed in the overall data set.

The linear relationship that predicts the relationship of standard deviation to mean in Great Basin projectile points is a particularly powerful middle range metric that allows us to examine variability independent of absolute size and measurement type across many different categories and collections. This is important because the statistical tests for equivalence of variance between two samples (e.g., Bartlett's or O'Brien's homogeneity of variance tests) are strongly conditioned by sample size. Since our data were summary statistics from collec-

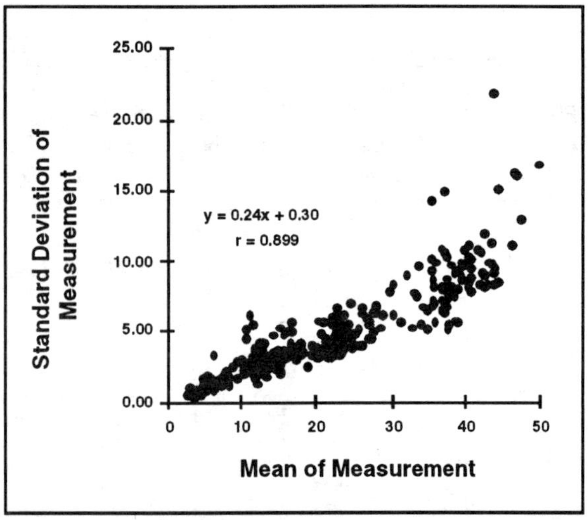

Figure 10.3 Mean and standard deviation for metrical variables of Great Basin projectile points calculated by section.

tions of unequal size, these tests yielded results that largely mirrored sample sizes. Differences in the variance between any two collections were almost always significant when sample sizes were large, and insignificant when they were small. In short, with these variance tests, "Large samples make everything significant!"

The 19% of overall variability that is not explained by variation in the mean can be parsed out by point type, section, and measurement type using the standardized residuals that remain when the scalar effect of the mean is subtracted from the original values of standard deviation. Large positive standardized residuals tell us that a particular measurement is more variable than the Great Basin average (Tables 10.2 – 10.5). Likewise, large negative measurements indicate less than average variability. These residuals directly bear on predictions for Great Basin projectile points derived above from culture transmission theory.

Metrical Variation by Type

The residuals in Tables 10.2 and 10.3 are consistent with many implications arising from the hypothesis that, in the case of Great Basin projectile points, the social component of cultural transmission will vary in force directly with technical complexity, its signature being reduced behavioral variation. As expected, Humboldt Concave Base, whose simple form and temporal persistence suggest cultural transmission with minimal social contribution, is more variable than any other Great Basin point type (mean residual = 0.74). Beyond this, complexity of shape correctly predicts overall metric variability for all other Great Basin dart points, the simple forms of which (Elko Corner-notched: 0.23; Gatecliff Contracting Stem: 0.59) are metrically variable and exhibit mean residuals above zero, and the complex shapes (Elko Eared: -0.28; Gatecliff Split Stem: -0.24; Large Side-notched: -0.60) are not and exhibit mean residuals below zero. Complexity of shape, however, not does predict variability in arrow points within the Desert Series, where the simpler Cottonwood Triangular shape is less variable overall (-0.26) than its more complex Desert Side-notched counterpart (-0.16). Indeed, Cottonwood Triangular, which is among the simplest of all Great Basin point forms, is also among the least variable. Perhaps this is because, being neckless, Cottonwood Triangular cannot be measured for neck width, which, following thickness, is the most variable of all point dimensions considered (Table 10.3). The neckless Cottonwood Triangular shape, then, is predisposed to be "less variable" than the necked Desert Side-notched shape. The greater variability of Desert Side-notched points relative to Cottonwood Triangular points, however, is mostly due to axial length, reflecting differences in basal treatment (e.g., straight, concave, notched, etc.) traditionally used to define variants within the basic Desert Side-notched type (Baumhoff & Byrne 1959). This suggests the cultural transmission of Desert Side-notch basal morphology contained greater experimental or individual stylistic contribution than characterized the other dimensions of that form (e.g., maximum length, maximum width), which are markedly less variable, presumably reflecting relatively greater social contribution to their transmission. In this case at least, the social and experimental contributions to transmission seem to differ by dimension.

Independent of the above, the hypothesized relationship between technical complexity, emphasis on social transmission, and reduced behavioral variability further predicts observed differences in arrow point and dart point variability. As expected, arrow points - which represent the more complex weapons system, are less variable overall (-0.15) than dart points (0.07). It is quite thinkable here that the Cottonwood Triangular shape is neckless precisely because neck width is inherently variable (i.e., that this neckless form facilitated efficient cultural transmission relative to necked forms, which frequently proved too difficult to maintain within reasonable tolerances through social transmission). Shape complexity, however, is a clearly more powerful determinant of variability than weapon system, since complex dart shapes are less variable (-0.37) than arrow points taken as a whole (-0.15). This suggests that the dart shapes characterized here as "complex" are not the locus of individual cultural expression of the kind Weissner (1983; Chapter 9) has termed "assertive style," which should be highly variable. The Desert Side-notched arrow point, on the other hand, which is more variable in basal mor-

Table 10.2 Standardized Residuals by Projectile Point Type.

	Desert Side-Notched	Cotton-wood	Rose-gate	Elko Corner-Notched	Elko Eared	Gate-cliff Contract-ing Stem	Gate-cliff Split Stem	Hum-boldt	Large Side-Notched
Mean	-0.16	-0.25	-0.04	0.23	-0.28	0.59	-0.24	0.74	-0.60
S	0.51	0.46	0.31	0.92	0.57	1.76	0.92	1.18	0.66
Minimum	-1.31	-1.24	-0.57	-1.13	-1.80	-1.32	-2.31	-0.76	-2.48
Maximum	1.05	0.56	0.99	3.72	1.49	8.00	1.61	4.33	0.58
Cases	48	36	48	48	48	48	48	40	42

phology than its more simply-shaped counterpart (Cottonwood Triangular), may contain such assertive stylistic expression in this dimension (see also Barton, Chapter 8).[6]

Metrical Variation by Section

The results are more mixed when we parse out the standardized residuals by section (Table 10.3). Elko Corner-notched points, which are time-sensitive in the central and western, but not northern or eastern, Great Basin, display patterns of variance that only partly conform to the argument that time-sensitivity is a function of variability-reducing social transmission. In accord with the hypothesis, Elko Corner-notched is less variable in the central Great Basin (East Central: -0.39; West Central: -0.29), where the type is time-sensitive, than in the Bonneville section of eastern Great Basin (0.19), where it is not time-sensitive. At the same time, the type is everywhere *more* variable in the western Great Basin (South Lake: 0.25; Reno: 0.96; Lahontan: 1.54), where the type is time-sensitive, than in the Tonopah section of the eastern Great Basin (-0.12) and northern Great Basin (North Lake: -0.32), where it is not time-sensitive. In short, these data provide no clear evidence that time-sensitivity is connected with variability-reducing social transmission as we have hypothesized.

That time-sensitivity overall is so poorly correlated with metric variability in these sections might be due to the presence of multiple transmission systems. We noted earlier that heavily weighted social transmission–indirectly biased social transmission in particular, will depress variation within, and exaggerate variation between, adjacent populations. Hence, where the weight of social transmission is large, sectional variability should be extremely low if intra-sectional boundaries (i.e., between transmission systems) are absent and extremely high if they are present. Culture transmission theory further predicts such boundaries should form where there is substantial opportunity for culture contact and movement between environmentally dissimilar sections (Boyd & Richerson 1987). This might explain the usually high metrical variation in the spatially contiguous Lahontan, Reno, and South Lake sections, which are situated along the Sierra Nevada-Cascade rim where Great Basin peoples routinely encountered representatives of the very differently organized adaptive systems of California. This is certainly an area of demographic impermanence and fluidity. Penutian-speaking groups are believed to have entered California through the South Lake district (Shipley 1978), the Lahontan section is the hypothesized location of major prehistoric population replacement (Aikens & Witherspoon 1986), and Washo groups ethnographically centered around Lake Tahoe are regarded by many as remnants of a formerly much larger Hokan population that once occupied the Reno section (Heizer & Baumhoff 1962). Given that South Lake and Lahontan section projectile points are highly variable in general, it is potentially significant that the Humboldt form, which is highly variable and nowhere regarded as time-sensitive in the Great Basin, is in fact relatively invariable (metrically at least) in the continuous Lahontan and South Lake sections, where the type was first identified and is particularly common (Heizer & Clewlow 1968). This suggests socially dominated cultural transmission of the Humboldt form in these sections and raises the possibility it might constitute a good chronological marker in these sections.

Metrical Variation by Measurement

As hinted very early above, when parsed by measurement, the residuals are at odds with many traditional

Table 10.3 Standardized Residuals for Projectile Point Types by Measurement.

Point Form	Maximum Length	Axial Length	Maximum Width	Basal Width	Neck Width	Thicknes	Mean	Standard Deviation
Desert Side-notched	-0.54	-0.02	-0.42	-0.21	0.05	0.18	-0.16	0.51
Cottonwood	-0.59	-0.33	-0.26	-0.23	-	0.09	-0.26	0.46
Rosegate	-0.12	-0.03	-0.17	0.04	-0.02	0.08	-0.04	0.31
Elko Corner-Notched	0.46	0.46	-0.32	0.22	0.23	0.31	0.23	0.92
Elko Eared	-0.60	-0.40	-0.58	-0.25	-0.11	0.23	-0.28	0.57
Gatecliff Contracting Stem	0.79	0.84	-0.16	1.11	0.72	0.25	0.59	1.76
Gatecliff Split Stem	-0.40	-0.13	-0.93	-0.06	0.03	0.06	-0.24	0.92
Humboldt	1.24	1.36	0.16	0.62	-	0.33	0.74	1.18
Large Side-Notched	-1.33	-1.08	-0.73	-0.61	0.13	0.00	-0.60	0.66
Mean	-0.10	0.10	-0.37	0.08	0.15	0.17		
Standard Deviation	1.56	1.50	0.55	0.68	0.42	0.24		

Table 10.4 Standardized Residuals for Projectile Point Types by Region.

Point Form	North Lake	South Lake	Lahon-tan	West Central	East Central	Reno	Tono-pah	Bonne-ville	Mean	Standard Deviation
Desert Side-notched	-0.47	0.05	0.23	-0.08	-0.36	-0.24	-0.10	-0.32	-0.16	0.51
Cottonwood	-0.61	-	0.21	-0.24	-0.47	-0.30	-0.61	0.17	-0.26	0.46
Rosegate	-0.02	0.07	0.03	-0.14	-0.06	0.17	-0.23	-0.13	-0.04	0.31
Elko Corner-Notched	-0.32	0.25	1.54	-0.29	-0.39	0.96	-0.12	0.19	0.23	0.92
Elko Eared	-0.79	-0.06	-0.40	-0.46	-0.11	-0.16	-0.01	-0.27	-0.28	0.57
Gatecliff Contracting Stem	-0.01	0.21	3.56	0.05	-0.26	0.11	0.50	0.58	0.59	1.76
Gatecliff Split Stem	-0.86	-0.01	0.43	-0.22	-0.07	0.02	-0.64	-0.53	-0.24	0.92
Humboldt	0.36	-0.11	-0.19	1.02	2.33	1.58	0.54	0.40	0.74	1.18
Large Side-Notched	-0.71	-0.26	-0.80	-0.44	-0.66	-	-0.75	-0.40	-0.60	0.66
Mean	-0.40	0.02	0.53	-0.11	-0.04	0.25	-0.16	-0.06		
Standard Deviation	0.79	0.45	1.82	0.59	1.06	0.96	0.80	0.55		

intuitions about metrical variability in different projectile point dimensions (Table 10.5). In particular, maximum length (-0.10) turns out to be less variable overall than basal width (0.08), which is at odds with the commonly held view that because length is more subject to post-manufacture alteration (i.e., resharpening and repair) than basal width, it should be more variable. Underlying the traditional view, of course, is the assumption that production standards (represented by basal width) are inherently more uniform than discard standards (represented by maximum length). Our data suggest just the opposite: that discard standards are more uniform than production standards, and that metrical variability diminishes with life span age of Great Basin projectile point populations.

The measurement residuals are instructive in two other respects. First, it is quite clear that thickness is consistently variable, across all types, weapon systems, and sections. Since thickness is presumably consequential to structural integrity, one is inclined to speculate that Great

Table 10.5 Standardized Residuals for Regions by Measurement.

Region	Maximum Length	Axial Length	Maximum Width	Basal Width	Neck Width	Thick-ness	Mean	Standard Deviation
North Lake	-1.04	-0.83	-0.60	0.01	0.00	0.13	-0.40	0.79
South Lake	-0.11	0.03	-0.38	0.26	0.20	0.12	0.02	0.45
Lahontan	1.02	1.32	-0.02	0.38	0.35	0.10	0.53	1.82
West Central	-0.44	-0.26	-0.39	0.14	0.16	0.21	-0.11	0.59
East Central	-0.02	0.17	-0.44	-0.12	0.08	0.12	-0.04	1.06
Reno	0.56	0.74	-0.20	0.05	0.05	0.26	0.25	0.96
Tonopah	-0.33	-0.19	-0.73	-0.11	0.25	0.21	-0.16	0.80
Bonneville	-0.34	-0.14	-0.22	0.07	0.08	0.25	-0.06	0.55
Mean	-0.10	0.10	-0.37	0.08	0.15	0.17		
Standard Deviation	1.56	1.50	0.55	0.68	0.42	0.24		

Basin point makers were simply less able to control it. Perhaps there are technical reasons for this, but it is tempting to argue it reflects point makers working with plan view, i.e., two-dimensional, templates. It is of significance in this regard that, by varying thickness as needed, San metal arrow point makers in the Kalahari attempt to maintain overall point length, width, and plan view outline, regardless of the wire gauge with which they happen to be working, even though this yields unduly thin and easily bent points when lighter wire gauges are used (Weissner 1983:261). In both the San and Great Basin cases, point makers seem to be acting as though point shape were a two-dimensional, not three-dimensional, problem. Second, our Great Basin data indicate that, of all the measures examined, maximum width is by far the most stable (-0.37), suggesting it was the most salient in the minds of point-makers and the one most strongly dominated by social transmission.

CONCLUSION

We have tried to demonstrate here a simple method for exploring the consequences of important evolutionary processes in the archaeological record. Perhaps the most innovative aspect of our proposal is its emphasis on variability rather than central tendency. This follows directly from basic evolutionary theory where variation plays a central role and is more tractable to theoretical analysis than central tendency. We have outlined a process by which one might set about measuring the relative weights of different components of cultural transmission that maintained distinctive projectile point shapes in the Great Basin. In the course of this, we demonstrated quite convincingly that metrical variation in projectile points is strongly scaled by size and that this relationship is linear. The metric obtained here is eminently well suited to measuring relative variability in other collections of Great Basin projectile points. Further, it points the way to similarly structured studies of projectile points from other regions and of other artifact classes. Perhaps the most important aspect of this study is that we used culture transmission theory to learn something we did not already know about the archaeological record. We established the basic character of metrical variability in Great Basin projectile points and made some progress toward linking the remaining variability to fundamental evolutionary processes.

ACKNOWLEDGEMENTS

We thank R. Azari for advice and assistance with statistical methodology, Susan Harris for help in data entry and data quality control, and David Hurst Thomas for providing the basic data.

END NOTES

1 The mechanics of the directly biased transmission, however, preclude habitat specific biases (e.g., "Use Elko Eared points in tall timber, Elko Corner-notched points in scrublands.") and extend only to generalized biases of the sort commonly imagined to guide rational choice, i.e., that permit discrimination between better and worse outcomes (Boyd & Richerson 1985:155-157).

2 Preferences are acquired by the same process. Thus, preference for models with high social standing would result in the acquisition of house forms made by high-status individuals *and* the preferences high-status individuals themselves use to rank potential models, which are

likely to attach greater than average importance to social status. In this way, house-building and social standing could increase in importance together: the initial preference for social standing increases the importance of social standing by reinforcing positive feedback, increasing at the same time the variant of house-building initially displayed by individuals of high social standing. If the effect is strong enough, this results in populations in which the preferences, preferred traits, and the indirectly-acquired behavioral traits are very strong, highly-correlated, and frequently exaggerated.

3 Pruett-Jones (1992) suggests that copying will conform to frequencies given by, $p = 1 - (k/f)$, where p is the frequency of copying, k is the cost of trial and error, and f is the fitness benefit of making the "right" choice.

4 It is important to note here that while cultural transmission is traditionally discussed principally in terms of its effects on production behavior, especially in the case of projectile points, there is every reason to think it was equally important in determining discard behavior, including the standards that caused objects to be classified no longer useful. Accordingly, that most of the points we recover archaeologically are extensively reworked discards does not negate the importance of cultural transmission in understanding the variability they display. Data bearing on this point are presented below (see Metrical Variation by Measurement).

5 As shown in Figure 10.2, geographical realities required that we divide Thomas's floristically-defined Central Great Basin section into an East Central section and a West Central section, and his Lake section into a North Lake section and a South Lake section. Following Thomas, we assigned Dirty Shame Rockshelter to the North Lake section and Freightor's Defeat to the East Central section, although both sites fall slightly outside these sections as floristically defined. Further, following existing convention (e.g., O'Connell and Inoway 1994), we treat the South Lake section as part of the western Great Basin, which it most closely matches in terms of point chronology. Likewise, we treat the Tonopah section as part of the eastern Great Basin because the two sites representing that section are clearly in the eastern Great Basin, even though the section as whole is not.

6 Rosegate Series as identified by the Thomas key comprises two recognized types - Rose Springs and Eastgate, which, if separated, might clarify some of these ambiguities. Eastage is generally conceded to be more formally redundant and carefully made, suggesting greater emphasis social transmission. If that is so, and if point shape is a more important determinant of the weight of social transmission than weapons system, then the simple Rose Springs form should be more variable than the more elaborate Eastgate alternative and approach simple dart point forms in this respect. If, on the other hand, weapons technology is the more important determinant the weighting of social transmission, then, as in the Desert Series, the simpler Rose Springs form might be quite invariable, though that need not be so.

REFERENCES CITED

Aikens, C. M., & Y. Witherspoon

1986 Great Basin Numic Prehistory: Linguistics, Archaeology, and Environment. In *Anthropology of the Desert West: Essays in Honor of Jesse D. Jennings* (C.S. Condie & D.D. Fowler, eds.), pp. 7-20. Salt Lake City: Anthropological Papers No. 110. University of Utah.

Baumhoff, M. A., & J.S. Byrne

1959 Desert Side-notched Points as a Time Marker in California. *Archaeological Survey Reports* No. 48, pp. 32 - 65. Berkeley: University of California.

Beck, C.

1984 *Steens Mountain Surface Archaeology: The Sites.* Unpublished Ph.D. dissertation, Department of Anthropology, University of Washington.

Bettinger, R. L.

1987 Archaeological Approaches to Hunter-gatherers. *Annual Review of Anthropology* 16: 121-142.

1980 Explanatory/predictive Models of Hunter-gatherer Adaptation. *Advances in Archaeological Method and Theory* 3: 189-255. New York: Academic Press.

1991 *Hunter-Gatherers: Archaeological and Evolutionary Theory.* New York: Plenum.

Binford, L. R.

1977a General Introduction. In *For Theory Building in Archaeology* (L. R. Binford, ed.), pp. 1-10. New York: Academic Press.

1977b Forty-seven Trips: A Case Study in the Character of Archaeological Formation Processes. In *Stone Tools as Cultural Markers: Change, Evolution, and Complexity* (R. V. S. Wright, ed.), pp. 24-36. Canberra: Australian Institute of Aboriginal Studies.

1978a *Nunamiut Ethnoarchaeology.* New York: Academic Press.

1978b Dimensional Analysis of Behavior and Site Structure: Learning From an Eskimo Hunting Stand. *American Antiquity* 43: 330-361.

1979 Organization and Formation Processes: Looking at curated technologies. *Journal of Anthropological Research* 35: 255-273.

Boyd, R., & Richerson, P. J.

1985 *Culture and the Evolutionary Process.* Chicago: University of Chicago Press.

1987 The Evolution of Ethnic Markers. *Cultural Anthropology* 2: 65-79.

Clewlow, C. W.

1967 Time and Space Relations of Some Great Basin Projectile Point Types. *Archaeological Survey Reports* No. 70: 141-150. Berkeley: University of California.

Conover, W.J., Mark E. Johnson & Myrie M. Johnson

1981 A Comparative Study of Tests for Homogeneity of Variances, with Applications to the Outer Continental Shelf Bidding Data. *Technometrics*: 23: 351-361.

Ericson, J. E.

1982 Production for Obsidian Exchange in California. In *Contexts for Prehistoric Exchange* (J. E. Ericson & T. K. Earle, eds.), pp.129-148. New York: Academic Press.

Dunnell, R.

1980 Evolutionary Theory and Archaeology. *Advances in Archaeological Method and Theory* 3: 35-99.

Fenenga, Franklin

1953 Weights of Chipped Stone Points: a Clue to Their Functions. *Southwestern Journal of Anthropology* 9: 309-323.

Goddard, I.

1903-1904 Life and Culture of the Hupa. *Publications in American Archaeology and Ethnology* 1(1). Berkeley: University of California.

Grayson, D. K.

1986 Eoliths, Archaeological Ambiguity, and the Generation of "Middle-Range" Research. In *American Archaeology: Past and Future*, (D. J. Meltzer & D. D. Fowler, eds.), pp. 77-133. Washington: Smithsonian Institution Press.

Heizer, Robert F. & M. A. Baumhoff
1961 Archaeology of Two Sites at Eastgate, Churchill County, Nevada: I Wagon Jack Shelter. *Anthropological Records* 20: 119-138.
1962 *Prehistoric Rock Art of Nevada and Eastern California*. Berkeley: University California Press.

Heizer, Robert F., Martin A. Baumhoff, & C.W. Clewlow, Jr.
1968 Archaeology of South Fork Shelter (Nv-El-11), Elko County, Nevada. *Archaeological Survey Report* No. 71, pp. 1-58. Berkeley: University of California.

Heizer, Robert F., & C.W. Clewlow, Jr.
1968 Projectile Points From the Site NV-CH-15, Churchill County, Nevada. *Archaeological Survey Report* No. 71, pp. 59-88. Berkeley: University of California.

Heizer, R. F., & T. R. Hester
1978 Great Basin. In *Chronologies in New World Archaeology* (R. E. Taylor & C. W. Meighan, eds.), pp. 147-199. New York: Academic Press.

Hester, T. R.
1973 Chronological Ordering of Great Basin Prehistory. *University of California Archaeological Research Facility Contributions* No. 17. Berkeley: University of California.

Holmer, R. N.
1978 *A Mathematical Typology for Archaic Projectile Points of the Eastern Great Basin.* Unpublished Ph.D. dissertation, Department of Anthropology, University of Utah.
1986 Common Projectile Points of the Intermountain West. In Anthropology of the Desert West: Essays in honor of Jesse D. Jennings (C. Condie & D. Fowler, eds.). *Anthropological Papers* 110: 89-116.

Kroeber, T.
1976 *Ishi In Two Worlds*. Berkeley: University of California Press.

Lanning, Edward P.
1963 Archaeology of the Rose Spring Site Iny-372. *Publications in American Archaeology and Ethnology* 49: 237-336.

O'Connell, J. F.
1967 Elko Eared/Elko Corner-notched Projectile Points as Time Markers in the Great Basin. *Archaeological Survey Reports* 70: 129-140.
1975 *Prehistory of Surprise Valley*. Palo Alto: Ballena Press Anthropological Papers No. 4.

O'Connell, J. F., & C. M. Inoway
1994 Surprise Valley Projectile Points and Their Chronological Implications. *Journal of California and Great Basin Anthropology* 16: 162-198.

Orans, M.
1975 Domesticating the Functional Fragon: An Analysis of Piddocke's Potlatch. *American Anthropologist* 77: 312-328.

Pruett-Jones, S.
1992 Independent Versus Nonindependent Mate Choice: Do Females Copy Each Other. *American Naturalist* 140: 1000-1009.

Shipley, W. F.
1978 Native Languages of California. In *California, edited by R. F. Heizer*, pp. 80-90. *Handbook of North American Indians, vol 8* (W.C. Sturtevant, ed.). Washington: Smithsonian Institution Press.

Simpson, G.G., A. Roe, & R.C. Lewontin
1960 *Quantitative Zoology*. New York: Harcourt and Brace.

Sugden, R.
1986. *The Economics of Rights, Co-operation and Welfare.* Oxford: Blackwell.

True, D. L.
1970 Investigation of a Late Prehistoric Complex in Cuyamaca Rancho State Park, San Diego County, California. *Archaeological Survey Monograph*. Department of Anthropology, University of California, Los Angeles.

Thomas, D. H.
1981 How To Classify the Projectile Points from Monitor Valley, Nevada. *Journal of California and Great Basin Anthropology* 3: 7-43.
1986 Contemporary Hunter-gatherer Archaeology in America. *American Archaeology: Past and Present* (D. J. Meltzer & D. D. Fowler, eds.), pp. 237-276. Washington: Smithsonian Institution Press.

White, L.
1959 *The Evolution of Culture*. New York: McGraw-Hill.

Weissner, P.
1983 Style and Social Information in Kalahari San Projectile Points. *American Antiquity* 48: 235-276.

Wilde, J. D.
1985 *Prehistoric Settlements in the Northern Great Basin: Excavations and Collections Analysis in the Steens Mountain Area, Southeastern Oregon.* Unpublished Ph.D. dissertation, Department of Anthropology, University of Oregon.

11

Transmission Theory in the Study of Stone Tools: A Midwestern North American Example

Michael J. Shott
University of Northern Iowa

ABSTRACT

Selectionism has to do with the differential transmission of traits through generations, however conceived. Popular in archaeology in recent years, selectionism remains better known for programmatic statements than for empirical applications. Boyd and Richerson's transmission models only weakly account for variation and, presumably, selection through time in metric attributes of points in Late Woodland and Emergent Mississippian contexts at the Range site in midwestern North America. Social transmission should increase with population size and social complexity. Range's occupation spans several cycles in these properties without corresponding variation in metric values and their dispersion predicted by theory. Selectionism and other interpretive traditions all require closer attention to how the record formed before we can distinguish between their often competing predictions.

INTRODUCTION

In a past variously considered golden or dismal, archaeologists gladly wore the processual label. Today, hardly anyone self-identifies as a processual archaeologist. Indeed, the field is crowded with what some are pleased to call archaeolog*ies*, identified chiefly by the many ways that they are not processual. To qualify as an archaeological ontology, it has become necessary to define oneself in opposition to processualism. If it ever was a unified school of thought, processualism now serves mostly as a syllabus of errors or sins, to be chronicled in opposition to the favored alternative. Its fate has been to pass by degrees from promise to whipping boy.

Today we have an ism for every liking, which is both blessing and curse. It is a blessing because archaeology legitimately is too big to be contained in any one ontology. It is a curse because the competing alternatives are embraced enthusiastically but evaluated poorly or not at all against the archaeological record.

Partly as a result of this enthusiasm, archaeology suffers from a serious bandwagon affliction–gender, post-processualism, behavioralism, processualism itself–whose objects have formed a parade stretching back at least 30 years. Too eagerly we jump on bandwagons to ride them for their runs. Selectionism derives from a biological theory over a century old, itself a bandwagon of sorts because its enthusiastic reception and subsequent triumph, at least in natural sciences, depended on the values of bourgeois Victorian England. Despite its pedigree, selectionism may be the latest archaeological bandwagon to judge from its rapid rise in popularity.

Bandwagons, alas, tend to run out of fuel and to end up abandoned by the side of the road. The path that archaeology has traversed in recent decades is marked by such abandoned wrecks. Like auto wrecks, some bandwagons would remain roadworthy if respected and properly maintained. There is a big difference between joyriding and responsible ownership, to invoke a bourgeois sensibility.

SELECTIONISM

Scholars long have recognized apparent parallels between change over time in culture and biology. Basalla (1993: 43), for instance, notes in both the absence of ultimate goals and the impetus to change from "pressures of existing circumstances". But Basalla, a historian, accepts the analogy only to a point. He cites Kroeber (1993: 44), one of anthropology's near-mythical ancestors, to the effect that "In the made world. . .there are no boundaries [as between species], no inherent prohibitions against the mating of different species of artifacts." Basalla argues, that is, for important differences in the mechanisms of cultural and biological change, differences that might be explored using selectionist theory.

As used here, selectionism denotes the various recent applications of evolutionary theory to archaeological problems. None of this is to be confused with a crude biological determinism, and there is no reason to equate selection in material-culture traits with individual inclusive fitness or other sociobiological concepts (Boyd & Richerson 1985: 12-15; cf. Abbott et al. 1996). Anthropology treats of people in culture and thus of contexts governed as much by meaning as by practical reason.

To follow Abbott et al. (1996: 34-35) advisedly, selectionism rests on the premises that (1) behavior and therefore technology are parts of the human phenotype, (2) this phenotype changes through the agency of selection, (3) technology consists of discrete, competing variants or alternatives whose fates are analogous to reproductive success, (4) variants may wax and wane at random or change through time in a consistent, directed pattern. The former pattern identifies stylistic variants, the latter either functional variants directly subject to selection or those 'sorted' by their necessary link to selected variants. Direction is identified with selection; (5) evolutionary explanation consists of identifying the selective cause of directed change.

Most selectionists probably locate selection and evolution at the level of individuals, both people and traits. Abbott et al.'s first premise seems at least to imply as much. Yet some of the most interesting recent work in paleobiology concerns how and why selective change occurs at several levels (e.g., Depew & Weber 1985) and not just at what some might call the level of the fetishized individual. There seems no reason then to confine selection to individuals, whether people or things. Technology and other "phenotypic" traits may consist of discrete variants such as Abbott et al.'s (1996: 33) 800-odd tractor makers, and selectionism seems well suited to analysis of some discrete traits (e.g., Neiman 1995). But technological traits often vary continuously, and selectionism seems less well suited to their study.

Abbott et al. (1996: 35) distinguish selectionism from processualism, in part by perhaps eliding the important differences between the latter and adaptationism. (Certainly crude adaptationism [Gould & Lewontin 1979] is an easier mark than is processualism). To them, processualism assumes that adaptation drives change and that technology, by changing, is 'reflexive' of adaptation. In contrast, selectionism assumes that selection or drift drive change and that the differing fates of technological variants are determined by the differing values they return to users. This seems a rather labored distinction that may amount to little difference. Nevertheless, Abbott et al. (1996: 35) rightly celebrate selectionism for its equal regard to neutrality, sorting, and adaptation-via-selection as sources of change. Of course, these interests are not entirely original; the archprocessualist himself has written on drift (Binford 1963).

Like other approaches, selectionism serves admirably as polemic, and professes to be empirically grounded. It is ironic, therefore, that the recent spate of selectionist literature is far more programmatic than applied. To put it bluntly, selectionists have talked much but done little. This may be a good time for an agnostic approach. Selectionism needs empirical applications, not more programmatic statements, and it might use them even from those uncommitted in advance to selectionism's tenets.

Transmission Models

Despite its central tenets, there are nearly as many selectionisms as there are selectionists to advocate them. One important part of broader selectionist theory is the body of transmission theory devised by Boyd and Richerson (1985). Selectionists do not agree that transmission theory is important; Dunnell (1992: 214), for instance, dismissed it as "a theory in search of an application". Yet others have adopted it enthusiastically in the study of persistence and change in material culture (e.g., Bettinger 1991, Eerkens & Bettinger 1994).

Boyd and Richerson (1985) devised three chief transmission models to account for the selection of traits over time. 'Transmission' is defined as the persistence of existing variants, their gradual modification, and/or the production of new variants, all subject to selection.

1. Guided variation (Boyd & Richerson 1985: 95). A user approximates the average value of existing variants and modifies them by trial and error. Moore (1978) proposed a similar model, specifically for point production.
2. Frequency-dependent adoption (Boyd & Richerson 1985: 208-209). A user simply adopts whatever variant is most common. To invoke a simplistic metaphor, this is the same process by which burger companies entice prospective customers by telling them that many millions already have been served, so the product must be good.
3. Indirect bias (Boyd & Richerson 1985: 249-252). Users rank sets of variants by socially determined preference, and adopt all variants of different traits that are part of the set, whatever their actual value. This is tantamount to convincing people that Michael Jordan knows men's briefs because, after all, he knows basketball.

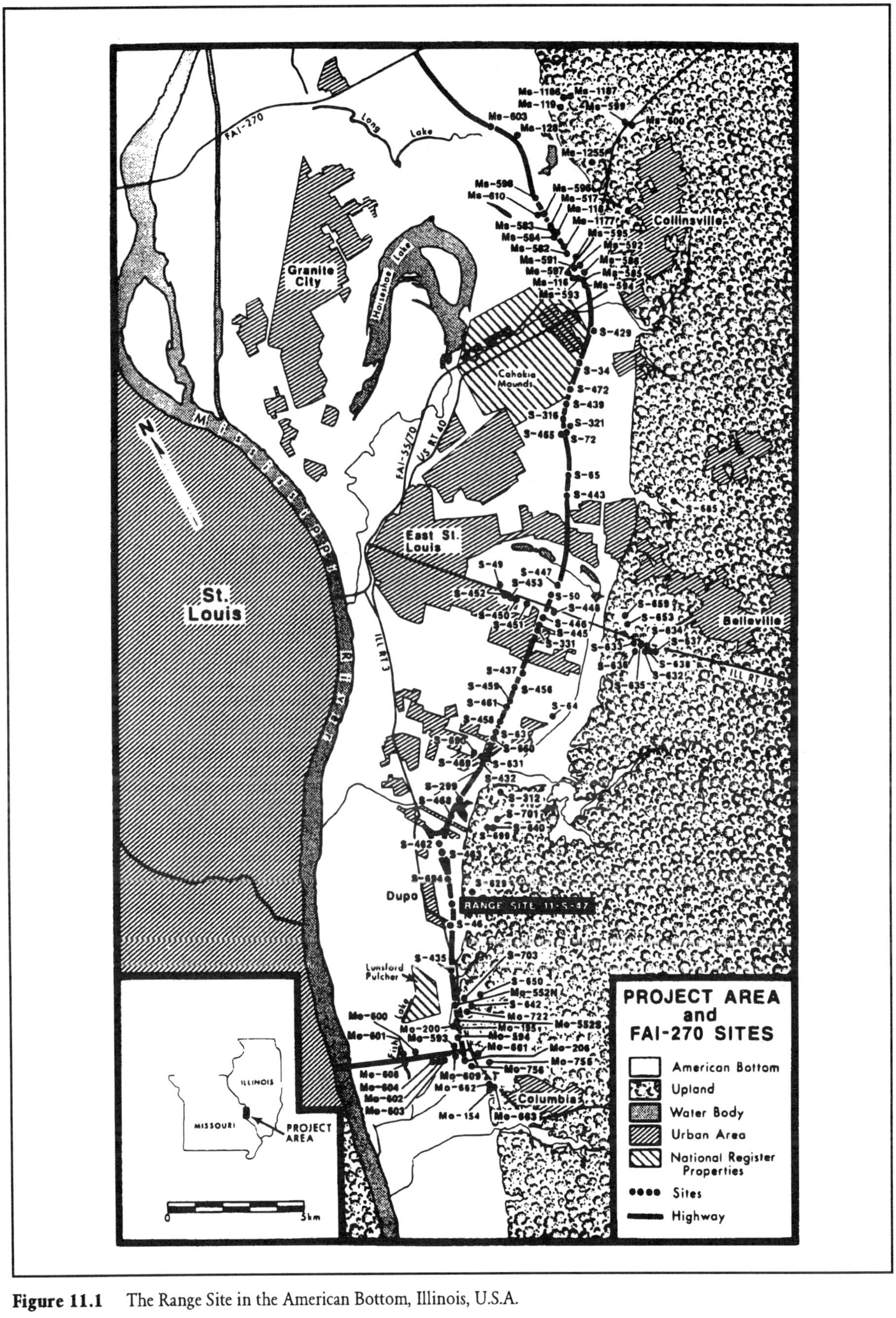

Figure 11.1 The Range Site in the American Bottom, Illinois, U.S.A.

The latter two models can be grouped as processes governed substantially by social factors, and hence can be called 'social transmission.' To Eerkens and Bettinger (1994), guided variation should obtain best where society is small and both society and technology are simple. Here, social considerations and constraints are weak, and guided variation's trial-and-error basis is sensible because different variants of simple technologies can easily be ranked. In contrast, the risk of 'experimental error' (Eerkens & Bettinger 1994: 4) rises with technological complexity.

Study Area

Transmission theory can be tested against first-millennium A.D. archaeological data from the North American Midwest. The American Bottom near St. Louis, Missouri, U.S.A. marks the upper extent of the lower Mississippi Valley (Fig. 11.1). Locus of the Cahokia polity, it was densely settled during at least the last millennium of prehistory. The American Bottom stretches about 150 km along the Mississippi River, is as wide as 16 km in its northern half, and narrows markedly in its southern half (Bareis & Porter 1984: 3). In the 1970s, the University of Illinois and other institutions began an enormous program of salvage excavation ahead of highway construction (Bareis & Porter 1984). Among this noteworthy project's results is a local cultural sequence of remarkable detail and resolution over the past 2,000 years.

Located in the southern half of the American Bottom, the Range 'site' is an extremely complicated deposit of occupations that spanned Middle Woodland to late prehistoric times (Fig. 11.1). Most occupation dates from the Late Woodland and Emergent Mississippian periods, including the successive Patrick (A.D. 600-700), Dohack (A.D. 750-850), and Range (A.D. 850-900) phases. Kelly et al. (1987, 1990) report on these phases; additional data were obtained in consultation with project archaeologists. Comparable data on the Emergent Mississippian George Reeves Phase were not available for study.

During the study interval Range witnessed a succession of occupations by horticulturalists. Using seriation and contextual evidence, Kelly (1990; see also Kelly et al. 1987: 417-435, Kelly et al. 1990: 281-302, 531-549) inferred a sequence of occupations, each possessing substantial spatial integrity. Judging from each occupation's size, assemblage, and settlement plan, Kelly classified them variously as homesteads, hamlets, or villages, in ascending order by size. Table 11.1 lists in order those for which sufficient relevant data exist. For the Patrick Phase, Kelly (1990: 79-86) adduced a cycle of progressive fusioning of homesteads and then hamlets into a succession of villages. This culminated in a fissioning of the P5 or P6 settlement into several smaller units, one of which was excavated and denoted as P7. The cycle repeated in the Dohack Phase with a series of villages "strongly reminiscent" (Kelly 1990: 92) of the Patrick Phase P5 and P6 villages. The Range Phase marked the onset of another fission cycle, with the hamlet R4 formed from the remnants of the later Dohack Phase villages and the start of another cycle represented by the subsequent R2 village. Thus, Range witnessed a series of occupations over about 300 years that varied in duration, population size, scale and perhaps strength of social integration, and that spanned several cycles of fusion and fission.

Table 11.1 Sequence and Type of Occupations at Range. Source: Kelly (1990).

Occupation	Type
P3	homestead
P4	hamlet
P5 or P6	villages
P7	village, fissioned
D2	village
D4	village
D1	village
R4	hamlet
R2	village

Obviously these occupations are inferred, not revealed; any study based upon them is derivative in the sense that it assumes their validity. Doubts can be entertained about any analysis, but the Range sequence seems well grounded. Anyway, it was defined independent of data used and the questions raised in this study. Equally as obvious, patterns of variation between occupations are complicated by variation in size and duration of the occupations. On present evidence, this variation cannot be controlled and is ignored. Finally, analysis assumes that the occupations are ordered correctly, but need not assume that they form an unbroken continuum; indeed, there is an interregnum between the Patrick and Dohack phases as understood at present that might be filled by occupations found at the nearby Sponemann settlement (Kelly 1990: 87-88).

The Category 'Point'

The various Range site occupations yielded many 'points.' 'Point' is not a revealed kind but a constructed or inferred category. There is little doubt that what archaeologists routinely identify as hafted bifaces were indeed hafted. But flakes as well as bifaces may have been hafted and used as points (Odell 1988) and many hafted bifaces may have been used as knives or other tools instead of, or in addition to, as points. Obviously, many if not most points were fashioned from perishables and the time and effort needed to canvas assemblages for what may have been flake tools is prohibitive. Thus, focus is on hafted bifaces identified as points. To follow standard usage, these henceforth are called 'points' without qualification, which nevertheless always should be remembered. Among others, this treatment excludes the broad stemmed bifaces called 'Sugar Loaf knives' (Kelly et al. 1990: Plate 4.3).

Mixing

The American Bottom boasts perhaps the most finely resolved local cultural sequence anywhere in the world for the A.D. 600-1400 interval. The sequence is defined principally by ceramic seriation but also radiocarbon dating and contextual evidence. Biface typology has contributed less to development of the sequence, although the constructed type sequence is corroborated generally by the sequence of forms found in deeply buried stratified sites in the region (Jolly & Roberts 1974). Fair confidence attaches to the constructed sequence of point types.

Most if not all specimens were found in features that could be assigned to well defined phases in the American Bottom chronology. However, early types regularly occur in later American Bottom occupations due to heirloom effects, recycling, and incidental secondary deposition attending the excavation and filling of features (Mehrer 1995: 32-33). At major American Bottom sites like Range, features cannot be treated as random traps or sinks nor their fills as representative of settlement-wide patterns (Mehrer 1995: 66-67). It is questionable to assume that all points found in a feature were used and discarded at about the same time. Yet the only available choices are to treat all points as coeval or to selectively ignore specimens that belong to earlier types, leaving aside the sometimes problematic character of type assignment. The former practice risks inviting confusion by failure to separate specimens made, used and therefore subject to cultural selection at different times in the past. The latter invites us to assume whatever trend or pattern that we should wish to document empirically. It is entirely possible that some types were used for very long periods, longer than those we ordinarily suppose, and that 'early' and 'late' types may have enjoyed concurrent use. Certainly the evidence from stratified sites suggests that many types were used over very long periods (Jolly & Roberts 1974: Fig. 41), although again this may owe in part to stratigraphic mixing. At one extreme, we could ignore many specimens in use during a period of interest and thereby hopelessly bias the data under study; at the other, we could be forced to conclude from the occasional discovery that Woodland cultures used fluted bifaces. Nevertheless, points and other stone tools are not the only artifacts subject to mixing; indeed, mixing of ceramics was common in Range site features (Mehrer 1995: 73), an especially troubling observation considering the role that ceramics play in grounding the seriation upon which the cultural sequence is constructed and in assigning features to occupations.

Ultimately, the only way to resolve this problem is to excavate many more settlements that have large and representative tool assemblages but that were occupied during one phase only. No mixing can occur then because there was nothing left behind earlier to mix. This requirement might have the salutary effect of drawing some interest to areas outside the American Bottom, where neither people nor their goods were so concentrated in time and space.

EXPECTATIONS AND CRITIQUE

Eerkens and Bettinger (1994) have proposed ways to test transmission theory using biface metric data, and the series of Range occupations differing in particulars of the inferred social context provide a suitable test. By their reasoning, the guided-variation model should increase within-population variation in points. However, if selective factors like performance requirements are rigorous, then variation might be minimized as selection favors a narrow range. Moreover, if those performance requirements themselves change over time (as they may have in the American Bottom during the study interval [Shott 1996]), then the rigorously constrained variants will change correspondingly in value. In archaeological deposits that accumulate over considerable periods, this will produce wide variation even though selection at any one point favors a narrow range of values. Such apparent smearing of selection effects can make it difficult to distinguish the effects in material culture of different transmission models.

Eerkens and Bettinger (1994) link frequency-dependent and indirectly biased transmission as forms of socially determined transmission. They should obtain as population rises and society grows complex, because individual action is constrained in these circumstances by limits on time to experiment and by pressure to conform. Under these circumstances, social transmission

Table 11.2 Dispersion Measures by Settlement Type.

Type	Maximum Length s.d.	Maximum Length c.v.	Shoulder Width s.d.	Shoulder Width c.v.	Shoulder Width s.d.[1]	Neck Width s.d.	Neck Width c.v.	Thickness s.d.	Thickness c.v.	Thickness s.d.[1]
homestead	1.05	33.62	0.31	20.13	2.25	0.38	40.43	0.22	50.00	7.89
hamlets	1.06	30.64	0.46	30.46	4.40	0.77	60.16	0.28	52.83	5.80
villages	0.78	25.32	0.42	28.19	4.44	0.66	54.54	0.22	45.83	6.13

1. power-transformed variable.

should gradually replace guided variation. To Eerkens and Bettinger (1994), social transmission reduces within-population variation. Why time is more limited in larger, more complex societies and why conformity is enjoined more there both are unexplained. The former seems arguable but plausible, but social pressure to conformity is no weaker in "simpler" societies nor opportunities to resist it weaker in 'complex' ones.

Like guided variation, social transmission makes little allowance for the effects of rigorous performance requirements (i.e., external selective factors), although it models very clearly some kinds of social barriers or resistance to selection. Thus, it offers a way to model our understanding of such barriers or resistance. Weak or no performance requirements should allow social transmission to reduce variation, as Eerkens and Bettinger propose, although it may produce several modes corresponding to different groups or classes, not a single mode.

Besides selection, raw material supply and flaking properties can bear on metric attributes of stone tools, although archaeologists are perhaps overinclined to ascribe variation to this source. Chert types used by American Bottom knappers were many and varied, and their treatment is well standardized by now (e.g., Jackson et al. 1992: Appendix A). There were no significant differences in metric attributes by chert type in an earlier study that included data used here (Shott 1996), and none are assumed here.

Dispersion Measures

Properly, selectionism is as interested in variation as in modes, as interested in dispersion as in central tendency in continuous variables. In metric data, therefore, standard deviation and other measures of spread are as informative as is the mean. Coefficient of variation is a scale-free measure of dispersion relative to mean. Spread-versus-level plots (Norusis 1993: 188-189) also test and control for scale-dependence in dispersion measures, and suggest power-transformations sufficient to remove it. In the Range data, shoulder width and thickness exhibit such scale-dependence, and were power-transformed accordingly. Finally, box-and-whisker plots display dispersion in summary fashion. By convention, box limits mark the 25th and 75th percentiles or 'hinges' of a distribution, whiskers the distribution of cases that are not outliers, open circles cases that lie between 1.5 and 3 box-lengths from hinges, and asterisks cases that lie beyond those points (Norusis 1993: 185-186). The horizontal line inside the box marks the median.

Eerkens and Bettinger usefully suggest that dispersion also can be measured as the residual to the regression of standard deviation upon mean. In their Great Basin point data, standard deviation and mean correlate strongly ($r^2 = .81$) (Eerkens & Bettinger 1994: 8). At Range, the comparable result is lower but still significant ($r^2 = .62$). Standardized residuals are not used here, but pattern much as do reported measures of dispersion.

Dispersion by Settlement Type. Table 11.2 compiles dispersion figures by settlement type for Range site Patrick, Dohack and Range Phase occupations that have sufficient point samples. The small P7 village is separated from other villages because Kelly (1990: 86) considers it the product of fissioning of a larger, earlier settlement, presumably P5 or P6. Therefore, P7 may lack the size and apparently differs in social context from other Range village occupations in ways that may bear on the selection and transmission of metric values. For what it is worth, adding P7 to other villages does not significantly change the result. There are many more villages than either other type and correspondingly many more points from villages, a difference in sample size that is ignored. Occupations are classified in part by size, but an occupation's settlement type does not covary simply with the size of its point assemblage.

By Eerkens and Bettinger's reasoning, the progression from homestead to hamlets to villages involves greater reliance on social transmission and therefore decreasing metric variation. Maximum length substantially fits this pattern but this result is problematic because length is influenced as much by each specimen's history

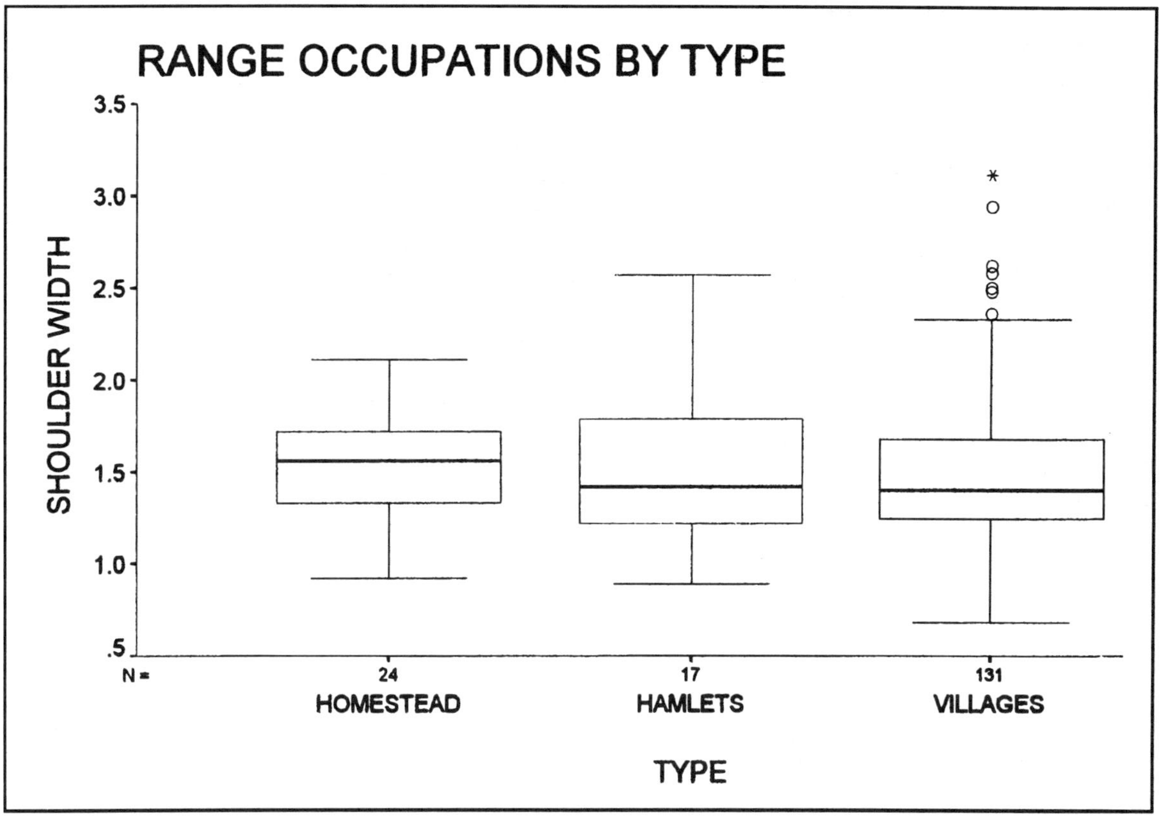

Figure 11.2 Dispersion in shoulder width by settlement type.

of use as it is by that specimen's original design. Other variables do not fit the pattern, either failing to trend consistently or trending in the direction opposite of expectations (e.g., power-transformed shoulder width). There is a slight decline in the median of variables like shoulder width that may register regional metric trends (Shott 1996), but box-and-whisker plots show increasing, not decreasing, dispersion from homestead to villages (e.g., Fig. 11.2). Grouped by settlement type, population dispersion does not pattern consistently with Eerkens and Bettinger's expectations.

Dispersion by Order in Time. According to Kelly (1990) the Patrick Phase sequence of P3 to P6 is a time series of occupations that describes a cycle of increasing population and social integration. The positions in this series of P5 and P6 are uncertain so they are listed simply in number order. By Eerkens and Bettinger's (1994) reasoning, a cycle of this nature should witness a progressive shift from guided variation to social transmission. Kelly (1990: 86) suggests that P7, the latest Patrick Phase occupation, reversed the trend or completed the cycle by being a product of fission of the P5 and/or P6 community. Thus, it may have witnessed reversion to greater reliance upon guided variation. The cycle apparently repeated from occupations D2 (village) to R4 (hamlet).

Table 11.3 shows no comparable cyclical trend in metric dispersion according to this sequence of occupations. A table of this nature defies easy reading; to aid comprehension, Figure 11.3 graphs dispersion in shoulder width. Villages P5 and P6 have more, not less, dispersed distributions than do their smaller P3 and P4 predecessors. Dispersion is low for most variables in the P7 occupation even though it represents a context in which dispersion-promoting guided variation presumably grew more common. In most variables, there is some tendency for dispersion to decline as expected from D2 through D4 to D1. Yet fissioned village P7 and hamlet R4 bracket these occupations and they should register greater guided variation and therefore more dispersion. They do not.

Time Dependence and Dispersion. Significant time-dependent metric trends were found in a different American Bottom data set in an earlier study (Shott 1996). Several Range occupations are dated. Within each, dates were pooled (they were concordant in all cases), averaged and calibrated using CALIB 3.0.3 (Stuiver & Reimer 1993). The means of haft variables but not maximum length in the dated occupations decline weakly with time (e.g., for shoulder width r = .74, p = .10; for thickness r = .69, p = .13), indicating that time-dependence characterizes Range occupations.

Time trends partly register the introduction of bow and arrow technology and the progressive decline of darts, but also may register constantly changing performance requirements of arrows. Certainly, time-dependence persists even if analysis is confined to arrows specimens (Shott 1996: 294-295), so it cannot be explained entirely by the technological change.

In this study, time-dependence must be considered both because arrow specimens might usefully be distinguished from darts and because declining metric attributes imply a declining range of potential variation. In effect, the range of metric variation diminishes regularly with time, which means that dispersion of metric values is increasingly constrained with time. This scale-dependence must be controlled to study patterning in dispersion through time.

Using archaeological and ethnographic unknowns, discriminant analysis successfully distinguishes dart from arrow points for over 90% of arrows and about 85% of darts (Shott 1997). To accommodate broken specimens for which length and other measurements are impossible, classification functions can be defined on one to four metric variables (Shott 1997: 93-95). In general, shoulder width and thickness are the most useful discriminators (Shott 1997: 94).

Classification invokes an either/or decision, forcing a specimen into one group or the other. One could argue that, applied to archaeological unknowns, classification always will finely split a data set, making the dart-arrow transition appear gradual no matter its actual form and rate. In fact, it is the continuous metric trend so evident in American Bottom data that determines the form of classification results. If the dart-arrow transition occurred abruptly and involved a discontinuous shift in point metric values, then classification would faithfully record the transition in these terms.

Moreover, some might object that classification functions are defined on specimens that clearly are either dart or arrow, and may be poorly suited to classify 'transitional' specimens. Indeed, dart and arrow technology are not qualitatively different, and some specimens may seem transitional (e.g., Shott 1997: 93). These may be misclassified but there are bound to be errors in any classification function, especially one involving darts and arrows with their partly overlapping ranges of metric values (Shott 1997: Fig. 3). Fortunately, the recently defined function performs very well on arrows, slightly less well on darts. Thus, most specimens identified as arrows almost certainly were arrows.

Finally, it is possible that the pattern of gradual replacement of dart by arrow technology (Shott 1996), characterized by diminishing numbers of darts specimens through the Late Woodland, is the false product of data contaminated by earlier-period specimens incorporated by mixing or recycling into later contexts. To sustain this argument, however, requires explaining how such mixing should have affected Late Woodland contexts like Mund, Rosewood and Patrick Phase occupations but not

Table 11.3 Dispersion Measures by Time-Ordered Occupations.

Occupation	Maximum Length		Shoulder Width			Neck Width		Thickness		
	s.d.	c.v.	s.d.	c.v.	s.d.[1]	s.d.	c.v.	s.d.	c.v.	s.d.[1]
P3	1.05	33.81	0.31	19.83	0.29	0.38	40.77	0.22	50.03	7.89
P4	1.11	27.56	0.45	25.28	0.14	0.64	33.96	0.32	49.16	3.11
P5	0.71	23.19	0.48	31.67	0.47	0.69	56.62	0.24	48.74	4.19
P6	1.24	35.59	0.57	34.62	0.80	0.71	45.65	0.34	58.45	12.57
P7	0.35	11.86	0.13	11.68	0.15	0.15	21.43	0.15	32.96	3.61
D2	0.57	19.71	0.27	19.12	0.42	0.23	29.60	0.10	25.06	3.31
D4	0.32	11.67	0.21	15.66	0.19	0.15	20.24	0.14	30.88	3.28
D1	0.71	23.30	0.23	16.96	0.24	0.47	47.55	0.08	19.93	2.91
R4	0.83	27.65	0.34	26.67	0.43	0.99	14.41	0.14	30.42	7.13
R2	0.69	23.18	0.32	20.34	0.14	.[2]	.	0.12	26.77	3.23

1. power-transformed variable
2. n = 1

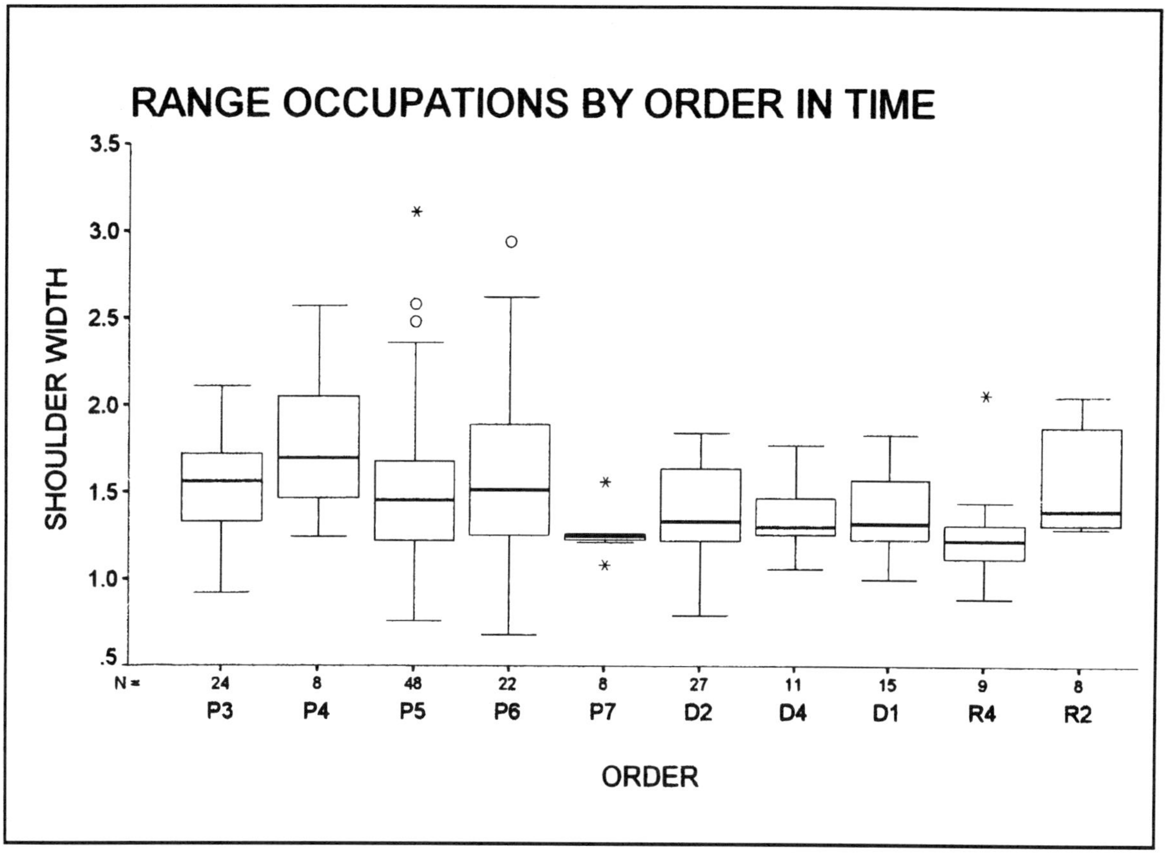

Figure 11.3 Dispersion in shoulder width by occupation ordered in time.

have affected Emergent Mississippian and Mississippian Phase contexts. In this case, it requires explaining why such mixing occurred during the Late Woodland Patrick Phase and somehow stopped once the Emergent Mississippian Dohack and Range Phases began.

Table 11.4 shows dispersion measures for shoulder width and thickness by occupation, but confined to specimens identified as arrows. As above, shoulder width and thickness are the most useful variables in distinguishing darts and arrows. Again, patterning in shoulder width is equivocal and more comprehensible in thickness, although standard deviation of the power-transformed variable for villages very slightly exceeds the figure for hamlets. Power-transformed thickness dispersion mirrors the fusion-fission cycle especially from occupations D2 to R2, where large villages have low values, the R4 hamlet a considerably larger value, and R2 a low one again. On balance, there is mixed support for transmission theory in these data.

Summary

This exercise does not prove that Eerkens and Bettinger (1994) are wrong, merely that their predictions do not always bear out. Indeed, the data pattern as they would predict in some respects, although they do not consistently. More importantly, Eerkens and Bettinger have taken their expectations to the archaeological record, for which they deserve credit. Their own data, from very different time-space contexts, do in fact pattern as their reading of selectionist theory would predict.

THE NEED FOR FORMATION THEORY

Formation theory models archaeological properties of cultural organization and practice. Because these properties are contextual, complex to put it mildly, and perhaps simply arbitrary, Dunnell (1992: 80-81) argued that they can be neither described nor explained scientifically. To him, science explains only directly observed phenomena, "not some secondary construct" like behavior, which must be inferred. Dunnell thus would reject out of hand the validity of formation theory. But no observation is direct; all require and produce secondary constructs or inferences. Conclusions, like predictions, are theory-bound. To explain the past Dunnell's selectionism invokes behaviors and their consequences that are

Table 11.4 Dispersion in Range Site Occupations, Arrows Only.

		Shoulder Width		Thickness	
Occupation	Type	original s.d.	transformed s.d.	original s.d.	transformed s.d.
P3	homestead	.255	.300	.172	8.53
P4	hamlet	.220	.129	.162	2.74
P5	village	.248	.485	.127	3.82
P6	village	.312	.881	.234	13.78
P7	village	.134	.150	.153	3.44
D2	village	.261	.426	.101	3.31
D4	village	.172	.182	.143	3.28
D1	village	.183	.228	.079	2.91
R4	hamlet	.149	.400	.118	7.21
R2	village	.205	.112	.083	3.09
mean	homestead	.255	.300	.172	8.53
mean	hamlet	.185	.264	.140	4.97
mean	village[1.]	.231	.386	.128	5.03

1. P7 excluded

just as secondary and no more directly observable than the objects of other archaeologists' inferences. Indeed, Dunnell's disciples have argued the need to infer behavior to test their evolutionary theory (e.g., O'Brien et al. 1994: 260).

Therefore selectionism, like other theoretical approaches, would benefit from the development of suitable formation theory (Shott n.d.). For instance, we assume blithely that pit fill, including points, is undifferentiated material structured only by quotidian concerns and processes. But there certainly are social dimensions to the use of pits (DeBoer 1988) and there may be similar dimensions to artifact deposition in them (Hill 1995). Even if ordinary activities determined the contents of pit fill, the problem of mixing of earlier and later items must be controlled. Mixing can occur in many kinds of archaeological deposits, not just American Bottom settlements. Natural scientists have dealt with similar problems blissfully free of the problematics introduced by symbolic meaning (Berger & Heath 1967); until we develop comparable theory, which is a kind of formation theory, any use of pit-fill data is subject to doubt. Also, Boyd and Richerson's transmission models are usefully operationalized by Eerkens and Bettinger, but nevertheless remain unscaled. Eerkens and Bettinger (1994: Tables 1-3) report results that pattern as expected but other data could produce similar patterning at different scales. Until we develop the necessary formation theory we cannot explain possible scale differences, compare data sets that differ considerably in scale, nor distinguish with confidence between significant and incidental variation.

But formation theory is not confined to deposition in features or metric scale effects. Cultural transmission models are elegant and precise but the material consequences of different models can be subtle enough to defy distinction in the archaeological record, raising the unpleasant specter of equifinality. Until we devise suitable formation theory to distinguish between finely distinguished alternatives, selectionist and otherwise, transmission theory will lack the empirical grounding that alone can transform an appealing prospect into an operational body of theory. Then we are free to attribute patterning found to any explanation we please. If selectionism accomplishes nothing more than to encourage the development of such theory, it will be worth the effort. Until then we argue by accommodation, which is an extended exercise in cleverness that accomplishes little and that comes at the price of our scientific integrity because it encourages the easy reception of bandwagons without rewarding their tedious but critical empirical grounding. Formation theory is independent of general theory (Shott n.d.), and selectionists as well as archaeologists of other theoretical stripes may use it, once developed, in pursuit of their goals.

CONCLUSION

'Processualism,' 'selectionism' and the like are labels that reflect some legitimate distinctions but too often erect contrived boundaries within archaeology. Their use implies a totalizing commitment to particular ontological views, which is needlessly limiting. All serious archaeologists should be concerned with the issues that selectionism raises, making all of us selectionists to some degree. Better still to abandon obsession with labels and to devote ourselves instead to the serious study of the past, using selectionist and other theories.

There is plenty to criticize in all approaches. If the purest processualism is rigid and ahistoric, some postprocessualist views are equally doctrinaire. The important distinction is not between selectionists and others, be they postprocessualists, behavioralists or even processualists, no matter how unfashionable the latter. It is between dogmatism of whatever persuasion on the one hand, and constructive engagement with the archaeological record on the other. Such engagement requires commitment to the notion that there was a single past, not many; that it is imperfectly but substantially knowable despite its contextual meaning and complexity; that we do not reveal the past nor fabricate it, but construct it; that, therefore, knowledge is theory-bound and not free invention and is subject to, yes, scientific reasoning; and that we are not objective but should aspire to this impossible goal (Clark 1993).

Such an archaeology can accommodate many constructive isms. We should resist labelling and respect any approach like selectionism that might help understand the past. None is a panacea and none is totalizing. If postprocessualism is no more than the "fuller exploration of process through . . . historical context" (Preucel 1995: 148), then we should be postprocessualists although this definition can make us behavioralists or selectionists just as easily. If it is a commitment to constructive engagement as outlined above, a much more accurate characterization of processualism than its equation with adaptation, then we should be processualists and glad of it.

ACKNOWLEDGMENTS

John Kelly kindly provided the data used here, based on investigations sponsored by the Illinois Transportation Archaeology Program at the University of Illinois (Thomas Emerson, Director). Geoffrey Clark and C. Michael Barton invited me to contribute to the volume and ably guided the chapter through the editorial process. All deserve my thanks, but I alone am responsible for the results.

REFERENCES CITED

Abbott, Alysia L., Robert D. Leonard & George T. Jones
1996 Explaining the Change from Biface to Flake Technology: A Selectionist Application. In *Darwinian Archaeologies* (H. Maschner, ed.), pp. 33-42. New York: Plenum.

Bareis, Charles J. & James W. Porter
1984 *American Bottom Archaeology: A Summary of the FAI-270 Project Contribution to the Culture History of the Mississippi River Valley.* Urbana: University of Illinois Press.

Basalla, George
1993 On the make: Desire is the mother of invention, and failure its guide. *The Sciences* 33: 40-45.

Berger, Wolfgang H. & G. Ross Heath
1968 Vertical Mixing in Pelagic Sediments. *Journal of Marine Research* 26: 134-143.

Bettinger, Robert L.
1991 *Hunter-Gatherers: Archaeological and Evolutionary Theory.* New York: Plenum.

Binford, Lewis R.
1963 'Red Ochre' Caches from the Michigan Area: A Possible Case of Cultural Drift. *Southwestern Journal of Anthropology* 19: 89-108.

Boyd, Robert & Peter J. Richerson
1985 *Culture and the Evolutionary Process.* Chicago: University of Chicago Press.

Clark, G. A.
1993 Paradigms in science and archaeology. *Journal of Archaeological Research* 1: 203-234.

DeBoer, Warren R.
1988 Subterranean Storage and the Organization of Surplus: The View from Eastern North America. *Southeastern Archaeology* 7: 1-20.

Depew, David J. & Bruce H. Weber
1985 *Evolution at a Crossroads: The New Biology and the New Philosophy of Science.* Cambridge: MIT Press.

Dunnell, Robert C.
1992 Archaeology and Evolutionary Science. In *Quandaries and Quests: Visions of Archaeology's Future* (L. Wandsnider, ed.), pp. 209-224. Carbondale: Southern Illinois University Press.

Eerkens, Jelmer & Robert L. Bettinger
1994 Variability in Great Basin Projectile Points. Paper presented at the 59th Annual Meeting of the Society for American Archaeology, Anaheim.

Gould, Stephen J. & Richard Lewontin
1979 The Spandrels of San Marco and the Panglossian Paradigm: A Critique of the Adaptationist Programme. *Proceedings of the Royal Society of London, Series B* 205: 581-595.

Hill, J.D.
1995 *Ritual and Rubbish in the Iron Age of Wessex: A Study on the Formation of a Specific Archaeological Record.* Oxford: BAR British Series No. 242.

Jackson, Douglas K., Andrew C. Fortier & Joyce A. Williams
1992 *The Sponemann Site 2: The Mississippian and Oneota Occupations.* American Bottom Archaeology, FAI-270 Site Reports Volume 24. Urbana: University of Illinois Press..

Jolly, Fletcher & Ralph G. Roberts
1974 Projectile Point Sequence at the Williams Shelter (23PH34) in the Ozarks of South Central Missouri. *Central States Archaeological Journal* 21: 59-78.

Kelly, John E.
1990 Range Site Community Patterns and the Mississippian Emergence. In *The Mississippian Emergence* (B. Smith, ed.), pp. 67-112. Washington: Smithsonian Institution Press.

Kelly, John E., Andrew C. Fortier, Steven J. Ozuk & Joyce A. Williams
1987 *The Range Site: Archaic through Late Woodland Occupations.* American Bottom Archaeology, FAI-270 Site Reports Volume 16. Urbana: University of Illinois Press.

Kelly, John E., Steven J. Ozuk & Joyce A. Williams
1990 *The Range Site 2: The Emergent Mississippian Dohack and Range Phase Occupations.* American Bottom Archaeology, FAI-270 Site Reports Volume 20. Urbana: University of Illinois Press.

Mehrer, Mark W.
1995 *Cahokia's Countryside: Household Archaeology, Settlement Patterns, and Social Power.* DeKalb: Northern Illinois University Press.

Moore, Dan
1978 Simulation Languages for Archaeologists. In Simulati*on Studies in Archaeology* (I.Hodder, ed.), pp. 11-19. Cambridge: Cambridge University Press.

Neiman, Fraser D.
1995 Stylistic Variation in Evolutionary Perspective: Inferences from Decorative Diversity and Interassemblage Distance in Illinois Woodland Ceramic Assemblages. *American Antiquity* 60: 7-36.

Norusis, Marija J.
1993 *SPSS for Windows Base System User's Guide Release 6.0.* Chicago: SPSS, Inc.

O'Brien, Michael J., Thomas D. Holland, Robert J. Hoard, & Gregory L. Fox
1994 Evolutionary Implications of Design and Performance Characteristics of Prehistoric Pottery. *Journal of Archaeological Method and Theory* 1: 259-304.

Odell, George H.
1988 Addressing prehistoric hunting practices through stone tool analysis. *American Anthropologist* 90: 335-356.

Preucel, Robert W.
1995 The Postprocessual Condition. *Journal of Archaeological Research* 3: 147-175.

Shott, Michael J.
1993 Spears, Darts, and Arrows: Late Woodland Hunting Techniques in the Upper Ohio Valley. *American Antiquity* 58: 425-443.

1996 Innovation in Prehistory: A Case Study from the American Bottom. In *Stone Tools: Theoretical Insights into Human Prehistory* (G.Odell, ed.), pp. 279-309. New York: Plenum,

1997 Stones and Shafts Redux: The Metric Discrimination of Chipped-Stone Dart and Arrow Points. *American Antiquity* 62: 86-101.

n.d. Status and Role of Formation Theory in Archaeological Practice. Unpublished ms. on file, University of Northern Iowa.

Stuiver, Minze & Paula J. Reimer
1993 Extended ^{14}C Database and Revised CALIB Radiocarbon Calibration Program. *Radiocarbon* 35: 215-230.

Part IV

Case Studies & Applications in Paleoanthropology & Archaeology

Introduction

The five essays in this section consist of case studies and applications of Darwinian approaches to a range of topics in paleoanthropology (Clark) and archaeology (Webb and Rindos, Diehl, Neiman, Bird). In what is probably the least empirically grounded of the papers, Geoff Clark (Chapter 12) explores aspects of early hominid sociality using an eclectic approach that has more in common with those of Gowlett (Chapter 3) and Mithen (Chapter 4) than with those of most American workers. Clark is interested in developing strong inferences about Pliocene hominid social organization, a tricky subject even in much later time ranges. After a discussion of conceptual and methodological atomism, and a brief critique of Gould and Lewontin's classic essay (1978) on developmental constraints, he combines data drawn from the ethology of extant hominoids with Foley's 'behavioural cladistic' approach (e.g., 1989) to model early hominid patterns of dispersal at sexual maturity, local group size and composition, sexual selection and mating practices, cognitive evolution and the social context of learning in order to create a scenario for early hominid social organization. While firmly grounded in neo-Darwinian evolutionary theory (NDT), the scenario is acknowledged to be untestable in any direct way. Like Mayr (1982), Clark believes that satisfactory explanations for pattern in the earliest phases of human evolution can only be built from circumstantial evidence - intertwined strands of 'facts' linked together by strong inference and tied to central NDT concepts warranted in biological evolution in general.

In Chapter 13, Esmée Webb and the late David Rindos adopt an evolutionary perspective on the initial human colonization of the Sahul landmass and critique existing models grounded in the assumption that the rate of adaptation to new environmental conditions determines the rate of spread of those adaptations. While they consider the initial human colonization of Sahul to have been relatively rapid, they argue that rapid colonization was not a consequence of rapid and successful adaptation, but rather of the high mobility inherent in maladaptive subsistence strategies that resulted in inefficient extractive methods. Although developed to explain the colonization of Sahul, the model is tested against the chronologically late and archaeologically distinctive evidence for the initial human colonization of the New World.

Michael Diehl (Chapter 14) uses principles of behavioural ecology in combination with macrobotanical data from late Archaic and Hohokam sites in the Tucson Basin to evaluate three competing models for late Archaic (Cienega Phase) subsistence in the area. These foragers are characterized (1) as logistically organized, long-distance hunters who adopted agriculture to support long-distance hunting expeditions extending as far north as the Mogollon Rim (Wills), (2) as sedentary horticulturalists dependent upon agricultural crops and a few species of wild annuals, differing little in their subsistence and landuse practices from later Hohokam farmers (Huckell), and (3) as forager-horticulturists who shifted relatively sedentary residential locations between resource patches according to season of availability (Roth). Niche width statistics calculated on macrobotanical remains from late Archaic and Hohokam sites in the basin, and on energy budgets required to engage in long-distance hunting, indicate little support for the Wills and Huckell models, but considerable support for Roth's characterization of Cienega Phase adaptations as seasonally sedentary forager-horticulturists.

An historical archaeologist 'poaching' (as he puts it) in the Maya lowlands, Fraser Neiman (Chapter 15) develops an ingenious application of Darwinian theory that focuses on the 'wasteful advertising' manifest in Maya monument construction and the competitive ability it signals amongst Maya elites to arrive at a novel interpretation of the processes behind the well-documented but poorly understood Maya 'collapse' (AD 760-900). Arguing that Maya monuments were the result of selection for wasteful advertising in the context of elite competition, and using terminal monument dates from 69 sites in conjunction with geoscience evidence (lake cores) supporting a population-resource imbalance provoked by soil erosion, Neiman is able to show that, as the collapse proceeded, monument-constructing elites entered into political competition with elites at sites on average about 65 km distant. This apparently had the effect of prolonging competition at the latter sites (i.e., they tend to have later terminal monument dates). A surprising conclusion from the research was that Late Classic Maya polities were quite small, on average only about 50 km in diameter. The cessation of monument construction across the Maya lowlands was, as Morley (1920) proposed long ago, a single time- and space-transgressive 'event', and not a palimpsest of two downward trends in an ongoing cycle, as proposed by Marcus (1993).

In the last contribution, Douglas Bird (Chapter 16) offers a tightly-focused application of optimal foraging theory in support of the Darwinian notion that economic decision-making behaviours have been shaped by natural selection and, in the extent to which they contribute to reproductive success, will tend to persist in a population at the expense of alternatives. He illustrates the approach with a fine-grained study of intertidal prey choice, patch switching, processing and transport strategies from the Meriam Islands, in the Torres Strait between Australia and New Guinea. Cast in a simple, yet elegant, deductive framework that allows for precise, quantifiable prediction of the material correlates of various kinds of economic behaviour, the research shows that Meriam shellfish processing decisions are consistent with the hypothesized goal of maximizing resource delivery rates, but that the model does not predict the archaeological consequences of Merian shellfish gathering. These inconsistencies only become apparent when the predictions generated by quantifiable hypotheses are matched against observed behaviour and its material signatures in an ongoing ethnographic context.

REFERENCES

Foley, R.

1989 The evolution of hominid social behaviour. In *Comparative Socioecology* (V. Standen and R. Foley, eds.), pp. 473-494. Oxford: Blackwell Scientific.

Gould, S. and R. Lewontin

1978 The spandrels of San Marco and the Panglossian paradigm: a critique of the adaptationist programme. *Proceedings of the Royal Society* 205: 581-598.

Marcus, J.

1993 Ancient Maya political organization. In *Lowland Maya Civilization in the Eighth Century A.D.* (J. Sabloff and J. Henderson, eds.), pp. 111-171. Washington: Dumbarton Oaks.

Mayr, E.

1982 *The Growth of Biological Thought.* Cambridge, MA: Belknap Press.

Morley, S.

1920 *The Inscriptions of Copan.* Washington: Carnegie Institute of Washington Publication No. 219.

12

Aspects of Early Hominid Sociality: an Evolutionary Perspective

G.A.Clark
Arizona State University

ABSTRACT

Despite nominal acknowledgment, archaeologists have been slow to take advantage of neo-Darwinian evolutionary theory (NDT) in the course of their research. While perhaps understandable for those working in later prehistory, where quasi-historical explanations for process can sometimes be invoked, archaeologists working in 'deep time' have also shown a marked reluctance to use it. This essay asks why this is so, and goes on to apply NDT concepts in an effort to reconstruct aspects of early hominid sociality. Local group size and composition, sex-based dispersal at maturity, sexual selection and mating practices, kinship, enculturation, primate cognitive abilities and the social context of learning are addressed using an eclectic approach in which non-human primate data, sociological data on modern human mating practices and human paleontology play important roles.

We can see the seeds, the origins, of everything we know about our culture in the distant past. This means that every aspect of our culture can benefit from some understanding of the biology from which it sprang (Bonner 1980: 186).

INTRODUCTION

Neo-Darwinian evolutionary theory (NDT) and its modern derivatives currently offer the only generally acknowledged conceptual framework within which to understand human evolution (Foley 1987). The central tenet of NDT is natural selection - Darwin's 'descent with modification' - whereby organisms better adapted to their particular ecological circumstances enjoy greater reproductive success and consequently increase the relative frequency of their genotypes at the expense of those of their less-well-adapted conspecifics. Since the genes are invisible, natural selection operates on the phenotype, and in the extent to which those aspects of the phenotype that confer reproductive success are encoded genetically, it is to that extent that they will tend to spread throughout a population, causing shifts in their relative frequencies and in those of their competitors. The end result of evolution by natural selection is a constellation of functionally interrelated traits (Harvey et al. 1987). Understanding a particular evolutionary trajectory involves (1) identifying a set of phenotypic characters that can plausibly be argued to confer reproductive success, (2) demonstrating that those phenotypic properties actually existed in the past, (3) showing that they were subject to selection pressure, and (4) that the likely outcome of selection would have tended to favor them (Lake 1996). The phenotype is a generalized concept, however, and when phenotypic characters are expressed behaviorally, the relevance of the concept for archaeology and human paleontology becomes clear. If the phenotype has material correlates either in aspects of skeletal morphology or in the tangible properties of an archaeological record, questions of evolutionary process can be addressed and - at least in theory - subjected to empirical scrutiny.

Lest these introductory remarks be regarded as yet another promotion of 'the adaptationist program' (cf. Clark 1989, 1990; Foley 1989, Mithen 1991), it is important to point out that NDT also embraces various kinds of 'non-adaptive evolution' wherein natural selection - while of undeniable centrality - is not invoked as the cause of frequency shifts in the genotype. These are familiar to any biological anthropology student, and are nominally acknowledged but (usually) subsequently ignored. They include genetic drift, allometry, the physics of growth fields, pleiotropy, differential fecundity and phyletic inertia. Darwin, while he clearly considered selection to be the most important mechanism of heredity, acknowledged other kinds of 'descent with modification':

> As my conclusions of late have been much misrepresented, and it has been stated that I attribute the modification of species exclusively to natural selection, I may be permitted to remark that in the first edition of this work (i.e., *The Origin of Species*), and subsequently, I placed in a most conspicuous position - namely at the close of the Introduction - the following words: "I am convinced that natural selection has been the main, but not the exclusive means of modification." This has been of no avail. Great is the power of steady misrepresentation (1872: 395).

It is not my intent here to pursue these issues at any length. However, it is worth pointing out that, in the classic essay on this subject, Gould and Lewontin (1979) recommend decoupling adaptation and selection in efforts to explain aspects of form, function and behavior. They argue that 'adaptation' - the 'good fit' of organisms to their environments - conflates three hierarchical levels of change with different causes: (1) physiological adaptation - phenotypic plasticity that permits organisms during ontogeny to modify their form to accommodate changing environmental circumstances (not directly heritable, although the propensity to develop them may be), (2) cultural adaptation - the capacity to incorporate a large component of learning in the behavioral repertoire (heritable in a Lamarckian mode, although - again - the propensity to 'learn' almost certainly has a basis in brain evolution), and (3) adaptation arising from the conventional Darwinian mechanism of selection on genetic variation.

Gould and Lewontin (1979) add a fourth approach to evolution, important in these days of cladistic methodologies, taxonomic 'splitting' and the tendency to 'atomize' the organism into parts upon which optimizing selection supposedly operates. This is the phyletic inertia manifest in the *bauplan*, or basic design of organisms. Of German origin, the *bauplan* takes into account the extraordinarily conservative nature of the design of organisms, and suggests that their body plans are so completely integrated and replete with structural constraints that selectionist arguments can do little to explain them (Schindewolf 1950, Remane 1971, Grassé 1977). In other words, prior system states severely restrict possible evolutionary pathways to adaptation, although change - when it occurs - may be mediated by Darwinian selection on genetic variation. What is of primary interest to Gould and Lewontin, however, are the constraints themselves, rather than the mitigating effects of natural selection, which in the original formulation of this approach were argued (incorrectly) to be inconsequential.

In light of alleged parallels between models of biological and cultural evolution (e.g., Boyd & Richerson 1985, but see Clark 1993a), the concept of the *bauplan* has important implications for how paleoanthropologists conceptualize change. The kind of phyletic inertia exhibited by the *bauplan* is a too-often overlooked feature of all evolutionary trajectories. Moreover, the atomism engendered by the current emphasis on cladistic methodologies has shifted attention away from the total morphological pattern toward that manifest in anatomical peculiarities considered in isolation. As I have tried to point out elsewhere (Clark 1988, 1989), this is not a 'good thing', but as will become evident below, I would not want to try to push this argument too far. The atomism that plagues paleoanthropology today is a 'methodological' or 'classificatory' atomism which, carried to an extreme, would result in the creation of a new taxon for every fossil discovered. 'Conceptual' atomism is not only necessary but can be exceptionally productive, and a failure to atomize results in the practical impossibility of exploring *any* morphological or behavioral trait. I am indebted to John Alcock (pers. comm.) for pointing out that Gould and Lewontin's strictures on *baupläne* and developmental constraints are probably intended to discourage inquiry into the adaptive basis for human behavior, essentially because their ideological biases render them unable to accept the possibility that cultural evolution is largely epiphenomenal, thinly covering the evolved genetic-developmental mechanisms that generate what some think of as the 'uniquely human'. Finally, Alcock notes that Darwin's pluralism, of which Gould and Lewontin make so much, involved his semi-acceptance of bits and pieces of Lamarckian and blending inheritance mechanisms as well as the natural and sexual selection mechanisms which are the foundations of NDT. As David Queller (1995) points out in a brilliant and funny critique of Gould and Lewontin (1979), this is not the kind of pluralism likely to inspire modern evolutionary biologists.

Gould and Lewontin's (1979) critique of the adaptationist program had a chilling effect on the study of adaptation in evolutionary biology, although its impact in archaeology and paleoanthropology was much more attentuated (one might even say indiscernible, given the general absence of an evolutionary perspective in these disciplines). The essay initiated an era of recrimination and self doubt that, at the same time, caused evolutionary biologists to pay increased attention to new comparative methods rooted in phylogenetic systematics; an analytical framework that merges quantitative genetics, functional morphology, and natural history; and increasingly sophisticated study of adaptation in laboratory contexts and in nature. In fact, many of the criticisms that Gould and Lewontin (1979) levelled at ignorance of history and disregard for developmental constraints have been appropriated by evolutionary biologists to generalize and improve the adaptationist program. The essay, by making biologists more self-critical and by raising the standard of proof, ultimately enhanced and reenergized the study of biological adaptation (Feder 1997).

ON BUILDING STRONG INFERENCE

Understanding early hominid social behavior is by no means a straightforward endeavor. In addition to the numerous problems engendered by the exceptionally 'coarse-grained' time-space grid, a poor grasp of taphonomic and diagenetic process, and the fact that the kinds of activities likely to be identified in 'sites' are only a small part of a much larger organizational framework (most of which is likely to remain invisible), there is a kind of naïve empiricism that permeates much of this research, especially in its early phases. Work in east Africa in the 1970s on early hominid 'home bases' underscores some of the problems with these approaches. The objective of this research was to reconstruct "the basic features of early Pleistocene life" (Isaac 1969: 8) using what were assumed at the time to be primarily archaeological sites with a high degree of contextual integrity. In its most complete formulations, Isaac argued that a sexual division of labor, pair-bonding, food sharing, enculturation and social learning at a central place could be more or less directly tested against the particulars of the east African archaeological record (1984, Lovejoy 1981). Subsequent efforts to carry out these tests eventually led to the dismantling of the home base as originally conceptualized, and to a better understanding of the complexity of natural and cultural site formation processes in 'deep time' (e.g., Binford 1981, Bunn & Kroll 1986, Shipman 1986, Potts 1988). Although we are still in a 'pattern searching' mode, and far from consensus, the important point is that inferences about early hominid sociality cannot be directly evaluated against construals of pattern in the archeopaleontological record. In recent years there has been some recognition of this fact, and both Potts (1988) and Gifford-González (1991) have offered protocols designed to make the logic of inference more secure.

The scheme Potts (1988) proposes is hierarchical and involves three successively more inclusive levels of inference. First-order inference constitutes a demonstration that 'sites' actually exist, and that they are at least partly explicable by invoking hominid activity. The existence of sites is established by taphonomic inference whereby the bones and stones are shown to be anomalous concentrations in the paleolandscape that accumulated over a relatively short period of time (most of the 'high resolution' east African sites apparently formed over a minimum of 5-10 years), and that the agents of accumulation can be more or less successfully identified. From the perspective of the mid-1990s, these inferences are typically the most secure.

Second-order inference involves determination of the nature of 'exclusively hominid' activities. Whether hominids hunted or scavenged in order to obtain the bones brought to a site, how much meat or fat might have been extracted from these bones, and when hominids might have had access to carcasses are good examples. Inference here is also grounded in taphonomy (and 'actualistic' research), although there is as yet little sign of consensus. Whether uniformitarian assumptions about the behavior of present-day, non-human bone accumulators can to extended to their Pliocene ancestors is, of course, something of an inferential leap. In the end, Potts (1988) recommends evaluating the credibility of second-order inference on a case-by-case basis.

Third-order inference is the most problematic of all. It addresses regional scale process questions relating to how hominids organized their social landscapes. Were they hunting, scavenging, or both? What kinds of constraints operated under what conditions to emphasize one or the other meat procurement system? Who was doing the hunting, scavenging, gathering? Was there food sharing and, if so, what was shared and how was food sharing organized? In addition to the matrifocal unit, what was the composition of the local group? How did it change during the course of an annual round? At the level of generational replacement? What did supra-local mating networks look like? Was there adult pair-bonding and, if so, what was the basis for it and how long did it last? What form did kin-based adult co-residence patterns take? Most generally, how can we constrain choice about the forms that early hominid social organization might have taken?

Lake (1996) points out that, unlike first- and second-order inference, which are 'site-based', third-order inference requires knowledge of regional systems and, given the poor time-space resolution of the Plio-Pleistocene archaeopaleontological record, there are formidable (probably insurmountable) obstacles to subjecting any third-order inference to an empirical test. Consequently, all third-order inference is presently considered unreliable. Several researchers, notably Blumenschine and Potts, have initiated research with the aim of surveying a paleolandscape in the hope of identifying the particular topographic and ecological contexts of specific behaviors. This in turn should help facilitate reconstruction of more complex, inclusive behaviors like food sharing (Gibbons 1990).

While commendable, it is hard to be very optimistic about the outcome of these approaches. First and foremost, there are the enormous problems with temporal and spatial resolution alluded to above. Second, it is very likely that there is a lot of equifinality in the relationship between time-factored spatial form and the diverse processes that combined to create what we observe today. This means that, even given the highest resolution data likely to result from a paleolandscape survey, there is probably no relatively direct relationship between spatial form and spatial process. Equifinality is an almost insurmountable problem in recent historical contexts, let alone after hundreds of thousands or even millions of years have elapsed (Clark 1977). The gloomy implication of this is that, for the foreseeable future at least, third-order inference is likely to remain in the realm of more or less well-informed speculation.

This essay is, of course, about third order inference. Acknowledging its untestability in any direct way, I remain relatively optimistic that, by using an eclectic approach grounded in NDT, we can at least narrow the range of plausible models of early hominid sociality, and perhaps eliminate aspects of some of them altogether. The remainder of the chapter is an attempt to do this, using what we know about broad patterns in primate social organization, and making certain assumptions drawn from the literature of evolutionary psychology (e.g., sex-specific reproductive strategies, adult dispersal patterns). While some readers might take issue with these, I do not regard them as conflicting with the tenets of NDT as generally understood.

EARLY HOMINID SOCIOECOLOGY

Because of the likelihood that the socioecology of early hominids was very different from that of modern *H. sapiens,* there is currently a broad consensus that paleoanthropology must look beyond its traditional concerns with forager ethnography if the objective is to build strong inferences about ancestral hominid social systems (Wobst 1974, 1976; Foley 1988, Binford 1989). However, there is a diversity of opinion as to what aspects of non-hominid primate data are most relevant for modeling ancient human sociality, and also sharp differences as to how we might go about doing this (cf., e.g., Clutton-Brock 1989, Foley & Lee 1989, Hinde 1983, Maryanski & Turner 1991, Kano 1992, Steele & Shennan 1996). Most of this literature is relatively recent and has not had much of an impact so far on the research protocols of paleoanthropology, which tend to be very 'data driven', and characterized by pattern searches, rather than by more eclectic approaches to theory building (Clark 1988).

Since my own exposure to this literature is also relatively recent and, therefore, partial, I cannot claim to control it very well. However, it is possible to make some broad generalizations about the sexual division of labor, local and regional group size and composition, patterns of aggregation and dispersion, sexual selection, group stability and the structure of alliances based on the fact that pongid social systems are highly conservative and, while extremely variable, nevertheless exhibit certain general tendencies that might, for purposes of hypothesis building, be considered synapomorphies - characters shared by the Hominoidea but derived vis à vis the Old World monkeys (Foley & Lee 1989). Beyond invoking a 'cladistics of behavior' (which some might find objectionable), the logic of doing this is simply that derived characters (apomorphies) are useful in assessing genealogical links among taxa. Given the conservative nature of hominoid social systems, and the fact that we can contrast them with those of the Cercopithecoidea, we might be able to use them to identify aspects of ancient hominid sociality by arguing that the behavior of early hominids resembled more closely that of the great apes than that of the Old World monkeys. This kind of modeling can be done either 'from the top down', in terms of a concept of finite social space wherein the range of options for social interaction, relationships and structure is limited and specifiable (e.g., Foley & Lee 1989), or 'from the bottom up', starting with a comprehensive survey of hominoid social structure, and using network analysis to extract general patterns from it (e.g., Maryanski & Turner 1992, Maryanski 1993).

Hominoids as Large Mammals

Hominoids are large mammals, and as such have long lifespans, tend to mature 'late' (relative to smaller primates), have long gestation and lactation periods, long interbirth intervals, tend toward single births of altricial young, and have low basal metabolic requirements. In other words, they are '*K*-selected' vis à vis other, smaller primates. *K*-selection implies adaptation to fairly constant or predictable environments, density-dependent attritional mortality profiles and a relatively low juvenile death rate (compared to *r*-selected species), relatively stable population densities at or near carrying capacity, saturated communities (i.e., few or no ecological vacuums), and little or no colonization (or recolonization) over the short-term. Competition is relatively intense, and intraspecific competition in particular is an important factor in reproductive success. Primates in general, and apes and humans in particular, are *K*-selected. The life-history factors implied by *K*-selection have strong implications for the kinds of social organizations that might be reasonable to entertain for early hominids since there are systematic, although not necessarily straightforward, relationships between general body proportions, estimates of brain size, metabolic rate, niche width, mobility, predator-prey relations, thermoregulatory efficiency and, eventually, encephalization (Foley 1987). Put another way, much of human evolution might be explicable in terms of a large hominoid exploiting a relatively stable food supply, its stability enhanced by virtue of its breadth (Lewin 1993: 62, 63). While technology might have eventually allowed for more efficient exploitation of plant and animal resources, thus broadening (and stabilizing) the diet further, a stable environment with a predictable food supply is implied by *K*-selection itself. Thus it seems likely to predate what we would recognize as technology in archaeological contexts.

Generalizations About Hominoid Sociality

Although certain aspects of hominid behavior (e.g., subsistence) might better be modeled by looking to the social carnivores (e.g., Turner 1984, Stiner 1994), there is a strong general bias in favor of primate models to address hominid social organization. This is simply because the non-human primates, and particularly the African apes, are our closest living relatives and are thus likely to conserve in their behaviors traces of hominid ancestral conditions. Moreover, we know a great deal about the primates (relative to other orders); this is particularly true of the superfamily Hominoidea. Some of the salient hominoid life history characteristics are enumerated in Table 12.1; social and demographic features are given in Table 12.2. While these data, culled from Rowe (1996), are the best available to date, inspection of the tables shows pretty clearly the enormous range of variability in hominoid socioecology, demography, food distribution and foraging behavior, social organization and patterning in sexual activity within species, let alone genera. Thus it would be ill-advised to try to identify any particular hominoid taxon to use as a 'most likely' model for early hominids.

Cheney et al. (1986) identify three characteristics that set primates apart from other animals: (1) primates are extremely flexible and diverse in their social arrangements, and this flexibility would seem to provide a basis for complex reciprocal interactions of all kinds. (2) Primate social organizations are complex vis à vis those of other orders, and link together individuals of different ages, sex, dominance-rank and kin relatedness, forming temporary alliances, subgroups and long-term associations within and across universal categories of social geography. (3) Because of their intelligence and long life spans, primates form long-term social bonds. Interactions can be shaped in terms of anticipated outcomes, implying a certain amount of displacement or 'planning depth'.

Primate models make sense for other reasons, too - among them the often-noted similarities between apes and humans in terms of brain structure and cognitive abilities. It is possible to think of early hominids as chimp-like so far as their skulls are concerned, but with slightly larger brains (EQs in the region of 2.5, vs 2.0 for the common chimpanzee), better cranial balance, facial and dental adaptations for powerful mastication, and with smaller canines and incisors than chimps. Their brains differ from those of apes because of (1) expansion of the posterior parietal associative areas, (2) greater size and complexity of the frontal lobes (esp. the areas associated with speech), and (3) by an expansion of the temporal lobes. Holloway's studies of cranial endocasts suggest that early hominids had already undergone what he calls "an encephalization of behavioral structure" (Holloway 1975, 1983a, b). This means that their brains had expanded in those areas required for the development of the complex neural models that underlie learned behavior. In particular, the areas related to cognition, categorization, symboling, ordering, discrimination, cross-modal transfer (the ability to shift referential formats) and the verbal-auditory regions all appear to be more developed than those of contemporary apes. Moreover, this pattern appears to have been established very early on; according to Holloway, it is evident even in *A. afarensis* (Holloway & Kimbel 1986). While this view has not gone unchallenged (cf. Falk, 1983, 1986, 1987, 1991), it has led some workers to the conclusion that, since primate intelligence seems to have been selected for because of social factors (i.e., living in large groups with complex social interactions), primate intelligence and sociality probably co-evolved (see, e.g., papers in Byrne & Whiten 1988).

The Phylogenetic Context of Human Sociality

In an ingenious series of essays, Foley and Lee (1989, 1996; Foley 1989, 1992, 1996) use the concept of finite social space as a template against which to 'score' the frequency of extant primate social systems, and thus circumscribe their ranges of variability. What emerges from this is (1) that social variability in terms of male and female dispersal patterns at sexual maturity is very limited, (2) that not all primate social systems are close to one another in evolutionary space/time, and (3) that the distance between social states allows for assessment of the likelihood that some states were ancestral to others.

Table 12.1 Life History Characteristics of the Hominoidea (after Rowe 1996).[1.]

Family Genus/Species	Weaning (months)	F/M Sexual Maturity (months)	Estrus Cycle (days)	Mating Season	Gestation (days)	Age at 1st Birth (months)	Birth Interval (months)	Lifespan (years)
H. hoolock (hoolock gibbon)	23	84	28	rainy			36	42
H. agilis (agile gibbon)							38	32
H. klossii (Klossi's gibbon)	12			year-round	210		40	
H. lar (white-handed gibbon)	24	108/78	27		205	112	30	44
H. moloch (silvery Javan gibbon)								35
H. muelleri (Bornean grey gibbon)				year-round			36	47
H. pileatus (capped gibbon)								39
H. concolor (black gibbon)	18							36
H. gabriellae (golden cheeked)								46
H. leucogenys (Chinese white cheek)		48/-			200-212	54		28
H. syndactylus (siamang)					189-239	108	36-59	35
Gibbon Mean:	19.3	78/81	27.5	-	209	91.3	37.9	38.5
P. abelii (Sumatran orang)	42	84/114	30		260	144-196	72-144	>50
P. pygmaeus (Borneo orang)	42 (60-84)	84/114	30-35	year-round	223-267	144-180	32(c) 84-96(w)	
Orangutan Mean:	42(57)	84/114	31.2	-	252	166	99	54.5
G. gorilla beringei (mountain gorilla)			27-39	year-round		120-144	36-59	40-50
G. gorilla gorilla (western lowland)	52	78/120	25-42		237-285	108-132	48	50
G. gorilla graueri (eastern lowland)								
Gorilla Mean:	52	78/120	33	-	237-285	126	47.7	47.5
P. paniscus (bonobo)			35-40		240	168	54	40
P. troglodytes (common chimp)	48	135/156	36	year-round	240	168-180	60	53
Chimpanzee Mean:	48	135/156	37	-	240	171	57	46
H. sapiens sapiens (humans)	24	198/-	28	year-round	270	192-240	10-48	80-90

Table 12.2 Social and Demographic Characteristics of the Hominoidea (after Rowe 1996).

Family Genus/Species	Social Structure	Group Size	Home Range (hectares)	Day Range (meters)	F/M Body Weight (kg)	Habitat	Diet[2]
H. hoolock	MFG[1]	3.5	15-30	600	6.1/6.9	swampy	F,L,B
		3-6	300-400	300-1000		forest	
H. agilis	MFG	4.4	25-29	1335	5.9/-	swampy	F,L
		2-7		650-2200		forest	
H. klossii	MFG	3.4	7-11	1514	5.8	swampy	F,A
		2-6	34-54	885-2150		forest	
H. lar	MFG	5	12-53	1545	5.6/6.3	forest	F,L,I
		3-12		1490-1600			
H. moloch	MFG	3-4	17	1400	5.7	forest	F,L
H. muelleri	MFG	3.5	38	890	5.0-6.4	forest	F,L
		2-5		350-1520			
H. pileatus	MFG	4	15-50	833	7.5/9.1	forest	F,Fl,I
		2-6		450-1350			
H. concolor	MFG	2-5	300-500	90-750	4.5/9.0	forest	B,F,L
H. gabriellae					5.7/-	forest	
H. leucogenys	MFG	3.7			5.8/5.6	swampy	
		2-6				forest	
H. syndactylus	MFG	3.5	47	738-969	10.5/13.5	forest	L,F,Fl,A
		2-10	15-50				
Gibbon Mean:	MFG	4	69	1038	6.5/7.3	forest	F,L
P. abelii	DH[3]	1-3	200-1000	800-1200		swp/forest	F,L,Bk,A
P. pygmaeus	DH	1-3	42-777	305-800	37.0/77.5	swp/forest	F,L,Bk,A
Orangutan Mean:	DH	1-3	505	776	37.0/77.5	swp/forest	F,L,Bk,A
G. gorilla beringei	OH[4]	9	400-800	400	97.7/159.2		L,W,R,Fl
				100-2500			
G. gorilla gorilla	OH	3-21	800-1800	2300	71.5/169.5	forest	F,L,S,I
G. gorilla graueri	OH		3200		80.0/175.2	forest	F,L,Bk,R,I
Gorilla Mean:		10	1700	1350	83.1/168.0	forest	F,L,Bk
P. paniscus	MFFG[5]	50-200[6]	2200-5800	1200-1400	31.0/39.0	rainforest	F,L,B,A
P. troglodytes	MFFG	100	1250	3900	39.5/50.0	rainforest,	F,L,Fl,S,
		7-25 (f),	3900-7800			grassland,	I,A
		5-16 (m)				wdln/sav	
Chimpanzee Mean:	MFFG	15	3775	2850	35.2/44.5	-	F,L
H. sapiens sapiens	SM/P[7]	variable	variable	variable	55.0/68.2	variable	variable

1. MFG = monogamous family groups
2. F = fruit, L = leaves, B = buds, shoots; A = animal prey, I = insects, Fl = flowers, Bk = bark, W = wood, R = roots, S = seeds in descending order of importance
3. DH = distributed harem
4. OH = orthodox harem
5. MFFG = multimale, multifemale foraging groups
6. foraging groups are much smaller
7. SM/P = serially monogamous, polygynous

This in turn is the basis for constructing hypothetical trajectories of primate social evolution, the credibility of which can then be assessed against what we know about the living primates and landmark primate divergence events.

Focusing on the Old World anthropoids (Catarrhini), there is a significant evolutionary divide at the superfamily level between the Old World monkeys (Cercopithecoidea) and apes and humans (Hominoidea) in respect of a whole range of physiological, ecological and behavioral differences (summarized in Foley & Lee 1989, 1996; see also Maryanski 1992, 1993, 1996). Especially pertinent here are broad generalizations about dispersal at sexual maturity. Although there are exceptions, cercopithecoids are primarily characterized by male dispersal and female coresidence, whereas among the hominoids, the reverse is true - stable female kin-based coresidence is unknown. Both males and females can disperse, as in the monogamous gibbon (*Hylobates* spp.) and the solitary orangutan (*Pongo pygmaeus*). Among gorillas (*Gorilla* spp.) organized into harems, females disperse as well as most of the males, although some remain coresident with their fathers (and thus might eventually inherit the harem) (Harcourt & Stewart 1987). In the case of common chimpanzees (*Pan troglodytes*), females disperse while males remain coresident with their male kin, and form kin-based alliances of various kinds (Goodall 1986). Bonobos (*Pan paniscus*) also appear to follow a pattern of female dispersal, although male-female alliances would seem to form more regularly than those among males (Kano 1992).

That female kin-based coresidence is unknown among the hominoids is a striking conclusion, with numerous and far-reaching implications for subsequent aspects of hominid social organization (e.g., the strong tendency toward patrilocality evident in hunter-gatherer ethnography - Radcliffe-Brown 1930, Service 1962, Williams 1974, Martin & Stewart 1982). Without going into detail and compressing a lot of sophisticated argument, Foley and Lee (1989, 1996) suggest that, while the cercopithecoids continue to exhibit the mammalian conservative condition of female coresidence (an anthropoid plesiomorphy), the apes display a temporally-vectored series of more 'derived' states which, over time, depart more and more from the primitive ancestral condition. Using an cladistic approach as a heuristic device upon which to 'hang' extant primate social systems, they argue that, at the cercopithecoid-hominoid split at ca. 25 mya, female coresidence appears to have lapsed, probably in the context of monogamy, in which both sexes disperse. This pattern is retained in the sexually monomorphic gibbons, the first hominoid lineage to diverge (at ca. 17-15 mya). Set against a backdrop of ecologically mediated increases in overall body size, and more marked sexual dimorphism, the pattern characteristic of orangs (distributed harem) and gorillas (orthodox harem) emerges next, in the middle-late Miocene at ca. 12-8 mya. The interval corresponds to the estimated divergence time of the orangutan clade, as determined from the first appearance of *Sivapithecus* at ca. 12 mya, to the largely hypothetical divergence of the gorilla lineage at ca. 8 mya (estimates derived from DNA hybridization data summarized in Pilbeam [1986]). The ancestral pattern of both orangs and gorillas would have entailed greater male tenure in harems (because of increases in longevity - Clutton-Brock [1989]), and an increasing tendency for female dispersal to join successful males, thus further reducing the likelihood of a reversion to the earlier pattern of female coresidence. Finally, the fragmentation of forested habitats in the late Miocene resulted in a niche shift after ca. 6-5 mya among some African hominoids, marked by an extension of foraging ranges, which provoked the transition to the pattern of more marked territoriality found in common chimpanzees. The niche shift would have coincided with a shift from a pattern of female defense (as implied by harems) to one emphasizing territorial defense (as inferred from range expansion). It would have further reduced the likelihood of male dispersal, and would have tended to strengthen the tendency for male kin-based alliances (Fig. 12.1).

In sum, Foley and Lee (1989, 1996; Foley 1992) show from the pattern of sex dispersal at maturity among the living catarrhines that hominids - as hominoids - were much more likely to have evolved from a lineage in which male dispersal was rare or absent than from one in which it was relatively common. This means that any descent and alliance-based ties in hominids are more likely to be male-based than female-based. These findings are both anticipated (e.g., Wrangham 1979a, b; 1987) and corroborated by other workers (e.g., Maryanski 1992, 1993, 1996), sometimes using very different approaches.

Body Size, Sexual Dimorphism and Their Implications

A common finding of comparative zoological research is that morphological, behavioral and life-history variables tend to be highly correlated with body size. These allometric relationships are well established and general enough across mammal taxa that they can be described by a linear equation of the form $y = a(x)^b$, where y is the life-history variable, x is body size, and a and b are constants (usually the allometric coefficient and exponent respectively - see Harvey et al. 1987: 186). These relationships have a number of important implications for the evolution of hominid sociality since they link brain and body size to basal metabolic rates, gestation length, and

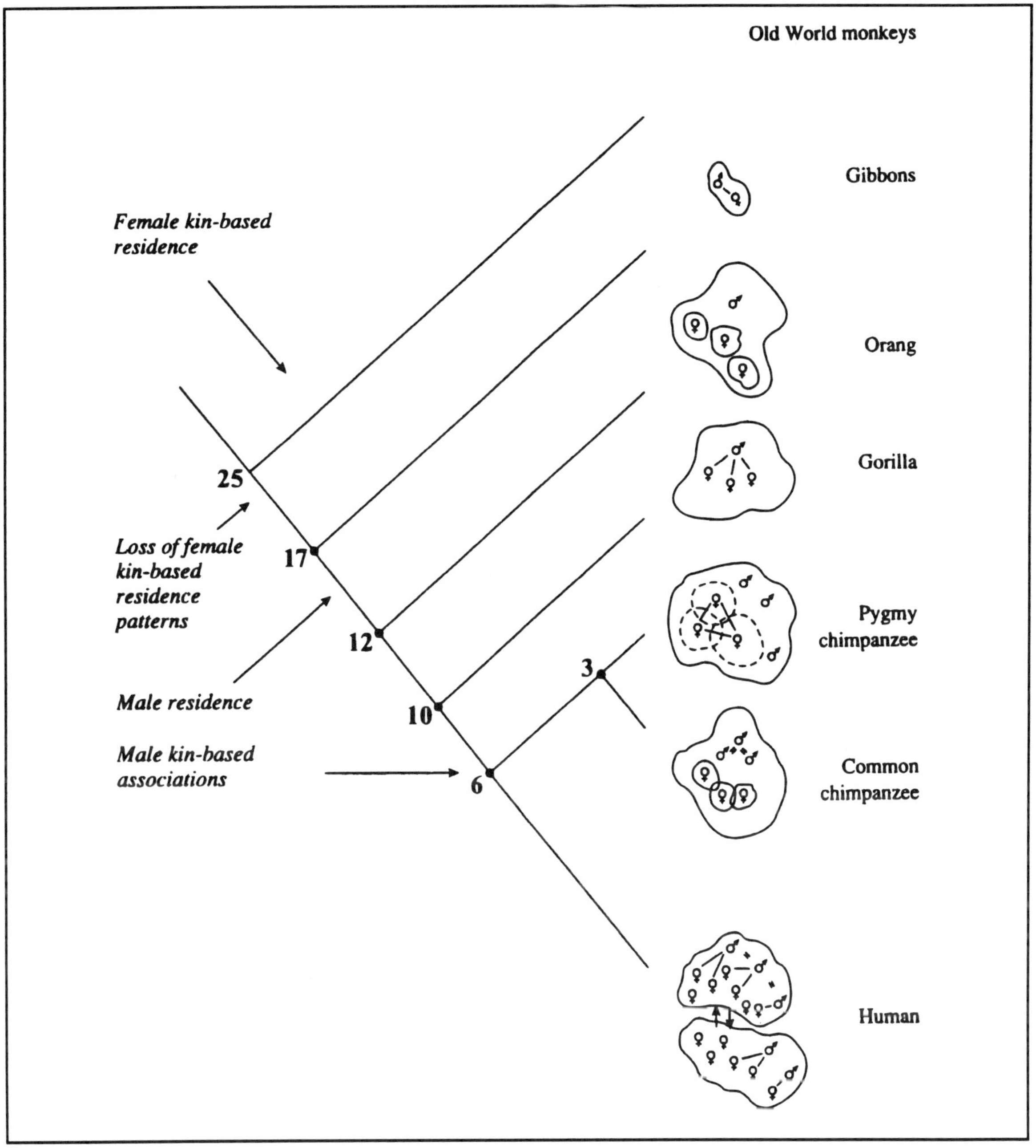

Figure 12.1 A cladistic phylogeny of catarrhine social systems (from Foley & Lee 1996: 56, used with authors' permission). Female kin-based co-residence is considered to be the ancestral catarrhine system as represented in the Old World monkeys, and lost in the hominoid clade. Male kin-based co-residence becomes established in the African ape/hominid clade. Numbers give approximate divergence dates.

precociality (primates are moderately precocial vis à vis other mammalian orders).

In a consensus rare in human paleontology, there is broad agreement that the earliest hominids were considerably more sexual dimorphic in terms of body size than the later ones. Body size dimorphism, in turn, has important implications for mating systems, models of male and female sexual selection, agonistic relationships of various kinds (esp. intermale competition, dominance hierarchies) and the sexual division of labor, among many other aspects of social organization. Although sexual dimorphism is also manifest in aspects of the dentition (esp. the canines), mandibular robusticity, craniofacial morphology in general, and limb bone proportions, pattern searches on these variables have so far tended to produce conflicting results (and, parenthetically, atomistic, hyper-selectionist, post-hoc explanations for the various anatomical parts involved). Only body size, weight and stature are strongly autocorrelated.

A focus on body size dimorphism has the additional advantage that there are a number of quasi-independent algorithms for its calculation, although - as in all efforts to deal with sparse hominid samples widely separated in space and time - taxonomic ambiguities and sample size effects inevitably determine how groups are constituted and what kinds of comparisons are reasonable to make. The major implication of body size sexual dimorphism is that monogamous species tend to have very low body weight dimorphism and polygynous species tend to have higher dimorphism (Clutton-Brock et al. 1977). While not without its problems, this association holds up pretty well in the Order Primates: monogamous species are always monomorphic, and polygynous species are dimorphic to various degrees. In some extreme cases (e.g., gorillas, orangs), males can weigh twice as much as females.

Sexual selection is usually invoked to explain sexual dimorphism. In the classic articulation of this view, Trivers (1972) argued that male/male competition to monopolize the reproductive potential of females selected for greater size, strength and agonistic behaviors of various kinds (some with material correlates) in males, and that in long-lived, *K*-selected species like the hominoids, males would have sought to maximize their reproductive potential by impregnating as many females as possible, while denying other males sexual access to females already impregnated. Extreme sexual dimorphism tends, therefore, to be strongly correlated with polygyny, and with the formation of harems (as, e.g., in gorillas, orangs). Harem defense is an expensive proposition, however (ask any Saudi prince!), and the extent to which it is successful is ultimately determined by local group size and composition, the periodicity of female sexual receptivity, and the intensity of male/male competition.

Selection operates differently on females because of the vastly greater costs that females incur in terms of the energy budget of reproduction (gestation, lactation etc.). Faced with the consequences of dalliance, it makes them more selective in regard to decisions with whom to copulate. However, size differences per se have an inconsequential effect on female reproductive success, essentially because they do not figure in interfemale competition. Small females are not at the same disadvantage as small males in the competition for mates (McHenry 1996).

In monogamous species, sexual dimorphism tends to be minimal or absent because of the inability of males to monopolize the reproductive potential of more than one female at a time. However, the converse does not hold true - some polygynous species are highly dimorphic and some are not (Frayer & Wolpoff 1985). Although the reasons for this are not clearly understood, the differences probably have something to do with combined effects of frequency and intensity of male/male interaction.

In a study of intermale competition aimed at determining its effects on canine dimorphism, monogamous, monomorphic primates like gibbons are shown to be characterized by low-frequency, low-intensity male/male competition, whereas moderately promiscuous, moderately dimorphic common chimpanzees typically exhibit a pattern of high-frequency, low-intensity intermale competition (Plavcan & van Schaik 1992). Chimps in general seem to be characterized by a complex network of many informal but relatively weak ties (a 'hang loose' community structure - Maryanski 1996: 71). Female ties are neutral to moderately antagonistic, except in the case of dependent female subadults. Mother/daughter ties weaken and disappear, however, as daughters mature and disperse, whereas mother/son ties tend to remain relatively strong since mothers and male progeny tend to coreside in the same foraging area. Ties between female siblings also weaken at maturity, and enduring voluntary associations among unrelated, adult females are seemingly non-existent (Maryanski 1996: 71-73). Taken together, these patterns suggest that changes in sexual dimorphism within lineages over time are probably explicable mainly in terms of changes in the nature of intermale competition, and have little or nothing to do with ecology, subsistence or interfemale competition except insofar as local demographic factors would have affected the intensity of competition among males. There is a clear difference in body size dimorphism between species with high and low frequencies of intermale competition among species with overall low intensity competition (McHenry 1996: 99).

Sexual Dimorphism in the Hominid Fossil Record

How do patterns in extant hominoid data compare with evidence for body weight dimorphism in the hominid fossil record? This question has been a central concern of much of McHenry's research (e.g., 1986, 1991a,b; 1992a,b; 1994) and his data on body weight dimorphism in extant hominoids and fossil hominids are reproduced here (Table 12.3). Table 12.3 shows that highly dimorphic, polygynous, harem-forming gorillas and orangs represent the modern hominoid extremes (2.1, 2.0 respectively), whereas the early hominid values (1.3-1.6) bracket those typical of chimps (1.4). Although taxonomic uncertainty advises caution in any attempt at more fine-grained comparison, McHenry points out that in at least

Table 12.3 Body Weight Dimorphism in Extant Hominoid Species and in Fossil Hominid Taxa (after McHenry 1996: 93).[1,2]

Species	Male	Female	Ratio
extant *Homo sapiens*	64.9	53.2	1.2
Pan troglodytes	54.2	39.7	1.4
Pan paniscus	47.8	33.1	1.4
Gorilla gorilla	157.9	75.4	2.1
Pongo pygmaeus	78.8	38.8	2.0
Hylobates syndactylus	11.3	11.3	1.0
Hylobates lar	5.5	5.2	1.1
Australopithecus afarensis	44.6	29.3	1.5
Australopithecus africanus	40.8	30.2	1.4
Australopithecus boisei	48.6	34.0	1.4
Australopithecus robustus	40.2	31.9	1.3
Homo habilis sensu lato[3.]	51.6	31.5	1.6
Homo habilis sensu stricto[3.]	37.0	31.5	1.2
Homo rudolfensis[3.]	59.6	50.8	1.2
early African *Homo erectus*[4.]	62.7	52.3	1.2
neandertal[5.]	60.1	51.8	1.2
early modern *Homo sapiens*[5.]	70.0	56.8	1.2

1. from McHenry (1991).
2. from McHenry (1992), except where indicated.
3. from McHenry (1994).
4. = *Homo ergaster.*
5. derived from stature estimates in Feldesman et al. (1990) by regression formulae for male and female given in Table 11 of Ruff & Walker (1993). Probably underestimates true values because these hominids were more robust than modern humans from whom the stature/weight formulae are derived.

two cases - the G trails of hominid footprints at Laetoli and the 'first family' site at Hadar (AL 333) - we probably have samples of the range of variability within single biological populations (1996: 92, 93). These are both *A. afarensis* sites, dated to 3.7-2.9 mya. If *A. afarensis* represents a single species, as Johanson and White (1979) originally argued and as many workers believe, the range of size variation is well above that seen in any modern human population. If there is an interval during which hominid body size dimorphism begins to approximate the modern condition (1.2), it is after 1.8-1.7 mya, with the appearance of *H. erectus.* This trend, which might have been much more gradual than commonly represented, is attributed to overall increases in *H. erectus* body size, possibly tied to increases in foraging range, and in the width of the food niche, as more high energy animal products are added to the diet (Shipman & Harris 1988, Potts 1988). Competition with baboons was likely throughout the early hominid range, and might also have led to (or accelerated) niche differentiation, and selection for larger body size (Wolpoff 1994).

A puzzling aspect of these data is that the increase in *H. erectus* body size would appear to have been proportionately greater in females than in males. Males are about 50% larger than their australopithecine counterparts, whereas females are about 70% larger, approaching the modern ratio (1.2). It has been suggested that this relative increase in female body size is associated with an increase in estimated neonatal brain size (166 g for *A. africanus*, 270 g from early *H. erectus*); the relatively greater increase in female body size might have been due to selection for the physiological mechanisms needed to give birth to larger-brained infants (McHenry 1994). I will return to this topic below.

In sum, this research shows clearly that body weight dimorphism in the earliest hominids is greater than that of later ones, but well below that of gorillas and orangs. In the extent to which these relationships are relatively direct, the dimorphism apparent in *Australopithecus* and *H. habilis* pretty much rules out both the long-term monogamous relationships of gibbons and the harem-organized polygyny of gorillas and orangs. Loosely organized, polygynous, kin-related, multi-male groups with relatively weak affective ties overall and a low levels of male/male competition would seem most probable. The Laetoli footprints (which represent one large, and two smaller individuals) and AL-333 would also imply the presence of adult females and juveniles. What is striking about McHenry's work (1994, 1996) is just how closely his conclusions converge with those of Foley and Lee (1989, 1996) and Maryanski (1993, 1996). Although the approaches are completely different, this convergence of results engenders a certain confidence in the overall ability of socioecology to explain aspects of human evolution that are not directly amenable to testing. However, one aspect of this scenario - polygyny - merits further scrutiny in light of a large body of evidence suggesting that the basic mating pattern in modern humans, at least, is not polygyny but serial monogamy (Fisher 1987, 1989, 1992). If this is so, how far back in our evolutionary heritage does it go? Can it be explained by an NDT conceptual framework, and how can it be reconciled with the early hominid polygyny inferred from hominoid socioecology?

Unpacking Polygyny

As a regulatory ideal, monogamy is rare both in nature and among humans. Of 1154 societies in the HRAF for which adequate data are recorded, more than 1000 (93%) recognize some degree of sanctioned polygyny (i.e., at least occasionally, males can mate with more than one female), and polygyny is the preferred choice in 70% of them (Murdock 1967). However, it is important to remember that polygyny refers only to male reproductive strategies, overlooking those of females, and that it conflates at least four modal types found widely among animals and paralleled in human society. (1) Resource defense polygyny tends to be found in species where food sources, 'sanctuaries' (hiding places), nesting spots or mating grounds are spatially clustered. If a male can succeed in establishing proprietary rights to one of these resource-rich patches by excluding other males, he can monopolize the reproductive potential of any females who might - for whatever reason - gather there. (2) Female defense polygyny occurs when a male can round up a group of females and forcibly prevent other males from copulating with them. (3) Male dominance polygyny depends upon variation in individual aspects of personality (strength, empathy, charisma, guile etc.), rather than resource control, to attract mates. (4) Search polygyny involves a strategy of identifying receptive females, mating, and then moving on (Emlen & Oring 1977, Frayser 1985, Flinn & Low 1986). These strategies are played out contextually. They are neither mutually exclusive, nor exclusive of other strategies, nor fixed and invariant over the male reproductive lifespan. The explanation for the prevalence of polygyny in general is quintessentially Darwinian: males form polygynous unions to spread their genes as widely as possible, whereas females benefit by resource acquisition, thus insuring a greater probability of the survival of their young. Where, then, and how, does monogamy fit in the picture?

Monogamy Defined

Because of the genetic advantages of polygyny for men and because polygyny is so widespread among humans, it is taken by many anthropologists to be the basic human (and hominoid) reproductive strategy. However, in practically all societies where polygyny is a possible reproductive strategy, only 5-10% of men actually have more than one wife simultaneously (Frayser 1985, van den Berghe 1979). So striking was this pattern that, after surveying 250 cultures worldwide, HRAF founder George Murdock remarked that:

> An impartial observer . . . would be compelled to characterize nearly every known human society as monogamous, despite the preference for and frequency of polygyny in the overwhelming majority (1949: 27, 28).

In the modern world, then, and for a variety of economic and other reasons, men tend to marry only one woman at a time. However, one of those 'other reasons' is potentially important for understanding the origins and implications of monogamy - a hominoid tendency for pair-bonding. Although rare in nature, affective, heterosexual relationships of varying intensity and duration would seem to be part of our evolutionary heritage (Hrdy & Whitten 1987). They are also well-documented among apes, where they play important roles in alliance formation and in 'courting' behavior strikingly similar to our own (Fisher 1992: 19-36).

The *Oxford English Dictionary* defines monogamy as "the condition, rule or custom of being married to only one person at a time". The definition does not imply that the partners in a monogamous relationship are remorselessly sexually faithful to one another. Zoologists James Wittenberger and Ronald Tilson use the term to refer to "a prolonged association and essentially exclusive mating relationship between one man and one female" (1980: 198). Fidelity is not part of this definition either. They add that, "By 'essentially exclusive', we imply that occasional covert matings outside the pair bond, (i.e., 'cheating') do not negate the existence of monogamy" (1980: 198). While I strongly suspect that 'monogamy' is a western cultural construct devoid of much analytical utility, the important point is that 'monogamy' implies neither life-long commitment nor exclusive sexual fidelity in the context of heterosexual pair bonds (see also Small [1993, 1995]; Palombit [1994a, b]; Manson [1997]). It is a reproductive strategy that makes Darwinian sense in certain circumstances (see below). Moreover, it can be relatively easily derived from polygyny (Fisher 1992: 62-64).

The conventional anthropological explanation for monogamy is essentially an economic one: humans endure it because men are unable to acquire the resources they need to accumulate a harem, and women tolerate it because they are unable to entice more than one man at a time to provide resources. Supporting this view is the ample evidence for polygyny among rich and powerful men (Betzig 1986, 1989; Betzig et al. 1988). However, Fisher 1989, 1992) points out that monogamy in conjunction with adultery provides reproductive advantages similar to those of polygyny - males have the opportunity to inseminate multiple females, females get access to resources beyond the provisioning capabilities of a single male. Other reasons that adultery might have been adaptive for females include 'insurance' (e.g., supplementary males might help with parenting chores if the primary male dies or deserts) and genetic improvement (e.g., the principal mate is not an adequate provider, or exhibits other deficiencies that are likely to affect the survival of her progeny).

Taking note of the high sex drive in female primates, Hrdy points out that female apes engage in a great variety of non-reproductive coitus. This suggested to her that a female chimp's pursuit of sexual variety would tend to create alliances by befriending males who might otherwise be inclined to kill a newborn, and to confuse paternity so that each male might act paternally toward her forthcoming child, if he had reason to suspect that it might be his own (1981, 1986). Applying this reasoning to humans, she concludes that the high female sex drive is an ancient evolutionary tactic designed to obtain supplementary paternal investment in progeny, and as insurance against infanticide (Hrdy 1981, 1986; Hrdy & Whitten 1987). From a female perspective, the credibility of this argument turns on the nutritional depletion that comes with pregnancy and lactation, the problems involved with bearing altricial young and the long period of dependency they entail, and the scheduling conflicts that arise from balancing the demands of childcare and subsistence. From a male perspective, what might be construed as unproductive paternal investment in provisioning could instead be attributed to mating effort, since there is a likelihood that provisioning would result in continued sexual access to the mother. Kleiman contends that monogamy occurs whenever more than a single individual is needed to rear the young (1977: 51). Ember and Ember agree, and add that:

> Heterosexual partnerships develop wherever the need of the mother to obtain her nutrition interferes with the care of the young. The duration of this bond is dependent upon parental care time (1979: 53).

That monogamy only arises under certain circumstances is an important insight because it provides an explanation for pair bonding in general and for increased parental investment on the part of males in particular. However, 'parental care time' would presumably set limits to the duration of the pair bond. It would not make Darwinian sense to continue it once the child was mature, so long as either partner was still reproductively active.

To deal with the many implications of monogamous pair bonding that follow more or less directly from this approach is a book-length subject. Particularly influential has been the work of the Lancasters (1983; Lancaster & Kaplan 1992) and of Hawkes (1990, 1996), who argue that the differential trade-offs that males and females make in regard to mating and parenting effort are manifest in many aspects of higher primate sociality, and probably underlie what is usually taken to be a unique human characteristic - the sexual division of labor. The complex, flexible and context-sensitive mating strategies of modern women exhibit strong modal tendencies rooted in remote antiquity, having been shaped by natural and sexual selection to reflect compromises whereby 'success' is determined by an individual woman's resources and options mitigated by conflicts of interest with men and with other women (Cashdan 1997).

Implications of Divorce Statistics

Citing robust statistical patterns derived from demographic data on divorce taken from the *United Nations Demographic Yearbooks,* Fisher (1987, 1989, 1992) found that, among women of reproductive age, divorce tended to peak around the fourth or fifth year of marriage, and that marriage, therefore, exhibited a cross-cultural pattern of decay. This suggested to her that humans tended to marry one person at a time and then, during the peak reproductive years, divorce, only to remarry again three or four years later. She points out that this pattern in modern demographic data corresponds almost exactly to the four-year forager birth interval, where constant nursing, high levels of exercise and a low fat diet combine to suppress ovulation until weaning has occurred, thus postponing the ability to become pregnant again for about that interval. This in turn implies that a four-year pattern of birth spacing was the norm in our evolutionary past (Lancaster & Lancaster 1983), before diets high in calories and fat, lack of exercise and limited nursing boosted the percentage of body fat needed to trigger ovulation, lowered the age of menarche, and reduced birth spacing to its modern average of two years or less (Frisch & Revelle 1970; Frisch 1978, 1989). Data on birth spacing among the apes also support the antiquity of this reproductive pattern. Among chimpanzees and gorillas, birth spacing is on the order of four or five years (Allen et al. 1982).

Fisher's research (1987, 1989, 1992) also uncovered strong, cross-cultural statistical evidence for a high incidence of philandering, from which she concluded that monogamy in conjunction with adultery is the primary modern human reproductive strategy, whereas polygyny (and polyandry) are opportunistic, secondary reproductive tactics that only show up in post-agricultural contexts with a concept of land tenure and institutionalized inequities in the distribution of wealth and power. This has been contested by behaviorists and culturists of various persuasions (although, in my opinion, not very successfully). In terms of this essay, I am not sure that it matters much, since the strategies themselves are complementary. The implication for early hominids, though, is that human pair bonds originally evolved to last only long enough to raise a single, dependent child to sexual and social maturity, and that that interval was approximately four or five years. This is the major conclusion of Fisher's book (1992). However, if the hominoid ancestral pattern is assumed to be polygyny, it is necessary to account for the greater paternal investment in child rearing that accompanies the shift to monogamy.

As already mentioned, the heterosexual partnerships implied by monogamy can be explained by time stress resulting from the need to balance child care demands and subsistence. I suggest that these conditions are less likely to have arisen in the extensive, resource-uniform, early Miocene forests to which our hominoid ancestors were adapted. The shift that selected for more enduring heterosexual pair bonds, therefore, probably coincided with the shift to the more heterogeneous, nutrient-rich resource patches found in savanna-margin environments of the late Miocene, and the extension of foraging ranges that it entailed. The credibility of this argument turns on how the niche shift to life on the ground is modelled. Resources for folivores like chimps in relatively continuous forest environments are themselves relatively continuously distributed, so there is no conflict between subsistence requirements and the rearing of young. Consequently, female chimps raise their young on their own. However, when hominids - as hominoids - began to exploit more open, forest-margin and savanna environments as these became increasingly available during the Middle and Late Miocene, they encountered temporally and spatially dispersed resource patches which are at once more nutritious (i.e., concentrations of high-energy seeds, nuts, tubers etc.) but also more irregularly distributed in the landscape, and thus less immediately available to nursing mothers. Satisfying individual subsistence needs and curating young at the same time would have provoked scheduling conflicts, selecting for the conditions under which more male provisioning (in exchange for increased sexual access) would also have resulted in more resource security for females and their progeny. This relationship was doubtless very tenuous early on (i.e., late Miocene) but would have tended toward an evolutionarily stable strategy (ESS) over the course of the Pliocene, characterized by increased aridity and by a progressive multiplication of nutrient-rich, forest margin habitats as the Miocene forests broke up and shrank in a mosaic pattern. Serial monogamy probably became fixed in the hominid lineage when brain size began to increase around the beginning of the Pleistocene (as a consequence of more meat-eating), and niche differentiation became even more marked than it had been previously. Bearing larger brained, more altricial infants is very costly energetically, and would have tended to incapacitate hominid moms for extended periods of time. In sum, the shift in the food niche that accompanied the transition to life on the ground also had an effect on hominid sociality, since it had consequences for the ability of hominid mothers to care for their young. If these inferences are at all logically secure, they suggest that hominids were characterized by serial monogamy ever since they became savanna-adapted bipeds (but probably not before).

Suppression of Estrus

Finally, a case could be made that estrus - temporary, phenetically obvious sexual receptivity - might also have disappeared at around this time (late Miocene). Estrus is characteristic of all the higher primates except humans. Lancaster (1979) has proposed that the episodic physiological and behavioral changes which accompany ovulation in females would have tended to disrupt local group structure, and would have amplified male/male competition for sexual access to females, thus precluding pair bonding. She reasoned from this that if the marked, cyclic, sexual activity associated with estrus were replaced by relatively continuous sexual interest under cortical control, it would have acted to alleviate stress and would have tended to promote more enduring pair bonding between males and females. The many functions of sex among the hominoids tend to support this scenario.

In an ingenious (although politically incorrect) corollary to this approach, Szalay and Costello (1991) suggest that the modern condition of concealed estrus (i.e., concealed ovulation) is not really 'concealment' at all, but rather a permanent and obvious display of continual sexual attractivity. They argue that the notion of concealed estrus cannot be reconciled with a Darwinian explanation for the distinctive somatic sexual attributes of modern human females, which make them more dimorphic in life than their pongid counterparts (similarities in skeletal structure notwithstanding). They focus on the phenetic contrasts between temporary estrus display (TED) - typical of apes - and permanent estrus display (PED) - typical of humans - and argue that the latter evolved from the former as a consequence of a shift from quadrupedality to bipedal erect posture and locomotion. They suggest that, since reproductive strategies are constrained by feeding and locomotion, and by the mechanical design constraints of bipedality, the hominoid female sexual signal system was transformed from the configuration seen in the ancestral estrus state (i.e., vulva-focused swelling) to the fatty buttocks and permanent, pendulous breasts seen in post-pubescent human females. These changes in the sexual signalling paraphernalia are explained by continuous sexual and natural selection, acting together and based on a successful somatic and behavioral female survival and reproductive strategy. The continuous attractivity of PED females, they argue, resulted in both feeding advantages (i.e., increased male provisioning in exchange for sex), and a probable reduction in the incidence of infanticide by potential 'fathers'. The loss of estrus in the hominid clade would, therefore, have coincided with the emergence of bipedality, and would date to the late Miocene.

While impossible to evaluate directly, the modern human female pattern of masked but relatively continuous sexual receptivity must have emerged at some point in our evolutionary history. As Szalay and Costello (1991) observe, under an NDT perspective the origin and transformation of morphology and physiology occur together with behavioral changes, and the latter cannot be decoupled from the somatic attributes of organisms. Despite well-intentioned caveats against post hoc accommodative argument, many crucially important aspects of human evolution simply do not fossilize. This means that we will have to develop other methods if we hope to explain them. Taken together, these observations on the suppression of estrus suggest that the human female sexual signal system is probably very ancient indeed. I think it probably came about as part of the adaptive shift to life on the ground. If the foregoing arguments about monogamy and pair bonding are credible, suppression of estrus would seem almost a necessity in order to counter the centripetal tendencies of intermale competition that it promotes.

Local Group and Foraging Range Sizes - Traditional Approaches

Non-human primates live in groups and, although they vary markedly in size and composition depending on season, aspects of local ecology, and demographic factors unrelated to density, these groups nevertheless exhibit some modal tendencies (Jewett & Clark 1987, Dunbar 1988). While there are advantages to living in relatively large groups (e.g., protection from predators, increased ability to locate resource patches), there are also drawbacks (e.g., stress from resource competition can provoke or accelerate fissioning). For example, Dunbar points out that the incidence of intragroup aggression is directly correlated with group size and the extent to which resources are clumped (1988: 113-115). In order for some degree of stability to prevail, living in large groups would also require social knowledge about other group members, although there is currently no consensus about the content of that knowledge, nor how extensive it must be (cf. Aiello & Dunbar 1993, Mithen 1994). What is important here, though, is that the causes and consequences of large group size are as likely to be applicable to hominids as to any other large terrestrial primate (Mithen 1996).

Early efforts to set limits on local group size estimates were 'site centered' and arose out of the 'home base' literature of the 1960s, and the forager ethnographies that framed it conceptually. The existence of a home base was postulated on the prolongation of pre-adult life and the greater dependency of human young upon adults, especially the mother. Although there are many uncertainties about the relationship between modern and ancient maturation rates (reviewed in Wolpoff 1996: 180-185), the eruption cycle of australopithecine teeth suggests that this period of dependency was on the order of 5-6 years (i.e., like that of modern African apes, ca. 30-50% faster than that of modern humans) (Smith 1986, 1989). There is some reason to believe that this is an underestimate. Inspection of hominoid life history data given in Table 12.1 would imply an early hominid maturation rate, mean age at sexual maturity and at first birth longer than those indicated by the dental eruption data (i.e., ca. 7-9 years). Whatever the case, these studies all strongly suggest that modern developmental patterns have 'slowed down' relative to those of apes, affecting the length of gestation, age at weaning and the onset of menarche. Primate studies indicate that the period of dependency is closely related to the acquisition, through learning, of the skills and behaviors required for physical and social survival in adult society. Young chimps become independent of the mother at about five or six years of age, and it seems likely that Pliocene hominids remained dependent for approximately that length of time. As originally conceptualized, the home base was seen as a constraint imposed on the mobility of primitive hominids by the necessity for this relatively long period of enculturation (Clark 1970). However, and despite a similar period of subadult dependency, none of the African apes produces anything like a home base. Nest sites are individual, change nightly and are never reused.

Analogies with different kinds of extant higher primates established a range of possibilities for early hominid social organization but, by the early 1980s, it became apparent that no pattern 'matched' the archeological record in relatively well-studied areas like Olduvai Gorge, itself the subject of critical scrutiny in terms of site contextual integrity and formation processes (Binford 1981). Because comparisons of possible social organizations are dependent in part upon ecological factors (which can be controlled to some extent) and in part upon demographic factors (which cannot), it became important to try to establish a range for local group sizes (LGS) and for foraging ranges (FR). Unfortunately, there are no established procedures for doing this. The problem is exacerbated by (1) the realization that only a small part of the hominid social landscape is likely to be recognizable archaeologically (and that only rarely), and (2) the near-certainty that early hominid group size and composition varied over the course of an annual cycle and at larger scales in response to changing ecological factors. This means that the exceptionally coarse 'grain' of the Pliocene archaeopaleontological record is likely to confound any attempt at isolating the full range of structural poses necessary to do landscape archaeology (see above).

J. Desmond Clark, using a chimp analogy, suggested groups composed of no more than two or three 'families' of adult females and juveniles, with perhaps three or four adult males more tenuously attached - a LGS on the order of a dozen individuals, possibly with a sex ratio favoring females (1970). By ransacking forager ethnographies, Birdsell proposed a modal LGS of about 25 individuals. Groups in his model are male-dominated, and expand and contract in response to the relative abundance or scarcity of seasonally available resources (1968). He also postulated a larger aggregate - the dialect tribe - composed of ca. 20 local groups, and numbering ca. 500 individuals. The dialect tribe expressed the maximum social 'reach' of any individual during his/her lifetime, and probably corresponded to the limits of the regional mating network based on the potential for face-to-face interaction. Interestingly, Kosse (1990) also sets the maximal face-to-face group at 500 ± 100 individuals, arguing that 500 is a 'cognitive threshold' in long-term memory - the limit set by problems in information processing in non-hierarchical forager contexts. Finally, Jewett and I, using a regression of floor area on population, derived LGS estimates in the 14-21 range for a series of relatively 'high resolution' Oldowan, Developed Oldowan and Acheulean 'living floors' in Bed I at Olduvai Gorge (Jewett & Clark 1987). No temporal trends were apparent. As is often remarked, these are 'magic numbers', invoked to describe the interactions and information processing capabilities of mobile foragers (Martin 1973, Gamble 1996). Although derived empirically by very different techniques, and only partly explicable theoretically, these figures tend to converge and are probably on at least the right order of magnitude (esp. the LGS estimates). It is important to keep in mind that flexibility in local group size and composition is a characteristic response to seasonal resource abundance and scarcity, not only among exant hunter-gatherers in a variety of habitats (e.g., Bushmen, Hadza, Mbuti, Ache, Agta) but also among higher primates in general (e.g., chimpanzees, baboons). It seems plausible to argue that early hominid group size and composition would have been affected by similar factors.

Local Group and Foraging Range Sizes - Recent Approaches

Recent approaches to estimating hominid group sizes use regression analyses on aspects of modern primate morphology to predict modern primate group sizes, thus creating a basis for evaluating the fossils (Dunbar 1992, 1993; Aiello & Dunbar 1993, Steele 1996). Most of these studies have to do with surrogates for braincase volume - relatively accessible for many fossil hominid taxa. Dunbar initiated this approach in 1992, when he developed a model based on a bivariate regression analysis of the relationship of group size and (scaled) neocortex size across extant primate genera, and used it to predict hominid group sizes given known cranial capacities for the fossils. Group sizes predicted for living non-human primate genera using his equations gave an overall encouraging $r^2 = 77.3$ when observed and expected mean group sizes were compared ($n = 35$ genera, both variables logged). He argued from this that strong correlations would indicate a causal relationship between the complexity of primate social life and the cognitive resources at the disposal of its members (see, e.g., Byrne & Whiten 1988). Aiello and Dunbar (1993) then took Dunbar's model and extended its predictive scope by interpolating steps to derive an expected 'cortex ratio' from measurements of the endocranial capacity of primate and hominid skulls expressed as a line (the Reduced Major Axis, or RMA) which they used to predict group size from cortex ratios across 35 primate genera. Fossils are accommodated by deriving a neocortex ratio from measurements of endocranial capacity, although Aiello and Dunbar do not test the RMA equation against primate species of known cranial capacities and known average group size.

Steele (1996: 230-252) provides a comprehensive evaluation of the statistical methodology underlying this research and, surprisingly (given Dunbar's initial results), finds that the confidence with which we can predict cortex ratios from endocranial capacities using the RMA line is very limited ($r^2 = 0.11$). He then goes on to provide a series of tests of the predictive capacity of the Aiello and Dunbar (1993) model, most of which are rather discouraging. Without reiterating Steele's arguments in detail, the poor fit is probably due mostly to two factors. The first is the confounding effect of using a curve based on generic averages (Dunbar 1992) to predict specific (and intraspecific) values for group size on the basis of cortex ratios of extinct hominids of known cranial capacity (Aiello & Dunbar 1993). Steele points out that Aiello and Dunbar assume that there is no significant decrease in the goodness-of-fit for the relationship between cortex ratio and group size when congeneric species are compared for these variables. This assumption is probably not warranted. The second factor is that neocortex ratio is expected to scale isometrically to available metabolic energy, calculated as the product of body mass and the specific metabolic rate (Armstrong 1985). It would seem that Aiello and Dunbar did not take sufficient account of the possible effects of energetic constraints on corticalization, and that some of the variance in the neocortex/brain volume ratio may be explained by variance in total metabolic energy budgets between primates of similar body sizes (Steele 1996: 247). Steele concludes that, while predictions of cortex ratio from absolute brain size in fossil hominid taxa will not be completely unrealistic, they can serve only as very broad indicators of trends in hominid social evolution (1996: 242, 243). As one of the commentators on Dunbar (1992) points out, if conventional least squares regression had been used to predict average group size for humans, the 95% confidence interval would cover a range of 23-446 individuals (Janson 1993).

While regarded as evidence for a woeful lack of precision, these values actually correspond quite closely to the results of traditional research on this subject, and also bracket those of other modern workers, using different approaches (e.g., Martins 1993, Dunbar 1993, Steele 1996). The differences are differences in scale determined by the constants built into the different statistical algorithms employed. Nevertheless, they are not 'all over the map' - there *is* a convergence of results. For example, Martins (1993) predicts ancestral human groups on the order of magnitude of foraging local bands (25-50 persons); Dunbar (1993) predicts groups intermediate in size between minimum and maximum bands (125-150 persons), and the method proposed by Steele (1996) as an improvement on Aiello and Dunbar (1993) yields human groups approaching the upper limits of forager social integration (ca. 300 persons). Since Dunbar (1993: 725) is almost certainly correct when he underscores the archaeological invisibility of much of the early hominid social landscape, it occurred to me that what is identifiable archaeologically is probably restricted to the smaller end of the hominid group size range. In other words, relatively high resolution archaeological sites probably correspond to the material remains of groups of 10-50 individuals, rather than groups of > 100 individuals. The material remains of larger social aggregates are either so dispersed across the landscape that they are unrecognizable as archaeological sites, or they simply have not been located yet. There is some empirical support for this notion in the regressions of floor area on numbers of hominids in Bed I sites at Olduvai Gorge but this research is also limited by what some might consider unwarranted assumptions about the size and shape of these 'sites' (Jewett & Clark 1987).

More problematic than estimates of local group size are foraging or 'home range' estimates. Put another way, LGS estimates converge more closely across different techniques than do those of foraging ranges. A solution proposed by Steele (1996: 248-250) to get at the latter by making use of the fact that home range area scales to group mass (the total biomass of a social group of animals - Grant et al. 1992) systematically underestimates foraging ranges when group mass is estimated using a primate model. Carnivore models would seem to provide a more realistic approximation of home range when compared against raw material transport data from the European Middle Paleolithic (Gamble 1993, 1996) but, as Steele points out, dietary niche, trophic level, body mass and group size all figure in these estimates, and the complex interactions among these variables are not well understood.

Site Settings

It is of some interest to note that all of the Pliocene hominid sites identified so far are located close to a permanent source of fresh water, most commonly at the confluence of a river and a (usually alkaline) lake. Because of a lack of evidence to the contrary, it is assumed that early hominids had no means of transporting or conserving water. Well-watered areas might also have been selected for their denser vegetation cover and consequent protection, and the prevalence of nutrient-rich ecotones. In view of the fact that the topography was characterized by relatively treeless, open savanna vegetation, permanent water sources probably attracted many creatures out of the necessity to slake their thirst, as is the case with east Africa today. If the biomass was greater per unit area in the vicinity of a lake or stream, as certainly seems likely, there would have been more food available in the form of nuts, seeds, rhizomes etc., and as scavengable carcasses (for the carnivores would be present in greater numbers too) than elsewhere on the savanna. An alternative (or complementary) explanation is simply that low energy, lake margin environments make for optimal preservation of fossils, and the sample is biased toward these localities for diagenic and taphonomic reasons.

Seasonal Aggregation and Dispersion

Early hominid environments consisted of open brush grasslands and woodland-parkland mosaics, often in gallery forest situations. Theirs was not a purely tropical adaptation, however, since the high veldt sites in the Transvaal lie well south of the Equator at about 26° S. lat., and at 1500-1800 m in elevation. These are latitudes characterized by marked seasonality, with nightime winter temperatures below freezing. Taking seasonality into account, and given a bimodal distribution in rainfall in both east and south Africa, many workers posit an aggregation-dispersion cycle for early hominids according to the season of the year. This model derives initially from Bushman and Hadza ethnographic accounts, but seems plausible enough in general, if not in detail. As it is usually presented, the largest aggregates (100-200 persons) would have formed during the wet season (October-December, March-May in Kenya, Tanzania) at ecotones rich in plant foods near reliable water sources. Variety and availability of plant foods would have been at a maximum, and scavengable carcasses would have been relatively plentiful. There might also have been a peak in sexual activity at this time, since local groups would have assembled from far and wide, with a resultant increase in potential mating opportunities. In contrast, the east African dry season (August-September) is characterized by resource depletion (esp. of seeds, which were probably a dietary staple). Fissioning into minimal groups accompanied by increased mobility would have been all but inevitable, as hominids spread out over the land in search of resource patches. That Pliocene hominids were periodically subject to dietary stress is indicated by evidence of malnutrition in the teeth (Tobias 1967; Smith 1989, 1992). Enamel hypoplasia - interrupted enamel development seen today at weaning in the children of the very poor - indicates periods of sickness, lack of essential vitamins and minerals, and/or starvation, possibly equated with the lean dry season.

CONCLUSIONS

By way of a summary, there are good reasons to think that Pliocene hominids were (1) *K*-selected, and that they exhibited life-history profiles broadly comparable to those of other *K*-selected mammals; (2) that they were characterized by female dispersal at sexual maturity, and that the roots of this pattern extended back in time beyond the hominid/pongid split; (3) that the female sexual signal system resembled that of modern humans, rather than that of apes; (4) that their social organizations were marked by numerous, although generally weak affective ties manifest in alliances of various kinds; (5) that a pattern of high frequency, low intensity intermale competition was the norm, and that interfemale ties were neutral to moderately antagonistic; (6) that serial monogamy with opportunistic auxiliary liaisons was the primary reproductive strategy for both sexes (but for different reasons); (7) that monogamous pair-bonding between adult males and females lasted only for the period of dependency of single-birth, altricial young, and that (8) that period of dependency was approximately 5-7 years; (9) that traces of this primordial mating pattern can be detected in modern human mating practices (esp. in divorce statistics); (10) that, given estimated maximum lifespans of ca. 20-25 years, the average number of relatively enduring unions was three or four (depending upon whether or not menopause - rare in nature - had evolved); (11) that Pliocene hominids exhibited a pattern of seasonal aggregation and dispersion mediated by the availability of surface water; (12) that wet season aggregations took place in the late Fall-early Spring, and numbered > 100 individuals, of both sexes and all ages; (13) that mating probably peaked during (and might have been largely confined to) wet season aggregations, which would have constituted the most varied genetic reservoirs, and (14) that average group size was minimal (< 20 individuals) at the end of the dry season (August-September), when hominids were maximally dispersed over the landscape.

Whether *AP3A* readers will regard these conclusions as credible or not will be determined by their sympathy (or lack thereof) with the eclectic approach taken here. At the beginning of the essay, I acknowledged that what Potts (1988) has called 'third order inference' is not directly amenable to empirical evaluation. Those readers who subscribe to strict empiricist biases will, therefore, probably dismiss the paper out of hand (see Clark [1993b: 211-213] for a definition of strict empiricism). However, it strikes me that satisfactory explanation in human evolution must be built from 'circumstantial evidence' - intertwined strands of 'facts' linked together by strong inference and tied to central NDT concepts (e.g., adaptation) warranted in biological evolution in general (Mayr

1982). If we continue to consider humans as somehow 'unique' in terms of their evolutionary heritage, we will continue to turn our backs on neo-Darwinian evolutionary theory, considered by some to be the most powerful metaphysical paradigm for describing and explaining 'reality' that humans have ever developed. I suggest that the reason that archaeology has been slow to take advantage of NDT is simply that most archaeologists are insufficiently aware of its explanatory potential. Coming out of an intellectual tradition dominated by social and cultural anthropology, and with a weak commitment to the materialist biases that underlie NDT, they usually have little exposure to biological evolution.

In mulling over what I had learned from trying to write this essay, I was left with three strong impressions. The first was that archaeologists interested in evolutionary process questions in 'deep time' can only ignore NDT at their peril. The time-space grid of the Pliocene archaeological record is so coarse-grained that recourse to pattern searches on it can explain almost nothing of interest beyond the level of the individual site (and that only with difficulty). The second was that 'culture', defined as complex learned behavior, is likely to have been utterly irrelevant in deep time, and is probably largely epiphenomenal as a determinant of human behavior in general. Even today, the tyranny of the genes lies just below the surface of the cultural veneer, and the notion that culture is deterministic in some fundamental sense is probably an illusion. As Dawkins put it, organisms can be thought of simply as machines engineered by natural selection to permit the survival of DNA (1995). All that really matters from a genetic point of view is that we live long enough to reproduce, and it is the contest of the genes for survival, rather than anything to do with organisms per se, that is driving evolution. While admittedly reductionist, and perhaps objectionable from the perspective of the methodological individual so important to NDT, Dawkins' perspective underscores the fact that NDT can be invoked to explain a wide range of phenomena usually regarded as 'cultural' when dealing with humans. Although, ignorant of genetics, Darwin focused on the effects of natural selection on the individual organism, Dawkins' genetic determinism is not at all incompatible with NDT as it is presently conceptualized. Finally, and in light of the behaviorist claim that NDT is so loose and all-encompassing that practically anything can be explained post hoc by invoking it, I would point out that the credibility of inference grounded paradigmatically in NDT can only be evaluated in terms of the specifics of the particular case. *Pace* Gould and Lewontin (1979), explaining human biological and cultural evolution is not something that can be done convincingly without partitioning it into its constituent parts. This means that the enormous explanatory potential of NDT will only begin to be realized by applications of its tenets and concepts in more circumscribed problem contexts.

"And do you, Rebecca, promise to make love only to Richard, month after month, year after year, and decade after decade, until one of you is dead?"

Drawing by Cheney © 1997 The New Yorker Magazine, Inc.

ENDNOTE

Data in Tables 12.1 and 12.2 (Rowe 1996) combine captive and wild studies but are heavily biased toward the former. They cannot be taken to indicate wild life history and demographic parameters.

ACKNOWLEDGMENTS

This essay was written while I was a visiting fellow at Clare Hall, Cambridge. I would like to take this opportunity to thank all those individuals connected with my stay who made it such an intellectually rewarding and enjoyable experience. Particularly important were other visitors and dining companions at Clare Hall, notably Randi and Gunnar Haaland, Rosemary Luff, Geoff Bailey, Helen Leach, Jim Council, Rob Finnegan, Henry Sullivan, Jim and Jill Evans and Dympna Callaghan - by no means all archaeologists. Of the archaeology and biological anthropology faculties, Paul Mellars, Geoff Bailey, Chris and Ann Chippindale, Marina Mosquera, William Davies and Terry Hopkinson were especially tolerant of my efforts to convey the ideas embodied in this essay, which remains very much 'a work in progress'. Obviously, Rob Foley's seminal publications on the general approach have been an inspiration. ASU colleague John Alcock gave the essay a good 'read' and I have sought to incorporate his thoughtful observations whenever possible. ASU primatologist Leanne Nash provided the helpful references incorporated in Tables 12.1 and 12.2, and in the bracketing discussion. I am grateful to Polly Wiessner for bringing the Szalay and Costello (1991) paper to my attention. Finally, *un agradecimiento* to my former student, Kathy Roler (Eastern New Mexico University) who, through the wonders of e-mail, engaged me in some serious discussion about what all of this might mean.

REFERENCES CITED

Aiello, L. & R. Dunbar
1993 Neocortex size, group size, and the evolution of language. *Current Anthropology* 34: 184-193.

Allen, L. et al.
1982 Demography and human origins. *American Anthropologist* 84: 888-896.

Armstrong, E.
1985 Allometric considerations of the adult mammalian brain with special emphasis on primates. In *Size and Scaling in Primate Biology* (W. Jungers, ed.), pp. 115-146 New York: Plenum.

Betzig, L.
1986 *Despotism and Differential Reproduction: a Darwinian View of History.* New York: Aldine de Gruyter.
1989 Causes of conjugal dissolution: a cross-cultural study. *Current Anthropology* 30: 654-676.

Betzig, L., M. Borgerhoff Mulder & P. Turke
1988 *Human Reproductive Behavior: a Darwinian Perspective.* New York: Cambridge University Press.

Binford, L.
1981 *Bones: Ancient Men and Modern Myths.* New York: Academic Press.
1989 Isolating the transition to cultural adaptations: an organizational approach. In *The Emergence of Modern Humans* (E. Trinkaus, ed.), pp. 18-41. Cambridge: Cambridge University Press.

Birdsell, J.
1968 Some predictions for the Pleistocene based on equilibrium systems among recent hunter-gatherers. In *Man the Hunter* (R. Lee & I. DeVore, eds.), pp. 229-240. Chicago: Aldine.

Bonner, J.T.
1980 *The Evolution of Culture in Arnimals.* Princeton: Princeton University Press.

Boyd, R. & P. Richerson
1985 *Culture and the Evolutionary Process.* Chicago: University of Chicago Press.

Bunn, H. & E. Kroll
1986 Systematic butchery by Plio-Pleistocene hominids at Olduvai Gorge, Tanzania. *Current Anthropology* 27: 431-452.

Byrne, R. & A. Whiten (eds.)
1988 *Machiavellian Intelligence: Social Expertise and the Evolution of Intellect in Monkeys, Apes and Humans.* Oxford: Clarendon Press.

Cashdan, E.
1996 Women's mating strategies. *Evolutionary Anthropology* 5: 134-143.

Cheney, D., R. Seyfarth & B. Smuts
1986 Social relationships and social cognition in nonhuman primates. *Science* 234: 1361-1366.

Clark, G.A.
1977 Review: *Spatial Analysis in Archaeology* (Hodder & Orton). *American Antiquity* 43: 132-135.
1988 Some thoughts on the Black Skull: an archeologist's assessment of WT-17000 *(A. boisei)* and systematics in human paleontology. *American Anthropologist* 90: 357-371.
1989a Alternative models of Pleistocene biocultural evolution: a response to Foley. *Antiquity* 63: 153-161.
1989b Paradigms and paradoxes in paleoanthropology: a response to C. Loring Brace. *American Anthropologist* 91: 446-450.
1992 A comment on Mithen's 'Ecological Interpretation of Palaeolithic Art'. *Proceedings of the Prehistoric Society* 58: 107-109.
1993a Review: *History and Evolution* (Nitecki & Nitecki, eds.). *American Journal of Physical Anthropology* 90: 129, 130.
1993b Paradigms in science and archaeology. *Journal of Archaeological Research* 1: 203-234.

Clark, J.D.
1970 *The Prehistory of Africa.* New York: Praeger.

Clutton-Brock, T.
1989a Female transfer, male tenure and inbreeding avoidance in social mammals. *Nature* 337: 70-72.
1989b Mammalian mating systems. *Proceedings of the Royal Society B* 236:339-372.

Clutton-Brock, T., P. Harvey & B. Rudder
1977 Sexual dimorphism, socionomic sex ratio and body weight in primates. *Nature* 269: 797-800.

Darwin, C.
1859 (1872) *The Origin of Species.* London: John Murray.

Dawkins, R.
1995 God's utility function. *Scientific American* 267 (Nov.): 80-85.

Dunbar, R.
1987 Demography and reproduction. In *Primate Societies* (B. Smuts et al., eds.), pp. 240-249. Chicago: University of Chicago Press.
1988 *Primate Social Systems.* London: Croom Helm.
1992 Neocortex size as a constraint on group size in primates. *Journal of Human Evolution* 22: 469-493.
1993 Coevolution of neocortical size, group size and language in humans. *Behavioral and Brain Sciences* 16: 681-735.

Ember, M. & C. Ember
1979 Male-female bonding: a cross-species study of mammals and birds. *Behavior Science Research* 14: 37-56.

Emlen, S. & L. Oring
1977 Ecology, sexual selection and the evolution of mating systems. *Science* 197: 215-223.

Falk, D.
1983 Cerebral cortices of East African early hominids. *Science* 222: 1072-1074.
1986 Evolution of cranial blood drainage in hominids. *American Journal of Physical Anthropology* 70: 311-324.
1987 Hominid paleoneurology. *Annual Review of Anthropology* 16: 13-30.
1991 3.5 million years of hominid brain evolution. *Seminars in the Neurosciences* 3: 409-416.

Feder, M.
1997 Review: *Adaptation* (Rose & Lauder, eds.). *Science* 277:189.

Fisher, H.
1987 The four-year itch. *Natural History* (Oct.): 22-33.
1989 Evolution of human serial pairbonding. *American Journal of Physical Anthropology* 78: 331-354.
1992 *Anatomy of Love.* New York: Fawcett Columbine.

Flinn, M. & B. Low
1986 Resource distribution, social competition and mating patterns in human societies. In *Ecological Aspects of Social Evolution* (D. Rubenstein & R. Wrangham, eds.), pp. 221-237. Princeton: Princeton University Press.

Foley, R. A.
1987 *Another Unique Species: Patterns in Human Evolutionary Ecology.* Harlow: Longman Scientific.
1988 Hominids, humans and hunter-gatherers: an evolutionary perspective. In *Hunters and Gatherers I: History, Evolution and Social Change* (T. Ingold et al., eds.), pp. 207-221. New York: Berg.
1989 The evolution of hominid social behaviour. In *Comparative Socioecology* (V. Standen & R. Foley, eds.), pp. 473-494. Oxford: Blackwell Scientific.
1992 Evolutionary ecology of early hominids. In *Evolutionary Ecology and Human Behavior* (E. Smith & B. Winterhalder, eds.), pp. 131-164. New York: Aldine de Gruyter.
1996 The adaptive legacy of human evolution: a search for the environment of evolutionary adaptiveness. *Evolutionary Anthropology* 4: 194-203.

Foley, R. & P. Lee
1989 Finite social space, evolutionary pathways and reconstructing hominid behavior. *Science* 243: 901-906.
1991 Ecology and energetics of encephalization in hominid evolution. *Philosophical Transactions of the Royal Society, Series B* 334: 223-232.
1996 Finite social space and the evolution of human social behaviour. In *The Archaeology of Human Ancestry* (J. Steele & S. Shennan, eds.), pp. 47-66. London: Routledge.

Frayer, D. & M. Wolpoff
1985 Sexual dimorphism. *Annual Review of Anthropology* 14: 429-473.

Frayser, S.
1985 *Varieties of Sexual Experience: an Anthropological Perspective.* New Haven: HRAF Press.

Frisch, R
1978 Population, food intake and fertility. *Science* 199: 22-30.
1989 Body weight and reproduction. *Science* 246: 232.

Frisch, R. & E. Revelle
1970 Height and weight at menarche and a hypothesis of critical weights and adolescent events. *Science* 169: 397-399.

Gamble, C.
1986 *The Palaeolithic Settlement of Europe.* Cambridge: Cambridge University Press.
1993 Exchange, foraging and local hominid networks. In *Trade and Exchange in Prehistoric Europe* (C. Scarre & F. Healy, eds.), pp. 35-44. Oxford: Oxbow Monograph No. 33.
1996 Making tracks. In *The Archaeology of Human Ancestry* (J. Steele & S. Shennan, eds.), pp. 252-277. London: Routledge.

Gibbons, A.
1990 Paleontology by bulldozer. *Science* 247: 1407-1409.

Gifford-González, D.
1991 Bones are not enough: knowledge and interpretative strategies in zooarchaeology. *Journal of Anthropological Archaeology* 10: 215-254.

Goodall, J.
1986 *The Chimpanzees of Gombe.* Cambridge, MA: Belknap Press.

Gould, S. & R. Lewontin
1979 The spandrels of San Marco and the Panglossian paradigm: a critique of the adaptationist programme. *Proceedings of the Royal Society* 205: 581-598.

Grassé, P.
1977 *Evolution of Living Organisms.* New York: Academic Press.

Harcourt, A. & K. Stewart
1987 The influence of help in contests on dominance rank in primates: hints from gorillas. *Animal Behaviour* 35: 182-190.

Harvey, P., R. Martin & T. Clutton-Brock
1987 Life histories in comparative perspective. In *Primate Societies* (B. Smuts et al., eds.), pp. 181-196. Chicago: University of Chicago Press.

Hawkes, K.
1990 Why do men hunt? Some benefits for risky strategies. In *Risk and Uncertainty* (E. Cashdan, ed.), pp. 145-166. Boulder: Westview Press.
1996 Foraging differences between men and women. In *The Archaeology of Human Ancestry* (J. Steele & S. Shennan, eds.), pp. 283-305. London: Routledge.

Hinde, R. (ed.)
1983 *Primate Social Relationships.* Oxford: Blackwell.

Holloway, R.
1975 Early hominid endocasts: volumes, morphology, and significance for hominid evolution. In *Primate Functional Morphology and Evolution* (R. Tuttle, ed.), pp. 393-415. The Hague: Mouton.
1983 Cerebral brain endocast pattern of *Australopithecus afarensis* hominid. *Nature* 303: 420-422.
1983 Human brain evolution: a search for units, models and synthesis. *Canadian Journal of Anthropology* 3: 215-230.

Holloway, R. & W. Kimbel
1986 Endocast morphology of Hadar hominid AL 162-28. *Nature* 321: 536-537.

Hrdy, S. B.
1981 *The Woman that Never Evolved.* Cambridge, MA: Harvard University Press.
1986 Empathy, polyandry, and the myth of the coy female. In *Feminist Approaches to Science* (R. Bleier, ed.), pp. 41-60. New York: Pergamon Press.

Hrdy, S. B. & P. Whitten
1987 Patterning of sexual activity. In *Primate Societies* (B. Smuts et al., eds.), pp. 370-384. Chicago: University of Chicago Press.

Isaac, G. Ll.
1969 Studies of early culture in East Africa. *World Archaeology* 1: 1-28.
1984 The archaeology of human origins: studies of the Lower Pleistocene in East Africa 1971-1981. *Advances in Old World Archaeology* 3: 1-87.

Janson, C.
1993 Primate group size, brains and communication: a New World perspective. *Behavioral and Brain Sciences* 16: 711-712.

Jewett, R. & G. A. Clark

1987 Observations on estimating local group size at Olduvai Gorge, Tanzania. In *Coasts, Plains and Deserts: Essays in Honor of Reynold J. Ruppé* (S. Gaines, ed.), pp. 117-127. Tempe: Arizona State University Anthropological Research Paper No. 38.

Johanson, D. & T. White

1979 A systematic assessment of early African hominids. *Science* 203: 321-330.

Kano, T.

1992 *The Last Ape: Pygmy Chimpanzee Behavior and Ecology.* Palo Alto, CA: Stanford University Press.

Kleiman, D.

1977 Monogamy in mammals. *Quarterly Review of Biology* 52: 39-69.

Kosse, K.

1990 Group size and societal complexity: thesholds in the long-term memory. *Journal of Anthropological Archaeology* 9: 275-303.

Lake, M.

1996 Archaeological inference and the explanation of hominid evolution. In *The Archaeology of Human Ancestry* (J. Steele & S. Shennan, eds.), pp. 184-206. London: Routledge.

Lancaster, J.

1979 Sex and gender in evolutionary perspective. In *Human Sexuality* (M. Katchadourian, ed.), pp. 232-256. Berkeley: University of California Press.

Lancaster, J. & H. Kaplan

1992 Human mating and family formation strategies. In *Topics in Primatology, Vol. 1* (T. Nishida et al., eds.), pp. 21-33. Tokyo: Tokyo University Press.

Lancaster, J. & C. Lancaster

1983 Parental investment: the hominid adaptation. In *How Humans Adapt: a Cultural Odyssey* (D. Ortner, ed.), pp. 117-131. Washington, DC: Smithsonian Institution Press.

Lewin, R.

1993 *Human Evolution: an Illustrated Introduction.* Oxford: Blackwell Scientific Publications.

Lovejoy, C. O.

1981 The origin of man. *Science* 211: 341-350.

Manson, J.

1997 Primate consortships: a critical review. *Current Anthropology* 38: 353-374.

Martin, J.

1973 On the estimation of the sizes of local groups in a hunting-gathering environment. *American Anthropologist* 75: 1448-1468.

Martin, J. & D. Stewart

1982 A demographic basis for patrilineal hordes. *American Anthropologist* 84: 79-96.

Martins, E.

1993 Comparative studies, phylogenies and predictions of co-evolutionary relationships. *Behavioral and Brain Sciences* 16: 714-716.

Maryanski, A.

1992 The last ancestor: an ecological network model on the origins of human sociality. *Advances in Human Ecology* 1: 1-32.

1993 The elementary forms of the first protohuman society: an ecological-social network approach. *Advances in Human Ecology* 2: 215-241.

1996 African ape social networks. In *The Archaeology of Human Ancestry* (J. Steele & S. Shennan, eds.), pp. 67-90. London: Routledge.

Maryanski, A. & J. Turner

1991 Network analysis. In *The Structure of Sociological Theory* (J. Turner, ed.), pp. 540-572. Belmont, CA: Wadsworth.

Mayr, E.

1982 *The Growth of Biological Thought.* Cambridge, MA: Belknap Press.

McHenry, H.

1986 Size variation in the postcranium of *A. afarensis* and extant species of Hominoidea. *Journal of Human Evolution* 15: 149-156.

1991a Sexual dimorphism in *A. afarensis. Journal of Human Evolution* 20: 21-32.

1991b Femoral lengths and stature in Plio-Pleistocene hominids. *American Journal of Physical Anthropology* 85: 149-158.

1992a Body size and proportions in early hominids. *American Journal of Physical Anthropology* 87: 407-431.

1992b How big were early hominids? *Evolutionary Anthropology* 1: 15-19.

1996 Sexual dimorphism in fossil hominids and its socioecological implications. In *The Archaeology of Human Ancestry* (J. Steele & S. Shennan, eds.), pp 91-109. London: Routledge.

Mithen, S.

1991 Home bases and stone caches: the archaeology of early hominid activities. *Cambridge Archaeological Journal* 1: 277-283.

1994 Technology and society during the Middle Pleistocene. *Cambridge Archaeological Journal* 4: 3-33.

1996 Social learning and cultural tradition. In *The Archaeology of Human Ancestry* (J. Steele & S. Shennan, eds.), pp. 207-229. London: Routledge.

Murdock, G. P.

1949 *Social Structure.* New York: Free Press.

1967 *Ethnographic Atlas.* Pittsburgh: University of Pittsburgh Press.

Palombit, R.

1994a Dynamic pair bonds in hylobatids: implications regarding monogamous social systems. *Behaviour* 128: 65-101.

1994b Extra-pair copulations in a monogamous ape. *Animal Behaviour* 47: 721-723.

Pilbeam, D.

1986 Hominoid evolution and hominoid origins. *American Anthropologist* 88: 295-312.

Plavcan, J. & C. van Schaik

1992 Intrasexual competition and canine dimorphism in anthropoid primates. *American Journal of Physical Anthropology* 87: 461-477.

Potts, R.

1988 *Early Hominid Activities at Olduvai.* New York: Aldine de Gruyter.

Queller, D.

1995 The spaniels of St. Marx and the Panglossian paradox: a critique of a rhetorical programme. *The Quarterly Review of Biology* 70: 485-489.

Radcliffe-Brown, A. R.

1930 The social organization of the Australian tribes. *Oceania* 1: 1-63, 206-246.

Remane, A.
1971 *Die Grundlagen des natürlichen Systems der vergleichenden Anatomie und der Phylogenetik.* Königstein-Taunus: Koeltz.

Rowe, N.
1996 *The Pictorial Guide to the Living Primates.* East Hampton, NY: Pogonias Press.

Schindewolf, O.
1950 *Grundfragen der Paläontologie.* Stuttgart: Schweizerbart.

Service, E.
1962 *Primitive Social Organization.* New York: Random House.

Shipman, P.
1986 Scavenging or hunting in early hominids: theoretical framework and tests. *American Anthropologist* 88: 27-43.

Shipman, P. & Harris, J.
1988 Habitat preference and paleoecology of *A. boisei* in eastern Africa. In *Evolutionary History of the 'Robust' Australopithecines* (F. Grine, ed.), pp. 343-382. New York: Aldine de Gruyter.

Small, M.
1993 *Female Choice: Sexual Behavior of Female Primates.* Ithaca: Cornell University Press.

1995 Making a monkey out of human nature. *New Scientist* 146: 30-33.

Smith, B. H.
1986 Dental development in *Australopithecus* and early *Homo. Nature* 323: 327-330.

1989 Dental development as a measure of life history in primates. *Evolution* 43: 683-688.

1992 Life history and the evolution of human maturation. *Evolutionary Anthropology* 1: 134-142.

Smuts, B.
1987 Sexual competition and mate choice. In *Primate Societies* (B. Smuts et al., eds.), pp. 385-399. Chicago: University of Chicago Press.

Steele, J. & S. Shennan
1996 Introduction. In *The Archaeology of Human Ancestry* (J. Steele & S. Shennan, eds.), pp. 1-42. London: Routledge.

Steele, J.
1996 On predicting hominid group sizes. *The Archaeology of Human Ancestry* (J. Steele & S. Shennan, eds.), pp. 230-252. London: Routledge.

Stiner, M.
1994 *Honor Among Thieves.* Princeton: Princeton University Press.

Szalay, F. & R. Costello
1991 Evolution of permanent estrus displays in hominids. *Journal of Human Evolution* 20: 439-464.

Tobias, P.
1967 *The Cranium and Maxillary Dentition of A. (Zinjanthropus) boisei.* Olduvai Gorge, Vol. 2. Cambridge: Cambridge University Press.

Trivers, R.
1972 Parental investment and sexual selection. In *Sexual Selection and the Descent of Man 1871-1971* (B. Campbell, ed.), pp. 136-179. Chicago: Aldine.

Turner, A.
1984 Hominids and fellow-travelers: human migration into high latitudes as part of a large mammal community. In *Hominid Evolution and Community Ecology* (R. Foley, ed.), pp. 193-218. London: Academic Press.

van den Berghe, P.
1979 *Human Family Systems: an Evolutionary View.* Westport, CT: Greenwood Press.

Williams, G.
1974 *Sex and Evolution.* Princeton: Princeton University Press.

Wittenberger, J. & R. Tilson
1980 The evolution of monogamy: hypotheses and evidence. *Annual Review of Ecology and Systematics* 11: 197-232.

Wobst, H. M.
1974 Boundary conditions for paleolithic social systems: a simulation approach. *American Antiquity* 39: 147-178.

1976 Locational relationships in paleolithic society. *Journal of Human Evolution* 5: 49-58.

Wolpoff, M.
1996 *Human Evolution.* New York: McGraw-Hill.

Wrangham, R.
1979a On the evolution of ape social systems. *Social Science Information* 18: 334-368.

1979b Sex difference in chimpanzee dispersion. In *The Great Apes* (D. Hamburg & E. McCown, eds.), pp. 481-489. Menlo Park, CA: Benjamin Cummings.

1987a The significance of African apes for reconstructing human social evolution. In *The Evolution of Human Behavior: Primate Models* (W. Kinzey, ed.), pp. 51-71. Albany: SUNY Press.

1987b Evolution of social structure. In *Primate Societies* (B. Smuts et al., eds.), pp. 282-296. Chicago: University of Chicago Press.

13

The Mode and Tempo of the Initial Human Colonisation of Empty Landmasses: Sahul and the Americas Compared

R. Esmée Webb
Yamaji Language Centre, Geraldton, Australia

David J. Rindos
20 Herdsman Parade, Wembley, Australia

ABSTRACT

The *tempo* and *mode* of the initial human colonisation of *terra incognita* has long been a controversial topic in archaeology. Two opposing models have been used to explain the earliest evidence for human occupation of 'Sahul' (a eustatically enlarged Pleistocene landmass that incorporated Tasmania and New Guinea): that people adapted quickly to new environments and hence spread rapidly over the region, or that they found adaptation to new conditions difficult and hence spread slowly, initially around the coast. The assumption that the rate of adaptation to new conditions determines the rate of spread underlies both hypotheses.

We present here an alternative model of colonisation that reconsiders these issues. We suggest that initial human movement through Sahul would have been inherently fast. However, this rate of spread would *not* have been a consequence of rapid adaptation, but of maladapted subsistence strategies that resulted in inefficient extractive methods and hence low carrying capacities. Thus, we reverse the normal argument and posit that *minimising* adaptiveness, as long as human survival is possible, will *maximise* the rate of human spread over an empty landmass.

INTRODUCTION

Humans are now found on every habitable continent and all islands large enough to support viable breeding populations. However, their settlement of many parts of the world is a relatively recent event, in evolutionary terms. It is generally accepted that the genus *Homo* originated in Africa (Mellars & Stringer 1989; Bräuer & Smith 1990; Mellars 1990). Hence the earliest archaeological remains found anywhere outside that continent are presumably evidence of human migration. The fact that humans were not originally endemic to Europe, Asia, Australia or the Americas has been instrumental in fuelling a long-standing debate on the mechanics of migration, especially of humans into new worlds (e.g. Lee 1966; Kirk & Szathmary 1985; Mascie-Taylor & Lasker 1988; Hill & Serjeantson 1989; Anthony 1990; Beaton 1991; Rindos & Webb 1992; Cavalli-Sforza et al. 1993; Clark 1994; Young & Bettinger 1995).

This controversy is not really about the timing of migration events, however. Few, if any, scholars would be willing to maintain a model for human migration based on simple diffusion. Under such a model, migration would be controlled *solely* by non-cultural processes, and the arrival of humans into a new area would merely result from their presence in an immediately adjacent area at a slightly earlier period. Hence, given time and proximity, human populations would inevitably spread across the globe and the specific time at which a particular region was initially colonised would merely be a consequence of its accessibility. This model forms the null hypothesis for migration studies. It has never been considered, perhaps because it is self-evidently true. People *did* reach different regions of the world at different times, but that fact does not explain how they got there.

Yet, even under this null hypothesis, we must accept that a precondition for colonisation is a viable human population in proximity to unoccupied land. A standing population is also a requisite if a boundary is to be crossed or a frontier settled. Yet this population, too, had a history of migration and adaptation. At some time in the past, the source region was itself a frontier. That a standing population capable of further migration exists in a given region means that it had previously spread and adapted, via culturally mediated behaviour, from other, often different, conditions. That a population spread in the past does not, of course, mean that it will continue to do so in the future. Yet at least some populations did continue to spread. Was this merely matter of chance? Or could it be the case that certain cultural adaptations enhance the probability that a given population will disperse? Even taking the random movement of human populations as a given, it would seem clear that unless the appropriate behaviours for survival in new regions exist in a colonising population, the settlement of such regions will not be successful. It would also seem that the mean rate at which such adaptations occur will determine the absolute time taken for dispersal and the successful colonisation of new regions. Hence, even under a simple diffusion model, we cannot totally avoid the temporal problem raised by the process of cultural adaptation to local conditions. In earlier treatments of human migration, it has been just these parameters that have received most attention.

Precisely how (*mode*) and how quickly (*tempo*) people were able to colonise previously unoccupied territory after their initial arrival has been viewed as critical to understanding the cultural adaptability of past human populations (e.g. Davidson & Noble 1992). It has generally been assumed, almost as an article of faith, that humans somehow design their cultures to be well-adapted, perhaps even optimally adapted, to local environments. Hence, if the environment changes, or if a new environment is encountered during migration, people must alter their behavioural economics to re-establish the best culturally mediated adaptation to the local environment.

Many archaeologists consider that people had 'difficulty' making these kinds of cultural alterations and hence that human adaptation to radically changed environmental conditions, due either to migration to new regions or to changing climate, would be 'slow'. Other researchers hold that humans are capable of adapting their cultures 'quickly' to new conditions and that the challenges presented by migration or changing climate would be met rapidly by means of new, culturally mediated, adaptations. We examine these two approaches to the nature of cultural adaptation here in the context of migration to new, previously unoccupied, landmasses.

When the archaeological record is considered as a whole, the continents naturally appear to have been colonised sequentially from Africa. Predictably, since it is the closest to Africa and linked to it by land, Eurasia was the first region outside Africa to be occupied, apparently by archaic *Homo sapiens*, at least 750,000 years ago (Roebroeks 1994, Roebroeks & von Kolfschoten 1994, Carbonell et al. 1995, Dennell & Roebroeks 1996). Although the timing of the initial human colonisation of Sahul is the subject of ongoing debate (Allen 1989, 1994; Bowdler 1993, Smith et al. 1993, Roberts & Jones 1994; Roberts et al. 1994a, 1994b; Allen & Holdaway 1995, Chappell et al. 1996), it is generally accepted that a viable population of anatomically modern humans (*Homo sapiens sapiens*) was established there by at least 40,000 BP (Habgood 1989, 1991; Kramer 1991, Brown 1993). It appears that the Americas were the last habitable landmass to be colonised, also by anatomically modern humans, probably around 11,500 BP, although this date remains controversial (Carlisle 1988; Meltzer 1989, 1995; Bonnichsen & Turnmire 1991, Dillehay & Meltzer 1991).

Meaningful discussion of the nature of the original colonisation of Eurasia is effectively impossible at present due to the time at which it took place, the dearth of precise information about the ages of the known archaeological sites and the socio-economics and biology of the first occupants. Further confusion is added by the difficulties encountered in providing reliable calendrical dates for the oldest radiometric ages obtained in Eurasia. Precise temporal control is obviously an essential precondition for the discussion of the *mode* and *tempo* of the 'first horizons' of human occupation of any region. Moreover, analysis of the evidence for the far more recent initial human colonisation of Sahul and the Americas is not simple. Discussion of Sahulian colonisation is complicated by the fact that it apparently occurred around 60,000 BP (Roberts & Jones 1994; Roberts et al. 1994a, 1994b), a period that lies beyond the effective limit of radiocarbon (^{14}C) dating (Aitken 1990, Chappell et al. 1996). However, the model of Sahulian colonisation presented below can be tested with reference to the more recent colonisation of the Americas, which occurred well within the limits of ^{14}C dating. As will be seen, the New World provides a case study in colonisation that is not subject to the inherent problems of chronometric precision found in other parts of the world.

SAHULIAN COLONISATION

Humans are unquestionably intrusive into the endemic fauna of Sahul. Moreover, they could only have reached the region by over-water dispersal from Sundaland, the island-filled area of shallow sea that stretches from Indochina to Sulawesi. Clearly, at some time in the past, at least one founder population reached Sahul from Sundaland and survived. Radiometric dates obtained from archaeological sites throughout Sahul (Smith et al. 1993), show that this event had considerable antiquity, although it is unlikely to have taken place as early as has recently been claimed (Fullagar et al. 1996). Two alternative scenarios, derived from opposing preconceptions regarding the intrinsic adaptability of human culture, have been proposed to explain the *tempo* and *mode* of this initial colonisation.

The 'Fast' Model

Birdsell (1953, 1957, 1977) suggested that once people reached Sahul they would, of necessity, have adapted rapidly to their new environment. They would, therefore, have quickly radiated out to fill all the available ecological niches, else they would not have survived to leave archaeological traces of their presence. He developed mathematical models for population growth rates, which suggested that a biologically viable founding population could have multiplied to 'saturate' the entire continent within about 5000 years of first arrival.

This model of how humans colonised Sahul is in fundamental accord with biogeographic theories arguing that an invading species arriving in a new environment in which it encounters few, if any, competitors is likely to colonise it rapidly and successfully. The endemic Sahulian fauna is markedly depauperate in the large-bodied mammalian carnivores that would have been gatherer-hunters' chief competitors (Webb n.d. a). Thus, people should have been able to spread rapidly through the region upon arrival.

Descriptions of the colonisation of Sahul that are in keeping with the Birdsellian model have also been given by Mulvaney (1975), Shawcross (1975), Jones (1979, 1989, 1992) and Allen (1989, Anderson 1991). They presented archaeological evidence that supports Birdsell's theoretical model quite strongly.

The 'Slow' Model

On the other hand, Bowdler (1977, 1990a), Horton (1981) and Lourandos (1987) all considered the timescale implicit in Birdsell's model unrealistically short. They argued that he ignored the need for the colonising population to develop highly specialised cultural adaptations in order to survive in Australia's unique environment, especially the deserts of the continental interior.

Bowdler (1977) formulated her 'coastal colonisation hypothesis' specifically to make allowance for the problem of slow inherent rates of cultural adaptation. She assumed, reasonably, that the initial colonisers must have had some sort of maritime technology or they would never have reached Sahul. She therefore argued that they would have been 'pre-adapted' to the exploitation of coastal resources and would initially have been restricted to littoral environments. Later, as the original strictly coastal populations increased in number, population pressure would have forced people to venture up the major river systems, such as the Murray-Darling, where they would have been able to practice a more or less 'littoral' lifestyle. Only after the resource base in these areas had reached saturation and people had had the opportunity to develop new efficient extractive techniques would they have begun to spread into the interior deserts.

This hypothesis met early criticism (White & O'Connell 1982) on the cogent grounds that the sites Bowdler used as evidence that the initial colonists were coastally adapted, were not in fact located on or near where the coast would have been at the time when they were occupied. Bowdler (1977) had not allowed for the eustatic effects of temperate latitude glaciation. Glacially lowered sea levels would mean that many sites with old ^{14}C dates, that are situated close to the present coastline, in fact lay far from the ocean when they were first occupied. Since these sites are not in fact littoral (Hallam 1987), it is not surprising that the food remains recovered from them show little or no evidence that their occupants exploited coastal resources.

Bowdler (1990a) subsequently admitted that her original hypothesis was an inductive generalisation drawn from the data available in the early 1970s, when no sites older than 12,000 BP were known in central Australia. Hence, her argument that 'desert adaptations' would have been slow to develop might appear to have been conclusively negated by the discovery of sites in the central desert whose earliest archaeological levels are as old as those in sites located on the periphery of the continent (Smith 1987, 1989; Jones 1987).

However, the situation is not that simple and the problem of the *mode* and *tempo* of the initial human colonisation of Sahul remains unresolved. Not least because the earliest archaeological evidence for occupation of the Sahulian coastline is probably unrecoverable. It presumably lies inundated on what is now the continental shelf. The impossibility of recovering the earliest evidence for human occupation of Sahul leads to an apparent stalemate. The known 'coastal' sites are not in fact located on the then coastline, while sites known from the continental interior are inherently likely to be younger than the initial occupation of the continental margin, but the ages of the truly coastal sites are unknowable because their locations are unidentifiable. Hence, it is impossible to substantiate the date at which people first colonised Sahul, especially from the perspective of those who espouse a model of coastal colonisation. Thus, it is not surprising that Rindos and Webb (1992) concluded that some of the questions we would like to ask regarding the human capacity to adapt that are posed by the settlement of Sahul may remain unanswered forever.

Further Considerations

Debate on the *mode* and *tempo* of the initial human colonisation of Sahul has been hampered by a fundamental confusion in the minds of some participants regarding the nature of these concepts. The *tempo* (rate) at which people colonised an area must be carefully differentiated from the rate at which the colonising population developed new, culturally mediated, adaptations to their new environment - what can be called the 'settling in process'. We have already argued that rapid spread may be totally disconnected from rapid cultural change and/or adaptation to local conditions (Rindos & Webb 1992). We showed that the greater the degree of cultural adaptation to a given environment developed by a given population, the higher will be the realised carrying capacity (*K*) of their environment, and, hence, the slower will be their expected rate of spread.

We suggest that human colonisation of virgin territory can minimally be recognised to have occurred when an archaeological site can be identified in a previously uninhabited area. The *tempo* of that colonisation is a function of the degree of spatial and temporal separation between the new site in pristine territory and sites in the adjacent source area. It reflects the speed with which movement over the intervening space must have occurred. The *mode* of that colonisation, on the other hand, relates to the factors that governed the rate at which movement occurred. Before we can consider such rates, we must consider the measurement of archaeological time.

TEMPORAL LIMITATIONS OF RADIOMETRIC 'TIME'

The cultural chronologies of preliterate societies can only be determined stratigraphically or chronometrically. However, both these dating methods are imprecise. Stratigraphic schemas are severely limited because they can only create relative sequences. An age in years must be assigned by other means to the strata identified. On the other hand, chronometric techniques yield 'age estimates' that are actually probability statements, rather than dates in real years (Shott 1992). Thus, they are limited in ways not apparently appreciated by some archaeologists.

Until very recently, ^{14}C was the only radiometric technique that had been used to date the Australian archaeological record. However, ^{14}C has a comparatively short half-life of approximately 5570 ±30 years. Hence, its decay curve becomes asymptotic at about eight half-lives or 40,000 BP (Aitken 1990, Chappell et al. 1996). Therefore, ^{14}C samples close to that age or older all tend to yield similar, equally unreliable, dates. As a result, until very recently many Australian archaeologists argued strenuously that people first reached Sahul about 40,000 BP (e.g., Allen 1989, 1994; Allen & Holdaway 1995, Bowdler 1993). That idea is finally beginning to crumble as repeatable series of concordant luminescence dates (both TL and OSL) become available from sites in northern Australia. It now appears, for example, that people first reached Deaf Adder Gorge in western Arnhem Land in the Northern Territory about 60,000-55,000 BP (Roberts et al. 1994b), a time that lies beyond the limits of ^{14}C dating.

The probabilistic nature of radiometric 'time' is of greater importance to any discussion of the *tempo* and *mode* of the initial human colonisation of Sahul, than are discussions of the reliability of specific techniques or arguments over precisely when the first people arrived (Webb, n.d. b). As noted above, all chronometric techniques produce age estimates that are in fact probability statements. 'Dates' are expressed as mean values that have a 2/3 probability of accurately reflecting the 'true age' of the sample dated within the stated limits of error (Aitken 1990). For most of the chronometric techniques used in archaeology that error is at least ± 10%, although in conventional ^{14}C assays it can be as low as ± 1%, providing the sample is carefully pretreated and a high precision counter is employed (Gillespie 1991).

Because radiometric age estimates are probability statements, when comparing 'dates' to determine whether the events to which they relate were synchronous or diachronous, it is incorrect to consider just their mean ages. The *ranges* indicated by their standard deviations *must* be compared (Long & Rippeteau 1974, Ward & Wilson 1978, Rick 1987; Buck et al. 1991, 1994a, 1994b; Ward 1994, Webb, n.d. b). Only if there is no overlap between the ranges of two dates at 3σ, that is their mean ages are separated by at least three times the value of their combined standard deviations (ANU ^{14}C manual), is it likely that the events they date occurred at different times. We have called the period when age ranges overlap and it is impossible to determine whether the events they date occurred synchronously or diachronously, the *Radiometrically Instantaneous Interval* (RII) (Webb, n.d. b; Webb & Rindos, n.d.).

Obviously, the *RII* will vary in duration depending upon the precision of the radiometric technique employed and the errors on the dates being compared. Older dates usually have larger standard deviations, while those techniques able to date back beyond the range of ^{14}C are all inherently less precise. Thus, the further back in time the events being analysed occurred, the longer the *RII* is likely to be. This fact imposes serious limitations on the kinds of questions we can ask about long past events. For example, it is impossible to discriminate between two opposing hypotheses about the rate at which past events occurred, if the dates on which the alternative scenarios are based yield timescales that fall within the range of the *RII* for the period under contention. This is the problem faced by archaeologists wishing to discuss the initial human colonisation of Sahul. Given the length of the *RII* at >40,000 BP, it is impossible to discriminate between the 'fast' and 'slow' models outlined above. Both predictions of the rate of human movement through the continent fit the available timescale, as we shall now show.

Determining the *Tempo* of Sahulian Colonisation

Initially, we will illustrate the discriminatory limitations inherent to the *RII* concept with reference to the conventional 'date' for initial human entry into Sahul of 40,000 BP. In order to present a 'best case' scenario, we will also assume that the radiometric dates in question have standard deviations of only ± 1%, although such precision is unlikely to be achieved in reality on dates of such antiquity. Under the argument presented above, the *RII* on two dates of 40,000 + 400 BP will be >2400 years. Hence, in order to be certain that two sites radiometrically dated to about 40,000 BP were *not* occupied simultaneously, their mean ages would have to be separated by >2500 years. This degree of temporal separation has serious implications for the testability of hypotheses about the rate at which humans spread through Sahul and, hence, the speed at which they adapted to their new environment. As the following hypothetical example will demonstrate.

Let us assume that two archaeological sites have been found and radiometrically dated, one on the present coast of northwestern Australia and the other in highland Tasmania, and that we have reason to believe they represent the earliest human presence in both regions. Their ages can, in theory, be used to measure whether the rate at which people spread through the continent was 'fast' or 'slow'. The distance between the two sites is about 5000 km in a direct line that passes through the continental interior, the route implicit in the 'fast' model of Sahulian colonisation. The distance between the two sites passing along the present Western Australian coastline is about 7000 km. If the colonists had moved around the coasts of Queensland and New South Wales the journey would have been a little shorter. Calculating the distance around the coast is complicated by the fact that 40,000 years ago the seas around Sahul would have been about 50 m below their present level (Chappell 1993) and the coastline commensurately enlarged. The coastal route is fundamental to the 'slow' model of Sahulian colonisation.

Several implications follow from the distances and dates involved in this hypothetical example. First, in order to be certain that the two sites were *not* occupied simultaneously, their mean ages would have to be separated by >2500 years. Therefore, conversely, if their mean ages were separated by <2500 years, the sites would appear to have been occupied simultaneously and people would appear to have moved from one to the other 'instantaneously'. Second, as noted above, 5000-7000 km of country separates our two hypothetical sites, but only if their mean ages were separated by >2500 years could we suggest that they were occupied diachronously. That is an *average* rate of spread of 2-3 km/yr, although we are not assuming that people moved every year. The most likely scenario would be that successive generations would be expected to strike out and find new territory for themselves. That is, we suggest a 'hiving off' rate of 40-60 km every 20 or so years. A similar rate of spread has been suggested for Folsom people, on purely archaeological grounds (Amick 1996), and for anatomically modern people on theoretical grounds (Young & Bettinger 1995). However, such comparatively slow progress would allow more than enough time for the 'off-shoot' population to adapt their culture to any change in resources they encountered in their new environment. Third, if spread were >2-3 km/yr, then the colonising population would appear to have moved across the continent 'instantaneously', because the mean ages of our hypothetical sites would be separated by <2500 years. Thus, we are faced with a conundrum. Because radiometric 'dates' are probability statements, at 40,000 BP the *RII* is so long that it is impossible to discriminate between the 'fast' and 'slow' models of Sahulian colonisation. Hence, people could have spread quite slowly through the region and yet their rate of spread would still appear to have been 'instantaneous' in archaeological terms.

The scale of this problem becomes even clearer when we consider the actual ^{14}C dates that relate to the initial human colonisation of Sahul (Webb & Rindos, n.d.). As noted above, the ^{14}C decay curve becomes asymptotic at about 40,000 BP (Aitken 1990, Chappell et al. 1996). Therefore, it is not surprising that some of the approximately 100 published ^{14}C dates >25,000 BP from archaeological sites in Sahul have infinite ages and, hence, an *RII* of infinite duration. Moreover, and more critically, most of the dates >35,000 BP have standard deviations of ±2-5%. These dates yield an *RII* of >5000 years, an interval so long that it is impossible to distinguish the order in which the sites from which they come were occupied. Instead, the entire region, from New Ireland in the north to Tasmania in the south, appears to have been occupied 'instantaneously', over a period of about 5000 years. Thus, we cannot presently discuss the *tempo* of Sahulian colonisation because the rate at which people spread through the region cannot be determined from the available radiometric dates.

Furthermore, it is unlikely that technical developments or the addition of new early dates based on longer-lived radioisotopes will alter that conclusion, because all the techniques able to surpass the '^{14}C barrier', such as luminescence, have inherently large margins of error, usually ± 10%. Thus, the dates they yield also have large *RII* values. This will compound, rather than solve, the problem of determining the *tempo* of Sahulian colonisation. Moreover, initially at least, the addition of dates based on other radioisotopes may create new interpretive problems. For example, much ink has been spilled in the last few years questioning the validity of the luminescence dates from Arnhem Land (Bowdler 1990b, 1991; Hiscock 1990, Allen 1994, Allen & Holdaway 1995), because they are so much older than the ^{14}C dates from the same sites. That impression is misleading. The dates yielded by these techniques are not strictly comparable. Luminescence and ^{14}C measure 'time' in different ways and are applicable to different periods in the past (Frankel 1990).

Determining the *Mode* of Sahulian Colonisation

Not only is it apparently impossible to determine the *tempo* of Sahulian colonisation, but the length of the *RII* at the time when colonisation occurred disposes of the problem the colonists are supposed to have had adapting their culture to Sahulian conditions, at least insofar as that argument has been developed to date. As we have just shown, the *RII* at the time of initial colonisation was sufficiently long that the immigrant population would have been able to develop specialised desert economies, if such were necessary, while exploring and populating the continent as fast or as slowly as they wished, yet the process of adaptation remains archaeologically invisible. They might also have adapted to new regions much more rapidly, but the resultant archaeological record would not be visibly different from that generated by 'slower' adjustment. Thus, both modes of adaptation can easily be accommodated within the temporal limitations inherent to radiometric dating. Hence, it would appear that the long persistence of the debate among Australian archaeologists over the *mode* and *tempo* of Sahulian colonisation has been fuelled by the fact that the questions they are asking are fundamentally unanswerable, given the nature of the data on which the dispute is based.

Therefore, Rindos and Webb (1992) adopted a more theoretical approach, based on the relationship between *r*, the rate of population increase per generation, and *K*, the carrying capacity of a given environment (Pianka 1970), as mediated by cultural behaviour. Our approach took the dating problems outlined above into account. We argued that the better able people were to extract food resources from the environment in which they found themselves, the higher its carrying capacity would be for humans and, hence, the more slowly any excess population would be forced to hive off. Such *K*-maximising populations would, therefore, spread slowly, not least because they would also be *r*-minimisers. Conversely, populations with *maladapted* subsistence strategies that led to low food extraction rates and hence low carrying capacities would have to hive off their excess population more quickly, because they would also have to be *r*-maximisers to allow for the high infant mortality implicit in *K*-minimisation. Hence, they would be forced to spread rapidly, if thinly, over the landscape and would be advantaged when dispersing into previously unoccupied territory. Thus a *K*-minimising population has a dispersal advantage over a *K*-maximising population that is expressible as the square of the difference in *r* (Pianka 1970). It is *r*-selected in the sense that it is culturally preadapted to spread. Such populations succeed as colonisers precisely because they are forced to hive off their excess population frequently by the inefficiency of their cultural relationship to their environment, and their rate of population increase.

Therefore, we concluded that populations that successfully accomplished pristine colonisations could be expected to have *K*-minimising cultural adaptations because, as *r*-maximisers, they would out-compete alternative strategies by populating uninhabited land more rapidly (Rindos & Webb 1992). Hence, we showed that it is possible to discuss the conjoined problems of the *mode* and *tempo* of the initial human colonisation of empty landmasses, although speaking in terms of 'fast' and 'slow' rates of adaptation to new environments leads to a stalemate. We further showed that the greater the degree of pre-adaptation a population had to the new environments in which they found themselves, the more slowly they would spread, precisely because they would know how to extract the maximum amount of food from their environment. Hence, their excess population could hive off slowly into new areas because they would be able to support a much greater number of people in a given area, than would a 'maladapted' group.

Thus, the *mode* by which human colonisation of a region took place is independent of the *tempo* at which it occurred. As Young & Bettinger (1995) recently demonstrated, the *mode* of initial colonisation by anatomically modern humans appears to have been inherently *fast* world-wide. Determining how rapidly it took place is more difficult, however.

TOWARDS A GENERAL MODEL OF COLONISATION

It is generally assumed by both Australian and American archaeologists that because the colonisation of new environments, particularly on new continents, was difficult, then the rate at which people were able to move into virgin territory would have been *slow*, not least because the costs of mobility are high in terms of group size and viability (Perlman 1985). The argument is frequently advanced that totally new, highly specific, adaptations would have been required before people would have been able to move into such new regions. We believe this argument is incorrect and inverts well-established biogeographic models. Therefore, we proposed the alternative, testable, model outlined above (Rindos & Webb 1992).

We argued that the less familiar the resources of an area, and/or the less able people were to extract maximal energy efficiency from those resources, the fewer the number of people that area would be able to support. Hence, such areas could be defined as having low carrying capacities (*K*) for their human occupants. Moreover, in regions previously unoccupied by humans, low population densities would be a likely consequence of a lack of cultural adaptation to locally available food resources. Yet, if such regions have identifiable archaeological records, then their colonising populations clearly survived. Thus, a mode of cultural adaptation that yields low population densities has important consequences in a colonising context.

The pristine colonisation of new regions will be favoured by those cultural behaviours that result in an *explorative mode* of subsistence, but such behaviours did not have to be optimally efficient. Indeed, quite the reverse (Perlman 1985). The less 'efficiently' a population was able to extract energy from its environment, the greater its potential rate of dispersal. Although extractive 'inefficiency' or 'maladaptation' fails to build up dense local populations, it is the 'optimally efficient' or 'best adapted' strategy for the colonisation of virgin territory because it spreads people over the landscape rapidly, if thinly.

This conclusion has major implications for the formulation of general models regarding the *tempo* and *mode* of initial human colonisation of *terra incognita*. Moreover, and most importantly, it can be tested against the archaeological record. The converse of our model suggests that people's familiarity with the food resources they encountered in a previously unoccupied area would increase their extractive efficiency and thereby increase *K*. The same would be true of any new behaviours leading to increases in extractive efficiency. However, culturally mediated behaviours that lead to locally high population densities, reduce the probability that populations practising them will disperse rapidly. However, cultural behaviours that create locally low population densities are preadapted to dispersal. Hence, it is likely that their bearers will successfully colonise unfamiliar, uninhabited regions.

Thus, it would appear that an alternative hypothesis exists to the standard arguments that disagree about the rate at which humans are capable of generating those *K*-maximising strategies that supposedly bring about 'optimal' cultural adaptations to local resources. Moreover, it should be possible to recognise these alternative strategies in the archaeological record. A colonising population that adopted an adaptive mode involving *K*-maximising strategies could be expected to build up relatively high local population densities, to spread concomitantly slowly across the landscape and to leave a commensurately dense archaeological record. Conversely, colonising populations that adopted *K*-minimising strategies would have locally low population densities. Hence, they would spread rapidly and generate a very light, uniform archaeological signature, particularly at the earliest sites occupied.

However, as we have already seen, the Australian archaeological record presents numerous problems to researchers wishful to study the rate of spread of the initial colonists, problems that interfere with our ability to distinguish between the 'fast' and 'slow' modes of colonisation. If, as we have proposed, the optimal *mode* of colonisation entails an inherently 'fast' rate of spread, then the *tempo* of colonisation must also be considered. We argued above that, given the time when Sahul was colonised and the imprecision inherent to radiometric dating, it was possible for people to spread through the region exceedingly slowly and yet for that spread to appear to be 'instantaneous'. Thus, this study has shown that analysis of the *tempo* of Sahulian colonisation is impossible, although insights have been gained into its *mode*.

Therefore, we will now consider the cultural record of the New World to see whether it supports our contention that a *K*-minimising, that is relatively maladapted, extractive strategy was the optimal *mode* by which human groups successfully colonised uninhabited regions.

COLONISATION OF THE NEW WORLD

It is generally accepted that the first American colonists were anatomically modern humans, almost certainly descended from a parent population living in far eastern Asia (Kirk & Szathmary 1985, Bonnichsen & Turnmire 1991, Hoffecker et al. 1993). Despite persistent claims to the contrary (Bryan 1969, 1978; Carter 1980, Guidon & Delibrias 1986, Carlisle 1988; Dillehay 1988a, 1988b; Dillehay & Collins 1988; Adovasio et al. 1990, 1992; Guidon & Arnaud 1991, Adovasio 1993, Guidon et al. 1996), it is also generally accepted that the American landmass was first colonised by humans after the last glacial maximum, <15,000 BP, far more recently than Sahul. By then, the variance on most ^{14}C dates is quite small. Hence, in theory, the *RII* is only about 1000 years long. If our assumption is correct that a *K*-minimising adaptive *mode* characterises colonising populations and that the *tempo* of such subsistence strategies is inherently *fast*, then we would expect to find that the Americas were colonised quickly, probably within the limits of an *RII* of <1000 years. This, in fact, appears to have been the case (Taylor 1991; Haynes 1992, 1993; Meltzer 1995, Young & Bettinger 1995, Batt & Pollard 1996, Taylor et al. 1996), as we shall now show.

Whereas the colonisation of Sahul required the successful negotiation of a series of water barriers, human migration to the New World would have been comparatively easy. Faunal records of species movements across the Bering Strait in both directions convincingly demonstrate that it formed not a barrier but a bridge between Asia and the New World that facilitated terrestrial migration throughout the Quaternary (Kurtén & Anderson 1980, Hopkins et al. 1982, Marshall et al. 1982, Webb & Barnosky 1989). Hence, once people had successfully colonised northeastern Asia, they would have been able to migrate into northwestern North America without difficulty (Hallam 1977, Hopkins et al. 1982, Laughlin & Harper 1988, Beaton 1991). They simply had to walk into Alaska from Siberia. This could have been accomplished with least risk during periods of eustatically lowered sea level when the Bering land bridge was fully exposed, since the strait is <100m deep. Besides moving through the interior of North America, people could also have moved along the western coastal margin, either on foot or by boat (Fladmark 1979, Gruhn 1988, Ackerman 1989). They could also have simply walked across the ice that forms each winter in the Bering Strait, as modern Inuit do.

Although the Bering Strait was clearly not a barrier to mammalian colonisation of the New World, major obstacles to movement by non-volant terrestrial thermophilous species, including humans, existed further to the south and east in North America. During glacial periods large ice sheets formed on the Canadian shield and in the Rocky Mountains. These ice masses probably fused during the last (Wisconsinan) glacial maximum forming an impenetrable barrier to terrestrial migration during the period 20,000-15,000 BP. However, it is also highly probable that an open 'ice-free' corridor, trending north-south through Canada to the east of the Rockies, existed for lengthy, albeit uncertainly dated, periods throughout the beginning of the Wisconsinan (AMQUA 1978, Porter 1988). Hence, North American archaeologists generally assume that people would have been unable to penetrate south of the ice sheet once it fused, but that they could have easily moved down the corridor when it was open. That scenario is biogeographically naive. Conditions within the corridor are unlikely to have been favourable to the passage of humans or other large mammals for some time after it opened, because of the time lag between its deglaciation and colonisation by the plants that would have supported the herbivores that could provide a subsistence base for humans (Wendorf 1989, Mandryk 1992).

Thus, it is clear that humans faced fundamentally different problems when attempting to colonise the New World from those they faced during their migration to Sahul. Once people reached Sahul, having successfully crossed a major water barrier, they faced no further geophysical barriers. In contrast, entry into the far northwest of North America posed no real physical or ecological problems to terrestrial mammals (Hopkins et al. 1982, Marshall et al. 1982, Webb & Barnosky 1989). Instead, late Pleistocene migrants to the Americas encountered their first major barrier far to the southeast. It was only after they had penetrated *south* of the Wisconsinan ice sheet that people would truly have found themselves in a New World (Wright 1987, 1991). Determination of the *mode* and *tempo* of that initial colonisation is critically dependent on when they reached the more southerly parts of North America, apparently after 15,000 BP. The existence of a possible, earlier, colonisation of the landmass depends on finding securely dated archaeological evidence that humans were on the continent before 20,000 BP.

The 'early' or 'pre-Clovis' and 'late' hypotheses regarding the timing of initial human colonisation of the New World, will now be examined in terms of *K*-minimising modes of cultural adaptation and the model of pristine colonisation already outlined for Sahul.

The 'Early' Colonisation Model

As noted above, claims have repeatedly been made by a few committed researchers that there is good evidence that people were in the New World before the Palaeoindians, or Clovis people, arrived. Unfortunately, and despite ardent championship, none of the finds on which these claims are based has won general acceptance in the wider archaeological community because the sites from which they came have failed to produce indubitable humanly-flaked stone artefacts from securely stratified and reliably dated contexts. Critics who have examined the material in question have repeatedly thrown doubt on either the ages of the sites and/or the artefactual status of the flaked stones found at them (e.g., Dincauze 1984, 1991; Owen 1984, Lynch 1990, Jelinek 1992).

Nonetheless, a handful of sites refuse 'to go away' (Adovasio 1993), although it is noteworthy that different researchers champion different sites and many once popular sites are no longer championed! Of the allegedly pre-Clovis sites in North America, only Meadowcroft rockshelter in Pennsylvania continues to bedevil the literature (Adovasio 1993). However, most researchers accept that the oldest ^{14}C dates from even this site are misleading. The other sites with 'staying power' lie in South America. The two most often proposed as evidence that people reached the New World before the last glacial maximum are Monte Verde in southern Chile (Dillehay et al. 1992; Dillehay & Collins 1988, 1991) and Pedra Furada in Brazil (Guidon & Delibrias 1986, Guidon & Arnaud 1991, Bahn 1993, Guidon et al. 1996). Deposits dated to >30,000 BP at Monte Verde yielded pieces of wood that were originally thought to have been humanly-shaped. However, Dillehay himself is now sceptical of the authenticity of this material, although he still believes that the site was occupied about 13,000 BP. Guidon claims to have recovered amorphous pebble tools associated with hearths and 'non-natural structures' from a sequence of deposits ^{14}C dated to 50,000-30,000 BP. Not only are ^{14}C dates of such antiquity inherently unlikely to be reliable, but doubts have been thrown on the authenticity of the artefacts and the 'hearths' (Meltzer et al. 1995). Resolution of the numerous problems posed by this site must await its detailed publication.

Thus, at these sites, major doubts remain about the age and authenticity of the earliest material recovered. Furthermore, no antecedents for these sites are known in North America or far eastern Asia (Yi & Clark 1983, Bonnichsen & Turnmire 1991). Moreover, given the known high failure rate of the much later Polynesian dispersal into the Pacific (Bellwood 1978), colonisation of the South American mainland directly from Asia appears unlikely.

However, let us suppose for a moment that the data from these sites are acceptable, what implications would their alleged ages have for the *tempo* of human colonisation of the New World? ^{14}C dates in the region of 30,000-50,000 BP with standard deviations of ± 1% would yield a theoretical *RII* of 2000-3000 years. Travelling along the western seaboard of the Americas, the distance from the Seward Peninsula, the closest point to eastern Siberia, to Tierra del Fuego is approximately 21,000km. Thus, in order for people to have spread through the Americas within an *RII* of about 3000 years, they would have had to move at an *average* rate of 7-10km/yr. While this rate of spread is far higher than that calculated for Sahul, it is not unreasonable (Young & Bettinger 1995), particularly if the colonists were boat-borne and moved straight down the coast. Moreover, several researchers accept that Folsom people, at least, were equally mobile (Kelly 1983, Kelly & Todd 1988, Boldurian 1991, Hofman 1991, Amick 1996). Actually, the hypothetical rate of human spread at >30,000 BP was probably rather slower than has just been estimated because the actual errors on the available ^{14}C dates are rather higher than we gave above, <5%. Such errors yield an *RII* of 10,000-15,000 years and a rate of spread of <2/3 km/yr, a figure similar to that proposed for Sahul. Thus, we would argue that the ages of these early sites, supposing they can be validated, suggest that human settlement of the New World, supposing it occurred before the last glacial maximum, would still have appeared to have been 'instantaneous'. However, the 'artefacts' associated with this putative colonisation tell us little about its *mode*. They are so amorphous they could conform to any conceivable adaptational hypothesis.

It is the nature of these 'early' sites themselves that bear upon the model of colonisation advanced here. Under our hypothesis, we would expect that a founder population arriving in the New World would already have had and/or rapidly generated cultural traditions that led to *K*-minimising dispersal-adapted behaviour. Hence, if the earliest evidence from either Monte Verde or Pedra Furada stands up to further investigation, but a substantial number of other sites with similar archaeological records are *not* found further north, our hypothesis would have to be rejected. We would expect culture to generate such *K*-minimising behaviours randomly within colonising demes and that, once generated, such behaviours would leave clear signatures in the archaeological record.

If they do not, then Monte Verde and Pedra Furada become suspiciously unique.

The 'Late' Colonisation Model and the Clovis Tradition

The conservative, but generally accepted, scenario for the colonisation of the New World argues that humans first entered North America *after* the Wisconsinan glacial maximum, <15,000 BP. These initial colonists were the 'Clovis' people (Bonnichsen & Turnmire 1991). Advocates of the 'late' colonisation model point out that the initial colonists would have had to possess technologies for survival under extremely cold conditions before they could have penetrated first into the high Arctic and then across the Bering Strait. The few archaeological sites known in eastern Siberia appear to have been first occupied quite recently (Yi & Clark 1985). They have produced 'Upper Palaeolithic' type artefacts that resemble those dated to <30,000 BP elsewhere in Eurasia (Gamble 1986) and are likely antecedents for Clovis technology (Bonnichsen 1991). Hence, proponents of the late colonisation model argue that the apparent lack of earlier archaeological evidence from Siberia mitigates against a pre-Clovis colonisation of the Americas (Yi & Clark 1983).

Little is known about the socio-economy of the people who made Clovis artefacts, because few food remains and little evidence for food processing or shelters have yet been recovered from the rare occupation sites that have been identified and excavated. Moreover, many museum collections are composed entirely of artefacts, chiefly bifacially flaked projectile points, recovered from the present land surface. These artefacts not only lack any spatial or stratigraphic context, but they were also collected a long time ago, often by amateurs who possibly overlooked the less distinctive flake tools with which they were associated. Acheulean bifaces were selected similarly by early collectors. When occupation sites are excavated, the assemblage recovered usually comprises a much wider range of artefacts as well as flaking debris. It is likely, therefore, that the apparent predominance of projectile points in the Clovis tool kit relates as much to their identifiability as to their frequency in the archaeological record. Unfortunately, projectile points were first identified in the carcasses of extinct megaherbivores in the arid southwest of North America. This association, together with the paucity of occupation sites and, hence, the lack of other data on Clovis subsistence strategies, led to the long-held belief that Clovis people were big-game hunters. This description of Clovis socio-economics is now being questioned (Meltzer 1993) because more recent data suggest that Palaeoindian people pursued broadly based economies, like the first Australians. Fortunately, both big-game hunting and generalised subsistence strategies are congruent with the model of pristine colonisation proposed here, since the people pursuing them would be unlikely to extract maximal energy from an unfamiliar environment. Thus, analysis of what little is known about Clovis land use patterns lends strong support to our belief that a *K*-minimising strategy is the most profitable response for a population colonising an unfamiliar region to adopt.

Determining the *Mode* of Palaeoindian Colonisation - Clovis Artefacts

Like so many of the prehistoric 'social groups' recognised by archaeologists, Clovis people are really identified by their lithic technology, which has been found all over North America south of the Wisconsinan ice margin (Bonnichsen & Turnmire 1991). The excavation of occupation sites has shown that the Clovis tool kit comprises a variety of simple edge-retouched flake tools and manufacturing debris, as well as the ubiquitous and highly distinctive bifacially flaked projectile points. The flake tools are made on blanks struck from amorphous cores of locally available raw material often of comparatively poor quality. Distinct regional patterns have been identified in Clovis projectile point morphology, that distinguish 'eastern' Clovis from 'western' Clovis for example (Meltzer 1988). Nonetheless, the artefacts comprising Clovis assemblages demonstrate considerable stylistic homogeneity wherever they are found, suggesting that their makers belonged to a single extended social group that spread rapidly all over the continent soon after initial entry (Kelly & Todd 1988). A few blades or blade-like flakes, and even the occasional blade core, have been described from some of the larger Clovis assemblages (Stanford 1991), but flake tools predominate. This fact is of great significance since the east Asian Palaeolithic technologies from which Clovis is presumed to derive are all fully lamellar (Bonnichsen 1991).

The 'loss' of blade-making technology in North America during the Palaeoindian period may be a consequence of the difficulty the first colonists had in finding raw material of suitable quality for blade production, due to their unfamiliarity with resource locations, but this is unlikely to be the whole explanation. Modern replication experiments suggest that the bifacially flaked projectile points characteristic of Clovis, Folsom and related technologies, such as the 'fishtail' points of Central and South America, were 'energetically expensive' and difficult to make (Frison & Bradley 1980, Mason 1981). Flenniken (1978), a skilled modern flintknapper, experienced a 37% failure rate when attempting to drive off the basal thinning flutes that 'finish' a Folsom point. Fluting also occurs on Clovis points, but not as consistently as on Folsom points, which are *always* fluted. The results of this replication experiment are totally consistent with data from the few known Palaeoindian quarry sites, where 60% of the points recovered had been abandoned unfinished due to faults arising during manufacture (Frison & Bradley 1980, Mason 1981). Clearly, very strong selective pressure, probably related to some superiority of fluted points as projectiles, must be invoked to explain the maintenance of this labour intensive tradition for so long over such a wide area.

Clovis points were usually made from high quality raw materials with good flaking properties (Kelly & Todd 1988). They were percussion flaked and are clearly and easily distinguished from those of subsequent traditions, which were pressure-flaked. The deliberate selection of good raw material for making Clovis points may, in part, be a reflection of the fact that they were difficult to make (Tankersley 1994), but it is also consistent with their assumed importance in Palaeoindian subsistence strategies. The 'cost' of point manufacture can be seen in the fact that they were often reworked if they broke during manufacture or on impact, suggesting that it was 'easier' to retouch an old point than make an entirely new one, providing the resultant armature was long enough to be effective. Some points appear to have been reworked repeatedly, only being discarded when they were too short to haft (Webb, pers. obs.). Others were resharpened while still hafted, probably for use as knives (Meltzer 1993). Overall, a higher percentage of artefacts seem to have been reworked during the Palaeoindian period than in later cultural horizons, suggesting that the Palaeoindians needed to conserve raw material more than later people. Perhaps because they had not discovered many good sources of supply, but also because they appear to have been highly mobile (Shott 1986, 1989; Boldurian 1991, Hofman 1991, Amick 1996). Hence, they needed a transportable tool kit.

Evidence of their mobility is seen in the fact that Clovis points made from high quality 'exotic' raw material are frequently found hundreds of kilometres from the nearest known source of the material from which they are made (Kelly & Todd 1988). In a settled society with a developed economy similar distribution patterns might be interpreted as evidence that the raw material was being traded over the intervening distance. However, it is unlikely that trade is the explanation in this case (Amick 1996). If Clovis people were the first occupants of North America, then it is unlikely that there would be anyone for them to trade with. Nor, given the essential similarity of their tool kit wherever it is found, is it easy to determine what, if anything, they would have had to trade. It is much more likely in the Palaeoindian context that the distance between a raw material source and the place at which an artefact made from that material was discarded indicates the extent of the territory exploited by its makers in their search for game (Boldurian 1991, Hofman 1991). It could even be argued that Palaeoindian hunters placed as much emphasis on the quality of their projectile points as they did on the type of game they were used to kill.

Apart from projectile points, the Clovis assemblage contains few items with specialised functions. Instead, it mainly comprises artefacts of long-term, general utility that were highly transportable and could be made on any type of rock. This tool kit permitted its makers to penetrate into regions where they only knew of a limited number of raw material sources. It suggests that Clovis people were highly mobile, supporting our contention that they pursued *K*-minimising socio-economic strategies.

Determining the *Mode* of Palaeoindian Colonisation - Clovis Sites

The spatial distribution of Clovis sites also suggests that the first Palaeoindians practised *K*-minimising strategies, rather than utilising their environment intensively. Most artefact scatters are restricted both in area and in the numbers of pieces recovered from them, suggesting that occupation was transient. Particularly on the Plains (Hofman 1994), localities were only occupied once, for a fairly brief period, before being abandoned. Where more than one occupation can be detected, for example at Debert in Nova Scotia (MacDonald 1968) or Bull Brook in Massachusetts (Byers 1954, Grimes 1979), the artefacts appear to comprise a number of separate scatters, but it is impossible now to determine whether they represent contemporaneous occupations by more than one social group or successive occupations by the same or different social groups. Significantly, these 'aggregation' sites are all located in northeastern North America, and have yielded little evidence that their occupants were big-game hunters (Meltzer & Smith 1986). This settlement pattern suggests that *generally* Clovis people were thinly scattered and rarely returned to any given locality.

What little else is known archaeologically about the Clovis people reinforces our suggestion that they used their environment differently from later peoples. For example, artefacts from the entire range of the tool kit have been found at many quarry sites (Funk 1976), suggesting that Clovis people lived at quarries and carried out normal subsistence activities there, while raw material was being extracted and fashioned into artefacts. In contrast, later peoples tended to visit quarries solely to extract raw material that was then removed to be finished elsewhere.

As noted above, the association early noted between Clovis points and the extinct megafauna, mainly mastodon, mammoth, horse, camel and bison, led to the belief that Clovis people comprised small groups of highly mobile big-game hunters (Bonnichsen et al. 1987, Belovsky 1988). Although this scenario is now questioned (Meltzer 1993), it is undoubtedly true that most of these animals would have been familiar to immigrants from far eastern Asia and might have permitted some groups of colonists to follow a pattern of niche exploitation in which they focused primarily on hunting megaherbivores. In which case they would undoubtedly have spread rapidly over the North American landmass because megaherbivores were almost certainly rare in the North American biota (Graham & Mead 1987). Large animals need large territories in which to forage (May 1974, Brown & Maurer 1989, Owen-Smith 1988). Hence, they were unlikely to have been encountered frequently or in great numbers by Palaeoindian people. Thus it is not surprising that the megaherbivores found on Clovis, and Folsom (Jodry & Stanford 1993), sites usually comprise only a few individuals. Only after bison herd size increased in the Holocene (Epp 1988), was it possible for Archaic hunters to slaughter them in large numbers (Frison 1978). Subsistence strategies based on the use of rare, but high-value, food resources are typical of *r*-selected organisms (Pianka 1970). They clearly do *not* maximise the potential carrying capacity of the environment of the demes that practice them.

Evidence for specialised food activity sites is really rather rare in the Clovis period. In northeastern North America (Dincause 1988, Meltzer 1988, Tankersley & Isaac 1990) and South America (Dillehay et al. 1992, Gibbons 1996, Roosevelt et al. 1996), Clovis people appear to have pursued broadly-based opportunist economic strategies in which a wide variety of resources were exploited. Moreover, even on megafaunal 'kill sites', chiefly located on the High Plains and in the arid southwest of North America (Meltzer 1993), there is little evidence of the storage-related activities frequently found on later sites, such as breaking up large bones for the extraction of marrow or fat; while 'processing areas', such as stone heating or boiling pits, are completely unknown (Kelly & Todd 1988). Indeed, the megaherbivore bones found on most Clovis 'kill sites' are usually intact, suggesting that the carcasses were under-utilised (Hofman 1994). Bison may even have only been exploited seasonally (Epp 1988, Todd 1991, Todd et al. 1990). Haynes (1983, 1987) has gone further. His taphonomic analysis of several mammoth 'kill' sites suggests that the animals died naturally, although people may have scavenged from the carcasses *post mortem*. The lack of evidence that Clovis people stored food is part and parcel of an *r*-selected subsistence strategy in which *K*-maximisation is seen as irrelevant.

Therefore, like Meltzer (1993), we would invert Kelly and Todd's (1988) summation of Clovis subsistence strategies. Rather than being 'generalists in relation to large terrestrial faunal resources and opportunists in relation to all other food resources', Clovis people were probably opportunists in relation to large animals and generalists in relation to all other food sources. However, we would agree with Kelly and Todd (1988: 239) that the little available evidence for Clovis patterns of land use suggests that their economic strategy was characterised by:

> little emphasis on 'place', high residential mobility, variability in mobility based on local abundance of resources, emphasis on search-and-encounter hunting tactics, and lack of stored resources) and collector systems (complex, curated technology, repetitive use of sites, high logistical mobility and large territories.

Therefore, it should not be surprising that Clovis people were able to move with great speed across the American continent. Indeed, were people consciously to 'design' a subsistence strategy that facilitated colonisation, they could do no better than adopt the opportunistic hunting and generalised foraging pattern apparent at many Clovis archaeological sites.

Determining the *Tempo* of Palaeoindian Colonisation

Stanford (1991) commented that, if Clovis people were the first American colonists, then either they must have been very numerous or they were highly mobile, because they appeared to have 'exploded' over North America within a very few generations. Their rate of spread is demonstrated by the ^{14}C dates associated with their artefacts, which fall into the narrow time frame of 11,500-10,500 BP. The contextual reliability of all the available dates has recently been critically re-evaluated by Haynes (1992, 1993; Taylor et al. 1996). As a result, some assays have been rejected and the window through which Clovis people spread over North America has been closed even further, to 11,200-10,900 BP.

Before considering the implications such a tight chronology has for the *tempo* of Palaeoindian colonisation, let us look more closely at the dates themselves. Haynes' (1992, 1993) aim was clearly to exclude the outliers from this population of dates, hence he combined the dates available from individual sites, thereby minimising their standard deviations. However, the question we need to answer in order to determine the *tempo* of Palaeoindian colonisation is: could these sites have been occupied diachronously? Therefore, we need to consider the *ranges* at >1σ of the average ages calculated by Haynes (*ibid.*) from the conventional dates. He divided these dates into three groups: western Clovis dated between 11,650 ± 60 and 10,690 ± 50 BP (mean 11,000 ± 11 BP); Folsom dated between 10,930 ± 100 and 10,260 ± 140 BP (mean 10,700 ± 11 BP); and eastern Clovis dated between 11,050 ± 300 and 9550 ± 320 BP (mean 10,500 ± 55 BP). Under the constraints of the *RII* as outlined above, not only is it clear that these sites were probably occupied synchronously because their time spans overlap, but as we predicted these ranges yield an *RII* of <1000 years, suggesting that the *tempo* of Palaeoindian colonisation of North America was indeed *fast*.

The few reliable Palaeoindian data from South America corroborate that conclusion. For example, the numerous ^{14}C dates available from Pedra Pintada in the Amazon Basin yielded an average age for the earliest Palaeoindian level at that site of 10,450 ± 50 BP (Roosevelt et al. 1996), an age contemporaneous with eastern Clovis. Thus, it appears that the initial colonisation of the New World was so rapid that it truly was 'radiometrically instantaneous'. It occurred well within the short duration of the *RII* at 11,00-10,500 BP.

This interpretation of the early North American archaeological record revives Martin's (1973, Mosiman & Martin 1975) 'bow-wave' model of New World colonisation, that has long been deeply unpopular with most archaeologists because it implies an even faster rate of spread than that inferred by Birdsell (1977) for Sahul. Martin (*ibid.*) hypothesised that people initially penetrated south of the Wisconsin glaciers about 11,500 BP and had reached Tierra del Fuego by about 10,500 BP. A rate of spread of 16km/year. We argued above that such rapid movement is totally congruent with our model of initial colonisation and quite incompatible with models that assume that cultural adaptation to unfamiliar country was a slow and difficult process. However, our model differs from Martin's in a crucially important way. Like Birdsell (1953, 1957, 1977), Martin (1973, Mosiman & Martin 1975) envisaged a numerous population that saturated the landscape. We believe that colonisation is most efficiently effected by populations that are few in number and thinly spread (Rindos & Webb 1992).

CONCLUSIONS AND CONJECTURES

Certain implications can be derived from the model of New World colonisation advanced above. If the initial Palaeoindian populations were highly mobile generalised foragers who opportunistically hunted a familiar, but comparatively rare, food resource: megaherbivores; then they are unlikely to have stayed in any given area long enough to have become fully familiar with local resources, such as lithic sources. Hence, the first colonists might well have exploited visible rock outcrops, even though they did not yield the best material for stone artefact manufacture, because they failed to locate better, but hard to find, sources. This might explain why Palaeoindian projectile points are rarely found in cave deposits. These natural shelters are comparatively rare and difficult to find in North America. Apparently, Clovis people did not use them.

We have already noted that Clovis people apparently transported, rather than traded, desirable raw materials over considerable distances in the form of finished artefacts. We also noted the 'lightness' of the Clovis archaeological record. It is almost exclusively restricted to artefact scatters and megafaunal remains. That sites were only occupied briefly is also congruent with our hypothesis that initial colonists are inherently mobile, as is the paucity of regional stylistic differences between spatially separated Clovis assemblages. If social groups frequently shifted the territory they exploited, cultural homogeneity would be maintained, because inter-group communication would remain randomised with respect to the totality of information known to all groups within the cultural tradition. Hence, Clovis people would not have been affected by those processes of transmission loss, status mimicry or isolation that enhanced local differences during the Holocene.

This review of Clovis material culture suggests that the Palaeoindians were a people who moved often over long distances. However, we are not suggesting that by so doing they were adapting to new ecological niches or developing new patterns of resource exploitation. Quite the reverse. We believe that Clovis economic strategies and artefact technology were fundamentally homogenous because they were determined by the unfamiliarity in detail Palaeoindians had with the territory through which they moved. By opportunistically exploiting large herbivores whenever possible, Clovis people appear to have been able to pursue similar hunting strategies in different locations. This had the effect of homogenising inter-regional environmental differences. Opportunistic exploitation whenever possible of a rare, widely-dispersed, but high-yield food resource led to an extractive *mode* that maximised human movement over unfamiliar territory.

Our model of pristine colonisation originally led us to posit that people who successfully accomplish initial settlement will have *K*-minimising, *r*-maximising cultures. They will, therefore, move with great speed across an empty landscape. In contrast to Sahul, the Palaeoindian colonisation of the New World occurred at a time when the *RII* was very brief. Nonetheless, Clovis people appear to have moved from North America to South America within its duration. Hence, the archaeological evidence for the initial colonisation of the Americas strongly supports the model we originally developed to explain the Sahulian data. Indeed, support has been found in the Americas far beyond what might have been anticipated when our model was first developed. The speed of human movement through the Americas and what little can be deduced about Clovis socio-economics from the known artefact assemblages both confirm our hypothesis that people with maladapted extractive strategies spread very rapidly across unfamiliar territory. However, it is important to stress that such strategies would only be selectively successful while unoccupied territory remained to be explored. Therefore, after an initial phase of explosive spread, we would predict that the selective pressures on the colonising population would change and extractive strategies that maximally exploited the resources in a given area would come to local dominance. Such 'intensified' economies would permit a greater number of people to live in a given area, particularly as r declined. These strategies would be maintained as long as countervailing pressures did not suppress them.

Thus, it appears likely that the subsistence patterns of colonising demes, such as we have described here for the New World, represent a mode of adaptation that is completely different from anything known from the ethnographic record. Therein may lie the difficulty many researchers have experienced in identifying and interpreting the earliest archaeological evidence from Sahul. The first colonists of any *terra incognita* may have had no sense of 'territory', because their adaptive mode required them to be 'technology oriented', not 'place oriented' people.

POSTSCRIPT

This paper was originally drafted while REW was a Visiting Fellow at the Glenn Black Archaeological Laboratory at Indiana University in 1993. She thanks the director, Christopher Peebles, and other members of the laboratory for the cordiality of her reception and the excellent research facilities with which she was provided, that were so conducive to her productivity while in Bloomington.

However, we were not satisfied with that expression of our ideas. When they were presented verbally at the Annual Meeting of the Australasian Society for Human Biology in Canberra later in 1993, they met with a mixed response. We, therefore, welcomed the opportunity to revise our ideas for this volume, despite the fact that REW was then based at Royal Holloway College, University of London, and DJR was fighting his dismissal from the University of Western Australia on the grounds that his performance while there had been 'substandard'.

Sadly, David died in his sleep from a stress-induced heart attack in early December 1996, before we had finalised a revision of this paper with which we were both satisfied. I have, therefore, completed it to the best of my ability, along the lines we had discussed and agreed, based on the last draft David sent me only a few days before he died. I hope that he would have approved this published version, which–like the volume itself–is dedicated to his memory.

REFERENCES CITED

Ackerman, R.E. (editor
1989 *Routes into the New World*. Pullman: Washington State University Press.

Adovasio, J.M.
1993 The ones that will not go away. In *From Kostienki to Clovis* (Soffer, O. & Praslov, N.D., eds.), pp. 199-218. New York: Plenum Press.

Adovasio, J.M., Donahue, J. & Stuckenrath, R.
1990 The Meadowcroft rockshelter radocarbon chronology 1975-1990. *American Antiquity* 55: 348-54.
1992 Never say never again: some thoughts on could haves and might have beens. *American Antiquity* 57: 327-31.

Aitken, M.J.
1990 *Science-based Dating in Archaeology*. London: Longman.

Allen, J.
1989 When did humans first colonize Australia? *Search* 20: 149-54.
1994 Radiocarbon determinations, luminescence dating and Australian archaeology. *Antiquity* 68: 339-43.

Allen, J. & Holdaway, S.
1995 The contamination of Pleistocene radiocarbon determinations in Australia. *Antiquity* 69: 101-12.

Amick, D.S.
1996 Regional patterns of Folsom mobility and landuse in the American Southwest. *World Archaeology* 27: 411-26.

AMQUA
1978 *The Ice-free Corridor and the Peopling of North America*. Edmonton: American Quaternary Association, Abstracts of the V Bienniel Meeting, Alberta.

Anderson, I.
1991 First Australians headed south in haste. *New Scientist* 132 : 3.

Anthony, D.W.
1990 Migrations in archaeology: the baby and the bathwater. *American Anthropologist* 92: 895-914.

Bahn, P.G.
1993 50,000-year-old Americans of Pedra Furada. *Nature* 362: 114-5.

Batt, C.M. & Pollard, A.M.
1996 Radiocarbon calibration and the peopling of North America. In *Archaeological Chemistry: Organic, Inorganic and Biochemical Analysis* (Orna, M.V., ed.), pp. 415-33. Washington, DC: American Chemical Society.

Beaton, J.M.
1991 Colonising continents: some problems from Australia and the Americas. In *op. cit.* (Dillehay, T.D. & Meltzer, D.J., eds.), pp. 209-30. Boca Raton: CRC Press.

Bellwood, P.
1978 *Man's Conquest of the Pacific.* Auckland: Collins.

Belovsky, G.
1988 An optimal foraging-based model of hunter-gatherer population dynamics. *Journal of Anthropological Archaeology* 7: 329-72.

Birdsell, J.M.
1953 Some environmental and cultural factors influencing the structuring of Australian Aboriginal populations. *The American Naturalist* 87: 171-207.
1957 Some population problems involving Pleistocene man. *Cold Spring Harbour Symposia on Quantitative Biology* 22: 47-69.
1977 The recalibration of a paradigm for the first peopling of greater Australia. In *Sunda and Sahul* (Allen J, Golson, J. & Jones, R., eds.), pp. 113-67. Sydney: Academic Press.

Boldurian, A.T.
1991 Folsom mobility and organization of lithic technology: a view from Blackwater Draw, New Mexico. *Plains Anthropologist* 36: 281-95.

Bonnichsen, R.
1991 Clovis origins. In *op.cit.* (Bonnichsen, R. & Turnmire, K.L., eds.), pp. 309-29. Corvallis: Oregon State University Press.

Bonnichsen, R., Stanford, D.J. & Fastook, J.
1987 Environmental change and developmental history of human adaptive patterns: the Palaeoindian case. In *North America and Adjacent Oceans During the Last Deglaciation* (Ruddiman, W. & Wright, H.E., eds.), pp. 403-24. Boulder: Geological Society of America.

Bonnichsen, R. & Turnmire, K.L. (editors)
1991 *Clovis: Origins and Adaptations.* Corvallis: Oregon State University Press, Centre for the Study of the First Americans.

Bowdler, S.
1977 The coastal colonisation of Australia. In *Sunda and Sahul* (Allen, J., Golson, J. & Jones, R., eds.), pp. 205-46. Sydney: Academic Press.
1990a Peopling Australia: the 'coastal colonisation' hypothesis re-examined. In *op. cit.* (Mellars, P.A., ed.), pp. 327-43. Edinburgh: Edinburgh University Press.
1990b 50,000 year-old site in Australia - is it really that old? *Australian Archaeology* 31: 93.
1991 Some sort of dates at Malakunanja II: a reply to Roberts et al. *Australian Archaeology* 32: 50-1.
1993 Sunda and Sahul: a 30kyr BP culture area? In *op. cit.* Smith, M.A. et al., eds.), pp. 60-70.

Bräuer, G. & Smith, F.H. (editors)
1990 *Continuity or Replacement: Controversies in* Homo sapiens *Evolution.* Rotterdam: Balkema.

Brown, J. & Maurer, B.
1989 Macroecology: the division of food and space among species on continents. *Science* 243: 1145-50.

Brown, P.
1993 Recent human evolution in East Asia and Australasia. In *The Origin of Modern Humans and the Impact of Chronometric Dating* (Aitken, M.J., Stringer, C.B. & Mellars, P.A., eds.), pp. 217-33. Princeton, NJ: Princeton University Press.

Bryan, A.L.
1969 Early man in America and the late Pleistocene chronology of western Canada and Alaska. *Current Anthropology* 10: 339-65.
1978 *Early Man in America: from a Circum-Pacific Perspective.* Edmonton: University of Alberta Department of Anthropology Occasional Papers No. 1.

Buck, C.E., Christen, J.A., Kenworthy, M.B. & Litton, C.D.
1994a Estimating the duration of archaeological activity using ^{14}C determinations. *Oxford Journal of Archaeology* 13: 229-40.

Buck, C.E., Kenworthy, M.B., Litton, C.D. & Smith, A.F.M.
1991 Combining archaeological and radiocarbon information: a Bayesian approach to calibration. *Antiquity* 65: 808-21.

Buck, C.E., Litton, C.D. & Scott, E.M.
1994b Making the most of radiocarbon dating: some statistical considerations. *Antiquity* 68: 252-63.

Byers, D.S.
1954 Bull Brook - a fluted point site in Ipswich, Massachussetts. *American Antiquity* 19: 343-351.

Carbonell, E., Mosquera, M., Rodriquez, Z.P. & Sala, R.
1995 The first human settlement of Europe. *Journal of Anthropological Research* 51: 107-14.

Carlisle, R.C. (editor)
1988 *Americans before Columbus: Ice-age Origins.* Pittsburgh: University of Pittsburgh, Anthropology Department Ethnology Monograph No. 12.

Carter, G.F.
1980 *Earlier than You Think.* College Station: Texas A & M University Press.

Cavalli-Sforza, L.L. Menozzi, P. & Piazza, A.
1993 Demic expansions and human evolution. *Science* 259: 639-46.

Chappell, J.
1993 Late Pleistocene coasts and human migrations in the Austral region. In *A Community of Culture* (Spriggs, M. et al., eds.), pp. 43-8. Canberra: Australian National University, Department of Prehistory, Research School of Pacific Studies, Occasional Papers in Prehistory No. 21.

Chappell, J., Head, M.J. & Magee, J.W.
1996 Beyond the radiocarbon limit in Australian archaeology and Quaternary research. *Antiquity* 70: 543-52.

Clark, G.A.
1994 Migration as an explanatory concept in Palaeolithic archaeology. *Journal of Archaeological Method and Theory* 1: 305-43

Davidson, I. & Noble, W.
1992b Why the first colonisation of the Australian region is the earliest evidence of modern human behaviour. *Archaeology in Oceania* 27: 135-142.

Dennell, R.W. & Roebroeks, W.
1996 The earliest colonisation of Europe: the short chronology revisited. *Antiquity* 70: 535-40.

Dillehay, T.D.
1988a How new is the New World? *Antiquity* 62: 94-7.
1988b The Palaeoindian debate. *Nature* 332: 150-2.

Dillehay, T.D., Calderón, G.A., Politis, G. & Beltrão, M.C.
1992 Earliest hunters and gatherers of South America. *Journal of World Prehistory* 6: 145-204.

Dillehay, T.D. & Collins, M.B.
1988 Early cultural evidence from Monte Verde in Chile. *Nature* 332: 150-2.

1991 Monte Verde, Chile: a comment on Lynch. *American Antiquity* 56: 333-341.

Dillehay, T.D. & Meltzer, D.J. (editors)
1991 *The First Americans: Search and Research*. Boca Raton, FL: CRC Press.

Dincauze, D.F.
1984 An archaeo-logical evaluation of the case for pre-Clovis occupations. *Advances in World Archaeology* 3: 275-323.
1988 Tundra and enlightenment: landscapes for northeastern Palaeoindians. *Quarterly Review of Archaeology* 1988(summer):6-8.
1991 Review of Dillehay, T.D. (ed.) *Monte Verde: a Late Pleistocene Settlement in Chile*. *Journal of Field Archaeology* 18: 116-9.

Epp, H.T.
1988 Way of the migrant herds: dual dispersion strategy among bison. *Plains Anthropologist* 33: 309-20.

Fladmark, K.R.
1979 Routes: alternate migration corridors for early man in North America. *American Antiquity* 44: 55-69.

Flenniken, J.J.
1978 Re-evaluation of the Lindenmeier Folsom: a replication experiment. *American Antiquity* 43: 473-480.

Frankel, D.
1990 Time inflation. *New Scientist* 127: 52-3.

Frison, G.C.
1991 *Prehistoric Hunters of the High Plains* (2nd. ed.). New York: Academic Press.

Frison, G.C. & Bradley, B.A.
1980 *Folsom Tools and Technology at the Hanson Site, Wyoming*. Albuquerque: University of New Mexico Press.

Fullagar, R.L.K., Price, D.M. & Head, L.M.
1996 Early human occupation of northern Australia: archaeology and thermoluminescence dating of Jinmium rockshelter, Northern Territory. *Antiquity* 70: 751-73.

Funk, R.E.
1976 *Recent Contributions to Hudson Valley Prehistory*. Albany: New York State Museum Memoir No. 22.

Gamble, C.S.
1986 *The Palaeolithic Settlement of Europe*. Cambridge: Cambridge University Press.

Gibbons, A.
1996 First Americans: not mammoth hunters, but forest dwellers? *Science* 272: 346-7.

Gillespie, R.
1991 The Australian marine shell correction factor. In *Quaternary Dating Workshop 1990* (Gillespie, R., ed.), p. 15. Canberra: Australian National University, Department of Biogeography and Geomorphology, Research School of Pacific Studies.

Graham, R.W. & Mead, J.I.
1987 Environmental fluctuations and evolution of mammalian faunas during the last deglaciation. In *North America and Adjacent Oceans during the Last Deglaciation* (Ruddiman, W. & Wright, H.E., eds.), pp. 371-402. Boulder: Geological Society of America.

Grimes, J.R.
1979 A new look at Bull Brook. *Anthropology* 3: 109-30.

Gruhn, R.
1988 Linguistic evidence in support of the coastal route of earliest entry intto the New World. *Man* 23: 77-100.

Guidon, N. & Arnaud, M.B.
1991 The chronology of the New World: two faces of one reality. *World Archaeology* 23: 167-78.

Guidon, N. & Delibrias, G.
1986 Carbon-14 dates point to man in the Americas 32,000 years ago. *Nature* 321: 769-771.

Guidon, N., Pessis, A-M., Parenti, F., Fontugue, M. & Guérin, Cl.
1996 Nature and age of the deposits in Pedra Furada, Brazil: reply to Meltzer, Adovasio and Dillehay. *Antiquity* 70: 408-21.

Habgood, P.J.
1989 The origin of anatomically modern humans in Australasia. In *op. cit.* (Mellars, P.A. & Stringer, C.B., eds.), pp. 245-73.
1991 Aboriginal fossil hominids: evolution and migration. In *Origins of Human Behaviour* (Foley, R.A., ed.), pp. 97-113. London: Unwin-Hyman.

Hallam, S.J.
1977 The relevance of Old World archaeology to the first entry of man into New Worlds: colonisation seen from the Antipodes. *Quaternary Research* 8: 128-148.
1987 Coastal does not equal littoral. *Australian Archaeology* 25: 10-29.

Haynes, C.V.
1992 Contributions of radiocarbon dating to the geochronology of the peopling of the New World. In *Radiocarbon after Four Decades* (Taylor, R.E., Long, A. & Kra, R.S., eds.), pp. 355-74. New York: Springer-Verlag.
1993 Clovis-Folson geochronology and climatic change. In *From Kostienki to Clovis* (Soffer, O. & Praslov, N.D., eds.), pp. 219-36. New York: Plenum Press.

Haynes, G.
1983 Age profiles in elephant and mammoth bone assemblages. *Quaternary Research* 24: 333-45.
1987 Proboscidean die-offs and die-outs: age profiles in fossil collections. *Journal of Archaeological Science* 14: 659-68.

Hill, A.V.S. & Serjeantson, S. (editors)
1989 *The Colonization of the Pacific: a Genetic Trail*. Oxford: Oxford University Press.

Hiscock, P.
1990 How old are the artefacts in Malakunanja II? *Archaeology in Oceania* 25: 122-4.

Hoffecker, J.F., Powers, W.R. & Goebel, E.
1993 The colonization of Beringia and the peopling of the New World. *Science* 259: 46-53.

Hofman, J.L.
1991 Folsom land use: projectile point variaibility as a key to mobility. In *Raw MaterialEconomics among Prehistoric Hunter-gatherers* (Montet-White, A. & Holen, S., eds.), pp. 335-55. Lawrence: University of Kansas Publications in Anthropology No. 19.
1994 Paleoindian aggregations on the Great Plains. *Journal of Anthropological Archaeology* 13: 341-70.

Hopkins, D.M., Mathews, J.V., Schweger, C.E. & Young, S.B. (editors)
1982 *Palaeoecology of Beringia*. New York: Academic Press.

Horton, D.R.
1981 Water and woodland: the peopling of Australia. *Australian Institute of Aboriginal Studies Newsletter* 16: 21-7.

Jelinek, A.J.
1992 Perspectives from the Old World on the habitation of the New. *American Antiquity* 57: 345-7.

Jodry, M.A. & Stanford, D.J. (editors)
1993 *Folsom Archaeology: Early Holocene Environments and Human Adaptations*. Washington, DC: Smithsonian Institution Press.

Jones, R.
1979 The fifth continent: problems concerning the human colonisation of Australia. *Annual Review of Anthropology* 8: 445-466.
1987 Pleistocene life in the dead heart of Australia. *Nature* 328: 666.
1989 East of Wallace's line: issues and problems in the colonisation of the Australian continent. In *op. cit.* (Mellars, P.A. & Stringer, C.B., eds.), pp. 743-82.
1992 The human colonisation of the Australian continent. In *op. cit.* (Bräuer, G. & Smith, F.H., eds.), pp. 289-301.

Kelly, R.L.
1983 Hunter-gatherer mobility strategies. *Journal of Anthropological Research* 39: 277-306.

Kelly, R.L. & Todd, L.C.
1988 Coming into the country, early Paleoindian hunting and mobility. *American Antiquity* 53: 231-244.

Kirk, R.L. & Szathmary, E. (editors)
1985 *Out of Asia: Peopling the Americas and the Pacific*. Canberra: Journal of Pacific History Supplement.

Kramer, A.
1991 Modern human origins in Australasia: replacement or evolution? *American Journal of Physical Anthropology* 86: 455-73.

Kurtén, B. & Anderson, E.
1980 *Pleistocene Mammals of North America*. New York: Columbia University Press.

Laughlin, W.S. & Harper, A.B.
1988 Peopling of the continents: Australia and America. In *op. cit.* (Mascie-Taylor, C.G. & Lasker, G.W., eds.), pp. 14-40.

Lee, E.S.
1966 A theory of migration. *Demography* 3: 37-57.

Long, A. & Rippeteau, B.
1974 Testing contemporaneity and averaging radiocarbon dates. *American Antiquity* 39: 205-215.

Lourandos, H.J.
1987 Pleistocene Australia: peopling a continent. In *The Pleistocene Old World* (Soffer, O., ed.), pp. 147-65. New York: Plenum Press.

Lynch, T.F.
1990 Glacial-age man in South America? A critical review. *American Antiquity* 55: 12-36.

MacDonald, G.F.
1968 *Debert: a PalaeoIndian Site in Central Nova Scotia*. Ottawa: National Museum of Man Anthropology Papers No. 16.

Mandryk, C.S.
1992 Human viability of the Late Quaternary ice-free corridor. *Abstracts of the XII Biennial Meeting of the American Quaternary Association*, p. 51. Davis: University of California at Davis.

Marshall, L.G., Webb, S.D., Sepkoski, J.J. & Raup, D.M.
1982 Mammalian evolution and the Great American Interchange. *Science* 215: 1351-7.

Martin, P.S.
1973 The discovery of America. *Science* 179: 969-74.

Mascie-Taylor, C.G. & Lasker, G.W. (editors)
1988 *Biological Aspects of Human Migration*. Cambridge: Cambridge University Press.

Mason, R.J.
1981 *Great Lakes Archaeology*. New York: Academic Press.

May, R.M.
1974 Patterns of species abundance and diversity. In *Ecology and Evolution of Communities* (Cody, M. & Diamond, J.M., eds.), pp. 81-120. Cambridge, MA: Belknap Press.

Mellars, P.A. (editor)
1990 *The Emergence of Modern Humans*. Edinburgh: Edinburgh University Press.

Mellars, P.A. & Stringer, C.B. (editors)
1989 *The human Revolution*. Edinburgh: Edinburgh University Press.

Meltzer, D.J.
1988 Late Pleistocene human adaptations in eastern North America. *Journal of World Prehistory* 2: 1-52.
1989 Why don't we know when the first people came to North America? *American Antiquity* 54: 471-490.
1993 Is there a Clovis adaptation? In *From Kostienki to Clovis* (Soffer, O. & Praslov, N.D., eds.), pp. 293-310. New York: Plenum Press.
1995 Clocking the first Americans. *Annual Review of Anthropology* 24: 21-45.

Meltzer, D.J., Adovasio, J.M. & Dillehay, T.D.
1995 On a Pleistocene human occupation at Pedra Furada, Brazil. *Antiquity* 68: 695-714.

Meltzer, D.J. & Smith, B.
1986 Paleoindian and early Archaic subsistence strategies in eastern North America. In *Foraging, Collecting and Harvesting* (Neusius, S., ed.), pp. 1-30. Carbondale: Centre for Archaeological Investigations, Southern Illinois University.

Morlan, R.E.
1991 Peopling of the New World: a discussion. In *op. cit.* (Bonnichsen, R. & Turnmire, K.L., eds.), pp. 303-7.

Mosimann, J.E. & Martin, P.S.
1975 Simulating overkill by PalaeoIndians. *American Scientist* 63: 304-313.

Mulvaney, D.J.
1975 *The Prehistory of Australia* . Melbourne: Penguin.

Owen, R.
1984 The Americas: the case against ice-age human population. In *The Origins of Modern Humans* (Smith, F.H. & Spencer, F., eds.), pp. 517-63. New York: Alan Liss.

Owen-Smith, R.N.
1988 *Megaherbivores: the Influence of Very Large Body Size on Ecology*. Cambridge: Cambridge University Press.

Perlman, S.M.
1985 Group size and mobility costs. In *The Archaeology of Frontiers and Boundaries* (Green, S. & Perlman, S.M., eds.), pp. 33-50. New York: Academic Press.

Pianka, E.R.
1970 On r- and K-selection. *The American Naturalist* 104: 592-7.

Porter, S.C.
1988 Landscapes of the last ice age in North America. In *op. cit.* (Carlisle, R.C., ed.), pp. 1-24.

Rick, J.W.
1987 Dates as data: an examination of the Peruvian preceramic radiocarbon record. *American Antiquity* 52: 55-73.

Rindos, D.J. & Webb, R.E.
1992 Modelling the initial human colonisation of Australia: perfect adaptation, cultural variability and cultural change. *Proceedings of the Australasian Society for Human Biology* 5: 441-454.

Roberts, R.G. & Jones, R.
1994 Luminescence dating of sediments: new light on the human colonisation of Australia. *Australian Aboriginal Studies* 2: 2-17.

Roberts, R.G., Jones, R. & Smith, M.A.
1994a Beyond the radiocarbon barrier in Australian prehistory. *Antiquity* 68: 611-6.

Roberts, R.G., Jones, R., Spooner, N.A. et al.
1994b The human colonisation of Australia: optical dates of 53,000 and 60,000 years bracket human arrival at Deaf Adder Gorge, Northern Territory. *Quaternary Science Reviews* 13: 575-583.

Roebroeks, W.
1994 Updating the earliest occupation of Europe. *Current Anthropology* 35: 301-5.

Roebroeks, W. & von Kolfschoten, T.
1994 The earliest occupation of Europe: a short chronology. *Antiquity* 68: 489-503.

Roosevelt, A.C., Lima da Costa, M., Lopes Machado, C. et al.
1996 Paleoindian cave dwellers in the Amazon: the peopling of the Americas. *Science* 272: 373-84.

Shawcross, W.F.
1975 Thirty thousand years and more. *Hemisphere* 19: 26-31.

Shott, M.J.
1986 Technological organisation and settlement mobility: an ethnographic examination. *Journal of Anthropological Research* 42: 15-51.

1989 Diversity, organisation and behaviour in the material record. *Current Anthropology* 30: 283-315.

1992 Radiocarbon dating as a probabilistic technique: the Childers site and late Woodland occupation in the Ohio valley. *American Antiquity* 57: 202-30.

Smith, M.A.
1987 Pleistocene occupation in arid central Australia. *Nature* 328: 710-1.

1989 The case for a resident human population in the Central Australian Ranges during full glacial aridity. *Archaeology in Oceania* 24:93-105.

Smith, M.A., Spriggs, M. & Fankhauser, B. (editors)
1993 *Sahul in Review*. Canberra: Australian National University, Department of Prehistory, Research School of Pacific Studies, Occasional Papers in Prehistory No. 24.

Stanford, D.J.
1991 Clovis origins and adaptations: an introductory perspective. In *op. cit.* (Bonnichsen, R. & Turnmire, K.L., eds.), pp. 1-13.

Tankersley, K.B.
1994 The effects of stone technology on fluted-point morphometry. *American Antiquity* 59: 498-510.

Tankersley, K.B. & Isaac, B.L. (editors)
1990 *Early Palaeoindian Economies of Eastern North America*. Greenwich, CT: JAI Press, Research in Economic Anthropology Supplement No. 5.

Taylor, R.E.
1991 Frameworks for dating the late Pleistocene peopling of the Americas. In *op. cit.* (Dillehay, T.D. & Meltzer, D.J., eds.), pp. 77-111.

Taylor, R.E., Haynes, C.V. & Stuiver, M.
1996 Clovis and Folsom age estimates: stratigraphic context and radiocarbon calibration. *Antiquity* 70: 515-25.

Todd, L.C.
1991 Seasonality studies and Palaeoindian subsistence strategies. In *Human Predators and Prey Mortality* (Stiner, M.C., ed.), pp. 217-38. Boulder: Westview Press.

Todd, L.C., Hofman, J.L. & Schultz, C.B.
1990 Seasonality of the Scottsbluff and Lipscomb bison bonebeds: implications for modeling Paleoindian subsistence. *American Antiquity* 55: 813-27.

Ward, G.K.
1994 On the use of radiometric determinations to 'date' archaeological events. *Australian Aboriginal Studies* 2: 106-9.

Ward, G.K. & Wilson, S.R.
1978 Procedures for comparing and combining radiocarbon age determinations: a critique. *Archaeometry* 20: 19-32.

Webb, R.E.
n.d.a Megamarsupial extinction: the carrying capacity argument. *Antiquity* (in press).

n.d.b Problems with radiometric 'time': dating the initial human colonisation of Sahul. *Radiocarbon* (in press).

Webb, R.E. & Rindos, D.J.
n.d. Problems with radiometric 'time': the initial human colonisation of Sahul was 'instantaneous'. *Journal of Archaeological Science* (in press).

Webb, S.D. & Barnosky, A.D.
1989 Faunal dynamics of Pleistocene mammals. *Annual Review of Earth and Planetary Sciences* 17: 413-38.

Wendorf, M.
1989 Diabetes, the ice-free corridor and the Paleoindian settlement of North America. *American Journal of Physical Anthropology* 79: 501-20.

White, J.P. & O'Connell, J.F.
1982 *A Prehistory of Australia, New Guinea and Sahul*. Sydney: Academic Press.

Winterhalder, B.
1986 Diet choice, risk and food sharing in a stochastic environment. *Journal of Anthropological Archaeology* 5: 369-92.

Wright, H.E.
1987 Synthesis: the land south of the ice sheet. In *North America and Adjacent Oceans During the Last Deglaciation* (Ruddiman, W. & Wright, H.E., eds.), pp. 479-88. Boulder: Geological Society of America.

1991 Environmental conditions for Palaeoindian immigration. In *op. cit.* (Dillehay, T.D. & Meltzer, D.J., eds.), pp. 113-35.

Yi, S. & Clark, G.A.
1983 Observations on the Lower Palaeolithic of northeast Asia. *Current Anthropology* 24: 181-202.

1985 On the Diuktai culture and New World origins. *Current Anthropology* 26: 1-20.

Young, D.A. & Bettinger, R.L.
1995 Simulating the global human expansion in the Late Pleistocene. *Journal of Archaeological Science* 22: 89-92.

14

Rational Behavior, the Adoption of Agriculture, and the Organization of Subsistence During the Late Archaic Period in the Greater Tucson Basin

Michael W. Diehl
University of Louisville

ABSTRACT

Recent descriptions of Late Archaic Period land use in the greater Tucson Basin have described the Cienega phase occupants of the region as either (1) logistically organized with very- long distance hunting forays that adversely affected the occupants of the montane areas of the Mogollon Rim, (2) fully sedentary with agricultural practices that were more or less identical to those of the later Hohokam, or (3) seasonally-sedentary forager-horticulturists. Using the interpretive framework of behavioral ecology and recently-excavated paleobotanical data, it is shown that the first two models are unlikely. Instead, Cienega Phase occupants of the Tucson Basin may best be characterized as seasonally-sedentary foragers, who made use of multiple residential locations throughout the year.

INTRODUCTION

Recent discussions of Late Archaic subsistence in the Tucson Basin have debated the nature of subsistence procurement and settlement. Competing characterizations describe the Cienega Phase occupants of the area as (1) hunters who adopted agriculture in order to support very-long-distance logistical forays through the use of cached maize, (2) sedentary farmers whose use of the land differed little from subsequent Hohokam occupants of the region, or (3) seasonally mobile forager-horticulturists who shifted their residential locations between different resource patches according to their season of availability.

This paper uses some of the principles of behavioral ecology in combination with macrobotanical data to assess the given descriptions of subsistence. It is organized into several sections. The first section describes models of Late Archaic subsistence and land use. The second section uses principles of behavioral ecology to derive archaeologically testable implications from the given models. The third section uses macrobotanical data to assess the merits of the given models. It is shown that one descriptive model, "seasonally mobile forager-horticulturists who shifted their residential locations between different resource patches according to the season of their availability," best fits the macrobotanical record.

MODELS OF LATE ARCHAIC SUBSISTENCE

This research examines descriptions of land use and resource procurement in the Greater Tucson Basin during the interval B.C. 1500 through A.D. 200. This interval has variously been labeled the "Cienega Phase of the Late Archaic Period" or the "Early Agricultural Period." This paper also considers the differences between Cienega Phase sites from the Greater Tucson Basin and subsequent Hohokam occupations. Sites that provided macrobotanical information used in this paper include; Clearwater Site (BB:13:6, Diehl 1996), Donaldson (EE:2:30, L.Huckell 1995), Los Pozos (AZ AA:12:90, Diehl 1997), Milagro (BB:10:46, Huckell et al. 1994), Santa Cruz Bend (BB:12:746, L. Huckell 1996), Stone Pipe (BB:13:425, Huckell 1996), Cienega Site (BB:9:143, Miksicek 1990), Julian Wash (BB:13:7, Miksicek 1996), Lonetree (AA:12:120, Miksicek & Bernard-Shaw 1990), Redtail (AA:12:149, Miksicek 1989), Square Hearth (Huckell 1996), West Branch, Cook Avenue Locus (AA:16:3, Swartz & Miksicek 1996), and the West Branch Site - Other loci (Miksicek 1986; see Table 14.1).

The foregoing sites were included in this study for several reasons. All are located within the Greater Tucson Basin, on or in close proximity to the floodplain of the Santa Cruz River or major, albeit intermittent, tribu-

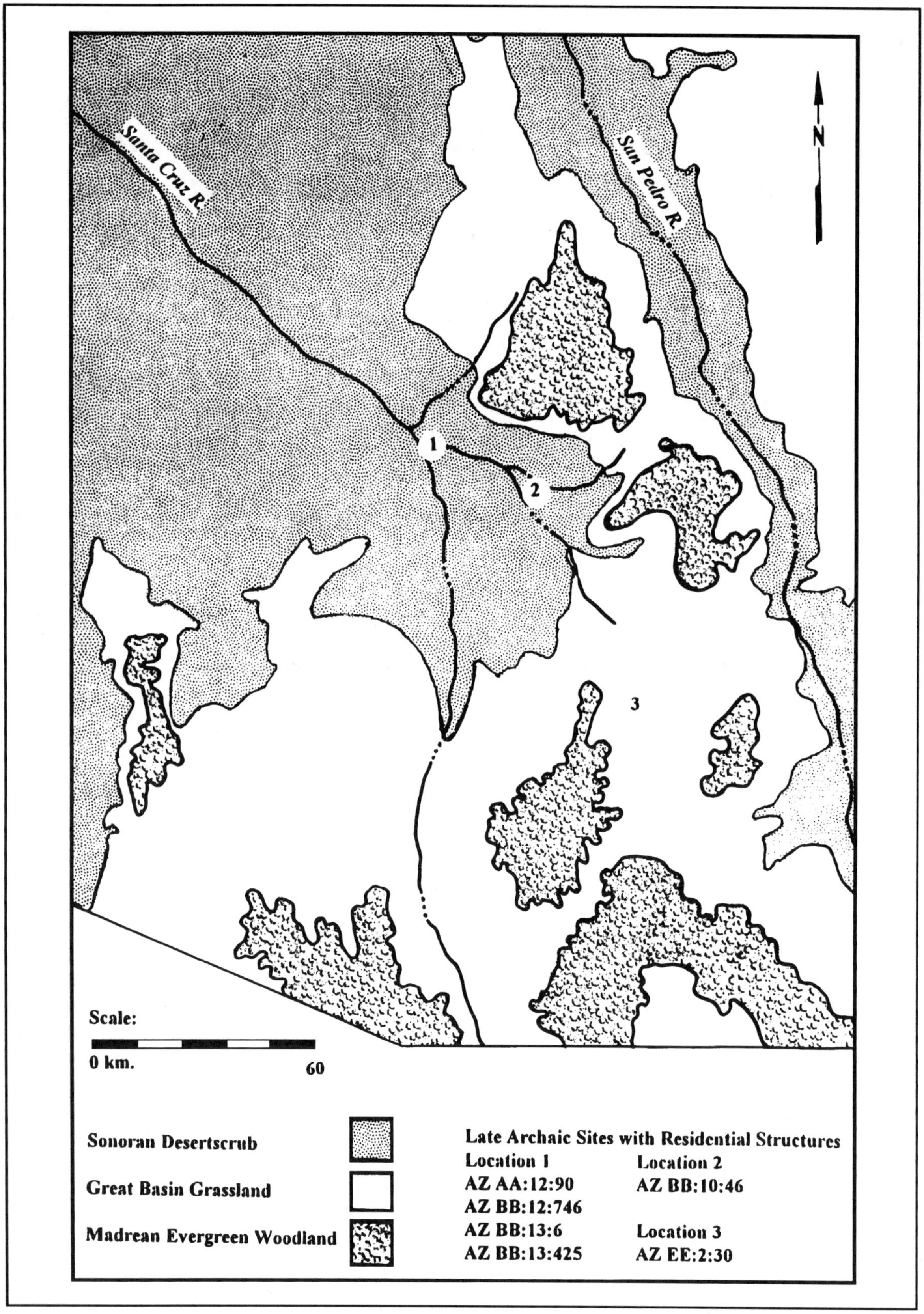

Figure 14.1 Late Archaic Sites in the Tucson Basin.

Table 14.1 General Chronology of Periods, Phase Names and Sites to which this Study Refers.

Site	Period	Phase	Approximate Date	Cultural affiliation
Clearwater	Late Archaic	Cienega	BC 800-AD150	Desert Archaic
Donaldson	Late Archaic	Cienega	BC 800-AD150	Desert Archaic
Los Pozos	Late Archaic	Cienega	BC 800-AD150	Desert Archaic
Milagro	Late Archaic	Cienega	BC 800-AD150	Desert Archaic
Santa Cruz Bend	Late Archaic	Cienega	BC 800-AD150	Desert Archaic
Stone Pipe	Late Archaic	Cienega	BC 800-AD150	Desert Archaic
Square Hearth	Early Ceramic	Agua Caliente	AD 150-550	Hohokam
Redtail	Early Ceramic	Canada del Oro	AD 750-850	Hohokam
Cienega	Classic Hohokam	Rincon	AD 900-1130	Hohokam
Julian Wash	Classic Hohokam	Rincon	AD 900-1130	Hohokam
Lonetree	Classic Hohokam	Rincon	AD 900-1130	Hohokam
West Branch	Classic Hohokam	Rincon	AD 900-1130	Hohokam

taries; the occupants of all sites had access to the same general suite of natural resources. Moreover, all sites are the loci of a full suite of domestic activities; all contain residential architecture, storage pits, and relatively complete arrays of stone and bone tools; in effect, these are most permanently and intensely occupied sites of their respective times.

Although there are a few other Archaic Sites such as Coffee Camp (AZ AA:6:18, Halbirt & Henderson 1993), these were not included in the study because the number of flotation samples were few, and because they lacked residential architecture.

Large Cienega phase sites that are located along the Santa Cruz River floodplain or major tributaries share in common several features that are considered unusual for highly mobile foragers. These include observable, shallow house depressions that include numerous post holes, usually one large central pit, and occasionally an intramural hearth (Gregory 1997; Mabry & Archer 1996). All contain abundant macrobotanical evidence of maize use, including charred kernels, cupules and cob segments. The co-occurrance of observable architectural (presumably residential) features, large central pits (usually interpreted as storage facilities), and domestic plants indicates that during the Cienega Phase, foraging was probably not the simple hand-to-mouth existence that previous generations of American Archaeologists assigned to Archaic Southwesterners (e.g. Cordell 1984:166-169; Irwin-Williams 1967; Sayles & Antevs 1941). The locations of Late Archaic sites that have obvious residential architecture and facilities, and their predominant environmental contexts, are illustrated in Figure 14.1.

CIENEGA PHASE HORTICULTURISTS AS LOGISTICALLY ORGANIZED LONG-DISTANCE HUNTERS

Three descriptive models of Late Archaic land use now compete for interpretive primacy. One model of Early Agricultural Period subsistence seeks to explain the adoption of agriculture in the low desert as a consequence of systemic relationships between foraging and horticulture. Wills (1996) compared the land use strategies of the occupants of the "Southern Basin and Range" biotic province with those of "cool temperate climates" (the Mogollon highlands and the Colorado Plateau). Wills (1996:238-240) suggested that the adoption of agriculture by occupants of the low desert (he cites sites from the Tucson Basin) occurred as a strategy for generating caches of stored food that could be used to support very-long distance logistical hunting forays. The transportation of agricultural surpluses to montane camps enabled low-desert farmers to out-compete the occupants of the Colorado Plateau and Upland Mogollon regions (Wills 1996:240; see Figure 14.2). The latter are said to have used the Mogollon Rim in a manner that is consistent with high residential mobility and "foraging" land use strategies. According to Wills, the strategy pursued by the low-desert farmers displaced that of the Colorado Plateau dwellers during the earliest centuries of farming in the Southwest.

> It is possible that the emergence of late Archaic settlements in southeastern Arizona and the corresponding trend toward specialized hunting tactics created economic problems for populations in other regions. With the development of surpluses from food production, desert populations were able to organize and support extremely specific hunting forays into distant upland areas during autumn months, selecting only the highest return species (Wills 1996:238-239).

Cienega Phase Horticulturists as Classically Sedentary Farmers

An alternative proposal characterized Cienega Phase occupants of the Greater Tucson Basin as essentially identical to later Hohokam farmers. Both groups were described as highly entrenched people who were predominantly dependent on agricultural crops and possibly a few species of important wild seed plants such as amaranth. This basic set of staples was supplemented by the addition of a few intensively harvested foods such as cactus fruit and animals. The floodplain villages in which these sedentary farmers lived were occupied year-round, and for numerous years. This is essentially the pattern of crop-dependent subsistence that is observed in subsequent Hohokam and Historic Period occupations in Southeast Arizona. Huckell (1995; Huckell et al. 1994) is an advocate of this model, and it is best conveyed using his own words:

> Domestic structures, agricultural products, storage pits, a thick and culturally rich midden deposit, and plants representing multiple seasons all suggest that late preceramic people were in residence at the Donaldson Site and probably Los Ojitos for a major portion of the year, if not the entire year, ... In my opinion, they may be considered villages inhabited both intensively during the course of a year and for periods of time best measured in decades, if not centuries. (Huckell 1995:131).

> In fact, the pattern of mixed farming and foraging established during the Late Archaic persisted *with few changes* until the arrival of the Spanish. (Huckell et al. 1994:3; emphasis mine).

Preliminary interpretations of the use of the Santa Cruz Bend site (Mabry & Archer 1996:2-132), and the Stone Pipe site (Swartz & Lindeman 1996:4-96) concur with this description. Moreover, Huckell's (1995) characterization is generally consistent with the results of some area surveys. Downum et al. (1986) surveyed and conducted limited test excavations in the Avra Valley, about 12 km. west of Tucson. They identified sites that included dense surface scatters and possible pithouses. These sites were said to be linked to the floodplain settlements as a component of a land use strategy that includes "permanent" floodplain settlements, and dune sites that are either "permanent base camps" or temporary "camps" (Downum et al. 1986:196) In other words, large floodplain sites such as Santa Cruz Bend are, from their point of view, unquestionably villages that were occupied year-round. The nature of the use of secondary camps in dune fields remains to be assessed.

Ciengea Phase Horticulturists as Seasonally Sedentary Foragers

Alternative models indicate that Cienega Phase foraging and horticulture involved a complex arrangement of residential and logistical moves that provided access to resources that were distributed in a spatiotemporally heterogeneous fashion. Forager-horticulturists shifted between seasonally occupied farming localities in the floodplains and *bajada* (foothills) and upland (montane or high plateau) hunting localities during other seasons.

Based on intensive survey data and some limited site excavations, Roth (1988:212; 1992:311-312) recognized two alternative models of Early Agricultural Period land use and subsistence. One model holds that the floodplain and bajada sites may be the material consequences of the existence of two groups of people engaged in fundamentally different settlement-subsistence strategies (Fish et al. 1990; Roth 1992). In this model, floodplain groups were wholly entrenched farmers who were relatively immobile in every sense of the word, and bajada groups were more transitory, engaging in a thoroughly mixed hunting-gathering with possibly a supplemental farming strategy.

Roth's preferred second model holds that the most intensively occupied sites were located on the floodplains because of the constraints and demands required to bring crops to fruition. Despite the mandates imposed by maize cultivation, these farmers are said to have made intensive use of some bajada sites by means of short-duration logistical forays to harvest cactus fruit and game (Roth 1995, 1992, 1988). This model is consistent with Huckell's, but differs to the extent that it does not require the year-round occupation of floodplain sites.

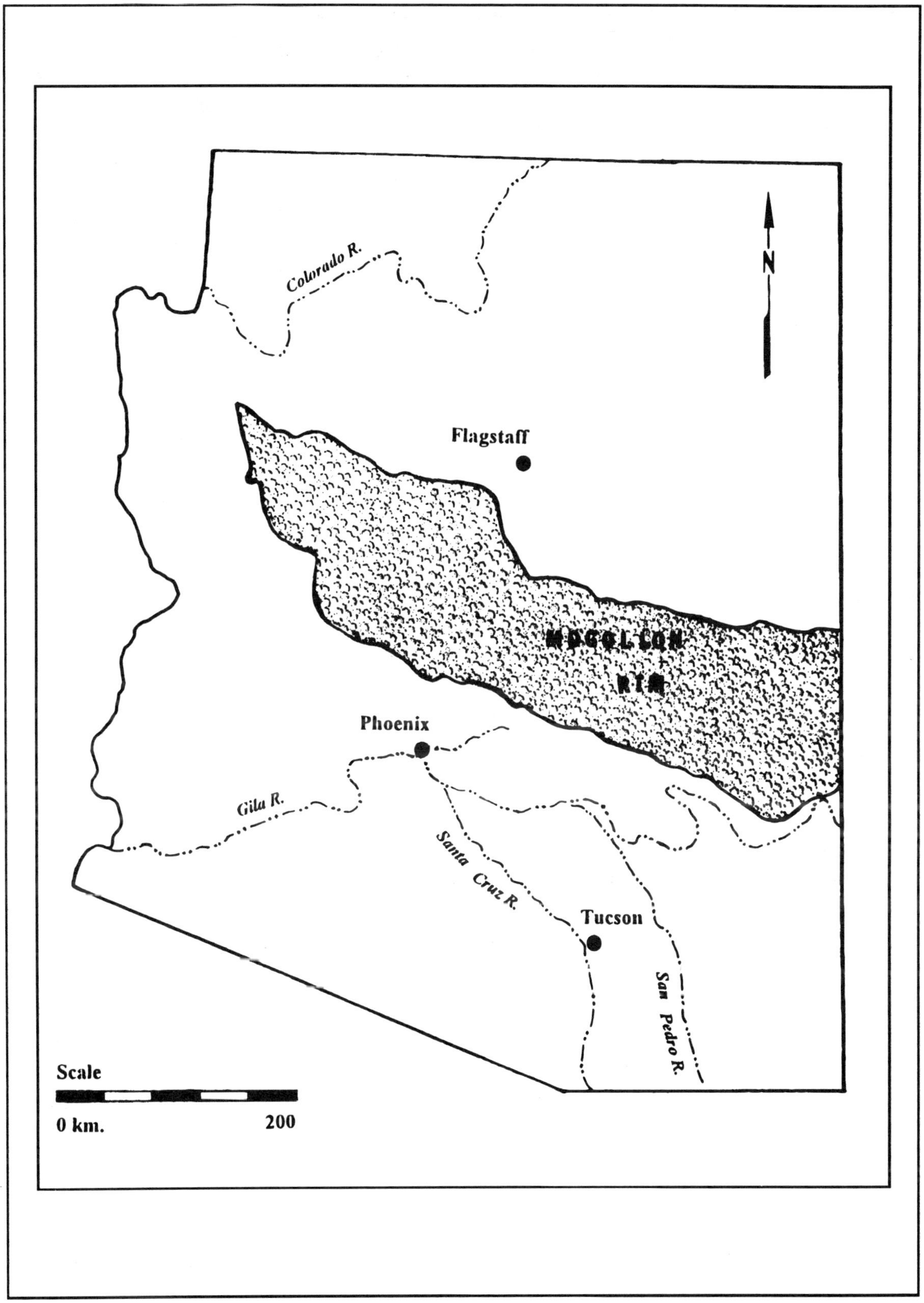

Figure 14.2 The Mogollon Rim in relation to the Tucson Basin.

Table 14.2 Ubiquities of Plant Taxa from Late Archaic Sites in the Tucson Basin.

	Clearwater (BB:13:6)	Donaldson (EE:2:30)	Los Pozos (AA:12:91)	Milagro (BB:10:46)	Santa Cruz Bend (BB:12:746)	Stone Pipe (BB:13:425)
No. of Contexts	15.00	9.00	72.00	15.00	41.00	45.00
Taxon						
Agrostis / Muhlenbergia	0.00	0.00	22.00	0.00	0.00	0.00
Artemesia	0.00	0.00	6.00	0.00	0.00	0.00
Astragalus	7.00	0.00	11.00	0.00	0.00	0.00
Bouteloua	0.00	0.00	4.00	0.00	0.00	0.00
Bromus	0.00	0.00	1.00	0.00	0.00	0.00
Carex	0.00	0.00	1.00	0.00	0.00	0.00
Chenopodiaceae	0.00	0.00	3.00	0.00	0.00	2.00
Chenopodium / Amaranthus	73.00	100.00	81.00	80.00	49.00	49.00
Cleome	0.00	0.00	1.00	0.00	0.00	0.00
Compositae	20.00	0.00	13.00	0.00	0.00	2.00
Cruciferae	7.00	0.00	11.00	0.00	0.00	0.00
Cyperaceae	0.00	89.00	0.00	0.00	0.00	0.00
Descurainia	60.00	0.00	63.00	53.00	56.00	38.00
Eleocharis	0.00	0.00	1.00	0.00	0.00	0.00
Eragrostis	0.00	0.00	7.00	0.00	0.00	0.00
Escholtzia	0.00	0.00	1.00	0.00	0.00	0.00
Fimbrystylis	0.00	0.00	10.00	0.00	0.00	0.00
Gramineae	27.00	50.00	46.00	13.00	44.00	62.00
Helianthus	0.00	0.00	3.00	0.00	0.00	0.00
Juncus	0.00	27.00	0.00	0.00	0.00	4.00
Kallstroemia	0.00	0.00	1.00	0.00	0.00	0.00
Malvaceae	0.00	27.00	1.00	0.00	0.00	0.00
Mentzelia	0.00	0.00	0.00	20.00	2.00	0.00
Panicum	0.00	0.00	11.00	0.00	0.00	0.00
Phragmites	0.00	0.00	1.00	0.00	0.00	0.00
Portulaca	7.00	50.00	8.00	0.00	10.00	0.00
Rumex	20.00	22.00	0.00	0.00	0.00	0.00
Salvia	0.00	0.00	1.00	20.00	5.00	2.00
Scirpus	13.00	0.00	0.00	0.00	0.00	0.00
Sporobolus	7.00	0.00	43.00	0.00	0.00	13.00
Suaeda	0.00	0.00	3.00	0.00	5.00	2.00
Trianthema	7.00	56.00	14.00	87.00	29.00	22.00
Vicia	0.00	0.00	1.00	0.00	0.00	0.00
Prosopis	13.00	33.00	40.00	13.00	29.00	36.00
Agavaceae	0.00	0.00	1.00	7.00	0.00	0.00
Atriplex	27.00	0.00	0.00	13.00	10.00	2.00

Table 14.2 Ubiquities of Plant Taxa from Late Archaic Sites in the Tucson Basin (cont.).

	Clearwater (BB:13:6)	Donaldson (EE:2:30)	Los Pozos (AA:12:91)	Milagro (BB:10:46)	Santa Cruz Bend (BB:12:746)	Stone Pipe (BB:13:425)
No. of Contexts	15.00	9.00	72.00	15.00	41.00	45.00
Taxon						
Celtis	0.00	0.00	0.00	0.00	0.00	2.00
cf. Labiatae	0.00	0.00	15.00	0.00	0.00	0.00
Euphorbiaceae	0.00	11.00	6.00	0.00	0.00	13.00
Juniperus	0.00	0.00	1.00	7.00	0.00	0.00
Leguminosae	0.00	33.00	4.00	0.00	2.00	25.00
Nicotiana	0.00	0.00	0.00	0.00	0.00	7.00
Quercus	0.00	0.00	0.00	7.00	0.00	0.00
Solanaceae	0.00	0.00	3.00	7.00	0.00	2.00
Solanum/Physalis type	7.00	0.00	1.00	0.00	2.00	0.00
Yucca	0.00	11.00	0	0.00	0.00	0.00
Cactaceae	0.00	0.00	6.00	0.00	2.00	0.00
Carnegia	53.00	11.00	58.00	27.00	34.00	42.00
Echinocereus	40.00	0.00	29.00	7.00	34.00	2.00
Mammillaria	0.00	0.00	7.00	7.00	0.00	0.00
Opuntia sp."platyopuntia"	7.00	11.00	10.00	7.00	5.00	0.00
Zea mays	87.00	100.00	88.00	0	83.00	84.00

BEHAVIORAL ECOLOGY: INTERESTS AND ASSUMPTIONS

How do we differentiate between the proposed models when evaluating their usefulness as descriptions of Late Archaic/Early Agricultural Period land use? The interest arena of behavioral ecology is suitable for assessing the viability of different resource procurement strategies. Others (Foley 1985; O'Connell & Hawkes 1994; Kelly 1995; Pianka 1978) have offered compelling testimony for the utility of the behavioral ecology perspective in archaeological research, and I shall dispense with the traditional review of successful efforts by behavioral ecologists who work in other areas of the world.

Behavioral ecologists study optimization (Foley 1985). It is assumed that optimization occurs as a consequence of rational choices between behavioral alternatives in such cases where one or more of the following conditions exist (Kelly 1995:54-57); (1) there is imminent threat of starvation, (2) specific nutrients are in short supply (not necessarily calories), (3) when there are time constraints on foraging or on other activities that are exclusionary to foraging activities, (4) when foraging exposes the forager to risks, or (5) when excess food or time may be used to enhance reproductive fitness. According to most theorists, human foragers who seek food optimize their rates of energy return when choosing between different classes of food (prey) or foraging areas (patches) (for an overview, see Kelly 1995:78-109). When optimization occurs one expects rational decisions to favor the use of the most efficient foraging strategies. To the extent that observable resource use strategies are observed to be inefficient, behavioral ecologists suppose that the conditions that demand optimization were not met, or that foragers behaved irrationally and unpredictably.

For this research several questions are important: (1) Did foragers optimize resource procurement? (2) If they did, are the given descriptions of Late Archaic subsistence practices consistent with the optimal use of resources? (3) Was long term resource use stable? (4) If not, what changes occurred and what might have motivated these changes? The first two questions are most suited to the evaluation of the very-long-distance hunting model offered by Wills (1996). The second two questions are well suited to the evaluation of the classically sedentary model offered by Huckell. The behavioral-ecological implications of the two models are discussed below.

Table 14.3 Estimated Energy Return Rates from Early Agricultural Period Plants.

Taxon (in rank order)	Gross Yield (Cal/kg)	Harvest Cost (Cal/kg)	Net Yield (Cal/kg) after Harvest	Process Cost (Cal/kg)	Net Yield after Processing (Cal/kg)	Processing Rate (kg/hr)	Net Rate of Return (Cal/hr)
Carnegiea	1100.00	68.00	1032.00	50.00	982.00	2.20	2160.00
Zea mays	3600.00	30.00	3570.00	500.00	3070.00	0.40	1228.00
Cheno-ams	3500.00	416.00	3084.00	600.00	2484.00	0.20	497.00
Descurainia	3400.00	416.00	2984.00	600.00	2384.00	0.20	477.00
Gramineae	3500.00	1153.00	2347.00	600.00	1747.00	0.20	350.00
Prosopis	3400.00	135.00	3265.00	900.00	2365.00	0.10	237.00
Other Seeds	2900.00	1500.00	1400.00	600.00	800.00	0.20	160.00

Behavioral Ecology and Implications for the Long-distance Hunting Model

Wills' (1996) model of Late Archaic land use asserts that hunters operating from low-desert base camps engaged in very long-distance hunting, and that such efforts impaired the subsistence efforts of hunters who were located in the Colorado Plateau. The Mogollon Rim and Uplands are said to be the spatial arena in which the strategies of the desert and plateau dwellers competed. Will's model may be tested for its concordance with the predictions of behavioral ecology by asking two questions: (1) Did desert dwellers optimize energetic returns? (2) If so, which of the available hunting strategies (models) are consistent with the optimization of resource procurement?

In this discussion, paleobotanical evidence is used to demonstrate that the Cienega Phase occupants of the Tucson Basin optimized resource procurement. An optimal rank order of plant use is derived from a review of research conducted in the botanical and ecological sciences. Examination of the ubiquities of macrobotanical remains from archaeological sites is consistent with the rank order that one would expect to observe if foragers optimized resource use. Estimates of the energetic costs and returns from very-long distance hunting and local hunting are proposed. A comparison of the estimates shows that foragers who optimized should have preferred a local hunting strategy over a very-long distance strategy. The very-long distance hunting model proposed by Wills (1996) is an energetically inefficient strategy for obtaining calories, and is an unlikely candidate as a description of Cienega Phase resource use.

Derived Energetic Return Rates from Southwest Plants

Fifty-three taxa of economically useful plants have been identified in flotation samples from Late Archaic / Early Agricultural Period, Cienega Phase sites. The "ubiquities" (see Minnis 1985, 1981; Popper 1988; Pearsall 1989) of plants from Tucson Basin Late Archaic sites are presented in Table 14.2. In the table, the ubiquities are the proportional frequency of the total number of analyzed archaeological features in which the taxon was present.

The different categories of plants may be ranked in order of their net energetic return rates. Table 3 lists the estimated rank-order of categories of plants found in Late Archaic archaeological sites. The calculation of the energy return rates is a complex process, and requires estimates of the intrinsic plant yields (kg/ha), the nutritional composition of their seeds, and the energy (Cal) required to harvest and process them with stone tools. This information was compiled by canvassing the literature on plant biology, arid lands resource management, ecosystems management, watershed management, and publications of various US government land management agencies[1]. A detailed discussion of that investigation and the subsequent reconstruction of energetic returns from plants vastly exceeds the capacity of the present venue. For details, readers are referred to the macrobotanical appendix in Gregory (1997).

Each of the plants discussed in Table 14.3 were specific targets of foraging, and that their distributions on the landscape were monitored. Monitoring promoted the harvest of productive patches when seeds ripened, and

Table 14.4 Averaged Ubiquities of Selected Resources from Late Archaic Tucson Basin Sites.

Taxon	Ubiquity
Zea mays	74.00
Cheno-ams	72.00
Descurainia	45.00
Gramineae	40.00
Carnegiea *	38.00
Prosopis	27.00
Other seeds	<<27

Note. * Marked resource is out of the predicted rank-order.

minimized the risk of the loss of ripe patches to competitors such as birds and rodents. The individual taxa that are not named in Table 14.3 are plants that may have been opportunistically acquired as especially productive patches were encountered, perhaps on returns from unsuccessful hunting forays or when targeted primary resources failed to meet expected yields. Rather than including these on a taxon by taxon basis, I have included a category "other seeds" that may represent a general estimation of the returns from opportunistic encounters.

The ubiquities of the plant taxa listed in Table 14.2 may be combined into a general score for Late Archaic sites that are situated in the low desert in and around the Tucson Basin (see Table 14.4). By inspection (Table 14.3 compared with Table 14.4), it is apparent that the derived rank-order of plants (which is based on the estimates of energy returns) closely matches the observed rank order of plants (by ubiquity, as observed in archaeological sites). It is suggested that the correspondence between the observed and predicted rank-orders of plants is consistent with optimization. A notable exception is the lack of fit between the predicted rank of saguaro and its actual use as indicated by its ubiquity. It is suggested that saguaro fruit, while nutritious, may have been identified less frequently than expected because the fruit were less storable than grain seeds, may have been rapidly consumed in other localities away from the site, or were proportionately less frequently subject to carbonization.

For the remainder of this discussion, it is assumed that the optimization of plant foraging is indicative of a general desire on the part of Late Archaic people to optimize all subsistence production. In other words, since Late Archaic forager-horticulturists optimized their use of plants, it is assumed that they optimized their use of animals as well. It should be noted that ongoing research on osteofaunal assemblages will allow analysts to assess whether or not the proportions of animal bones in assemblages are consistent with the inference that hunting, in addition to plant harvesting, was optimized.

Energetic Cost-Estimates of Upland Hunting

Wills (1996) proposed that inhabitants of the Sonoran Desert displaced hunters from the Colorado Plateau in an area where their resource strategies competed - the Mogollon Rim. The tactic that allowed them to successfully compete for preferred resources was the transportation and caching of maize. Wills' indicated that the use of maize to support long distance hunting forays was a strong motivating factor for the adoption and intensification of agriculture in the desert lowlands. The evidence from Early Agricultural Period sites indicates that, if Tucson Basin foragers were concerned with maximizing energy returns, then Wills (1996) model is unlikely.

For the sake of this discussion, the "Mogollon Rim" is defined as the nearest boundary of the Great Basin Conifer Woodland biological province (Brown 1994) on the south face of the Mogollon Rim. This definition was chosen because the "Pinyon-Juniper" zone is also one of the more ideal habitats for Mule deer; the latter are presumed to be among Wills' "highest return species,"since they are relatively large animals. Defined as such, the Mogollon Rim is roughly 150 to 200 km distant from Tucson, Arizona (see Brown & Lowe 1994). Between Tucson and the Mogollon Rim there are several mountain ranges, and we may assume that hunters skirted the base of these, rather than crossing them directly. The total distance traveled would have been roughly 200 km, from Tucson to the Natanes Mountains.

The energetic costs of transporting food would have imposed a tax on the amount of food that could be transported to logistical hunting stations in the Mogollon highlands. Following Kelly (1995) and Jones and Madsen (1989) it is assumed that a 20 kg load (containing 61,400 Cal after the costs imposed by harvesting and processing), could have been transported at a rate of three km/hour and at a cost of 125 Cal/km. The maximum round-trip distance that could be supported by 20 kg. of maize without a net energy loss is 491 km, assuming continuous movement without rest on a flat surface. If one adds some other reasonable qualifying conditions, for example a maximum of 12 hours of movement per day and an average cost of 50 Cal/hr when not moving towards the destination, then the entire round-trip would have required 133 hours of transportation (at 375 Cal/hr) and 133 hours of other activities (at 50 Cal/hr). The entire cost of the trip would be approximately 56,500 Cal. Subtracting the costs of the trip from the basket of transported maize, roughly 4,900 Cal. remain in the basket.

A hunter engaged in a long-distance foray to the Mogollon Rim would have to be extremely confident of success in a very narrow window of time. If we assume that hunting required 2,400 Cal/day, then roughly two days of hunting could be supported by the endeavor. It is suggested that a two-day window of availability risking net energy loss is a very risky strategy. Furthermore, although it is not central to this discussion, it is interesting to note that a person would have to grind maize for 80 hours in order to prepare the food for the hunter's trip. Thus, the maximum gross protein or energy return rate for a successful trip would be the edible nutritive component of an average Mule deer divided by 346 hours of human activity.

A second question may be even more important than the assessment of the gross costs of long-distance hunting. "What are the energy return rates, or rates of protein returned per expended Calorie, for Upland Mogollon hunting vs. local hunting?" An alternative would be to hunt close to the Tucson Basin, for example, in the nearby Santa Catalina Mountains, or to hunt for Mule deer along the low-desert watercourses. One could travel well-into the high elevations of the Santa Catalinas at the cost of a short (50 km 1-way) excursion. If all other factors in this exercise remain constant (for example, equal prey density at higher elevations in the Mogollon Rim and in the Santa Catalinas), then we may assume that local hunting would have offered energy or protein return rates that were minimally four times greater than the distant-hunting strategy proposed by Wills. Mule deer are common in the low desert (Turner & Brown 1994). Hence, unless there was extreme overpredation of preferred prey, a Cienega Phase hunter would likely have met with greater hunting success (higher return rates per hour or per expended Cal) within 50 km of his or her home.

Discussion

The foregoing discussion is admittedly hypothetical and the conclusion, that Wills' model of long-distance foraging by Cienega Phase hunters is unlikely, may be questioned on several grounds. The first possible objection is that the apparent concordance of the predicted rank-order of plant usefulness with the observed ubiquities of plants is spurious. It is possible that the estimates of the costs and returns from different plant resources have been inaccurately estimated, and that the true, predicted rank-order should be very different from the one obtained in this study. Although that possibility is regarded as unlikely, more actualistic research in this arena is needed. Alternatively, one might question the assumption that a correspondence between the predicted rank order of plants and the observed ubiquities is interpretively relevant. However, the assumption is plausible, and it provides a mechanism for testing models of prehistoric land use. An alternative objection is that other factors that influence hunting returns have not been assessed. For example, it is well-known that the abundance and prey-density of game will greatly affect the return rates of hunting. There exists one circumstance of two very specific conditions, in which Wills' model of very-long distance foraging would be preferred over local hunting. These are that (1) protein or fat, rather than carbohydrates, was the currency of optimization, and, (2) the density and capture rate of protein and fat in the Mogollon Rim greatly exceeded the density and capture rate of protein and fat in the low desert or the immediately surrounding mountains. Under such circumstances it is possible that the return rates (protein per expended Cal.) from very-long distance hunting strategies exceeded the return rates from local hunting strategies.

Finally, although Wills model was generated from (and rejected on the basis of) evidence that was recovered in the Tucson Basin, its plausibility increases for settlements that may have been located nearer to the Mogollon Rim. Future efforts at modeling Early Agricultural Period (Cienega Phase or its temporal equivalents) must account for the distances between resource use zones and residential sites. Attention to these details may lead to the recognition of a variety of land use strategies for the occupants of different areas in the low deserts of Arizona.

Behavioral Ecology and Implications for the Classically Sedentary Model

Huckell's model of "Early Agricultural Period" (Late Archaic) Period land use asserts that the subsistence practices of Cienega phase occupants of the Tucson Basin differed little from the subsistence practices of later Hohokam farmers. In other words, the proportional contribution of maize to the diet was the same in both cases. Huckell's description is, perhaps, reasonable given the relatively high maize ubiquity scores observed in Cienega phase sites.

However, when viewed through the lens of behavioral ecology there is reason to suggest that subsistence practices and the relative dependence on agriculture between the two time periods differed greatly. Specifically, the diversity of the diets of Cienega phase forager-horticulturists is greater than the diversity of the diets of Hohokam farmers. Behavioral ecologists' assumptions about the causes of changes in diet breadth imply that the return rates from one or more of the Hohokam resources

Table 14.5 Diet Breadths of Cienega Phase and Hohokam Period Occupants of the Tucson Basin.

Site	Occupation Period	Number of Taxa	Number of Analyzed Contexts
	Late Archaic/Early Agricultural Period Sites		
Clearwater (BB:13:6)	Cienega Phase	18.00	15.00
Donaldson (EE:2:30)	Cienega Phase	15.00	9.00
Los Pozos (AA:12:91)	Cienega Phase	42.00	72.00
Milagro (BB:10:46)	Cienega Phase	16.00	15.00
Santa Cruz Bend (BB:12:746)	Cienega Phase	17.00	41.00
Stone Pipe (BB:13:425)	Cienega Phase	20.00	45.00
	All Ceramic Period and Hohokam Sites		
	Early Ceramic Period and Hohokam Sites		
Cienega Site (BB:9:143)	Rincon Phase	13.00	16.00
Julian Wash (BB:13:7)	Rincon Phase	8.00	9.00
Lonetree (AA:12:120)	Early Pioneer	11.00	7.00
Lonetree (AA:12:120)	Early-Middle Rincon	11.00	3.00
Lonetree (AA:12:120)	All Rincon Phase	13.00	10.00
Redtail Site (AA:12:149)	L.Pioneer/E. Colonial	13.00	6.00
Redtail Site (AA:12:149)	Canada del Oro Phase	13.00	6.00
Square Hearth	Early Ceramic Period	9.00	18.00
West Branch, Cook Avenue Locus, All	Ricon Phase	6.00	10.00
West Branch, other loci	Early Rincon Phase	9.00	<14*
West Branch, other loci	Middle Rincon Phase	14.00	<69*
West Branch, other loci	L.Rincon/Tanque Verde	3.00	<5*

Note: * = authors presented ubiquity data based on number of sample

were so great that many of the low-quality plants used by Cienega phase foragers were ignored. In the following discussion I suggest that Cienega phase occupants of the Tucson Basin were substantially less dependent on agriculture than their Hohokam Period descendants.

BEHAVIORAL ECOLOGY AND DIET BREADTH

Behavioral ecologists have developed a set of assumptions that may be used to predict whether or not a specific resource will be included in the diet of foragers. One important assumption is that foragers optimize. Under such conditions, foragers should consider the search and pursuit costs (energy required to locate and capture prey) when deciding whether or not to use a given resource. Briefly summarized, a forager encountering a resource A should ignore that resource in favor of an as-yet-undiscovered resource B (suspected to be in the area), if the expected returns from B (even given the costs of continued search and pursuit) are higher than the immediate returns available from resource A (Kelly 1995:83). As higher ranked resources become available, low-ranked resources are dropped from the diet. As Kelly (1995:78) notes, there have been few actual tests of the diet breadth model however, the approach provides a powerful explanatory or predictive paradigm for explaining changes in diets.

The discussion that follows occurs in the context of two assumptions. First, it is assumed that Hohokam farmers optimized their diet structure. Second, it is assumed that the diet breadth model is usable as an interpretive device for explaining observed changes in diet breadth. In other words, this argument is not a test of the diet breadth model, but rather an interpretation of archaeologically observed changes in diet breadth, with recourse to the diet breadth model as a mechanism for generating an explanation.

The Diet Breadths of Cienega Phase and Hohokam Period Foragers in the Tucson Basin

A comparison of the macrobotanical assemblages from selected Hohokam Period and Cienega phase sites in the Tucson Basin indicates that the diversity of Cienega Phase assemblages is greater than the diversity of subsequent Hohokam assemblages. It is suggested that the differences are attributable to a general decrease in diet breadth that occurred during the Hohokam occupation of the Tucson Basin. For this discussion, diversity is measured by the number of observed taxa of potentially edible plants. The number of identified edible seed taxa from Cienega Phase and Hohokam (multiple phases) sites in the greater Tucson Basin are listed in Table 14.5.

What changes in the organization of subsistence were entailed by the decrease in diet breadth during the Hohokam occupations? Generally, inspection of the paleobotanical data offered in the cited sources indicate a replacement of wild plants with a few domesticated varieties, including maize, beans, squash, cotton, and possibly agave. The non-domesticants cheno-ams, tansey mustard, and various cacti (especially saguaro) are retained in the diets of Hohokam farmers; however, there is an extreme decline in the representation of wild grasses and other low-density wild annuals. Overall, agricultural plants are favored by Hohokam farmers, whereas Cienega Phase forager-horticulturists clearly incorporated a wide range of small-seeded, wild annuals in their diet.

One is led to the conclusion that Cienega Phase foragers were considerably more mobile than their Hohokam Period descendants. The use of a wider range of wild plants, whose distributions on the landscape were less predictable and not determined by deliberate plans, entails mobility while foraging. In contrast, the exclusive or primary use of agricultural plants enables high predictability in the timing and location of the ripe seeds, and entails reduced residential or logistical mobility; the need for searching is eliminated, and crops may be planted in close proximity to villages in an effort to reduce transportation costs. To the extent that Cienega Phase forager-horticulturists were more mobile, one must concede that their land use strategies may have been quite different from Hohokam farmers' land-use strategies (*contra* Huckell 1995), despite the incorporation of modest amounts of maize into their subsistence economy.

The observation that Hohokam farmers' subsistence practices were substantially different from Late Archaic forager-horticulturists should surprise few. Corroborating evidence in the arenas of architecture and ground stone tool technology support the interpretation offered in this report. Expansive irrigation systems, large granaries for maize, and a technology that included tools that were designed for processing maize (for example, large, vesicular basalt, partial- or full-trough *metates*) are common among Hohokam sites that were occupied after A.D. 500, and absent from Late Archaic / Early Agricultural Period sites.

Discussion

Although it is not incumbent upon a researcher to point out the potential errors in their own research, a few observations about the potential for error in this paper are appropriate. One might be tempted to criticize this study on the grounds that the apparent differences in the diversity of the Late Archaic and Hohokam macroplant assemblages are spurious. Two objections might be raised.

The first objection could be that the number of taxa represented in an assemblage is an inappropriate measure of diversity. Some paleobotanists might prefer the use of the Shannon-Weaver (S-W) diversity index (Popper 1988). One strength of the S-W index is that it attempts to compare assemblages using both the number of species that are represented and the quantity of seeds from each species. I avoided the use of the S-W index in this study because it imposes an arbitrary logarithmic transformation that has not been adequately justified by its proponents. As a consequence of the log transformation, there exists no scale by which to judge whether or not small differences in the S-W scores are indicative of great differences or only minor ones. Moreover, as is noted by Popper (1988), there exists no statistic by which one can assess whether or not different S-W scores are *significantly* different.

The number of cases that have been used to represent the Late Archaic Period and *all* of the Hohokam occupations is extremely small and is a potential source of other objections. Histograms (not shown) do not conclusively indicate the existence of any central tendencies, for example, such as a normal distribution of the number of taxa found at Cienega Phase and Rincon Phase settlements. If archaeologists are to observe statistically compelling differences between macroplant assemblages that were produced during different occupations, they shall require the excavation of scores of settlements from every occupation phase. In defense of the present study it is noted that the same objections could be raised in response to the models of Late Archaic land use that were heretofore considered. If one had a sufficient number of assemblages from Hohokam and Late Archaic sites, one might be able to compare the means of the number of taxa, by occupation phase.

LATE ARCHAIC HORTICULTURISTS AS SEASONALLY-SEDENTARY FORAGERS

From the preceding discussions it should be apparent that I would characterize Cienega Phase forager-horticulturists as "more mobile than classically-sedentary farmers and less mobile than very-long distance hunters." The results of this study accord reasonably well with Roth's characterization of the Late Archaic occupants of the Tucson Basin as seasonally-sedentary foragers. Unfortunately, the seasonally-sedentary model requires the elaboration of details that have not yet been specified. Where were the seasonal settlements? How many were used in a typical year? Which resources were obtained in their vicinities? Archaeologists will require the elaboration of these details if variants of the seasonally-sedentary model are to be assessed within the framework of behavioral ecology.

Hints of these details may be present in the limited array of non-floodplain Late Archaic sites that have been excavated. Coffee Camp (AZ AA:6:19, Hutira 1993) was a Late Archaic site at which maize and saguaro were not observed, but at which grasses and mesquite were strongly represented. Furthermore, numerous chipped-stone surface assemblages have been identified in the bajadas (Roth 1995, 1992) in biotic microclimates that are favorable for saguaro (Diehl 1997b); what was the nature of the occupation and use of these locations? Roth (1995) has identified upper bajada sites that may have been the locations of late fall residential base camps associated with hunting. Most importantly, what were the relationships between sites like Coffee Camp, or the bajada scatters, and the larger floodplain settlements? What are the implications of these relationships for the organization of Late Archaic subsistence? Future research will likely be structured in a way that will provide answers to these questions, so there is reasonable cause for optimism.

THE ADOPTION OF AGRICULTURE IN THE TUCSON BASIN

What motivated the adoption of agriculture in the Tucson Basin? According to Wills' (1996) model, agriculture was used to extend the range of logistically organized hunting efforts. From his point of view, the Tucson Basin was a naturally productive resource area in which foragers experienced little risk of subsistence shortfalls. The adoption of agriculture occurred to expand foraging opportunities, not as a response to resource stress. However, this study has demonstrated that Wills' very-long distance hunting model would have been both risky and energetically inefficient. Other strategies would have resulted in higher return rates (of Cal or units of protein per expended Cal). Although it is possible that the adoption of agriculture occurred as a mechanism to promote short-range hunting, an alternative motivation for the adoption of agriculture is proposed.

It should be noted that the energetic return rates from wild Tucson Basin resources are, with one exception, very low. The exception of course is the energy derived form the harvest of saguaro fruit. However, it has already been noted that saguaro may not have produced high yields per hectare, and the fruit may not have been suitable for long-term storage. Moreover, saguaro are very susceptible to frost damage, and the occurrence of even brief freezes may retard or halt the production of fruit for one ore more summers. All of the wild plants in the Cienega phase assemblages are resources whose specific timing and location of availability were unpredictable.

Only one resource used by Cienega Phase forager-horticulturists simultaneously offered high yields, relatively high energy returns, high storability, and high predictability. That resource was maize. Although costly to process, maize would have provided a reasonable energy return rate whenever it was available to grind. It is therefore suggested that the adoption of agriculture, at least in the Tucson Basin, occurred primarily as a risk minimization strategy. Wills (1996, 1988) has suggested that the minimization of risk promoted the adoption of agriculture by highland Southwesterners by ensuring access to highly valued foods such as pinyon nuts or meat. In contrast, it is suggested here that the occupants of the Tucson Basin valued maize principally because it minimized the risk of resource shortfalls that might otherwise be entailed by exclusive reliance on a broad spectrum of low-yield wild foods. Maize was grown in order to provide a large, stable, storable surplus of high-calory food that could be used in the season when other plants and animals were in short supply.

CONCLUSION

Debates over the organization of subsistence among prehistoric Southwesterners are common among area archaeologists. This paper offers resolutions for some of the disputes about the nature of Cienega Phase (BC 800-AD 150) subsistence practices and land use during the Late Archaic Period in the Greater Tucson Basin. Principles of behavioral ecology were used in combination with paleobotanical data in order to evaluate competing characterizations of Cienega Phase land use. It was shown that the model that best fits the extant macrobotanical record was provided by Roth (1992). In that description, Roth characterized the Cienega Phase occupants of the Tucson basin as seasonally sedentary forager-horticulturists. Finally, this paper also suggested that the adoption of agriculture in the Tucson basin occurred primarily as an effort to minimize the risk of seasonally-occurring productive shortfalls, rather than as a means for promoting other subsistence related activities.

ACKNOWLEDGMENTS

My thanks to Barbara Roth for comments on initial versions of this paper, and to Bill Doelle, David Gregory and Henry Wallace at the Center for Desert Archaeology for providing relevant Cienega Phase and Hohokam data from the greater Tucson Basin.

ENDNOTES

1. Estimates of the energetic return rates are based on a number of factors including harvesting and processing costs and rates. Although harvesting rates vary greatly between species, all energetic returns were limited by the technology used to prepare the plants as food. In particular, grinding seeds into flour has an equalizing effect that limits the energetic return rates of seed plants. For example, I estimated that Cienega Phase maize returned 200 kg/ha and could be harvested at rates of 5 kg/hour. Maize processing requires 1.25-1.7 hr/kg (M.Wright 1993:350-351) using modern varieties of flour kernelled maize, and large, efficient manos and metates. I assumed that maize grinding requires 350 Cal/hr (60 minutes times 5.8 Cal/minute; this is the same energetic cost used by Doelle [1978] based on a review of studies of African foragers). Owing to the use of a hard-kernelled maize (K.Adams 1994; Galinat 1988; Upham et al. 1988, 1989), and comparatively inefficient grinding tools, I assumed that Archaic maize was processed at a rate of .4 kg/hr.

Estimates of the returns from other plants are based on actualistic studies of the use of mesquite and saguaro by Native Americans who lived in the Sonoran Desert biotic province (Doelle 1976, 1978, 1980), Great Basin foragers' use of seed plants (Jones & Madsen 1989; Simms 1987), and African foragers use (harvesting and grinding) of seed plants (K. Wright 1994: 245). The estimates for "other seeds" use the same processing costs as K. Wright (1994), and assume higher harvesting costs owing to lower plant densities.

Where possible, nutritional data (Cal/kg) were obtained from primary sources especially Ensminger et al. (1990); when no information was available, the nutritional information provided by archaeologists and anthropologists was used. Tansey mustard (*Descurainia* sp.) was problematic to evaluate since there was no consistency between the estimates of the energy yields (Cal/kg) offered by this plant. The post-encounter return rate that I have offered in Table 2 is generous if compared with Kelly (1995: 80), yet less than one estimate of 4,600 Cal/kg (Roth 1988).

REFERENCES CITED

Brown, David E.
1994 Great Basin conifer woodland. In *Biotic Communities: Southwestern United States and Northwestern Mexico* (David E. Brown, ed.), pp.52-57. Salt Lake City: University of Utah Press.

Brown, David E., & Charles H. Lowe
1994 Biotic communities of the Southwest. Map supplement to *Biotic Communities: Southwestern United States and Northwestern Mexico* (David E. Brown, ed.), pp.52-57. Salt Lake City: University of Utah Press.

Cordell, Linda S.
1984 *Prehistory of the American Southwest.* New York: Academic Press.

Diehl, Michael W.
1996 Archaeological investigations of the Early Agricultural period settlement at the base of A-Mountain, Tucson, Arizona. *Technical Report No. 96-21.* Tucson: Center for Desert Archaeology.

1997 Macrobotanical and pollen remains and their Interpretation. In excavations in the Santa Cruz River floodplain: the early agricultural period component at Los Pozos (David Gregory, ed.). *Anthropological Papers, No. 21.* Tucson: Center for Desert Archaeology.

Downum, Christian E., Adrianne G. Rankin, & Jon S. Czaplicki
1986 *A Class III Archaeological Survey of the Phase B Corridor, Tucson Aqueduct, Central Arizona Project.* Tucson: Archaeological Series No. 168. Cultural Resource Management Division, Arizona State Museum.

Ensminger, Audrey H., M.E. Ensminger, James E. Konlande, & John R.K. Robson
1994 *Foods and Nutrition Encyclopedia*, 2nd ed. Boca Raton: CRC Press.

Fish, Suzanne K., Paul. R. Fish, & John Madsen
1990 Sedentism and settlement mobility Prior to A.D. 1000. In *Perspectives on Southwestern Prehistory* (Paul E. Minnis & Charles L. Redman, eds.), pp.76-91. Boulder: Westview Press.

Foley, Robert L.
1985 Optimality theory in anthropology. *Man* 20: 222-242.

Gregory, David A. (ed.)
1997 Excavations in the Santa Cruz River floodplain: the Early Agricultural period component at Los Pozos. *Anthropological Papers No. 21.* Tucson: Center for Desert Archaeology.

Huckell, Bruce B.
1995 *Of Marshes and Maize: Preceramic Agricultural Settlements in the Cienega Valley, Southeastern Arizona.* Anthropological Papers of the University of Arizona No. 59. Tucson: University of Arizona Press.

Huckell, Lisa W.
1996 Chapter 3.3. Macrobotanical remains. In *Archaeological Investigations at Early Village Sites in the Middle Santa Cruz Valley: Analyses and Synthesis* (Jonathan Mabry, ed.). Anthropological Papers No. 19. Tucson: Center for Desert Archaeology.

1995 Farming and foraging in the Cienega Valley: Early Agricultural period paleoethnobotany. In *Of Marshes and Maize: Preceramic Agricultural Settlements in the Cienega Valley, Southeastern Arizona* (Bruce B. Huckell, ed.), pp.74-97. Anthropological Papers of the University of Arizona, No. 59. Tucson: University of Arizona Press.

Huckell, Bruce B., Lisa W. Huckell, & Suzanne K. Fish
1994 *Investigations at Milagro, A Late Preceramic Site in the Eastern Tucson Basin.* Technical Report 94-5. Tucson: Center for Desert Archaeology.

Hutira, Johna
1993 Pits, pods and paradigms: Archaic subsistence at Coffee Camp. In *Archaic Occupation on the Santa Cruz Flats: The Tator Hills Archaeological Project* (Carl D. Hilbert & T. Kathleen Henderson, eds.), pp.333-343. Northland Research, Flagstaff. Submitted to USDI Bureau of Reclamation, Arizona Projects Office, Contract No. 3-PA-30-00740 Task 45. Copies available from the USDI Bureau of Reclamation, Arizona Projects Office, Phoenix.

Irwin-Williams, Cynthia
1967 Picosa: the elementary southwestern culture. *American Antiquity* 32: 441-456.

Jones, K. & D. Madsen
1989 Calculating the cost of resource transportation: a Great Basin example. *Current Anthropology* 30: 529-534.

Kelly, Robert L.
1995 *The Foraging Specturm: Diversity in Hunter-Gatherer Lifeways.* Washington: Smithsonian Institution Press.

Mabry, Jonathan B., & Gavin H. Archer
1996 The Santa Cruz Bend Site, AZ AA:12: 746 (ASM). Chapter 2 in *Archaeological Investigations of Early Village Sites in the Middle Santa Cruz Valley* (by Jonathan Mabry, Deborah L. Swartz, Helga Wocherl, Jeffery J. Clark, Gavin H. Archer & Michael W. Lindeman). Anthropological Papers No. 18. Tucson: Center for Desert Archaeology.

Miksicek, Charles H.
1996 Macrobotanical remains. In *A Rincon Phase Occupation at the Julian Wash Site, AZ BB: 13:7, ASM.* (by Jonathan B. Mabry), pp. 89-94. Technical Report No. 96-7. Tucson: Center for Desert Archaeology.

1990 Appendix C: botanical data. In *Rincon Phase Seasonal Occupation in the Northern Tucson Basin* (by Mary Bernard-Shaw & Frederick W. Huntington), pp.241-252. Technical Report No. 90-2. Tucson: Center for Desert Archaeology.

1989 Appendix F: supplemental plant information. In *Archaeological Investigations at the Redtail Site, AA:12:149 (ASM), in the Northern Tucson Basin* (by Mary Bernard-Shaw), pp.251-255. Technical Report No. 89-8. Tucson: Center for Desert Archaeology.

1986 Plant remains. In *Archaeological Investigations at the West Branch Site: Early and Middle Rincon Occupation in the Southern Tucson Basin*, by Frederick W. Huntington, pp.289-313. Anthropological Papers No. 5. Tucson: Institute for American Research.

Miksicek, Charles H., & Mary Bernard-Shaw
1990 Appendix F: supplemental plant information. In *Archaeological Investigations at the Lonetree Site, AA:12:120 (ASM), in the Northern Tucson Basin*, by Mary Bernard-Shaw, pp.245-252. Technical Report No. 90-1. Tucson: Center for Desert Archaeology.

Minnis, Paul E.
1985 *Social Adaptation to Food Stress: A Prehistoric Southwestern Example.* Chicago: University of Chicago Press.

1981 Seeds in archaeological sites: sources and some interpretive problems. *American Antiquity* 46: 143-152.

O'Connell, James F., & Kristen Hawkes
1994 Behavioral ecology, evolution and archaeology. Paper delivered at the annual meeting of the Society for American Archaeology, Anaheim.

Pearsall, Deborah M.
1989 *Paleoethnobotany: A Handbook of Procedures.* New York: Academic Press.

Pianka, Eric R.
1978 *Evolutionary Ecology, 2nd edition.* New York: Harper and Rowe.

Popper, Virginia S.
1988 Selecting quantitative measurements in paleoethnobotany. In *Current Paleoethnobotany* (Christine A. Hastorf & Virginia S. Popper, eds.), pp.53-71. Chicago: University of Chicago Press.

Roth, Barbara J.
1995 Late Archaic occupation of the upper Bajada: excavations at AZ AA:12: 84 (ASM), Tucson Basin. *Kiva* 61: 189-207.

1992 Sedentary agriculturists or mobile hunter gatherers? Evidence on the Late Archaic occupation of the Northern Tucson Basin. *Kiva* 57: 291-314.

1988 *Late Archaic Settlement and Subsistence in the Tucson Basin.* Ph.D. dissertation. Department of Anthropology, University of Arizona, Tucson.

Sayles, E.B. & E. Antevs
1941 *The Cochise Culture.* Medallion Papers No. 29. Globe: Gila Pueblo.

Swartz, Deborah L., & Lindeman, Michael
1996 The Stone Pipe Site, AZ BB:13: 425 (ASM). Chapter 4 in *Archaeological Investigations of Early Village Sites in the Middle Santa Cruz Valley* (by Jonthan Mabry, Deborah L. Swartz, Helga Wocherl, Jeffery J. Clark, Gavin H. Archer & Michael W. Lindeman). Anthropological Papers No. 18. Tucson: Center for Desert Archaeology.

Swartz, Deborah L., & Charles H. Miksicek
1996 Archaeobotanical remains from the Cook Avenue locus. In *Archaeological Data Recovery Project at the Cook Avenue Locus of the West Branch Site, AZ AA:16:3 (ASM)* (by Allen Dart & Deborah L. Swartz), pp.129-137. Technical Report No. 96-8. Tucson: Center for Desert Archaeology.

Turner, Raymond M., & David E. Brown
1994 Sonoran desert scrub. In *Biotic Communities: Southwestern United States and Northwestern Mexico* (David E. Brown, ed.), pp. 181-221. Salt Lake City: University of Utah Press.

Wills, Wirt H.
1996 Archaic foraging and the beginning of food production in the American Southwest. In *Last Hunters, First Farmers* (T. Douglas Price & Anne B. Gebauer, eds.), pp.215-242. Santa Fe: School of American Research Press.

1988 *Early Prehistoric Agricualture in the American Southwest.* Santa Fe: School of American Research.

15

Conspicuous Consumption as Wasteful Advertising: a Darwinian Perspective on Spatial Patterns in Classic Maya Terminal Monument Dates

Fraser D. Neiman
Monticello, Virginia

ABSTRACT

Spatial patterns in Classic Maya terminal monument dates offer clues about the processes behind the Classic "collapse" and the spatial scale of political competition that preceded it. A Darwinian theory of wasteful advertising delivers a simple model of how the collapse and political competition might map onto terminal monument dates. Modern statistical techniques provide the tools required to decompose spatial variation in terminal dates from a sample of 69 sites into the components required by the model. The large-scale trend in the cessation of dated monument construction, estimated using local robust regression (loess), supports recent evidence that the ultimate cause of the end of dated monument construction was an ecological disaster. The variogram characterizes spatial variation in residuals from this trend. The results suggest that as the collapse proceeded, monument-constructing elites, or individuals competing for that status, abandoned their original sites to enter political competition at sites on average 65 km distant, thus increasing competition, and prolonging monument construction at those recipient sites. This implies spheres of political competition in the Late Classic were small, and supports recent evidence from epigraphic sources they averaged about 50 km in diameter.

INTRODUCTION

Stone monuments have played a leading role in the archaeological study of Classic Maya society since archaeologists became aware of its existence. Over a century and a half ago, the remarkable monuments of Copan inspired John Lloyd Stephens to guess that he had discovered the remains of what we would today call a complex society, a place where "orators, warriors, and statesmen, beauty, ambition, and glory had lived and passed away" (1841:105). Inferences similar to Stephens' acquired real historical value when archaeologists learned how to date the monuments by reading the dedicatory dates inscribed on them. This development along with an accurate calibration between the Mayan and Gregorian calendars that was in place at the beginning of this century (Goodman 1905), made it possible to use monuments to chart the efflorescence and demise of Classic Maya civilization. Sylvanus Morley led the way in this endeavor by building histograms of monument dedicatory dates, at both site and regional scales (Morley 1920, 1938-39, 1946), a tactic still favored today (Jones 1991, Fash 1988, Mathews & Willey 1991). Morley's histograms allowed him to detect the relatively abrupt cessation of dated monument construction that became the famous Maya collapse. Since then region-wide frequency distributions of terminal monument dates–the dedicatory date of the latest known monument constructed at a site–have been used to address possible causes of the collapse (Hamblin & Pitcher 1980, Lowe 1985).

In the wake of the New Archaeology's interest in reconstructing ancient social and political organization, the traditional preoccupation with the distribution of monuments in time has been supplemented by a new interest in their distribution in space (Lowe 1985, Mathews 1985). Interest in the collapse has lead to a focus on the spatial distribution of terminal monument dates (Bove 1981, Whitley & Clark 1985, Kvamme 1990). The common motivation here is the notion that this evidence offers insight into two related topics. The first is the nature of the Classic Maya collapse itself. Different processes responsible for the collapse would appear to have different implications for spatial patterns in the timing of the cessation of dated monument construction. Second, different scenarios concerning the character of what was collapsing, in particular the spatial scale of political competition, yield different expectations about geographical patterns in terminal dates. Researchers have looked to spatial patterns in terminal dates to evaluate competing scenarios in both these areas (Marcus 1993).

My goals in this paper are both empirical and theoretical. On the empirical side, by building on previous work, I hope to advance our understanding of the historical processes responsible for spatial patterns in terminal dates and thereby contribute to our knowledge of ancient Maya history. Achieving this empirical goal requires the introduction of new conceptual tools, drawn from Darwinian theory. Hence I begin by sketching a set of ideas about the mechanistic evolutionary processes that cause conspicuous consumption, manifest in the case at hand by the apparently wasteful energy investment in the construction of dated stone monuments. The result, a Darwinian theory of wasteful social advertising, makes it possible to build a simple model of how historical processes related to the Maya collapse might map onto spatial patterns in terminal dates. It suggests what kinds of phenomena are measured by spatial variation in terminal dates. Expectations about this mapping indicate what sort of patterns we should be looking for in terminal dates and guide the choice of statistical tools likely to uncover them.

I then turn to evidence gleaned from terminal dates from 69 sites in the Maya Lowlands. Local, robust regression (loess) is used to identify the underlying regional trend in terminal dates. The trend offers an unappreciated and important clue to the ultimate cause of the cessation of dated monument construction. The residual variation is then examined for spatial pattern using a variogram, a simple spatial-analytic tool borrowed from geostatistics. While the results fall outside initial expectations, the larger theoretical framework indicates what kinds of historical processes created the patterns observed. The resulting inferences support characterizations of Maya polities that have begun to emerge over the last decade, spurred by the decipherment of Mayan hieroglyphic writing (Coe 1993). I conclude by suggesting how they can be evaluated objectively using additional neo-Darwinian theory and independent sources of evidence. The ultimate test of archaeological theory is its ability to deliver objective knowledge about what happened in history from a mute record. To the extent that this paper achieves its empirical goals, it is also an argument for the promise of Darwinian theory in archaeology.

MONUMENTS AND SOCIAL COMPLEXITY

The historical utility of dated stone monuments among Mayanists derives from the assumption that they can be used to monitor ancient phenomena that are variously glossed as "civilization" or "social complexity." Consider some examples, drawn from the recent literature on terminal dates. Justifying their use of terminal dates to document temporal dynamics of an hypothesized peasant revolt, Hamblin and Pitcher (1980: 246) observe that dated monument construction requires "literate forms of Classic Maya culture," including hieroglyphic writing, calendrics, mathematics and astronomy. Hence temporal patterns in terminal dates can be regarded as "quantitative indices of the presence of Maya elites, the carriers-producers of the civilization." Working in a similar tradition, but with an eye to spatial patterns, Bove (1981: 95) notes that the distribution of stone monuments is "highly correlated with a hierarchical system of settlement, a high degree of specialization, regionalization, and internal specialization" and can be regarded as a "response to increasing sociocultural evolution." Hence the cessation of monument construction is an "indicator of collapse" and the "demise of the elite." In the same vein, Whitley and Clarke (1985: 385) regard monuments as "indicators of 'elite' populations and attendant sociopolitical complexity."

This literature lacks a theoretical account of the causal mechanisms that might link the cessation of monument construction with societal "collapse." Instead the inferential connection is built on a covert empirical generalization that the occurrence of stone monuments on the one hand and "complexity" or "civilization" on the other are correlated across cultural contexts. This simple behavioral correlate, is the means by which the distribution of mute monuments in time and space has been given historical meaning since Morley's pioneering efforts. Its use is not, of course, unique to Mayanists. It underlies chronicles of the rise of civilization the world over. Perusal of this wider literature reveals variation in the precise way in which the behavioral meaning of monuments is cast. The most conspicuous source of variation is change in the interpretive fashions of archaeology. Over the last decade, as increased emphasis on social conflict, factionalism, and inequality displaces the functionalist neo-evolutionism of the New Archaeology, inferences tend to invoke "political power," in place of benign "social complexity." However recent post-processual approaches have failed to supply the missing etiological framework. Their proponents deny the desirability of attempting to build one, preferring instead to rely on their own covert assumptions about human nature (e.g. Hodder 1986, Shanks & Tilley 1987).

Recent work by Trigger (1990) is an apparent exception to archaeologists' inattention to theory in this area. Trigger's account of the link between monumental architecture and "complexity" or "power" begins with the contrast between Zipf's "principle of least effort" (1949) and Veblen's notion of "conspicuous consumption" (1899). Zipf argued that economizing and efficiency govern the production and distribution of goods necessary to sustain human life, while Veblen noted the ubiquity of wasteful spending to enhance social prestige in modern Europe. Trigger connects the two perspectives by defining conspicuous consumption as a violation of the least-effort principle. The definition is important because it raises the possibility that all artifacts and behavior, from pyramids to potlatches, that require significantly more energy than would be expected under the least-effort principle can be understood in the same conceptual framework as manifestations of a single set of causal dynamics.

The crucial question becomes why is conspicuous consumption a universal symbol of power? What is the causal connection between them? Trigger's answer has two parts. The first is that the energy expended often comes from other people's labor. When it does, conspicuous consumption signals the ability to control the behavior of those doing the work. But clearly this is not the crux of the issue. We still have no explanation for why, when people successfully invest energy in placing the labor of others at their disposal, they choose to spend the proceeds of their investment in a wasteful manner. The second part of Trigger's answer is that "monumental architecture and personal luxury goods become symbols of power because they are seen as embodiments of large amounts of human energy and hence symbolize the ability of those for whom they were made to control such energy to an unusual degree" (1989: 125). While superficially satisfying, this is no answer at all. It does not explain why the energy expenditure that denotes power is wasteful, that is violates the principle of least effort. Trigger's answer is little more that a reformulation of the original question in the form of an assertion.

The question goes unanswered because Trigger's account, like the empirical generalizations that preceded it, is crafted from common sense and not a larger mechanistic theoretical framework. Should archaeologists interested in figuring out what happened in history, whether in the Maya Lowlands or elsewhere in the world, worry about the lack of an answer? The price we pay for not doing so is the production of inferences about the past that are often impossible to evaluate objectively (Dunnell 1982, Neiman 1990), a point to which I will return at the end of this paper. It is the promise of objective evaluation that motivates the theoretical development that follows. There is no question that terminal monument dates are correlated with what archaeologists refer to as the "collapse" of Classic Maya civilization. But the foregoing argument does imply the need for greater precision regarding the social phenomena measured by these dates. Greater precision delivers richer expectations about how the historical process associated with the collapse maps onto spatial variation in terminal monument dates. Greater precision also specifies the statistical methods required to isolate relevant empirical patterns and opens the path to objective evaluation by the archaeological community.

A DARWINIAN THEORY OF WASTEFUL ADVERTISING

Temporal Dynamics

What aspects of social organization are we measuring when we look at terminal monument dates? Answering this question requires fundamental theory about the causes of conspicuous consumption. Trigger's treatment contains two clues for us. One is his suggestion that a history of natural selection operating on primate feeding strategies lies behind Zipf's empirical generalization about effort minimization in human behavior. If selection shapes feeding strategies of both human and non-human primates, and the predictive successes of Darwinian theory suggest it does (e.g. Broughton 1994, Kaplan & Hill 1992), then why is it not also responsible for variation in other aspects of human phenotypes, including conspicuous consumption? Placing conspicuous consumption in a Darwinian perspective immediately clarifies the precise sense in which the energy expenditure is wasteful extravagance. It is wasteful because it represents fitness costs that, unlike the time and energy expended on foraging, have no apparent compensatory fitness benefit. Consider dated Maya monuments. They are big chunks of stone in the form of altars, stairs, and stelae. They represent expenditures of resources to assemble and supervise the individuals who quarried, transported, assembled, and carved them. More importantly they required the training and maintenance of groups of individuals skilled in the use of a complex system of hieroglyphic writing, the accompanying iconographic repertoire, and in carving and painting both on stone. These expenditures represent the diversion of resources that members of the Maya elite might otherwise have invested directly in the maintenance of themselves and their kin, the acquisition of mates, and the production of children. In other words, the individuals responsible for the monuments appear to have suffered significant direct and opportunity fitness costs in order to construct them.

If a Darwinian perspective sharpens the sense in which conspicuous consumption is wasteful, it also renders it even more puzzling: how could selection possibly account for it? The answer I explore in the remainder of this paper was first suggested by the evolutionary biologist Amotz Zahavi (1975, 1987, cf. Kirkpatrick 1986)[1]. Graffen (1990a, 1990b) offers several elegant mathematical models that confirm the process envisaged by Zahavi actually leads to the outcome he claimed for it.

If investment in waste does not confer a direct selective advantage on the individual making it, where does the advantage lie? Zahavi's answer was to look for it in the responses waste engenders in other individuals who happen to observe it. Waste is a signal that conveys novel information about some hidden quality of the the sender, who incurred a fitness cost to send the signal, to a receiver. Selective advantage to the sender accrues because the signal causes the receiver to act in a way that benefits the sender. It is this beneficial effect that makes sending the signal worthwhile in the first place. Yet it must be in the fitness interest of the receiver to interpret the signal so as to benefit the sender. In this scenario, the investment in waste is smart advertising. It is smart in the technical Darwinian sense that it is engineered by natural selection so that both the sender's and receiver's strategies enhance their fitness, relative to alternative strategies, thus driving the signaling system to fixation.

Although signals may advertise a variety of qualities about the signaler, the one most relevant to Classic Maya dated monuments is competitive ability. By competitive ability, I mean the likelihood that individuals will win political contests with competitors, contests whose outcomes ultimately *do* determine access to mates and resources. An individual's ability to win political contests in turn is a function a variety of factors, among them physical size, physiological condition, fighting skill, and psychological cleverness, all of which contribute to an individual's ability to build and maintain coalitions comprised of kin and non-kin.

It is not hard to see why signaling competitive ability might be favored by selection on both sender and receiver: if the two parties differ significantly in competitive ability, communication allows both of them to avoid the fitness costs of a contest whose outcome is a foregone conclusion. Less obvious perhaps is the fact that *honest* signaling and accurate interpretation are in the interest of both parties. To see this, we must consider wasteful advertising from the perspective of both the sender and receiver. Senders should send accurate signals. There is little point in investing energy in a deterrent that can be counterfeited by inferior competitors. Hence there is continual selective pressure to invest in wasteful signals at a level that just exceeds the level that inferior competitors can afford. Conversely too large an investment in a wasteful signal guarantees that you will have nothing left for direct investment in yourself and your offspring.

There is a complementary pressure favoring the accurate interpretation of wasteful signals. Individuals fooled by counterfeit signals from inferior competitors suffer because they forego opportunities to enter political contests which they probably would win. Individuals who use interpretive rules that result in underestimation of the competitive ability of signalers will enter and lose many contests with superior competitors. It is to the signaler's selective advantage to send truthful signals and to the receiver's selective advantage to interpret them accurately. The result of this coevolutionary selection regime is the fixation of rules for sending and decoding wasteful signals. The rules produce a signaling system with two distinctive properties. First, advertising is on average honest. Individuals of greater competitive ability or quality advertise more and receivers correctly interpret the message. Honesty is a requirement if the signaling rules are to be selectively advantageous for both sender and receiver. Second, advertising is costly. Its costs scale with the amount of advertising. It is the cost of advertising, manifest in apparently wasteful extravagance, that guarantees its honesty. Natural selection is the key link between these two properties. Here at last is a causal account of the link between the wasteful extravagance or conspicuous consumption and competitive ability or power.

In our model the amount of wasteful advertising undertaken in a group of competitors is a function of three factors: the competitive ability of the worst advertiser, the variance in competitive ability in the group, and the payoff to a given investment in advertising. As the competitive ability of the worst competitor increases, so too will the amount of advertising of which she is capable. This in turn will drive up the amount of advertising necessary for superior competitors to distinguish themselves. Second, greater variance in competitive ability leads to more advertising, because there is more variation in political quality to distinguish and individuals higher up the political quality scale must advertise more. The third factor controlling the total amount invested in waste is the payoff to the signal. This is a function of the number of competitors that can be influenced by a given wasteful investment. As the signal reaches a larger audience, the payoff increases. The character of the audience is also an important determinant of payoff, independent of actual audience size. As the proportion of kin to the signaler in the audience increases, the payoff to advertising declines because kin are less likely to be competitors, other things being equal, given a history of selection for kin based-altruism (Hamilton 1964). In fact, selection may favor relatives who cooperate to pool their adverting resources across a nepotistic, kin-selected social coalition. Payoffs to advertising will increase when a larger proportion of the audience is comprised of strangers who have no other means of assessing accurately the competitive ability of signalers. For example, if audience members have a long history of personal acquaintance with the signaler which offers them a means of accurately assessing his competitive ability, or if the determinants of competitive ability are straightforward and available for easy inspection by audience members, then neither signaler nor receiver benefits from wasteful adverting. We can say that variation in kinship levels and prior knowledge of competitive ability alter the "effective" size of the audience.

Monument Design

Were dated Maya monuments the result of selection for wasteful advertising? We can begin to answer this question by looking for design characteristics that should be engineered into the monuments under such a selective regime. The monuments were engineered to be impossible to counterfeit. I have already alluded to the costs of supporting a literate elite, skilled carvers, labor and so forth that were required in order to erect them. These costs guaranteed the signal sent by the monuments was reliable[2]. Some members of the elite paid those costs directly. They were personally engaged in wasteful activities. For example, several sculptors' names, incised into the back of stelae, lintels, and altars carry the title *ahaw*, "lord." The son of the ruler of Naranjo was a painter of ceramic vessels (Stuart 1993). The development and transmission costs of other ancient scripts were paid for payoffs in the attestation and accounting of resource ownership (*e.g.*, Harrapan and Sumerian writing [Parpola 1994, Schmandt-Besserat 1979]). In contrast, current evidence indicates the Maya hieroglyphic tradition saw much less practical application. This raises the possibility that it owed its existence, although not its precise form, to selection for wasteful advertising.

The semantic content of the monuments, whether conveyed textually or iconographically, offers additional, if more ambiguous, clues. The ambiguity arises because the particular themes chosen do not require a cost which guarantees their truth in a Darwinian signaling system. On the other hand, if the juxtaposition of high manufacturing costs of the monuments with a particular iconographic message made the latter more credible in the brains of receivers, then we can expect those messages to have advertised political quality as well. The major themes of Classic texts and iconography fit this expectation. Among them is time depth in the elite status of individuals whose life histories are catalogued and transgenerational continuity in the elite status of their predecessors. Late Classic inscriptions at Copan trace continuity between powerful rulers back to the time when the site was in fact a tiny village (Mathews 1985: 53). This can be seen as an assertion that political quality of the kin-based coalition in question was high, and because temporal variance was low, quality was likely to remain high in the future. A second prominent theme is the participation of elites in self-inflicted bloodletting rituals (Schele & Miller 1986, Stuart 1988). In Darwinian perspective this practice, which has great time depth in Mesoamerica, is an advertisement of political quality. The ability to withstand blood loss without fainting on the spot or more long-term deleterious effects is a function physiological and nutritional status and hence an accurate indicator of fighting ability and resource access. The emphasis on battles, conquests, and the portrayal of rulers and their captive elite competitors in various postures of submission has a similar, if more obvious, message (e.g., Stuart 1993). These are precisely the kinds of themes that we would expect from political competitors who are advertising their political quality to intimidate rivals.

Our Darwinian theory of advertising gives us some insight into why the empirical generalizations and covert assumptions that archaeologists have traditionally used to translate conspicuous consumption into "social complexity" and "power" . It offers a causal explanation for why monumental architecture should scale with the range in political quality the payoffs to adverting. It suggests why Mayanists have been right in using terminal monument dates, which mark a downward spike in the amount of wasteful advertising effort in Maya social groups, as an indicator a "collapse." And finally, it indicates just what "collapse" means: a shrinkage of the range in competitive ability and a shift in audience size and composition, the latter a function of population density and the spatial scale over which political competition occurred within Maya groups

Spatial Variation in Wasteful Advertising

So far I have outlined the inferences we are entitled to make on the basis a drop in the amount of energy invested in advertising, signaled by the end of dated stone monument construction, within a single group of individuals. Termination of monument building means a decrease in our control variables: the range of political quality and the effective number of competitors within the group. But what causes spatial variation in the timing of monument termination among multiple groups of individuals distributed continuously across a region like the southern Maya Lowlands?

Consider the following model. We have a population of individuals, each assigned to a randomly chosen location in a region. Individuals sustain themselves and the social coalitions they lead by drawing on resources from a catchment falling within a certain radius of their location. They compete with other individuals for resources that fall within the portions of their catchment that overlap with the catchments of others. Resource catchments are thus also spheres of political competition. Individuals advertise within their catchments, according to the theory described above, in order to deter competitors. Their levels of resource expenditure on advertising are set by their condition, itself determined by resource abundance within their catchment and their past success against competitors, and by the payoff to advertising, which is controlled by the number of competitors they can reach with a given investment in waste.

Now each catchment is comprised of a finite number of resource patches. Resource abundance within each patch is a function of a region-wide parameter plus an error term randomly sampled from some probability distribution with mean 0. This implies that there are no spatial trends in resource abundance across the region and as a result the *expected* abundance of resources within each catchment (and patches within it) is identical. However the actual abundance will be different, because of random variation among patches. The expected difference between any two catchments in resource abundance is therefore a function of the amount of spatial overlap between them.

If we assume that the payoff to advertising is constant across catchments, then the expected difference between catchments in the level of advertising is controlled by the amount of overlap between them as well. We anticipate that difference will increase as a function of the distance between catchment locations, up to a distance equal to the circumference of the catchments. At this distance, catchment pairs no longer overlap and therefore differences in the level of advertising between them attain their maximal value and remain constant as inter-pair distance increases further.

The relevance of this simple model to the question of terminal monument dates becomes apparent if we add a temporal dimension. During successive time periods, we decrease the region-wide expected value for resource abundance. If we further assume that there is a threshold of energy investment in advertising at which construction of wasteful stone monuments is no longer adaptive, then the expected time at which investment falls below this level is the same for all catchments across the region. However, once again the actual times for each catchment will differ, because of random variation among resource patches within catchments. Once again the difference in the timing between pairs of catchments is expected to increase with the distance between them, up to the distance equal to the catchment circumference. The relevance of this simple model to making inferences about the spatial scale on which political competition occurred is apparent. Under the conditions described, an estimate of the distance at which there is no longer a detectable relationship between terminal monument date differences and geographical distance will also be an estimate of the spatial scale of political competition.

Perhaps the least realistic aspect of the model is the assumption of regional homogeneity in expected resource abundance, the timing of decrease in it, and ultimately the expected date at which different locations and their catchments cross the stone-monument building threshold. In the real world of the Classic Maya, there is likely to be a broad spatial trend in the expected crossing times whose contours are themselves determined by the causes of the collapse. Such a trend may confound our attempts to measure the spatial scale of terminal monument date difference that is determined by random variation among patches within catchments. By the same token, a regional trend, if it can be identified, might give us some insight into the causes of the collapse.

It is apparent that if we are to put our model of among catchment variation to work with real data, we are going to have to address two methodological issues. The first is the identification of the large-scale trend in the data. Once this trend is identified, we can remove it from the terminal monument dates and examining the local spatial variation that remains. The characterization of that local variation is our second methodological issue. I use robust local regression (loess) to estimate the trend. I then turn to the variogram to examine residual spatial variation in monument dates.

THE SPATIAL TREND AND COLLAPSE DYNAMICS

Data

The data on terminal monuments dates used in this study were gleaned from several sources. I started with Bove's (1981) sample of 47 Late Classic sites, and then corrected dates to reflect the results of more recent research. Terminal dates for an additional 22 sites were culled from several sources, as noted in Table 15.1. Terminal dates in the sample range over a 178 year period,

Table 15.1 Terminal monument dates from 69 sites used in this analysis. Fitted values are expected terminal dates, computed by loess, as described in the text. The residuals are the deviations of the actual terminal dates from these fitted values. The derivation of the mean annual rainfall estimates is described in Note 4.

Site	Monument	Long Count	Source	Date(AD)	Fitted Value	Residual	Rainfall
Aguacatal	Stela 1	9.16.0.0.0	Houston, n.d.	762	754	8	1600
Aguas Calientes	Stela 1	9.18.0.13.18	Mathews and Willey 1991	791	791	0	2682
Aguateca	New Stela	9.18.3.0.17	Houston, n.d.	793	789	4	2799
Altar de Sacrificios	Stela 15	9.17.0.0.0	Mathews and Willey 1991	771	790	-19	2734
Bonampak	Murals	9.17.15.12.10	Miller 1994	792	792	0	2632
Calakmul	Stela 64	9.19.0.0.0	Morley et al. 1983	810	833	-23	1671
Cancuen	Stela 1	9.18.10.0.0	Morley et al. 1983	800	795	5	3088
Caracol	Stela 17	10.1.0.0.0	Morley et al. 1983	849	810	39	2100
Chinkultic	Stela 1	10.0.15.0.0	Morley et al. 1983	844	855	-11	1462
Comalcalco	Brick	9.19.3.13.12	Houston, n.d.	814	820	-6	2471
Comitan	Stela 1	10.2.5.0.0	Morley et al. 1983	874	875	-1	1003
Copan	Altar L	9.19.11.14.5	Houston, n.d.	822	826	-4	1316
Dos Pilas	Bench 1	9.15.10.17.15	Houston 1993	742	789	-47	2735
El Caribe	Stela 2	9.17.10.0.0	Morley et al. 1983	780	791	-11	2682
El Cayo	New Orleans	9.18.1.12.16	Houston, n.d.	792	787	5	2738
El Chal	Stela 4	9.16.10.0.0	Houston, n.d.	761	804	-43	2501
El Choro	Altar 4	9.17.0.0.0	Mathews and Willey 1991	771	792	-21	2673
El Palmar	Stela 41	10.2.15.0.0	Morley et al. 1983	884	869	15	1652
El Peru	Stela 32	9.18.0.0.0	Houston, n.d.	790	801	-11	2672
Itsimte	Altar 1	9.17.5.0.0	Morley et al. 1983	775	805	-30	2288
Itzan	Stela 6	9.19.19.16.0	Mathews and Willey 1991	829	791	38	2675
Ixkun	Stela 5	9.18.10.0.0	Morley et al. 1983	800	806	-6	2601
Ixlu	Stela 2	10.2.10.0.0	Morley et al. 1983	879	819	60	2122
Ixtutz	Stela 4	9.17.10.0.0	Morley et al. 1983	780	801	-21	2738
Jimbal	Stela 2	10.3.0.0.0	Morley et al. 1983	889	836	53	1836
Jonuta	-	9.18.0.0.0	Marcus 1976	790	777	13	2589
La Amelia	Panel 1	9.18.17.1.13	Houston, n.d.	807	790	17	2722
La Florida	Stela 7	9.16.15.0.0	Morley et al. 1983	766	784	-18	2886
La Honradez	Stela 4	9.18.0.0.0	Morley et al. 1983	790	863	-73	1790
La Mar	Stela 2	9.18.15.0.0	Morley et al. 1983	805	790	15	2672
La Milpa	Stela 7	9.17.10.0.0	Morley et al. 1983	780	885	-105	1776
La Muneca	Stela 1	10.3.0.0.0	Morley et al. 1983	889	851	38	1617
Lacanha	Lintel 1	9.15.15.0.0	Morley et al. 1983	746	794	-48	2535
Laguna Perdita	Altar 1	9.15.11.2.17	Morley et al. 1983	742	807	-65	2149
Los Higos	Stela 1	9.17.10.7.0	Morley et al. 1983	781	785	-4	1699
Lubaantun	Altar 2	9.18.0.0.0	Bove 1981	790	800	-10	3103

Table 15.1 cont.

Site	Monument	Long Count	Source	Date(AD)	Fitted Value	Resi-dual	Rainfall
Machaquila	Stela 5	10.0.10.17.5	Mathews and Willey 1991	841	791	50	2844
Morales	Stela 1	9.16.5.0.0	Bove 1981	756	770	-14	2927
Mountain Cow	Altar 1	10.0.5.0.0	Bove 1981	835	820	15	2518
Naachtun	Stela 10	10.16.10.0.0	Morley et al. 1983	761	838	-77	1689
Naj Tunich	Drawing 29	9.17.0.6.3	Houston, n.d.	771	799	-28	3012
Nakum	Stela D	10.1.0.0.0	Morley et al. 1983	849	834	15	1933
Naranjo	Stela 32	9.19.10.0.0	Morley et al. 1983	820	835	-15	2010
Nimli Punit	Stela 14	9.18.10.0.0	Houston, n.d.	800	802	-2	3076
Oxpemul	Stela 7	10.0.0.0.0	Morley et al. 1983	830	833	-3	1593
Palenque	Tablet 96	9.17.13.0.7	Morley et al. 1983	783	798	-15	3681
Peidras Negras	Altar 3	9.19.0.0.0	Morley et al. 1983	810	785	25	2838
Poco Uinic	Stela 3	9.18.0.0.0	Morley et al. 1983	790	828	-38	1530
Polol	Stela 1	9.18.0.0.0	Morley et al. 1983	790	801	-11	2425
Pomona	Stendahl Panel	9.18.0.0.0	Houston, n.d.	790	781	9	3232
Pusilha	Stela E	9.15.0.0.0	Morley et al. 1983	731	794	-63	3217
Quen Santo	Stela 2	10.2.10.0.0	Bove 1981	879	859	20	1625
Quirigua	Temple 1	9.19.0.0.0	Morley et al. 1983	810	792	18	2220
Sacul	Stela 9	9.18.0.0.0	Houston, n.d.	790	805	-15	2667
Seibal	Stelae 18 & 20	10.3.0.0.0	Morley et al. 1983	889	792	97	2663
Tamarindito	Stairway 2	9.16.11.7.13	Mathews and Willey 1991	762	789	-27	2728
Tayasal-Flores	Stela 1	10.2.0.0.0	Morley et al. 1983	869	812	57	2182
Tikal	Stela 11	10.2.0.0.0	Morley et al. 1983	869	831	38	1882
Tila	Stela A	10.0.0.0.0	Morley et al. 1983	830	828	2	2500
Tonina	Monument 101	10.4.0.0.0	Morley et al. 1983	909	824	85	1741
Tortuguero	Stela 2	9.14.0.0.0	Morley et al. 1983	711	820	-109	3615
Uaxactun	Stela 12	10.3.0.0.0	Morley et al. 1983	889	834	55	1831
Ucanal	Stela 4	10.1.0.0.0	Morley et al. 1983	849	817	32	2333
Uxbenka	Stela 15	9.17.10.0.0	Houston, n.d.	780	799	-19	3065
Uxul	-	9.16.0.0.0	Morley et al. 1983	751	823	-72	1715
Xultun	Stela 10	10.3.0.0.0	Morley et al. 1983	889	850	39	1795
Xunantunich	Stela 1	10.1.0.0.0	Morley et al. 1983	849	835	14	2124
Yaxchilan	Lintel 10	9.18.17.13.14	Houston, n.d.	808	788	20	2711
Yaxha	Stela 31	9.18.5.16.4	Houston, n.d.	796	828	-32	2051

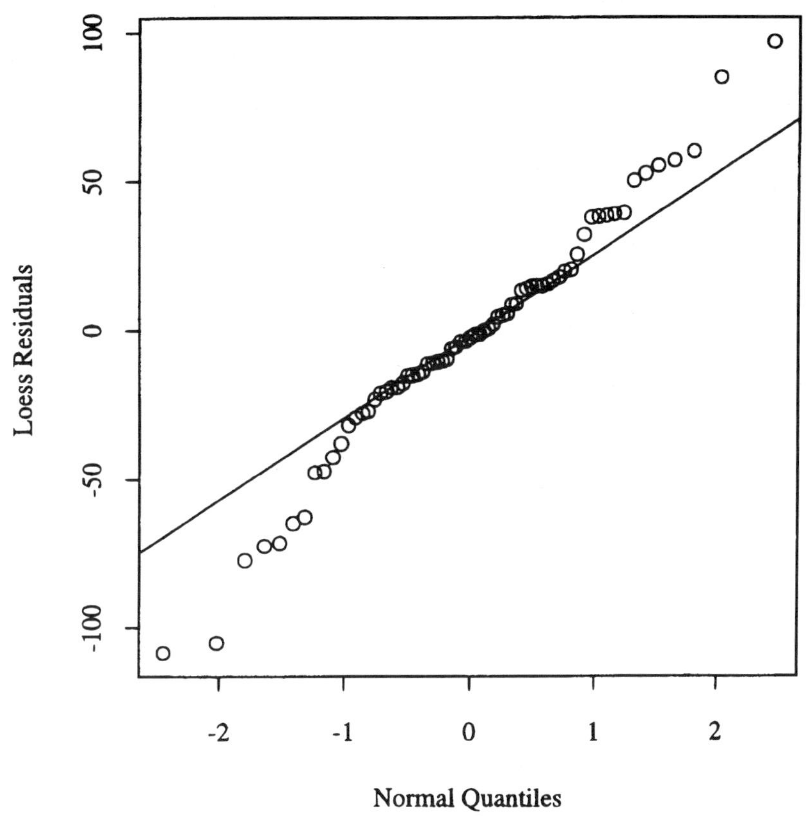

Figure 15.1 Normal probability plot of the terminal monument date residuals (actual terminal dates - expected dates, computed from loess). The line passes though the quartiles of the residuals and the reference Gaussian distribution. If the distribution of date residuals were Gaussian, all the points would fall on the line. The observed configuration indicates too much spread in the tails of the distribution. There are also 4 outliers, two at the top (Seibal and Tonina) and two at the bottom of the distribution (Tortuguero and La Milpa).

from AD 731 (Tortuguero) to 909 (Tonina). Geographical coordinates for the 69 sites were scaled from two recently published maps at 1:500,000 and 1:1,609,000 (Graham 1982, Garver 1989). Measurement error that infects the resulting estimates of inter-site distance is likely on the order of +/- 1km, but this is inconsequential given the scale of spatial pattern investigated here.

Trend Estimation

The question of how best to detect and estimate the broad geographical trend that likely underlies our data is a complex one. The trend in question will represent expected times that each site crossed the monument-building threshold. Actual crossing times will depart from expectation because of random among-patch variation within the political catchment which the site represents. Some clues about how the trend might be estimated can be had by considering just how those site-specific departures (i.e. the error term of the model) are likely to be distributed. We have no reason to believe that the resulting site-specific errors have any particular distribution, let alone a Gaussian or normal one. We can expect unique historical trajectories associated with some political catchments to produce actual terminal dates that greatly depart from expectation, resulting in an error distribution with long tails and perhaps some outliers. In fact, site groups in different areas of the study region may have different error distributions. On the other hand, our model does lead us to expect that the errors for sites that are geographically close to one another will tend to have similar values. This geographical dependence, or spatial autocorrelation, of the errors is precisely one of the phenomena we hope to investigate. These characteristics–stretched-out tails, regional heterogeneity in error variance, and autocorrelation–mean that the traditional parametric trend-surface approach to this problem, based on least-squares estimation, will not yield best unbiased

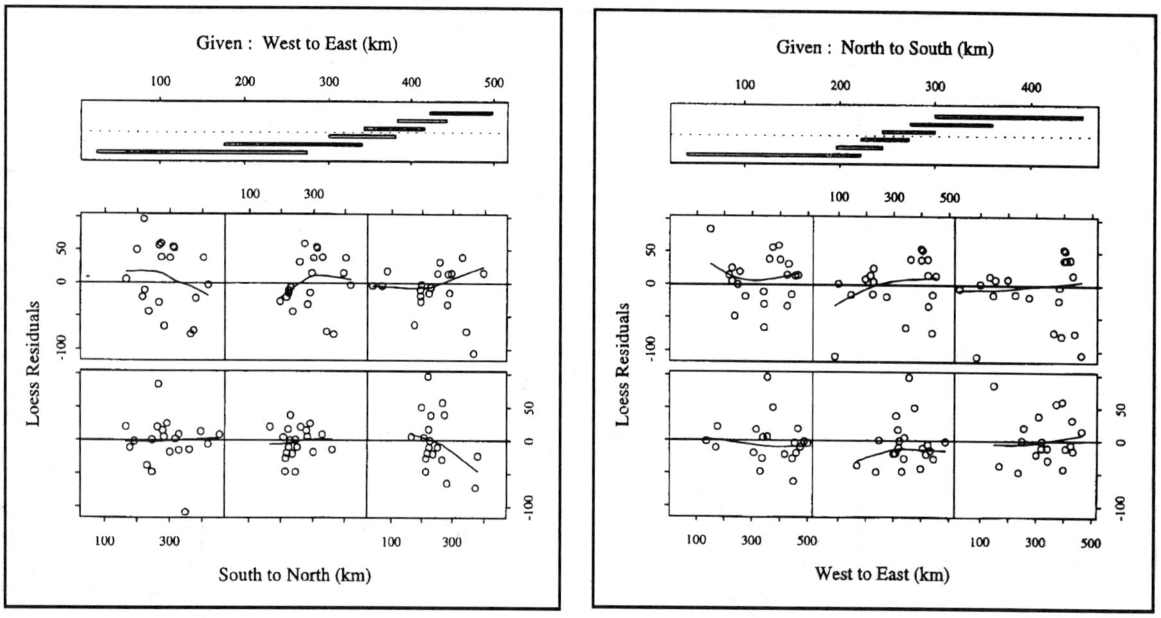

Figure 15.2 Conditioning plots of terminal date residuals. *A.* Residuals for six ranges of west-to-east coordinates, shown in the top panel, are plotted against the full range of south-to-north coordinates. *B.* Residuals for six ranges of north-to-south coordinates, shown in the top panel, are plotted against the full range of west-to-east coordinates. In both cases, loess curves fitted to the residuals fail to detect important large-scale trends, indicating that the loess fitted surface (*Figure 15.3*) captures these.

estimates nor will the usual statistical tests be meaningful. Hence we can expect poor results. The problem can also be considered from the perspective of the trend. There is no reason to suspect that the expected crossing times, which after all are a function of historical processes that varied across the region, will be well described by some single parametric surface, i.e. a linear, quadratic, or some higher-order polynomial trend surface globally fit to all the data points. Fitting such a global polynomial is likely to result in spurious inflections in the fitted surface.

What is required then is an estimation procedure that is local, in the sense that its estimates of expected crossing times for a given site are based on sites within a certain local neighborhood. Local estimation is also a way of coping with heterogeneous error distributions across the entire region. Our estimation procedure also needs to be robust so that the fitted surface is not unduly effected by long tails in the error distribution and outliers. One procedure that fits this requirement is loess, a technique of fitting a local regression model with robustness properties. Detailed accounts of loess can be found elsewhere (Cleveland et al. 1991, Cleveland 1993). Two parameters need to be specified to fit a loess model, α and λ. The first of these, α, specifies the size of the local neighborhood to be considered in fitting each point, while the second specifies whether the local fit is to be a linear or quadratic function of the predictor variable, in this case the site coordinates. Quadratic fitting performs better with curved trends and is used here. The expected terminal date for each is estimated from a weighted regression on its geographical coordinates, with the weights a function of proximity of other sites in the neighborhood to the site in question. Robustness is achieved by iteratively estimating the surface, while reweighting the influence of each point in the local neighborhood on the basis of its departure from the previous estimate until successive estimates converge.

The key question in using loess is how to choose a value for α. The technique used here is cross validation (Efron & Tibshirani 1993: 258-63). For each trial value of α, we proceed as follows. We fit the local regression surface 69 times, leaving one of the 69 sites out in turn. For each of the resulting 69 fits, we compute the expected value for the site we left out, then compute the deviation of the actual from the expected value. The 69 deviations can then be summarized in several ways. Squaring them, adding them up and taking their mean results in a mean squared prediction error (*MSPE*). I computed this badness of fit measure for successive trial values of α (from .2 to 1.0 in steps of .01). For the data in question, the minimum *MPSE* occurs at $\alpha = .90$. I used this value in the analysis reported below.

An examination of the residuals from the expected values makes possible an evaluation of the loess fit. The residuals are estimated in the standard way by subtracting the actual terminal dates from the expected values computed from the loess fit (Table 15.1, Figure 15.1). Plotting them against the corresponding quantiles of a

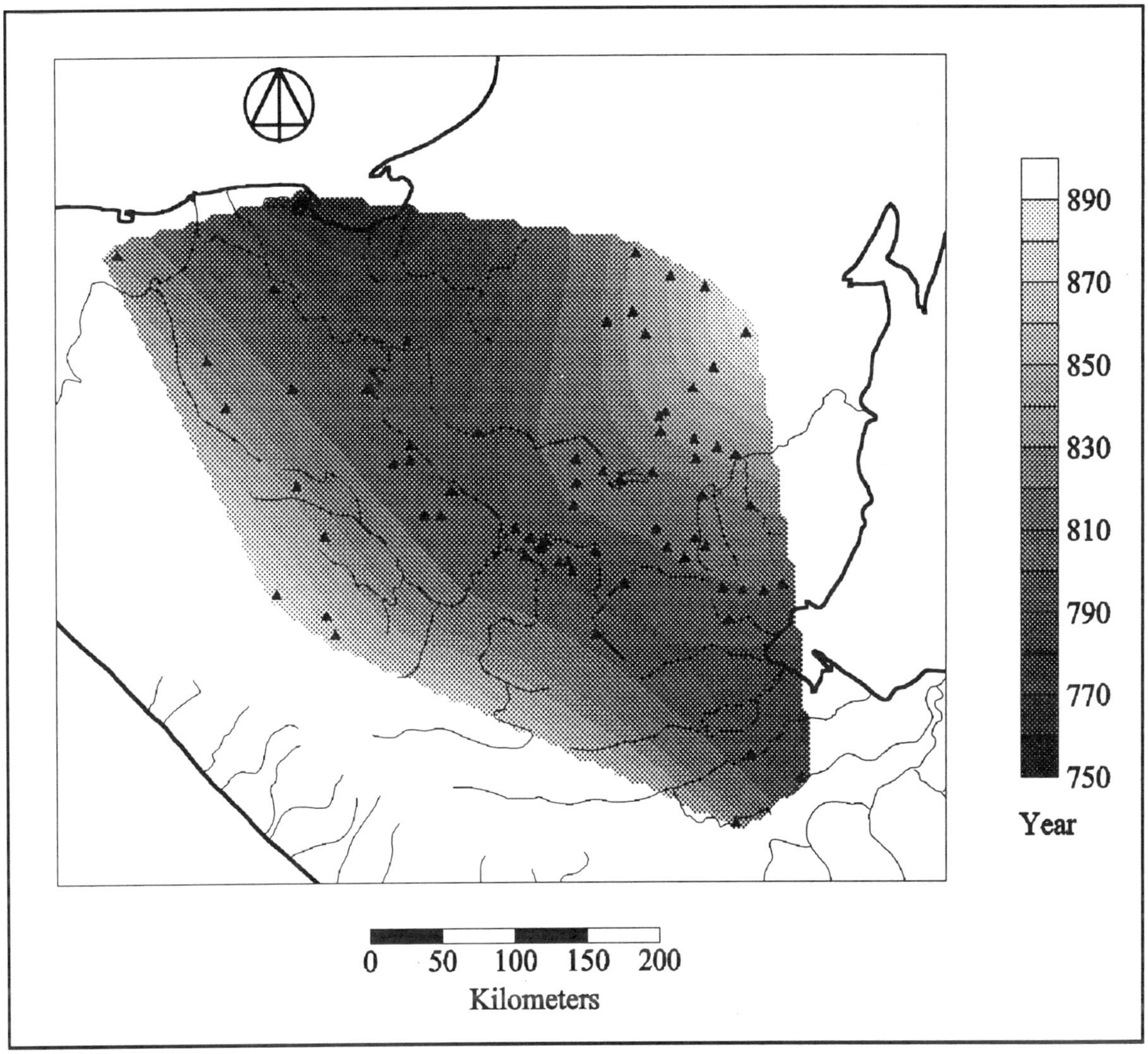

Figure 15.3 Contour plot of expected terminal monument dates fitted by loess. The value of the parameter ($\alpha = .90$), which determines the amount of smoothness in the surface, was chosen by crossvalidation.

standard normal distribution offers a helpful way to characterize their distribution. If the residuals follow a normal distribution, they all will fall on a single line. Adding a straight line to the plot, placed through the upper and lower quartiles, helps assess how well the residuals conform to this standard. It is obvious that the distribution of residuals has heavy tails at both ends–the departures from the fitted values here are too large. In addition, there appear to be outliers at each end of the residuals distribution. Both the heavy tails and outliers indicate the appropriateness of the loess fit, with its robustness features. There does not appear to be significant long-range dependence of the residuals on absolute spatial location, as demonstrated by conditioning plots of the residuals against their north-to-south and west-to-east spatial coordinates (Figure 15.2). Hence the loess surface appears to capture the large-scale trend in the data.

The final loess fitted surface is shown in Figure 15.3. A clear trend is apparent, with expected dates for crossing the monument-construction threshold varying widely across the region from c. AD 760 to 900. This means that the deterministic component behind the shrinkage in political-quality range and audience size, marked by the cessation of monument construction, is a time-transgressive process operating in different places in the region at times separated by nearly a century and a half. The surface is a valley whose floor runs from the Gulf of Mexico southeast to the Gulf of Honduras and whose sides slope up to the northeast and southwest [3]. This result can be used to evaluate collapse scenarios that have been prominent in the literature over the past several decades. There can be little doubt that proximate factors related to the cessation of dated monument construction were many and complexly interwoven. Among

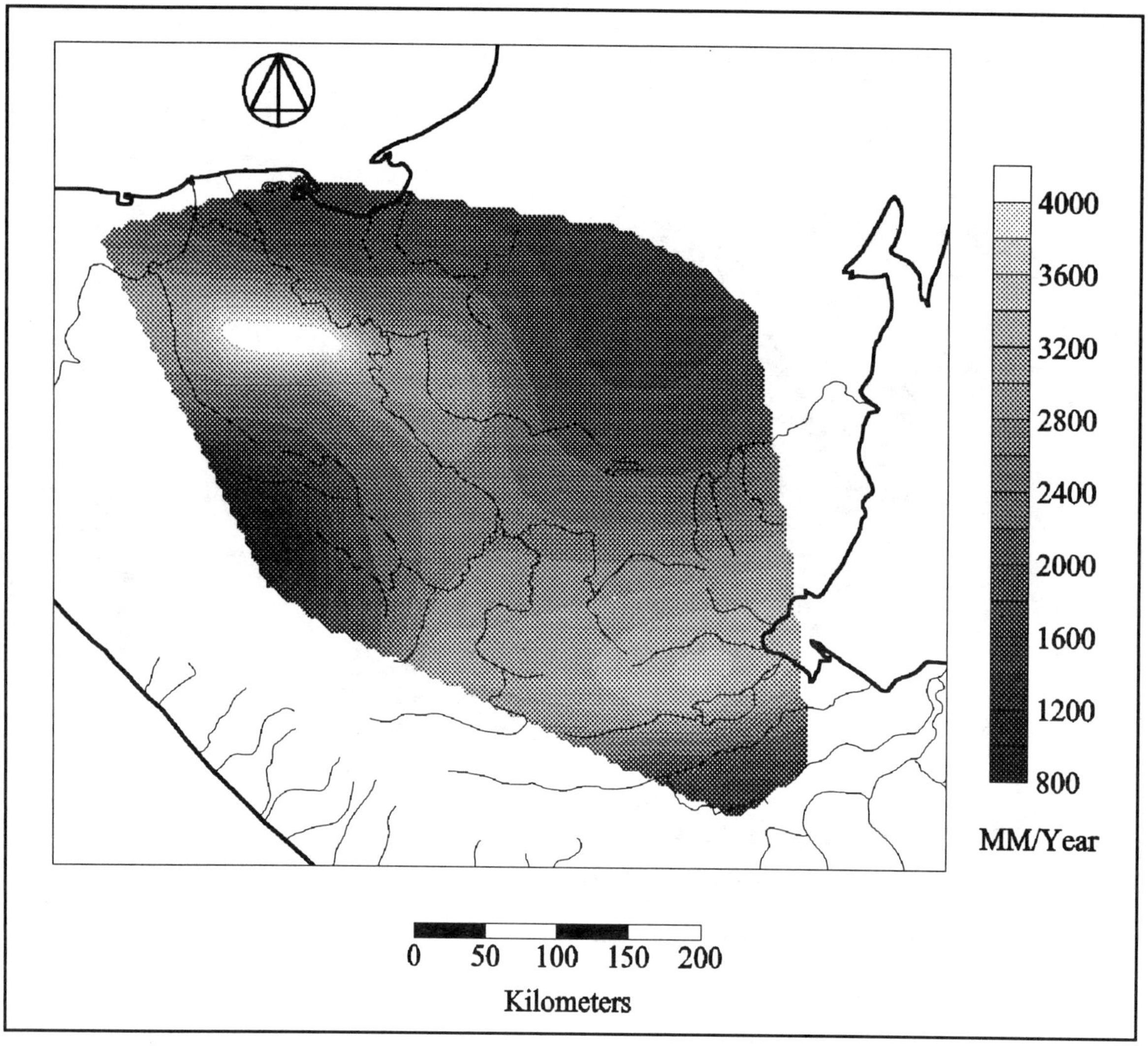

Figure 15.4 Contour plot of mean annual rainfall in the Maya Lowlands in millimeters (after Rice 1993).

these may have been increase in levels of political instability and inter-polity warfare, a possibility that has received increased attention among Mayanists in recent years (e.g. Demarest 1992, 1996). Explanations for the collapse that invoke warfare remain incomplete unless they reference changing ecological or social conditions that changed the costs and benefits of violence. The following discussion covers only ultimate causes whose implications for spatial pattern in terminal dates are reasonably clear. I consider three such causes here: invasion, drought, and long-term population-resource imbalance leading to an ecological disaster.

Invasion

Invasion explanations have enjoyed sporadic popularity over the past 30 years. Willey and Sabloff (1967) proposed that the collapse was caused by an invasion of the groups from the Gulf Coast Lowlands, perhaps ultimately of Mexican derivation. Sabloff and Willey imagined not only that invaders seized Seibal in the Peten, sending raiding parties out from this central base, but that this initial incursion eventually became part of a larger movement of diverse groups originating in the northwest into the Maya area. Invasion hypotheses were popular in the early 1970's. They were featured in the work of Sabloff (1973) and Adams (1973) and were favored among the majority of the participants in the School of American Research conference on the collapse (Culbert 1973). They continue to be proposed today. For example, Sabloff (1992) has resurrected a version of his earlier thesis, in which an incursion of Chontal Maya, now merchants not invaders, traveling up the Ucumacinta drainage from the Gulf Coast, had a hand in the collapse. In a similar vein, several workers have proposed incursions from the Yucatan, a region that appears to

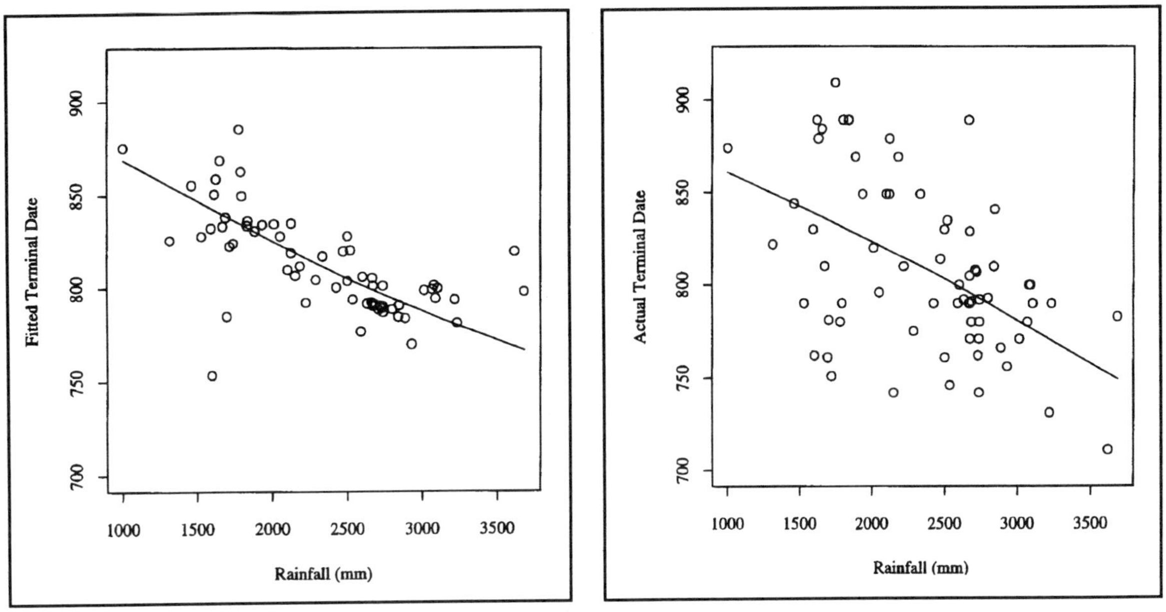

Figure 15.5 Scatterplots of fitted terminal dates (*A.*) and actual terminal dates (*B.*) as a function of mean annual rainfall. Fitted lines are from loess.

have come through the collapse relatively unscathed (e.g. Rice 1985, Chase & Chase 1982). In a recent review, Culbert has noted there is "increasing evidence that the southern Lowland Maya may have been under military pressure from the north" (1988: 100). Common to the recent versions of the external intrusion hypotheses, is the notion that the intruders were successful because the Maya had already become weak from internal causes (e.g. Culbert 1988: 79, Sabloff 1992).

The causal links between an incursion, however conceived, and the cessation of monument construction are poorly specified in the invasion literature (cf. Binford 1968). Under the wasteful advertising hypothesis, they can be sought in a reduction in the variance of political qualility within local groups or in a disruption of the transmission of skills required for dated monument construction. If outside intrusions were responsible for the cessation, what sort of spatial patterning in terminal monument dates might result? We can expect a clinal pattern whose configuration would be a function of the route followed by the intruders (Bove 1981). The clinal pattern recovered by loess from the terminal dates fits the pattern expected under Sabloff's various scenarios best. There is a tendency for fitted terminal dates to be earlier in the lower reaches of the Ucumacinta drainage. However, the time scale represented by the fitted surface fits neither invasion hypotheses. Both the incursions from the northern Yucatan and from the Gulf Coast are thought to be ninth-century phenomena. But the process mapped here appears to have been underway by the middle of the eighth century.

Drought

The notion that the collapse was the result of climatic forcing has recently received apparent support from evidence derived from sediment coring in Lake Chichancahab in northern Yucatan (Hodell et al. 1995). Temporal trends in oxygen isotopes from shell carbonates and the ratio of calcite to gypsum in sediments suggest that the interval between AD 800 and 1,000 was the driest from the mid-Holocene (c. BC 5000) to the present. The coincidence between this climatic event and the timing of the collapse has lead to the suggestion that a two-century long drought triggered the collapse. Although, as with the invasion hypothesis, proponents see internal stresses resulting from "an overall matrix of demographic and environmental pressures" as a necessary prior condition (Sabloff 1995: 357).

The connection between drought, specifically the agricultural crisis that it might precipitate, and the cessation of monument construction would be twofold. First we might expect a reduction in the range of political quality among competitors, caused by inability of individuals to produce or extract from others enough food to feed and maintain large cohorts of political allies and thus maintain former levels of political quality. Second, we might also expect a reduction in population size and thus the number of political competitors, caused by emigration and/or starvation in the face of lowered agricultural productivity.

The fitted surface indicates that the collapse had begun several decades before the onset of this climatic

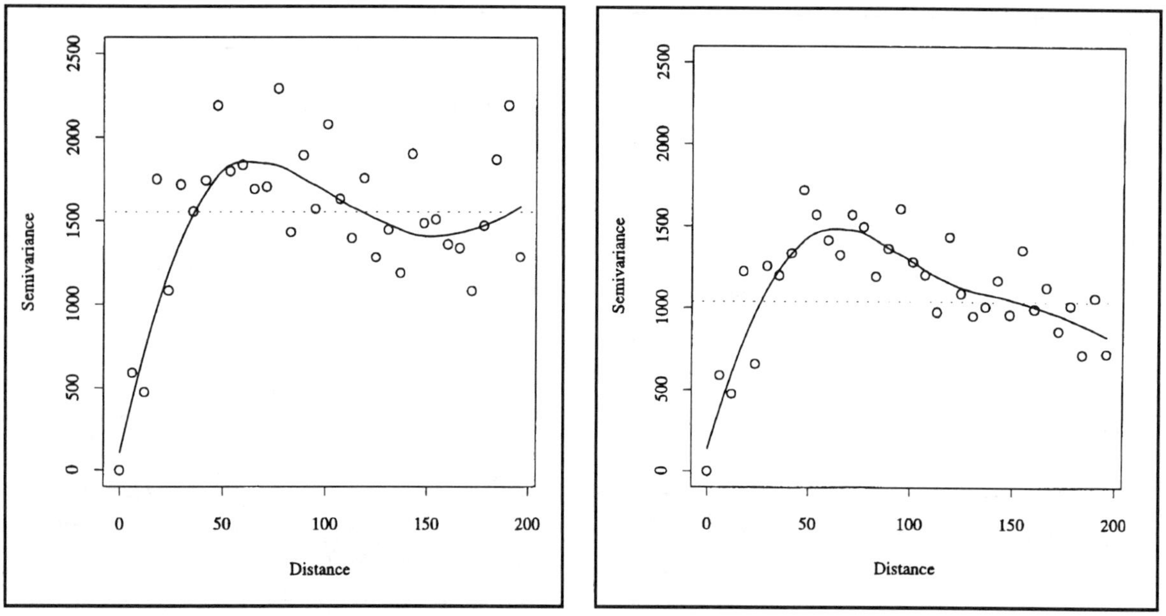

Figure 15.6 Variograms of the residuals from expected terminal monument dates. The curves were fitted by loess, with smoothing parameter ($\alpha = .65$) chosen by cross-validation. The horizontal line represents the total variance of the residuals and is the expected height of the variogram=s sill under the stationarity assumption. *A.* Variogram for all 69 residuals. *B.* Variogram for 65 residuals, 4 outliers excluded.

event. In addition the details of the deterministic component of the spatial pattern identified by loess do not support the drought hypothesis. The southern Lowlands exhibit considerable regional variation in annual rainfall, from 1000 mm to 4000 mm (Rice 1993). If drought were a significant factor in the cessation of monument construction, sites that receive the lowest rainfall should show the earliest effects of the drought and have the earliest fitted terminal dates. However, regional trends in rainfall and in terminal monument dates have precisely the opposite geographical relationship (Figure 15.4). Areas with the *highest* annual rainfall (> 2500 mm) have the earliest expected terminal dates. This is the swath extending from the Gulf of Honduras, westward across the southern Peten, into the Pasion and Ucumacinta primary drainages, and then northwest down the Ucumacinta to its mouth in the Gulf of Mexico. On the other hand, areas on the low end of the rainfall gradient, which bound this swath to the northeast and southwest, have the latest expected terminal dates.

A similar relationship between rainfall and the cessation of monument construction can be identified on a larger spatial scale. Northern Yucatan receives much less rainfall than the southern Lowlands. The mean annual figure for Lake Chichancahab is 1300 mm (Hobell et al. 1995) and this diminishes both to the north and northwest of the lake, where the Post Classic centers of Uxnal, Kabnah, Sayil, and Chichén Itzá lie. Yet these centers began their florescence as the southern Lowlands went into decline, after the initiation of the rainfall hiatus (Chase & Rice 1985, Andrews & Sabloff 1986, Sabloff 1995).

Ecological Disaster

The final collapse scenario considered here invokes a population-resource imbalance leading to an ecological disaster. Evidence in favor of this model has been accumulating from a variety of sources over the past two decades. Conspicuous among them is the record of anthropogenic environmental alteration recovered from lake sediment cores (recently reviewed by Rice 1993). The interval from BC 1000 to AD 900 witnessed dramatic increases in the amount of sediment and in silica concentrations, offering convincing evidence for soil erosion on a massive and destructive scale (Deevey et al. 1979, Leyden 1987, Rice et al. 1985). Increased levels of phosphorus attest to soil and ground-cover nutrient leaching. Increases in pollen from grasses and weeds associated with cultivation, and decline in arboreal pollen, point to deforestation wrought by a growing population seeking wood for fuel (Abrams & Rue 1988) and clearing land for swidden agriculture under shorter and ultimately unsustainable fallow cycles (Rice 1993: 42). Some archaeologists estimate that by *c.* AD 900 the Maya had cut down or significantly altered as much as 75% of the lowland forest (Whitmore et al. 1990) In some places, evidence of regrowth cannot be detected until *c.* AD 1500 (Brenner et al. 1990). That population growth continued unabated into the Late Classic, is documented by both archaeological surveys (Culbert 1988, Culbert & Rice 1990) and increased fertility levels in skeletal samples (Whittington 1991). Biological stress, particularly poor nutrition, even among members of the elite, is documented in Classic and Late Classic skeletal samples from Tikal (Haviland 1967), Altar de Sacrificios (Saul 1973), and Copan (Whittington 1989). There is evidence for an increase in childhood morbidity from Late to Terminal Classic times (Wright 1996)[4]. Taken together this evidence indicates that by the Late Classic, Maya populations found themselves in an increasingly barren and agriculturally unproductive landscape. That the result was a Malthusian catastrophe is suggested by the fact that Terminal Classic site densities across most of the southern Lowlands range between 15 and 60% of Late Classic levels, with some areas virtually abandoned well into the Post Classic (Culbert 1988: 84, 1995: 118,127, Culbert & Rice 1990).

Under this scenario, the cessation of monument construction was the predictable outcome of collapse in the range of political quality and the number of competitors, just as in the drought scenario. However in this case, the spatial trends would be determined by the intersection of locally variable population levels, edaphic factors, land use patterns, rainfall levels and the erosion and nutrient depletion that would result from their interaction. An accurate picture of variation in all these variables within the region during the Late Classic is unavailable. However, if local variation in population, soils, and land use were small, spatial variation in annual rainfall would be the major determinant of soil erosion and nutrient depletion rates, with those destructive processes proceeding most rapidly in areas receiving the highest rainfall. This seems unlikely. Yet surprisingly this expectation is met. It is implied in the previous discussion of drought: the southeast to northwest trending swath across the Lowlands with the highest rainfall levels also has the earliest fitted terminal dates. Even more surprising is the strength of the relationship. The rank correlation between rainfall levels and the fitted dates for the 69 sites in our samples is -.69 (Spearman's r)[5]. The relationship is so strong that it is even convincingly detectable using the raw terminal dates (Spearman's $r = -.44$) (Figure 15.5). This remarkable and unsuspected result indicates that spatial variation in annual rainfall is the most important determinant of the deterministic component of spatial patterning in terminal monument dates. It provides strong support for the idea that an erosion-driven ecological disaster was the principle, ultimate cause of the cessation of dated-monument construction.

To estimate the probability that these results are an artifact of sampling error, we cannot rely on standard methods that assume observations are independently sampled. Both the fitted date and rainfall values are spatially autocorrelated: sites that are geographically close have similar values. However, recasting the problem in terms of patterns of difference among fitted date and rainfall values makes statistical evaluation possible. By regressing the matrix of fitted date differences on matrices of geographical distances and rainfall differences among the 69 sites, we can partial the effects of distance (and hence spatial autocorrelation) out of the linear relationship between fitted-date and rainfall differences. Null-hypothesis distributions for estimated parameters can be obtained by randomly permuting the rows and columns of the fitted date matrix (Smouse et al. 1986, Smouse and Long 1992)[6]. Standardized regression coefficients, which measure effect sizes in standard deviation units, reveal that rainfall has over twice the effect on fitted dates as geographical distance, when each independent variable is held constant ($\beta_{distance} = .19$, $\beta_{rainfall} = .45$). In other words, a unit change in geographical distance brings with it only a .19 unit change in fitted date difference, while a unit change in rainfall difference is associated with a .45 unit change in fitted date difference. Permutation-based testing reveals the probability of observing effects of this size by chance alone is vanishingly small.

THE SPATIAL SCALE OF POLITICAL COMPETITION

The Variogram

Having estimated the deterministic regional trend in terminal monument dates, it now becomes possible to put our model of spatial variation in wasteful advertising to work to infer the average size of spheres of political competition on the eve of the cessation of monument construction. Under that model, the local pattern of residual variation around the expected terminal dates that comprise the loess fitted surface depends on the size of the areas over which elites compete with one another. Our model suggests that the spatial scale or distance over which there is a tendency for the residuals to exhibit statistical dependence will indicate the size of catchments within which Maya elites competed for resources and over which there was a payoff for wasteful advertising of political quality. The job of estimating that scale can be accomplished by computing a variogram for the terminal date residuals.

A variogram is a simple spatial-analytic tool borrowed from geostatistics (Davis 1986, Isaaks & Srivastava 1989). It is a plot of the dissimilarity between the values of a spatially distributed variable at points in space as a function of the distance between those points. In our case the point pairs are pairs of sites, while the spatially distributed variable is the deviation of actual terminal dates from values expected on the basis of the loess fit. The measure of dissimilarity used is called the semi-variance. It is half the mean of the squared difference between residual values at all pairs of sites that are certain distance or range of distances apart. The variogram is a plot of the semivariance against the successive distance classes over which the mean-squared differences were computed.

Local spatial dependence, in the absence of regional trends, will yield a variogram in which dissimilarity increases up to a given inter-site distance class and then remains unchanged over the remaining distance classes. This flat portion of the variogram is called the sill. A nice result from mathematical statistics says that when the process generating the data values contains no deterministic spatial trend, a condition known as stationarity, and only local spatial dependence, then the value of the semivariance at the sill is equal to the total variance in the data values. The distance at which the semivariance reaches its maximal value at the sill represents the range of distances over which terminal dates exhibit statistical dependence on one another or are spatially autocorrelated.

This set of theoretical relationships makes the variogram a useful tool with which to estimate the scale of political competition among monument building elites on the eve of the collapse. If our model of spatial variation in wasteful advertising approximates competitive dynamics on the eve of the collapse, then the semivariance of our date residuals will increase with increasing distance between sites up to a sill, which is equal to the total variance in the date residuals. After that point, the semivariance will remain flat across ever larger distance intervals. The distance at which the sill is detected, the range, is the average distance over which wasteful advertising and the political competition of which it was a part took place.

Before we can estimate the variogram, we need first to decide on the width of the distance classes over which we are going to compute the semivariance. Because interest focuses on the shape of the variogram, it is desirable to use relatively narrow distance classes. However, narrow classes are likely to yield a noisy point scatter when the semivariance estimates are plotted against the distance classes over which they were computed. Using wider distance classes may produce a better behaved point scatter, but may obscure interesting details of the semivariance-distance relationship. A way out of this dilemma is to use small distance classes for the semivariance estimates, and then smooth the resulting noisy scatterplot. Loess, with a smoothing parameter α again chosen by crossvalidation, recommends itself as a reasonable way to do this.

In computing the variogram, we are assuming that the process that generated the date residuals operated similarly at all sites. Examination of the distribution of residuals, suggests this is a reasonable assumption for most of the sites, but perhaps not for the four outliers (Figure 15.1). Hence it seems reasonable to estimate variograms for all the residuals and for a subset of the residuals, with the four outliers removed. I computed the semivariance of the date residuals for successive 6 km distance classes up to 200 km. It is customary to compute the variogram for distances up to half the smallest map dimension. Estimates of the semivariance for distances classes greater than this value depend not only on the distance between pairs of sites, but also on the absolute locations of the sites. For example, in a study area, *c.* 500 km on a side, estimates of the semivariance at distances greater than 250 km will increasingly be dominated by sites that fall close to the map edges. This dependence on location introduces unwanted bias into the semivariance estimates. I used loess to smooth the semivariance-by-distance scatterplots, with $\alpha = .65$ chosen by cross validation. The result is shown in Figure 15.6 .

It is clear that for both sets of residuals the variogram, as described by the loess curve, does not behave in the simple way that our model initially lead us to expect. Instead of increasing to a flat sill equal to the total variance in the data, the semivariance increases past the total variance, reaching a maximum at a distance of roughly 65 km. Thereafter it declines until it reaches a value that is roughly equal to the total residual variance, at about a distance of about 150 km[7]. This sort of pattern in a variogram is known as the "hole effect" in the geostatistical literature (e.g. Isaaks & Srivastava 1989: 156). It means that pairs of sites that are close together do tend to have more similar residual terminal dates than pairs of sites chosen at random, as we expected. However, pairs of sites that are 65 km apart have terminal dates that are less similar than pairs of sites chosen at random. This is a surprise.

What are we to make of this pattern? It indicates that our initial model of spatial variation in terminal monument dates is missing a key component. The Darwinian theory of advertising helps isolate just what this overlooked process might be. Recall that in the model terminal date thresholds are a function of two sets of quantities. The first in the range of political quality, which is determined by resource abundance within each competitor's catchment. This source of variation among sites should have been estimated by the loess fit and it should not appear in the residuals. Even if it does, it is difficult to envisage a process whereby long-term trends in population growth, rainfall, and soil erosion might result in sites that are 65 km apart having unusually different departures from the trend.

The second is the payoff to advertising. Here we find a more likely cause for the tendency of sites 65 km apart to depart from the loess trend in different directions: movement of individuals among resource catchments. As average resource abundance declined across the entire region, random variation among catchments prompted individuals based in areas with lower levels of resources to attempt move into adjacent catchments where conditions were more favorable. Such movement would result in an increase in the number of elite competitors in recipient catchments and a consequent increase in the effective size of the audience for advertising. Not only would absolute audience numbers increase, audience composition would shift as well, causing further increases in the payoff to advertising. New immigrant competitors would be unlikely to belong to local kin-based coalitions. In addition, immigrants and resident coalition members would be unfamiliar with one another. Hence members of each group would have no means of assessing the competitive ability of members of the other, save through wasteful social advertising. The result would be in increase in the payoff to wasteful advertising for all competitors within recipient political catchments. As resource levels continued to decline across the entire region, this local increase in the payoff to advertising resulted in the recipient catchments crossing the monument building threshold later that they would have, without recent additions to the pool of competitors. On the other hand, departure of emigrant competitors would hasten the crossing of the monument building threshold in donor catchments by reducing the effective audience size within them. We can expect this movement process would have been accompanied by increased levels of violence, as individuals in political catchments with more favorable resource levels attempted to defend them against the encroachments of outsiders. Thus spatial heterogeneity in the timing of resource abundance decline and the resulting movement process, documented in the variogram, point the way to a causal account for otherwise unexplained increases in the severity of warfare during the Late Classic that figure prominently in recent fashionable accounts of the collapse (e.g. Demarest 1996).

If the movement process is responsible for the conspicuous "hole effect" in the variogram, then the location of the peak in the semivariance at 65 km represents the mean distance over which such movements took place. In order to achieve the prolongation of monument construction, the immigrant competitors must have been new to the areas into which they moved. Hence 65 km also represents an *upper* limit on the mean diameter of spheres of political competition on the eve of the collapse. Movement during the collapse means that the price of an estimate of the *lower* limit is an additional assumption. The assumption is that the movement process merely altered the spatial distribution of residual variance, shrinking some residuals and stretching others, but leaving the total residual variance unchanged. Among the effects of the movement process on the variogram is an increase the rate at which the semivariance increases with distance. Hence without the movement process, the slope of the variogram from the origin would be lower. This fact, along with the constant variance assumption, means that the distance at which the loess fit to the variogram equals the total variance provides a lower limit on our estimate of mean size for political catchments - with a lower slope the equality would occur at a greater distance. This distance is roughly 35 km for both variograms. Taking the midpoint between our upper and lower limits on the scale of political competition yields a best estimate of 50 km.

The loess fit for the two variograms is very similar. However, there is much less variation about the curve fit to the 65-site data set from which the four outliers were excluded. Thus we have evidence that these four sites differ from the others, not only in their frequency distribution, but also in their pattern of spatial dependence. These are grounds for concluding that a different process is responsible for the cessation of monument construction at these four sites. For the two positive outliers, Tonina and Seibal, this seems like a reasonable inference, since elites based at these sites continued monument construction long after their adjacent competitors had ceased. Monument construction uniquely ceased at these sites at a time when there was there was no place else to go. The negative outliers, Tortuguero and La Milpa, may be sites at which the cessation of monument construction was related to ongoing social dynamics that predate the onset of the collapse in their respective areas. It is also possible that the known monuments at these two sites offer especially poor estimates of the timing of the cessation of monument construction.

The "Polity Size" Question

The question of the spatial scale of Classic political competition, discussed traditionally under the rubric of "polity size," is of particular interest because the past 20 years of research have given us, predictably, two opposing answers: "large" and "small"[8]. The "large" view, has its origins in Joyce Marcus' work in the 1970's and ultimately in the social-systems functionalism of the New Archaeology. Marcus proposed that by the Late Classic Maya polities were hierarchically organized into four or five regional states. Each polity was comprised of a four or five-tiered settlement hierarchy, whose elements Marcus thought could be defined on the basis of site size, patterns of emblem glyph occurrence, and a gloss of the inscriptions on two stelae, at Copan and Seibal respectively, as referring to four regional capitals (Marcus 1976, 1983). Such polities would have spanned 100 to 200 km in diameter. A similar view, based largely on estimates of site size and empirical generalizations about the political correlates of rank-size relationships, is espoused by Richard Adams (1990, Adams & Jones 1981). More recently Marcus (1993) has suggested that the Classic political hierarchy experienced temporal oscillations, as large regional states formed (c. AD 731), dissolved into their constituent provinces, and then reformed (c. AD 849). Among the initial attractions of Marcus' earlier model was its use of epigraphic evidence. However, the past two decades have seen major advances on the epigraphic front that have rendered Marcus' epigraphic interpretations suspect. Recent epigraphic research renders her reading of the Seibal and Copan monuments unsustainable, casting doubt on her reconstructions of political hierarchies (Houston 1992, 1993: 4-8, Stuart 1993).

The epigraphic revolution has produced a competing "small" alternative hypothesis. Its most quantitatively precise rendering is to be found in Peter Mathews' recent work (1985, 1991). It is based on the demonstration that emblem glyphs occur in the context of a personal title, which can be read as "ruler - literally *k'ul ahaw* or holy lord - of place X." This leads to the inference that they refer to political catchments under the control of individual political competitors. Mathews has mapped these units for the Late Classic. The mean nearest-neighbor distance between 32 Late Classic sites with identified emblem glyphs is 52 km (Houston 1993: 138) Our best estimate–50 km–of the mean size of political catchments from the variogram fits comfortably with this picture.

The variogram evidence clearly does not support Marcus' original regional state model. However, its relevance to the evaluation of the oscillation model is a little more opaque since the period over which the terminal date evidence accumulated encompassed, on her latest account, more than an entire cycle: from regional states, to fragmentation, to regional states, and then back to fragmentation. Marcus herself thinks the terminal date evidence is germane since she offers Bove's (1981) analysis in support of her more recent conclusions (Marcus 1993). However, the loess fitted surface may be of greater relevance to their evaluation than the variogram. The surface implies that the cessation of monument construction across the region was, as Morley (1920) proposed 75 years ago, a single time and space-transgressive event, and not a palimpsest of two downward trends in an ongoing political cycle.

CONCLUSION

Summary

How has our exploration of spatial patterning in terminal monuments added to our understanding of Classic Maya history? We have discovered a surprising, and surprisingly strong, relationship between mean annual rainfall and terminal monument dates. This lends support to the idea, warranted from a variety of other sources, that the deterministic component of the cessation of monument construction across the region was ultimately caused by an ecological disaster in which erosion played a leading role. Currently popular alternatives, drought and invasion, seem less likely in the face of the terminal date evidence. In addition our examination of the scale of spatial autocorrelation in the terminal date residuals indicates that as ecological conditions worsened, groups of individuals moved from lower to higher quality habitats, over an average of 65 km. Finally, we have seen how this pattern of movement fits nicely with the idea, derived from epigraphic sources, that spheres of political competition during the Late Classic averaged about 50 km in diameter.

Caveats

The inferential path leading from terminal monument dates to the causes of the Maya collapse and then to the size of spheres of political competition is a long one. There are obviously many opportunities for missteps. Hence our historical conclusions are provisional. It would be premature to suggest that this is the definitive analysis of terminal monument dates. As new monuments are discovered, we can expect changes in terminal dates for some sites that are included in the current sample. We can also look forward to the addition of new sites to the sample. A glance at the locations of the 69 sites reveals that the western half of the study area is more sparsely sampled than the eastern half. This is a defect that only additional research in this area can remedy.

Evaluation

Our conclusions are also provisional in the sense that they are open to evaluation. Toward the beginning of this paper, to motivate the development of a mechanistic theoretical framework on which the analysis is based, I suggested that the payoff would come in the form of the ability to evaluate the resulting inferences. Over the long term, the inferential process might thereby become less perilous. It is now time to redeem this promissory note and suggest just why we might expect mechanistic theory to deliver this kind of result.

Theory forges the links from which we can model the causal chain that connects monuments to the conditions and mechanisms responsible for them. An explicit mechanistic theoretical framework facilitates model construction - the valid deduction of precise empirical expectations that should hold if our initial ideas about the mechanisms and the historical contexts in which they operated are correct. Precision in empirical expectations is important because it lessens the chance that when there is agreement between them and the evidence, it is the result of chance convergence. Precision also facilitates recognition when agreement is lacking, paving the way for adjustments in our model.

The deductive chain makes manifest the causal role of key theoretical variables in producing empirical variation and motivates the attempt to resolve that variation into the multiple theoretical processes responsible for it. The attempt at resolution means we will probably learn something new. As long as dated monument construction was thought to be a simple correlate of Maya "civilization", it was possible to see the outlines of Maya polities of nearly any sort in terminal dates. Analysis consisted of fitting successively higher order global polynomials until polities emerged from their inflections or their residuals. The absence of theory guaranteed that the polities discovered matched the most *avant garde* interpretation at the time (Marcus 1976). As a result, the terminal date trend was incorrectly identified and its remarkable relationship with mean annual rainfall went unsought and unfound.

The deductive chain also means that when variation does not match initial modeled expectations, there is an audit trail with which we can begin to puzzle out what went wrong. In the case at hand, the possible causes of the departure of the variogram's shape from our initial expectations are clear, easing identification of the most likely overlooked culprit: variation in audience size caused by movement of competitors.

Mechanistic theory makes possible the deduction of multiple sets of expectations implied by the same mechanism's operation in *different* empirical domains. In a Darwinian framework, the domains that recur are the design properties of formal artifact variation, variation in physical traces of past behavior, and the distribution of that variation in time and space (Neiman 1990). In the case at hand, for example, this tactic delivers two complementary sets of expectations under the hypothesis of selection for wasteful advertising. One set details the properties engineered into the monuments and their semantic content (e.g. wastefully high manufacturing cost, low temporal variance in lineage political quality). The second describes the distribution in time and space of the cessation of monument construction under different collapse scenarios (e.g. the relationship with annual rainfall). If we grant that the ecological disaster scenario is the most likely on other grounds and use it to develop expectations regarding spatial patterns for terminal dates, agreement with the observed pattern can be used as independent evidence in favor of a history of selection for smart advertising. If we grant the design arguments establish that the monuments are the result of wasteful advertising, then the spatial distribution evidence can be used to evaluate different collapse scenarios. The second avenue is the one I took here. But both approaches are available and both take advantage of the existence of multiple expectations in different evidentiary domains to evaluate hypotheses about history in an objective fashion.

Another route to evaluation depends on the mutidimensional complexity of individual historical trajectories and the existence of multiple theoretical mechanisms to account for different dimensions of variation in a given historical case. We build multiple models, each involving the operation of different mechanisms, to account for different aspects of the same chunk of history. When two (or more) of these theoretically independent models contain a common key parameter, it becomes possible to deduce how variation in a particular parameter value implied by one model, might play out empirically in the context of the second. The two sets of empirical expectations can be made to yield independent estimates of variation in the common parameter. While this powerful line of evaluation has not been pursued here, it is worth considering briefly how it might be in the future.

Consider the question of shifts in audience size caused by movement that were detected in the variogram with the help of the theory of advertising. Darwinian theory offers a second, conceptually independent route to the estimation of audience size though the theory of cultural drift, the mechanism that controls the behavior of stylistic or selectively neutral cultural variation (Dunnell 1978, Neiman 1990). Selective neutrality here means that there are no deterministic forces like selection that favor the transmission or learning of one cultural variant (e.g. design motif) over alternative variants. Elsewhere I have shown that, under the selective neutrality hypothesis, among group variation in the diversity of alternative cultural variants along a dimension of stylistic variation is controlled by the product of the effective size of each group and the rate at which novel variants are introduced into it by immigration from outside (Neiman 1995).

Since we know that elite political competitors among the Classic Maya were personally responsible for carving hieroglyphs and iconographic motifs on monuments, the monuments themselves offer a place to look for relevant patterns of diversity. This raises the possibility of using diversity in alternative iconographic motifs (e.g. Proskouriakoff 1950) as a way of measuring variation in the number of immigrants among elites at different sites. Those immigrants represent an increase in the size of the local audience for wasteful advertising. If the movement hypothesis derived from the variogram is correct, then we should find that stylistic diversity among elites at sites with negative residual terminal dates should tend to be lower than diversity among elites at sites with positive ones since political competitors left the former the latter. It is theoretical clarity on the key role of variation in number of immigrants and in audience size and the overlap between them that makes this evaluation possible.

A similar tactic for evaluation is available when models built from entirely different theoretical frameworks contain common parameters (Kosso 1989). An example is the agreement between the estimate of the size of spheres of political competition based on the variogram and the estimate of polity size derived from Mathews' epigraphic work with emblem glyphs. Agreement lends objective support to both inferences. However in this case, a precise account of the theoretical independence is not possible because the theory that underwrites the mapping of emblem glyphs onto polities is opaque.

The quest for the means to evaluate our ideas about what happened in history is the motivation for building Darwinian models of historical processes in archaeology. Explicit theory about the mechanistic causes of history facilitates the evaluation of inferences about unique historical trajectories. We begin by hunting for agreement between precise, deductively modeled expectations and evidence. Explicit theory means there is an audit trail to follow if initial expectations are not met. Objectivity enters the evaluation process from independence between multiple sources of evidence implicated by the operation of a single mechanism and from independence among estimates of single parameters shared by different mechanisms. Building archaeological inference on simple behavioral correlates or covert, common-sense assumptions about human nature, and both processual and post-processual approaches do precisely that, does not guarantee that a particular inference is wrong. Nor can the use of causal theory guarantee correctness in a particular case. However, working with an explicit mechanistic theory that is the foundation for our scientific understanding of life *uniquely* raises the odds that, over the long term, more accurate inferences, and more powerful models with which to make them, will emerge. This paper is a modest attempt to further this process for ancient Maya history.

ACKNOWLEDGMENTS

An initial version of this paper was presented at the Symposium on Mesoamerican Archaeology at the Peabody Museum of Natural History, Yale University, May 1994, in honor of my Michael D. Coe, on the occasion of his retirement. Special thanks to Cary Carson who fundamentally shaped my interest in the problem of conspicuous consumption and my approach to it. I am also grateful for helpful comments on subsequent iterations from Bob Bettinger, Jim Boone, Michael Graves, Martha Hill, Bob Leonard, Garnett McMillan, Ann Ramenofsky, and Patrice Teltser. I owe an enormous debt to Steve Houston for helping me obtain up-to-date estimates of terminal monument dates. Steve also commented extensively on two drafts, enabling me to evade at least some of the pitfalls that await an historical archaeologist poaching in the Maya Lowlands.

ENDNOTES

1. Dunnell (1989) has offered a provocative alternative evolutionary model for wasteful energy expenditure. In Dunnell's account waste is a means by which groups achieve reproductive restraint and thus avoid extinction in stochastic environments (cf. Wynne Edwards 1962). His model, recently adumbrated by Graves and Ladefoged (1995), and the apparent group-selection mechanism that runs it deserve a more detailed examination than I can attempt here.

2. Unfortunately quantitative estimates of these costs are unavailable. Recent attempts to estimate the levels of human effort required to build Classic Maya residential structures (Abrams 1994) offer little guidance since they do not include sculptural and hieroglyphic carving and the skilled labor required to execute it.

3. The surface estimated by loess contrasts with the polynomial trend surfaces, fit by ordinary least squares to these data by previous researchers. Bove (1981) offered linear, quadratic, and cubic polynomial fits. But his interpretive efforts were concentrated on residuals from them, in which he glimpsed four or five polities. Kvamme (1990), on the other hand, fit a single quartic surface and saw the five polities directly in its inflections. Comparison of the location of inflections in Kvamme's fitted surface with actual data values suggests that former are to a significant extent artifacts the spatial distribution of data points and the global character of the polynomial trend surface approach. Bove's residuals suffer from similar defects, because they are computed from similar surfaces.

4. In her recent study of dental hypoplasias in 160 individuals from the Pasion drainage, Wright argues that "the relative frequency of stress events is shown to have been relatively constant over time," and suggests this finding does not support ecological models of the collapse (1996: 242). Her data (Table 2, Wright 1996: 241) belie this conclusion. I performed logistic regressions of hypoplastic tooth frequency on temporal period, examining the change from Classic to Late Classic and from Late Classic to Terminal Classic for each tooth separately.

Of the 14 teeth examined by Wright, none showed any statistically significant change (using a cutoff of $p < .10$) in hypoplastic frequency during the Classic-to-Late Classic transition. However, significant increases were detected on four teeth during the Late Classic - Terminal Classic transition (upper P3 and I2, lower P3 and P4).

5. Rainfall levels were estimated by digitizing 1000 *mm* rainfall contours in Rice (1993) and interpolating a 2.5 *km* grid over them using radial basis functions (Haigh 1992, 1993). The rainfall estimate for each site is the interpolated value for the grid node closest to it.

6. Geographical distance and rainfall are fixed in the usual regression sense. Thus to obtain a proper permutational test, geographical distance and rainfall matrices are held constant, while permuting the fitted date matrix. This protects the interrelations between distance and rainfall (Smouse & Long 1992: 192). After 1000 permutations, the 99th percentiles of the null hypothesis distributions for $\beta_{distance}$ and $\beta_{rainfall}$ lie at .14 and .10 respectively. Approximate z-values, obtained by dividing the standard deviation of the permuted null distribution into the parameter estimates, are in line with this result: $z_{distance} = 3.3$, $z_{rainfall} = 11.25$. I conducted the analysis on the square roots of all distances and differences to linearize the relationships among them.

7. This result isolates the source of recent controversy over the existence of spatial autocorrelation in terminal dates. Whitley and Clark (1985) divided the Maya lowlands into 17 quadrats, 80 km on a side. They averaged terminal dates within each quadrat, then tested the null hypothesis that the quadrat means were not spatially autocorrelated. They failed to reject this hypothesis. By contrast, Kvamme (1990) dispensed with the 80 km quadrats and examined individual site locations. He definitively rejected the null hypothesis that similarity among terminal dates was unrelated to inter-site distance. The variogram shows why these results diverge. By averaging over 80 km quadrats, Whitley and Clark insured they would not detect spatial dependence on a smaller scale. On the other hand, Kvamme's methods precluded an examination of just what the scale of spatial dependence might have been.

8. Grube and Martin (1994, 1995; Appenzeller 1994) have offered a third alternative, one that posits the existence of two large "superpowers", Tikal and Calakmul, each heading a suite of client states. Grube and Martin see this arrangement lasting from AD 690 to 740. Hence the terminal monument date evidence is too late to be useful in an evaluation of their scenario.

REFERENCES CITED

Abrams, E.M.
1994 *How the Maya Built Their World: Energetics and Ancient Architecture*. Austin: University of Texas Press.

Abrams, E.M. & D.J. Rue
1988 The causes and consequences of deforestation among the prehistoric Maya. *Human Ecology* 16: 377-396.

Adams, R.E.W.
1973 Maya collapse: transformation and termination in the ceramic sequence at Altar de Sacrificios. In *The Classic Maya Collapse* (T.P. Culbert, ed)., pp. 133-164. Albuquerque: University of New Mexico Press.

1990 Archaeological research at the Lowland Maya city of Rio Azul. *Latin American Antiquity* 1: 23-41.

Adams, R.E.W. & R.C. Jones
1981 Spatial patterns and regional growth among Maya cities. *American Antiquity* 46: 301-322.

Andrews, E. Wyllys,V, & J.A. Sabloff
1986 Classic to Postclassic: a summary discussion. In *Late Lowland Maya Civilization: Classic to Post Classic* (J.A. Sabloff & E.W. Andrews V, eds.), pp. 433-456. Albuquerque: University of New Mexico Press.

Appenzeller, T.
1994 Clashing Maya superpowers emerge from a new analysis. *Science* 266: 733-34.

Binford, L.R.
1968 Some comments on historical versus processual archaeology. *Southwestern Journal of Anthropology* 24: 267-275.

Bove, F.J.
1981 Trend surface analysis and the Lowland Classic Maya Collapse. *American Antiquity* 46: 93-112.

Broughton, J.M.
1994 Late Holocene resource intensification in the Sacramento Valley, California: the vertebrate evidence. *Journal of Archaeological Science* 21: 501-514.

Brenner M., Leyden B., Binford M. W.
1990 Recent sedimentary histories of shallow lakes in the Guatemalan savannas. *Journal of Paleolimnology* 4: 239-252.

Chase, D.Z. & A.F. Chase
1982 Yucatec influence in Terminal Classic Northern Belize. *American Antiquity* 47: 595-614.

Chase, A. & P.M. Rice (editors)
1985 *The Lowland Maya Post Classic: Questions and Answers*. Austin: University of Texas Press.

Cleveland, W.S.
1993 *Visualizing Data*. AT&T Murray Hill, N.J.: Bell Laboratories.

Cleveland, W.S., E. Gross & W.M. Shyu
1991 Local regression models. In *Statistical Models in S* (J.M. Chambers & T. Hastie, eds.), pp. 309-376. New York: Chapman and Hall.

Coe, M.D.
1993 *Breaking the Maya Code*. New York: Thames and Hudson.

Culbert, T.P.
1988 The collapse of Maya civilization. In *The Collapse of Ancient States and Civilizations* (N. Yoffee & G.L. Cowgill, eds.), pp. 69-101. Tucson: University of Arizona Press.

1995 *Maya Civilization*. Washington, D.C.: Smithsonian Books.

Culbert, T.P. & D.S. Rice
1990 *Precolumbian Population History in the Maya Lowlands*. Albuquerque: University of New Mexico Press.

Culbert, T.P. (editor)
1973 *The Classic Maya Collapse*. Albuquerque: University of New Mexico Press.

Davis, J.C.
1986 *Statistics and Data Analysis in Geology*. New York: Wiley.

Deevey, E.S., D.S. Rice, P.M. Rice, H.H. Vaughn, M. Brenner, & M.S. Flannery
1979 Maya urbanism: impact on a tropical karst environment. *Science* 206: 298-306.

Demarest, A.A.
1992 Ideology in ancient Maya cultural evolution: the dynamics of galactic polities. In *Ideology in Pre-Columbian Civilizations* (A.A. Demarest & G. Conrad, eds.), pp. 135-157. Santa Fe: School of American Research.

1996 War, peace, and the collapse of a Native American civilization: lessons for contemporary systems of conflict. In *A Natural History of Peace* (T. Gregor, ed.), pp. 215-248. Nashville: Vanderbilt University Press.

Dunnell, R.C.
1978 Style and function: a fundamental dichotomy. *American Antiquity* 43: 192-203.
1982 Science, social science, and common sense: the agonizing dilemma of modern archaeology. *Journal of Anthropological Research* 38: 1-25.
1989 Aspects of the application of evolutionary theory in archaeology. In *Archaeological thought in the Americas* (C.C. Lamberg-Karlovsky, ed.), pp.35-49. Cambridge: Cambridge University Press.

Efron, B. & R.J. Tibshirani
1993 *An Introduction to the Bootstrap*. New York: Chapman and Hall.

Fash, W.L., Jr.
1988 A new look at Maya statecraft from Copan, Honduras. *Antiquity* 62:157-169.

Garver, J.B.
1989 *The Land of the Maya: A Traveler's Map.* Washington, D.C.: Cartographic Division of the National Geographic Society.

Goodman, G.
1905 Mayan dates. *American Anthropologist* 7: 642-647.

Graffen, A.
1990a Sexual selection unhandicapped by the Fisher process. *Journal of Theoretical Biology* 144: 473-516.
1990b Biological signals as handicaps. *Journal of Theoretical Biology* 144: 517-546.

Graham, I.
1982 *Corpus of Maya Hieroglyphic Inscriptions, Volume 1, Part 3*. Cambridge: Peabody Museum of Archaeology and Ethnography, Harvard University.

Graves, M.W. & T.N. Ladefoged
1995 The evolutionary significance of ceremonial architecture in Polynesia. In *Evolutionary Archaeology: Methodological Approaches* (P. Teltser, ed.), pp.149-174. Tuscon: University of Arizona Press.

Haigh, J.G.B.
1992 Radial basis functions and archaeological surfaces. In *Computer Applications and Quantitative Methods in Archaeology: 1991* (G. Lock & J Moffett, eds.), pp. 157-162. Oxford: BAR International Series S577.
1993 Practical experience in creating digital elevation models. In *Computing the Past: Computer Applications and Quantitative Methods in Archaeology: CAA92* (J. Andersen, T. Madsen, & I. Scollar, eds.), pp. 67-74. Aarhus: Aarhus University Press.

Hamblin, R.L. & B.I. Pitcher
1980 The Classic Maya collapse: testing the class conflict hypothesis. *American Antiquity* 45: 246-267.

Hamilton, W.D.
1964 The genetical theory of social behavior, I,II. *Journal of Theoretical Biology* 7: 1-52.

Haviland, W.A.
1967 Stature at Tikal, Guatemala: implications for ancient Maya demography and social organization. *American Antiquity* 32: 429-433.

Hodder, I.
1986 *Reading the Past*. Cambridge: Cambridge University Press.

Hodell, D.A., J.H. Curtis, & M. Brenner
1995 Possible role of climate in the collapse of Classic Maya civilization. *Nature* 375: 391-394

Houston, S.D.
1992 Classic Maya politics. In *New Theories on the Ancient Maya* (E.C. Danien & R. Sharer, eds.), pp. 65-69. Philadelphia: University of Pennsylvania Museum.
1993 *Hieroglyphs and History at Dos Pilas: Dynastic Politics of the Classic Maya*. Austin: University of Texas Press.
n.d. Latest long count dates on monuments in the southern Lowlands. Ms. list in possession of the author.

Isaaks, E.H. & R.M. Srivastava
1989 *Applied Geostatistics*. New York: Oxford University Press,.

Jones, C.
1991 Cycles of growth at Tikal. In *Classic Maya Political History: Hieroglyphic and Archaeological Evidence* (T.P. Culbert, ed.), pp.102-127. Cambridge: Cambridge University Press.

Kaplan H. & K. Hill
1992 The evolutionary ecology of food acquisition. In *Evolutionary Ecology and Human Behavior* (E.A. Smith & B. Winterhalder, eds.), pp.167-202. New York: Aldine de Gruyter.

Kirkpatrick, M.
1986 The handicap mechanism of sexual selection does not work. *American Naturalist* 127: 222-240.

Kosso P.
1989 Science and objectivity. *Journal of Philosophy* 86: 245-257.

Kvamme, K.L.
1990 Spatial autocorrelation and the Classic Maya collapse revisited: refined techniques and new conclusions. *Journal of Archaeological Science* 17: 197-207.

Leyden B.W.
1987 Man and climate in the Maya Lowlands. *Quaternary Research* 28: 407-414.

Lowe, J.W.G.
1985 *The Dynamics of Apocalypse, A Systems Simulation of the Classic Maya Collapse*. Albuquerque: University of New Mexico Press.

Marcus, J.
1976 *Emblem and State in the Classic Maya Lowlands: An Epigraphic Approach to Territorial Organization*. Washington, D.C.: Dumbarton Oaks.
1983 Lowland Maya archaeology at the crossroads. *American Antiquity* 48: 454-488.
1993 Ancient Maya political organization. In *Lowland Maya Civilization in the Eighth Century A.D.* (J.A. Sabloff & J.S. Henderson, eds.), pp. 111-171. Washington, D.C.: Dumbarton Oaks.

Martin, S. & N. Grube
1994 Evidence for macro-political organization among Classic Maya Lowland States. Ms. in possession of the author.
1995 Maya superstates: how a few powerful kingdoms vied for control of the Maya Lowlands during the Classic Period (AD 300-900). *Archaeology* 48: 41-59.

Mathews, P.
1985 Early Classic monuments and inscriptions. In *A Consideration of the Early Classic Period in the Maya Lowlands* (G.R. Willey & P. Mathews, eds.), pp. 5-55. Albany: Institute of Mesoamerican Studies, Publication No. 10.
1991 Classic Maya emblem glyphs. In *Classic Maya Political History: Hieroglyphic and Archaeological Evidence* (T.P. Culbert, ed.), pp. 19-29. Cambridge: Cambridge University Press.

Mathews, P. & G.R. Willey
1991 Prehistoric polities of the Pasion region: hieroglyphic texts and their archaeological settings. In *Classic Maya Political History: Hieroglyphic and Archaeological Evidence* (T.P. Culbert, ed.), pp.30-71. Cambridge: Cambridge University Press.

Miller. M.E.
1986 *The Murals of Bonampak*. Princeton: Princeton University Press.

Morley, S.G.
1920 *The Inscriptions of Copan*. Washington: Carnegie Institute Publication 219.
1937-8 *The Inscriptions of Peten*. 5 vols. Washington: Carnegie Institute Publication No. 437.
1946 *The Ancient Maya*. Stanford: Stanford University Press.

Morley, S.G., G.W. Brainerd & R.J. Sharer
1983 *The Ancient Maya, Fourth Edition*. Stanford: Stanford University Press.

Neiman. F.D.
1990 *An Evolutionary Approach to Archaeological Inference: Aspects of Architectural Variation in the 17th-Century Chesapeake*. Ph.D. dissertation, Department of Anthropology, Yale University.
1995 Stylistic variation in evolutionary perspective: inferences from decorative diversity and interassemblage distance in Illinois Woodland ceramics. *American Antiquity* 60: 7-37.

Parpola, A.
1994 *Deciphering the Indus Script*. Cambridge: Cambridge University Press.

Proskouriakoff, T.
1950 *A Study of Classic Maya Sculpture*. Washington: Carnegie Institute Publication No. 593.

Rands, R.L.
1973 The Classic collapse in the southern Maya Lowlands: chronology. In *The Classic Maya Collapse* (T.P. Culbert, ed.), pp.43-62. Albuquerque: University of New Mexico Press.

Rice, D.S.
1985 The Peten Postclassic: perspectives from the central Peten lakes. In *Late Lowland Maya Civilization: Classic to Postclassic* (J.A. Sabloff & E.W Andrews V, eds.), pp. 301-344. Albuquerque: University of New Mexico Press.

Rice, D.S. & P.M. Rice
1993 Physical geography, environment, and natural resources. In *Lowland Maya Civilization in the Eighth Century A.D.* (J.A. Sabloff & J.S. Henderson, eds.), pp. 11-63. Washington, D.C.: Dumbarton Oaks.

Rice, D.S., P.M. Rice & E.S. Deevey
1985 Paradise Lost: Classic Maya impact on a lacustrine environment. In *Prehistoric Maya Environment and Subsistence Economy* (M. Pohl, ed.), pp. 91-105. Cambridge: Peabody Museum of Archaeology and Ethnography, Harvard University.

Sabloff, J. A.
1973 Continuity and disruption during Terminal Late Classic Times at Seibal: ceramic and other evidence. In *The Classic Maya Collapse* (T.P. Culbert, ed.), pp.107-132. Albuquerque: University of New Mexico Press.
1992 Interpreting the collapse of Maya civilization: a case study of changing archaeological perspectives. In *Metaarchaeology* (L. Embree, ed.), pp.99-119. Dordrect: Kluwer Academic Publishers.
1995 Drought and decline. *Nature* 375: 357.

Sabloff, J.A. & G.R. Willey
1967 The collapse of the Maya civilization in the southern Lowlands. *Southwestern Journal of Anthropology* 23: 311-336.

Saul, F.P.
1973 Disease in the Maya Area. In *The Classic Maya Collapse* (T.P. Culbert, ed.), pp. 301-324. Albuquerque: University of New Mexico Press.

Schele, L. & M. Miller
1986 *The Blood of Kings: Dynasty and Ritual in Maya Art.* Fort Worth: Kimball Art Museum.

Schmandt-Besserat, D.
1979 Reckoning before writing. *Archaeology* 32: 22-31.

Shanks, M. & C. Tilley
1987 *Re-Constructing Archaeology*. Cambridge: Cambridge University Press.

Smouse, P.E. & J.C. Long
1992 Matrix correlation analysis in anthropology and genetics. *Yearbook of Physical Anthropology* 35: 187-213.

Smouse, P.E., J.C. Long, & R.R. Sokal
1986 Multiple regression and correlation extensions of the Mantel test of matrix correspondence. *Systematic Zoology* 35: 627-632.

Stephens, J.L.
1841 *Incidents of Travel in Central America, Chiapas, and Yucatan.* New York: Harper and Brothers.

Stuart, D.
1988 Blood symbolism in Maya iconography. In *Maya Iconography* (E. Benson & G. Griffin, eds.), pp. 175-221. Princeton: Princeton University Press.
1993 Historical inscriptions and the Maya collapse. In *Lowland Maya Civilization in the Eighth Century A.D.* (J.A. Sabloff & J.S. Henderson, eds.), pp. 321-354. Washington, D.C.: Dumbarton Oaks.

Trigger, B.
1990 Monumental architecture: a thermodynamic explanation of symbolic behavior. *World Archaeology* 22: 119-131.

Veblen, T.
1899 *The Theory of the Leisure Class.* New York: Macmillan.

Whitmore, T.M., B.L. Turner II, B.L. Johnson, R.W. Kates & T.R. Gottschange
1990 Long-term population change. In *The Earth as Transformed by Human Action* (B.L. Turner, ed.), pp.25-39. Cambridge: Cambridge University Press.

Whittington, S.
1991 Detection of significant demographic differences between subpopulations of prehispanic Maya from Copan. *American Journal of Physical Anthropology* 85: 167-85.
1989 *Characteristics of Demography and Disease in Low Status Maya from Copan, Honduras.* Ph.D. dissertation, Department of Anthropology, Pennsylvania State University.

Whitley, D.S. & W.A.V. Clark
1985 Spatial autocorrelation tests and the Classic Maya collapse: methods and inferences. *Journal of Archaeological Science* 12: 377-395.

Wright. L.E.
1996 Intertooth patterns of hypoplasia expression: implications for childhood health in the Classic Maya collapse. *American Journal of Physical Anthropology* 102: 233-47.

Wynne Edwards, V.C.
1962 *Animal Dispersion in Relation to Social Behavior.* Edinburgh: Oliver and Boyd.

Zahavi A.
1975 Mate selection - a selection for a handicap. *Journal of Theoretical Biology* 53: 205-214.

1987 The theory of signal selection and some of its implications. In *Proceedings of the International Symposium of Biological Evolution* (P.V. Delfino, ed.), pp. 305-325. Bari: Adriatica Editricia.

Zipf, G.K.
1949 *Human Behavior and the Principle of Least Effort.* Cambridge: Addison-Wesley.

16

Behavioral Ecology and the Archaeological Consequences of Central Place Foraging among the Meriam

Douglas W. Bird
University of Washington

ABSTRACT

Archaeology is ultimately concerned with explaining variability in the patterns of past human behavior. This chapter illustrates how archaeologically falsifiable hypotheses derived from the neo-Darwinian framework of behavioral ecology can serve this goal. Most archaeological applications of this approach have focused on explaining differences in the material consequences of subsistence practices with reference to the general predictions of the encounter contingent prey-choice model. I examine how testing hypotheses grounded in this model illuminate its relevant assumptions and direct our attention toward new questions about human subsistence strategies, especially where patterns in subsistence remains are likely to be influenced by the processing and transport decisions of central place foragers. The argument developed here is that the same neo-Darwinian logic that provides predictions about prey choice also provides appropriate tools for generating testable explanations of variability in central place foraging strategies, resource transport, and patch choice. I illustrate the advantages of this approach with an analysis of shellfishing efficiency and shell deposition among contemporary Meriam shellfishers in the Torres Strait Islands. The results of this investigation demonstrate that Meriam processing decisions and the archaeological products of intertidal gathering are consistent with an hypothesized goal of maximizing resource delivery rate. This finding has ramifications well beyond the context of shellfishing, with implications for arguments about early hominid subsistence and settlement patterns drawn from interpretations of variability in large mammal remains.

INTRODUCTION

This chapter presents an attempt to link archaeological variability with the proposition that human behavioral capacities and propensities are shaped by natural selection. The main justification for this proposition is the assumption that where variations in behavioral patterns are at least partly heritable and time and energy are limited, those variants that increase an individual's genetic representation in future generations will persist in a population at the expense of alternatives. We need not assume that there is necessarily a close link between genetic and behavior variability or that selection acts on specific behaviors. If we suppose only that the underlying decision rules of behavior have been shaped by selection, then it should be possible to predict adaptive decision making with measurements of the fitness-related trade-offs an organism faces (Dawkins 1982: 18-29, Grafen 1991).

If selection shapes behavioral variation and behavior creates the archaeological record, then in principle, theoretically framed explanations of the evolution and maintenance of behavioral patterns should be testable with observations of variation in the material results of human activities. While controlling for factors such as non-anthropogenic site formation processes may make it difficult to operationalize this proposal, archaeological applications should not be impossible (O'Connell 1995). The goal of this chapter is to demonstrate how theory in behavioral ecology can assist in developing ideas about foraging strategies that can in turn generate archaeologically falsifiable explanations of behavioral variability. To do so, I briefly introduce behavioral ecology and some problems associated with explaining faunal assemblage variability that might be amenable to investigation from this perspective. I then illustrate some advantages of this approach with an analysis of Meriam intertidal prey choice, patch switching, processing/transport strategies, and their material correlates. Finally, I point out some general implications of this approach for archaeological interpretations of faunal remains.

BEHAVIORAL ECOLOGY AND FORAGING THEORY

Behavioral ecology is the subset of evolutionary ecological theory that seeks to understand behavioral variability as the outcome of selective processes within specific ecological and/or social settings (Krebs & Davies 1997, see Smith & Winterhalder 1992 for thorough review of application in the study of humans). Foraging theory, a subset of behavioral ecology, is used to direct inquiry into how resource acquisition decisions are affected by the trade-offs that individuals face in their allocation of time and energy to alternative courses of action. Foraging models analyze the relationship between four hypothesized components of foraging: (1) a goal, (2) a specific decision, (3) a currency, and (4) pertinent constraints (Stephens & Krebs 1986: 5-11).

Foraging models assume that subsistence behavior is ultimately shaped by selection; that natural selection designs organisms with the ability to balance the fitness-related opportunity costs of foraging against the benefits received. These costs and benefits are ultimately related to inclusive fitness in some way, but our aim should not necessarily be to search for correlations between measures of fitness and alternative foraging strategies. Such correlations are likely to be, in a direct sense, irrelevant to the question at hand because foraging is only one, sometimes minor component of fitness. Foraging models thus specify more immediate *goals* than fitness maximization–such as rate of energetic acquisition or time minimization–because these more immediate decision rules are more likely to closely track the (fitness-related) consequences of time or energy allocation decisions in particular and varying circumstances. This does not mean that selection does not design foraging strategies, it means that direct correlations between measures of fitness and alternative foraging decisions are likely to be of little consequence for addressing the question at hand (Hawkes 1993: 359; see also Cuthill & Houston 1997: 98-100).

For example, if we are interested in a functional (as opposed to causal, see Krebs & Davies 1993: 4-5) explanation for variability in field processing and transport strategies, we might begin by hypothesizing that the goal of field processing is to maximize the rate at which resources can be delivered to a central place, not to maximize fitness per se, because countless other decisions will have an influence on fitness outcomes. We can then specify the trade-offs in efficiency that a forager would face if she, for example, stopped collecting to cull a resource load of low utility parts by field processing. If her goal is to maximize delivery rate in terms of some specified *currency* (such as energy, protein, edible weight, etc.), what will serve that goal better: trading off the opportunity to collect and transport more resources by stopping to field process, or trading off the opportunity to increase the load quality with field processing by continuing to collect and transport more resources? This question incorporates a specific *decision* variable (in this case, field processing vs. collecting and transporting more resources) whose solution does not depend in some direct sense on fitness trade-offs, but rather on the immediate opportunity costs that foragers face. We assume that natural selection would design organisms to be effective at negotiating immediate trade-offs, and from there we can generate hypotheses with specific predictions about behavioral outcomes. Foraging models do not assume that other trade-offs other than those specified have no effect on behavioral outcomes, nor do they specify a mechanism for how such 'decisions' are made (e.g., cognized intent, instinctual program, learning, genetic transmission, socialization etc.).

The *constraints* of the model set limits on the feasible options available to the forager. Often the constraints are intrinsic characteristics of the prey or the forager. These include such things as prey distribution and the abilities of the forager (e.g., age, load size transport limits, incomplete knowledge of resource characteristics, etc.). The most successful foraging models stipulate very few constraints and specify currencies that are readily measurable (Stephens & Krebs 1986). Identifying these simple variables and analyzing their relationships allows the investigator to cast questions about behavior in economic terms to make specific testable predictions about how behavior should vary in specific settings.

Many people have argued that there is no reason to suspect that behavior should or would be optimal-that optimality models fail to capture the complexity of human behavior and are thus inappropriately simplistic because, especially among humans, decisions are the products of unique historical and social circumstances. However, optimality models from behavioral ecology are not intended to investigate the proposal that organisms are 'optimal'. An optimization framework simply generates hypotheses with specified variables and provides the tools for analysis of the predicted outcome of their relationships (Krebs et al. 1981: 5, Maynard Smith 1978, Smith

1987). Whether or not foragers exploit prey in a manner consistent with the hypothesized goal, currency, and constraints of a model is an empirical question. Conflicts between predictions and empirical observations can indicate that foragers face other constraints or have other goals than those hypothesized. Only by testing predictions of the models can we specify which of the model's assumptions are too simplistic and which variables may be inappropriate. Alternatives can then be hypothesized and novel predictions tested. Mismatches between observed and expected behavior do not indicate that the theoretical approach itself is flawed, rather that our hypotheses may be inappropriate.

Ethnoarchaeologists are just beginning to recognize the utility of applying the simplest form of foraging models in analyses of archaeologically significant behavior (O'Connell 1995, Simms 1992). The primary advantage in utilizing formal foraging models to inform ethnoarchaeological research lies in their ability to generate theoretically guided hypotheses with predictions about variability in behavioral phenomena long considered to be fundamental in human evolution (e.g., diet, processing and transport patterns, food sharing, and the sexual division of labor). Secondly, the assumptions which underlie foraging models are made explicit and are theoretically justifiable. Which assumptions are meaningful in a complex world can only be assessed by positing precise predictions against observations of food procurement and processing activities. While the models consist of a relatively narrow set of effective variables that will not necessarily reflect the complexity of foraging in the real world, their simplicity contributes a great deal to the level of precision at which the predictions can be assessed and the generality of their application to foraging issues (Levins 1966). Finally, operationalizing such models in an ethnographic setting provides a foundation to connect predictions about one aspect of behavior with another (e.g., relationships between foraging strategies and mate choice, see Hawkes 1993). Where at least some aspects of behavioral variation have direct archaeological consequences, this should provide a theoretically based link to variability in behavior that may be archaeologically indistinct otherwise (e.g., recognizing the differential effects of transport vs. trade, see Bird & Bliege Bird 1997: 53-54).

THE PROBLEM AND POTENTIAL SOLUTIONS

The familiar problem of interpreting variability in subsistence remains may serve to illustrate an archaeological avenue for applications of behavioral ecology. Two issues are particularly salient in debates about broad scale human subsistence patterns in the past: (1) assuming a simple relationship between resource choice and subsistence remains, how can we explain observable variability in the patterns of past diets, and (2) knowing that a simple relationship between patterned prey selection and food remains is unlikely, how can explanations of behavioral processes other than resource choice inform interpretations of faunal and plant assemblage variability. Conventional explanations of variability in the archaeological reflection of diet have often relied on factors related to synchronic or diachronic differences in climatic conditions, habitat familiarity, and/or population density. As others have pointed out (e.g., Edwards & O'Connell 1995), the problem with these constructions is that they provide us with no generally testable clues as to *why* particular sets of resources and changes in their frequency are observed in archaeological sites.

Archaeologists have long recognized that the *encounter contingent prey-choice model* from foraging theory may be useful for understanding shifts in the suite of resources exploited in the past (e.g., Beaton 1972, Earle & Christenson 1980, O'Connell & Hawkes 1981, Perlman 1980). This model provides a rate maximizing solution to the trade-off between the benefits of handling an encountered prey item versus the benefits of passing over an item to continue searching for other types of prey. The model assumes that, by handling prey, a forager will trade off search opportunities. The model generally uses energy as a currency and ranks prey types in order of their profitability (e/h or energy/unit time handling, where handling includes pursuit, harvesting, and processing). It predicts that overall return rates while foraging (E/T or total energy/unit time foraging, where foraging includes searching and handling) will be maximized when the highest ranked prey type is always handled on encounter and lower ranked prey are handled if, and only if, their e/h exceeds the expected E/T. Lower ranked prey are handled or passed over only as a function of changes in the encounter rate for higher ranked resources. The primary constraint of the model is that foraging occurs in a fine-grained manner such that search time is shared by all prey types (Smith 1991: 206-207). Where resources are patchy (search times for sets of different prey types are mutually exclusive), combined prey/patch models and marginal analysis of patch gain rates are more appropriate for analysis (see Stephens & Krebs 1986 for detailed treatment).

While a number of archaeologists have proposed that the qualitative predictions of the prey-choice model (that less profitable prey will be added to a diet with declines in the abundance of high ranked resources) are matched by variability in subsistence remains (e.g.,

Bettinger & Baumhoff 1982, Botkin 1980, Glassow & Wilcoxon 1988, Raab 1992), there are very few substantive tests of its specific quantitative predictions with archaeological data (but see Broughton 1994). One of the problems with testing the predictions of the model archaeologically is that any particular assemblage of subsistence remains is likely to be conditioned by a number of processes other than on-encounter prey choice. Two such processes include patterned variability in resource patch choice and differential resource transport. Of special archaeological importance is the observation that contemporary central place foragers in patchy environments face load constraints and often field process prey at or near procurement patches and differentially transport resources or resource parts to home bases (Binford 1978, Bunn et al. 1988, Emerson 1993, O'Connell et al. 1988, 1990; O'Connell & Marshall 1989, Savelle & Friesen 1996). Accurate reconstructions and explanations of subsistence changes are thus likely to require testable explanations of transport behavior and patch exploitation.

Problems related to recognizing and explaining archaeological patterning in prey choice are well illustrated in debates about the behavioral interpretation of Plio-Pleistocene hominid sites from east Africa (e.g., Binford 1983, Blumenschine & Madrigal 1993, Bunn 1986; Bunn & Kroll 1986, Isaac 1984, Monahan 1996, O'Connell et al. 1990, O'Connell 1997, Rose & Marshall 1996). These debates center on whether or not the sites in question represent central places to which early hominids transported food and hinge on anticipating the effects of differential body part transport to explain bone assemblage variability. While many of these studies have alluded to a host of factors (such as part value, travel distance, processing costs, and load size) that would influence the transport behavior of hominids (ancient or modern), until recently the predictions of hypotheses that model the relationship between such variables have remained untested with behavioral observations. As I illustrate below, analyses of central place foraging models from behavioral ecology (*sensu* Metcalfe & Barlow 1992, Orians & Pearson 1979) can anticipate the archaeological effects of processing/transport decisions and may provide insight into the conditions under which we will expect predictions of the prey-choice model to be falsified with archaeological data.

THE STUDY SITE

The Meriam are the indigenous Melanesian inhabitants of the three easternmost islands in the Torres Strait, the archipelago between the Australia's Cape York Peninsula and southern coast of Papua New Guinea. The Meriam Islands (Mer, Dauar, and Waier) lie 6 km west of the northern end of the Great Barrier Reef and 140 km southeast of New Guinea's Fly River Delta. The islands are small volcanic crests fringed by young, high-elevation reef flats. Mer, the largest (2.8 x 1.7 km) of the three islands, lies two km north of Dauar and Waier. Currently most of the Meriam population (about 400 individuals) lives in a village on Mer's north-west shore.

The Meriam Islands have been part of the British Commonwealth since the Torres Strait Islands were annexed by Queensland in 1879 (see Beckett 1987, Haddon 1908, 1935; and Sharp 1993 for details on Meriam history and ethnography). Until the 1970's intensive horticulture and fishing remained the primary source of subsistence for the Meriam. Since that time the importance of gardening has markedly declined with access to Australian welfare benefits; flour and rice now comprise the majority of the Meriam diet (Beckett 1987). Goods arrive on a barge every fortnight and are purchased in the local shop on Mer. However, marine resources acquired from foreshore, intertidal, and offshore habitats remain an important source of daily subsistence for most families. Per capita, on average the Meriam acquire 80 gm of protein and 400 kcal per day from fishing, turtle hunting, and intertidal gathering.

Intertidal Gathering and Data Collection

During 1993-95 R. Bliege Bird and I spent 14 months on Mer collecting observation data on Meriam marine subsistence. Data on time allocation to subsistence activities, food-sharing, and foraging efficiency were collected to address a broad set of questions concerning the determinants of Meriam foraging patterns (see Bliege Bird et al. 1995, Bliege Bird 1996, Bliege Bird & Bird 1997, Bird 1996, and Bird & Bliege Bird 1997). We also collected quantitative measures of factors that influence the behavioral determinants of archaeological variability resulting from *intertidal gathering*.

Two types of gathering, both focused on exploiting shellfish, take place in Meriam intertidals: *reef flat collecting* and *rocky shore harvesting*. Reef flat collecting takes place in the mid-sublittoral margin of the fringing reefs. This is by far the more important of the two gathering activities but occurs primarily only during low spring tides that expose the reef for one to two weeks per lunar cycle from March to October. Although a wide range of prey types are available to collectors, three shellfish resources made up over 94% of all edible flesh gathered from the reef flat: two types of large tridacnid clams (*Hippopus hippopus*, and *Tridacna squamosa/maxima*, which on average yield 150-400 gm of meat per specimen) and one medium sized conch (spider conch, *Lambis lambis* which yields 25 gm of meat per specimen). Collectors gather these prey as they walk parallel to the shoreline through the mid-sublittoral zone. In many respects reef flat collecting provides an ideal situation for predictive tests of the prey-choice model: search costs while foraging are distributed among all reef flat resources in the mid-sublittoral and prey items of a particular types are almost always encountered singly. As such, decisions to handle an item of a particular type are made on-encounter, irrespective of the search time required for collecting that specific prey type or item.

Rocky shore harvesting takes place in shell beds found mostly in the upper-littoral of the fringing reef. While a wide array of possible mollusc resources are found in rocky shore patches, the Meriam limit their selection to two prey types: small filter feeding sunset clams (*Asaphis violascens,* sunset clam, which averages about 9 gm of meat per specimen) and small nerites (specifically *Nerita undata,* waved nerite, which yields only about 2.5 gm of meat per specimen). *Asaphis* are excavated from the sandy, gravelly substrate beneath volcanic cobbles while *Nerita* are plucked from the rocks that overlie the bivalve beds. As mentioned above, this activity is rare relative to the reef flat collecting.

Methods: Measures from four data sets are used in the analysis below (see Bird 1996 for detailed description of methods). First, we collected quantitative data on intertidal gathering efficiency and patterns of processing and transport through detailed focal individual follows. Follows occurred both during randomly selected observation periods and opportunistically when a researcher was present during foraging episodes that included intertidal gathering. During follows we recorded the time that a focal forager allocated to each aspect of foraging (traveling, searching, and post-encounter handling of each prey item) and the total bulk and flesh yield acquired from each prey type. Records are available for 91 intertidal gathering follows of 50 different individuals incorporating 138 observation hours with samples from every month of the year. Second, data on handling costs for shellfish that are rarely collected (or rarely field processed) and flesh/gross weight ratios were obtained through a series of processing experiments where we timed active foragers while processing loads of specific prey types and counted and weighed the bulk and processed products. Data were obtained from 51 different loads of 10 prey types processed by 7 different foragers. Third, data on contemporary depositional patterns were obtained by sampling the remains of shellfish that 6 randomly selected households accumulated over a period of four months. Each household was given a container and each week a researcher counted and weighed all shells deposited by household members. Finally, quantitative estimates of the densities and distributions of intertidal prey were obtained though a series of reef flat surveys where researchers recorded the number and location of all mollusc prey encountered. Data were recorded for 12 corridor surveys on Mer's fringing reef (most of which extended from the shore to the reef edge and ranged from .2 to 1.1 km in length). Each corridor included four transects (two researchers walking up and back) at 10 meters apart. Assuming total coverage within each corridor, 127,200 m^2 of fringing reef were surveyed.

SHELLFISH PREY CHOICE

The prey-choice model discussed above entails two main predictions: (1) foragers encountering resources in a fine-grained manner will pass over those prey types whose post-encounter profitability (*e/h*, kcal/hr. harvesting and processing) is exceeded by the rate of nutrient gain expected from searching for and handling other prey (*E/T*, kcal/hr. searching and handling), and (2) those prey types that will on average maximize overall foraging efficiency (where $e/h > E/T$) will be harvested in proportion to their abundance (i.e., encounter rate). Archaeologically we would predict that: (3) those resources that if handled on-encounter will reduce mean overall foraging efficiency will be absent from middens, and (4) those prey types whose profitability exceeds overall foraging efficiency in a fine-gained patch will be found in middens in proportion to their abundance in the patch. I tested predictions 1-4 with data from both Meriam intertidal gathering activities and contemporary shell deposits.

Selectivity: Comparisons between observed collecting rates and expected encounter rates of Meriam shellfish provide a measure of "selectivity", an estimate of the probability that an item of a particular shellfish prey type will be handled on-encounter. Details on calculating these comparisons are provided in Bird (1996). Briefly, collecting rates are simply the average number of specimens of each prey type collected by focal adults per hour spent in search. To estimate expected encounter rates I assumed that reef flat collectors walk at 3 km/hour (a reasonable walking speed for foragers, see Jones & Madsen 1989). If foragers can spot all resources within five meters to each side, they would then search the reef flat at 30,000 m^2/hour. By multiplying the estimated search rate by the average resource densities obtained from the reef flat survey described above, I can provide an independent measure of variability in expected rates of encounter with different prey.

Table 16.1 presents a comparison of the predicted encounter rates (λ_e, the number of specimens encoun-

Table 16.1 Observed adult collection rates, predicted encounter rates, and selectivity index for reef flat prey.

Prey type	λ_c	λ_e	$\lambda_{e\,(flesh)}$ kg	$\lambda_{e\,(shell)}$ kg	SI
Hippopus	5.38	3.18	0.580	5.01	1.69
Tridacna spp.*	2.06	2.93	0.870	5.12	0.73
Trochus (lg)	0.39	0.41	0.004	0.02	0.96
Lambis	18.51	21.10	0.530	5.73	0.88
Cypraea	1.45	3.90	0.011	0.04	0.37
Trochus (sm)	0.00	2.34	0.006	0.03	0.00
Strombus	5.49 (1.53)**	16.64	0.010	0.07	0.33 (0.09)**
Tridacna crocea	0.02 168.62	10.433	61.37	<0.01	

λ_c = observed number of specimens collected per hour spent searching (n = 58 follows)
λ_e = predicted number of specimens encountered per hour (if foragers walk at 3 km/hr and encounter items in proportion to reef flat survey results)
$\lambda_{e(flesh)}$ = predicted weight of edible flesh encountered per hour
$\lambda_{e(shell)}$ = predicted weight of shell encountered per hour
SI = selectivity index (λ_c / λ_e)
* *Tridacna* spp. includes *T. squamosa* and *T. maxima*
** The lower figure is calculated with the removal of the single follow when a substantial amount of *strombus* was collected.

tered per collecting hour based on the resource survey measures), the actual average collecting rates (λ_c, the number of specimens collected per searching hour during focal follows), and selectivity index (SI) values for adults. The SI provides simply a ratio of λ_c per unit λ_e for each prey type. A prey type with an SI value equal to one would be collected at rate in proportion to its relative abundance across Mer's fringing reef as estimated from the reef flat survey. Although Meriam foragers do not necessarily search the reef flat in the systematic and random fashion of survey transects, the SI is instructive in two respects. First, it provides a quantitative and comparative measure of the extent to which foragers are selective within a given patch, independent of the measures of foraging effort. Second, it provides a baseline from which we can begin testing hypotheses about the determinants of intertidal prey selection.

Observed prey choice: Figure 16.1 demonstrates a good match between predictions 1 and 2 and observed prey choice among adults while reef flat collecting. Those resources whose profitability (*e/h*, or kcal/hour handling)[1] is significantly greater than mean overall foraging efficiency (*E/T* = 1491.5 ± 173.2 kcal/hour searching and handling, n = 47 follow)[2] should always be harvested on-encounter (the SI should be near 1). The highest ranked reef flat resource, *Tridacna gigas* (giant tridacnid clam), is only available while reef flat collecting for a few days out the year when diurnal low spring tides fully expose the subtidal reef edge. As predicted, when available this prey type was never passed over and has a post encounter return rate (*e/h* = 13064.8 ± 2375.4, n = 4 follows) nearly an order of magnitude greater than overall mean foraging efficiency.

Hippopus, the highest ranked resource on most follows, is associated with the highest SI value and post-encounter return rates (*e/h* = 6858.9 ± 464.0, n = 34 follows) significantly greater than the mean overall foraging efficiency (paired t-test, DF = 33, t = 12.3, $p < .01$). Foragers increase their harvesting rate for *Hippopus* by concentrating their efforts when possible in the boundary of the mid-sublittoral fringe (hence the SI value is greater than 1.0). While across the entire intertidal certain prey are more likely to be found in the mid-sublittoral zone, within that zone encounters with prey are fine-grained. By avoiding the upper littoral and sublittoral fringe, foragers increase their chances of encountering *Hippopus* without significantly decreasing encounters with other reef flat prey types.

Tridacna maxima/squamosa is also significantly more profitable on encounter (*e/h* = 4417.6 ± 709.0, n = 16 follows) than expected overall return rates (paired t-test, DF = 15, t = 5.2, $p < .01$). As anticipated, the SI for this resource is also high.

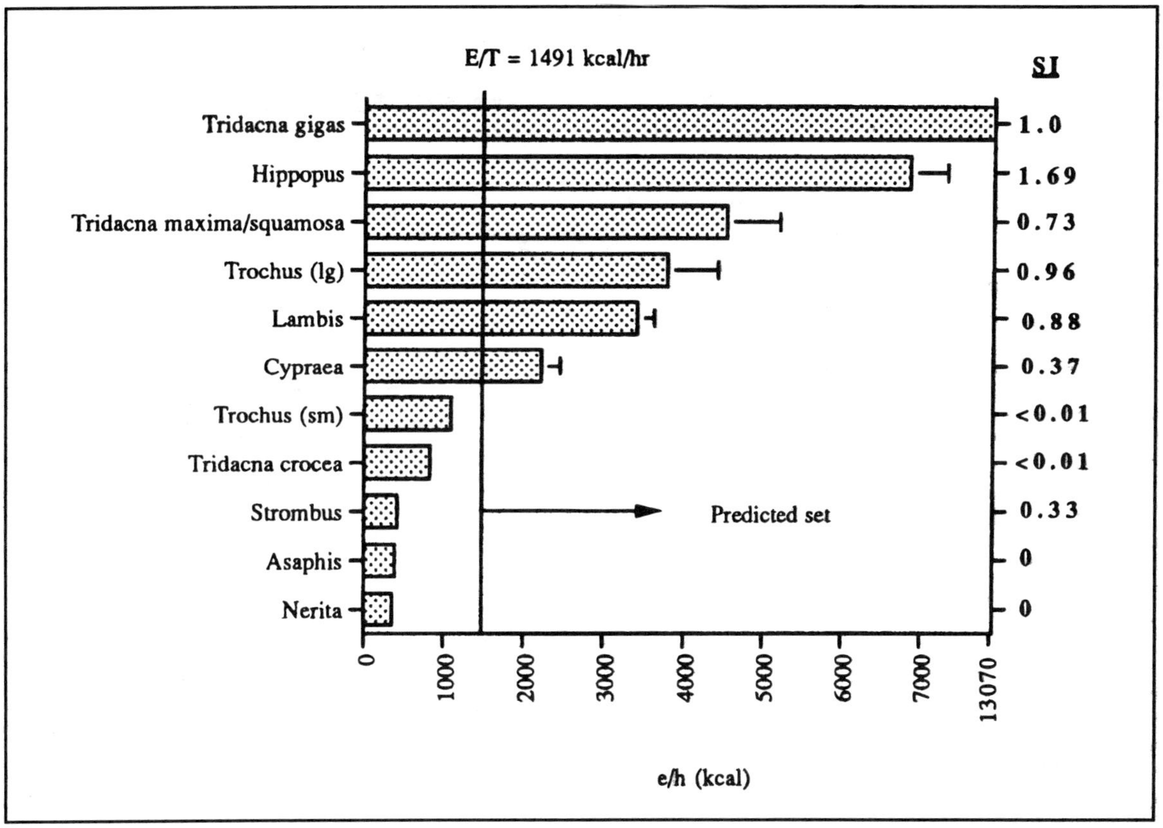

Figure 16.1 Adult foraging efficiency: post-encounter return rates (*e/h*) vs. mean reef flat collecting return rate (*E/T*). elh = k/cal/hour handling each prey type; *E/T* = kcal/hour searching and handling all prey types; SI = selectivity index, llecting rate/predicted encounter rate (see Table 16.1); Error bars = SE.

Lambis specimens were only very rarely passed over (SI = .88) and as anticipated by the model are also associated with *e/h* (3411.9 ± 205.6, n = 23 follows) significantly greater than the overall mean foraging return rate (paired t-test, DF = 22, t = 10.0, p < .01). Some specimens were not handled on-encounter during 2 follows where foragers who did not field process ran out of space in their container, and during 5 follows when only tridacnids *(Hippopus* and *Tridacna* spp.) were harvested.

Large *Trochus niloticus* (*e/h* = 3784.5 ± 638.1, n = 3 experiments) and *Lambis* are similarly ranked and, as expected, large *Trochus* were never observed to have been passed over (although due to the small sample size of processing experiments the significance of differences between *e/h* and *E/T* is not available). As with *T. gigas*, encounters with large *Trochus* occurred only in sublittoral zones at extreme astronomical low tides during a few follows in May-July.

Cypraea tigris (tiger cowrie) are somewhat problematic. The analysis suggests that while overall return rates will on average increase if this prey type is exploited (*e/h* = 2214.4 ± 239.2, n = 3 experiments), the SI value is well below what we might expect if it were consistently taken on-encounter. In concurrence with the predicted SI value, during follows foragers occasionally passed over this prey type and occasionally collected it. When asked why *Cypraea* were sometimes not taken, foragers usually indicated that this prey type can sometimes cause digestive problems. Thomas (1994: 22) has noted that Islanders in the Central Gilbert Islands of Kiribati completely avoid this resource when reef flat collecting. Subsequent nutritional analysis by Thomas (in prep.) has demonstrated that the *Cypraea* from Kiribati have varying levels of a toxic lipid of unknown origin. While nutritional analyses have not been conducted on Meriam *Cypraea*, it may be that prey choice on Mer is influenced by similar factors.

In brief, none of the resources that Meriam adults generally exploit on the reef flat while gathering (except possibly *Cypraea*) have post-encounter return rates that are significantly less than the overall foraging return rate. These findings are similar to other quantitative assessments of the prey-choice model among human foragers (e.g., Hawkes et al. 1982, Hill & Hawkes 1983, O'Connell & Hawkes 1981, Smith 1991).

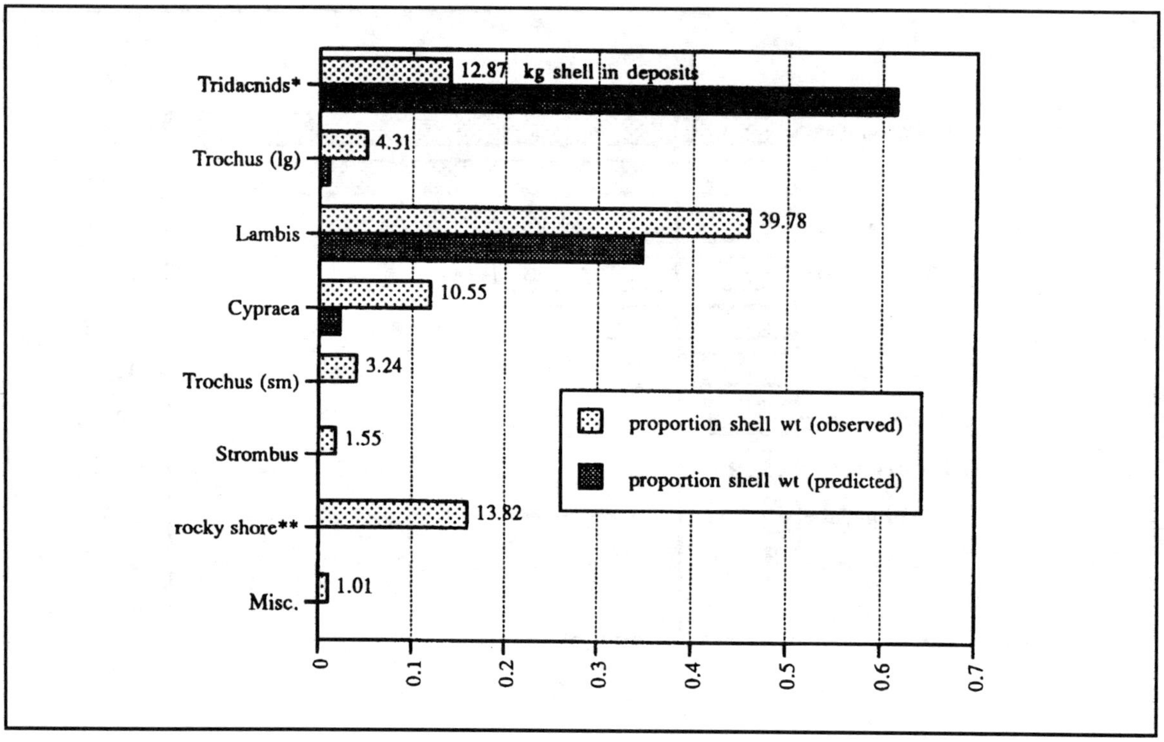

Figure 16.2 Observed frequency of each prey type in the shell deposits vs. frequency predicted by prey choice model for reef flat collecting.

Figure 16.1 also demonstrates a more robust test of the model's predictions. Those resources (again, except possibly *Cypraea*) which are systematically passed over by adults (where SI values are <.5), have on-encounter return rates that are lower than the mean return rate while reef flat collecting. While not enough processing experiments are available to demonstrate statistically significant differences between *e/h* for these resources and average reef flat collecting *E/T*, the differences are substantial.

For small *Trochus* (specimens <200 gm gross weight), the selectivity index is 0 (i.e. small *Trochus* were always passed over by adults), and as expected these small, abundant specimens fall outside the optimal set with on-encounter profitability (*e/h* = 1094.0 ± 54.7, n = 2 experiments) 400 kcal/hr below the average overall RFC return rate.

Tridacna crocea (boring clam) is the most abundant option on the reef, but is associated with an adult selectivity index value of nearly zero. This is anticipated by the model given that the *e/h* (554.1 ± 214.0, n = 3 experiments) in each processing experiment was less than half that of the mean adult *E/T*. On average, extracting a *T. crocea* specimen from its coral matrix took three times as long as it took to harvest a *T. maxima/squamosa* specimen and yielded significantly less flesh. During foraging follows I observed adult foragers collect only one *T. crocea* specimen, which was not deeply embedded in the coral. Harvesting this particular specimen was probably no more expensive than harvesting smaller specimens of *T. squamosa/maxima*.

Strombus luhuanus (black lipped conch) presents an interesting problem. This is the lowest ranked reef flat prey type (*e/h* = 425.9 ± 1.7, n = 3 experiments) and is associated with a harvesting rate that is surprisingly high (nearly that of *Hippopus*) producing an SI index value of .33, almost as high as *Cypraea*. However, the high harvesting rate is the result of only one follow when substantial amounts of this prey type were collected (8.2 kg bulk). This follow included a male forager whose maternal aunt (who usually cooked for the family) did all of the processing upon his return. In this case, the decision to exploit *Strombus* was made irrespective of its processing costs. If I remove this follow from the analysis, the adult collecting rate drops from 5.49 to 1.53 specimens per hour collecting. At that rate, *Strombus* has an SI value of .09, a value well anticipated by its low *e/h* rank.

Also included in Figure 16.1 are measures of post-encounter return rates for prey types in rocky shore resource patches. While Meriam foragers do occasionally harvest rocky shore prey types (*Asaphis violascens* and *Nerita undata*), they never do so when rocky shore patches are encountered *while* reef flat collecting (some rocky shore patches extend right out to the midlittoral). This is predicted by the model given that foragers handling these resources will gain energy at one fifth the rate of that expected from searching for and handling reef flat resources (rocky shore $e/h = 211.3 \pm 8.9$, $n = 13$).

THE ARCHAEOLOGICAL REFLECTION OF PREY CHOICE

Despite the good fit between the prey-choice model predictions and foraging observations, Figure 16.2 demonstrates that predictions 3 and 4 are falsified archaeologically. *Contra* prediction 3, some resources that fall outside the predicted range of prey choice make up a substantial part of the shell deposits; and contra prediction 4, those intertidal prey types that the prey choice model predicts will be taken on encounter are not necessarily found in contemporary deposits in proportion to their abundance on the reef flat.

Especially salient relative to predictions 3 and 4 is the unanticipated lack of *Hippopus* and *Tridacna squamosa/maxima* shells and the prevalence of rocky shore shellfish remains in the deposits. Of the resources predicted to be taken on encounter, only *Lambis* is encountered more often than tridacnids (*Hippopus* and *Tridacna squamosa/maxima*). We would expect that if *Hippopus* and *Tridacna squamosa/maxima* were represented in the deposits in proportion to their abundance, they should make up approximately 20% of the deposits by number of specimens[3]. and over 60% of the deposits by shell weight[4]. And yet combined the tridacnids are obviously underrepresented in the sample: they make up less than 1% of the number of shells and only 15% of the shell weight in the deposits. On the other hand, if we assume fine-grained encounters for all intertidal shellfish, the prey-choice model predicts that rocky-shore resources should not be present in the deposits. As Figure 16.2 demonstrates, *Asaphis* and *Nerita* combined make up 16% of the shell deposits by weight and over 70% by number.

Patch Differences, Field Processing, and Shellfish Delivery

Two explanations to account for these unexpected archaeological patterns are worth investigating. First, foraging strategies may often be conditioned by predictable variability in *intra-patch* return rates (again, where resource patches are defined as sets of prey types with mutually exclusive search times). Second, and potentially more relevant to current trends in zooarchaeological interpretation, foraging goals other than energetic capture rate while foraging may condition processing and transport decisions. I review analyses of these hypotheses only briefly here. More detailed treatments are presented in Bird (1996) and Bird and Bliege Bird (1997).

Patch choice: Although exploiting rocky shore patches on-encounter while reef flat collecting will, *on average*, reduce the reef flat collecting return rate, it does not necessarily follow that rocky shore harvesting will always be a poor option relative to overall intertidal gathering efficiency. If reef flat gains diminish with patch residence time or group composition, we would expect foragers to switch to rocky shore patches when the returns from traveling to and harvesting from the rocky shore are equal to the marginal returns for reef flat collecting, even if reef flat resources are still potentially available (Charnov 1976, Stephens & Krebs 1986: 38-45). Two observations suggest that there are occasions when this might be the case. First, during 6 out of 10 rocky shore harvesting follows, the midlittoral was inundated with 5-10 cm of water, which decreases collectors' ability to spot reef flat prey. When foragers switched patches they either moved *from* the rocky shore to collect reef flat prey when the tide exposed the reef flat (one follow), or moved *to* the upperlittoral after reef collecting when the tide flooded the midlittoral (5 follows). I suspect that if foragers had remained to collect reef flat resources while the tide had continued to rise in the midlittoral, their marginal energetic return rate would quickly have been exceeded by the returns to be gained by moving to the rocky shore resource beds. Data to test this are not yet available but could be obtained with experimental measures of reef flat collecting efficiency at different tidal heights.

Second, on three of the four remaining rocky shore follows that foragers chose to exploit rocky shore patches when the midlittoral was *not* inundated, children below the age of three were present. Only on rare occasions do women participate in reef flat collecting with small children or infants. Doing so entails carrying the child and resource loads while traveling across the reef searching for shellfish; encounter rates and field processing efficiency are presumably reduced as a result. However, rocky shore harvesting is a stationary activity and small children may affect a mother's efficiency less than while reef flat collecting. Furthermore, very small children may also be quite effective at rocky shore harvesting. Testing to see if women's marginal returns while reef flat collecting with small children are less than the expected foraging return rate for traveling to and harvesting from rocky shore patches with their children would require more data on mother-child 'team rates' (*sensu* Hawkes et al. 1995).

Field processing: While gathering in the intertidal, foragers often face load constraints: containers can be easily filled with bulk shellfish within an hour and foragers can quickly exceed a load size that they can effectively carry. In order to continue collecting, foragers can either trade-off foraging time (collecting and transporting more bulk shellfish) with time spent field pro-

Table 16.2 Summary of Predictions and Results from Tests of the Metcalfe-Barlow Central Place Foraging (CPF) Model (Bird & Bliege Bird 1997).

Prey type	number of loads	Z_t (min)	Z_d (meters)	number of loads fully field processed when MTFD exceeded Z_d (matched CPF prediction)	number of loads transported with *no* field processing when MTFD exceeded Z_d (failed CPF prediction)	number of loads transported with *no* field processing when MTFD did not exceed Z_d (matched CPF prediction)	% of loads field processed or transported in bulk as predicted by CPF model	frequency in deposits relative to prey-choice model predictions (see Figure 16.2)
Hippopus	43	2.9 ± .3	71.4 ± 8.5	37	5	1	.88	underrepresented
Tridacna spp. *	30	4.8 ± .8	120.8 ± 18.8	26	3	1	.90	underrepresented
Lambis	52	8.0 ± 1.5	200.8 ± 38.6	22	22	8	.58	proportional
Cypraea	17	18.9 ± 3.9	475.5 ± 97.2	3	6	8	.65	overrepresented (?)
Asaphis	7	96.7 ± 23.9	2418.5 ± 598.8	0	1	6	.86	overrepresented
Nerita	6	214.2 ± 51.2	5355.7 ± 1279.4	0	0	6	1.00	overrepresented

z_t = predicted round trip time beyond which field processing will increase the rate at which a forager can deliver edible flesh to a central locale
z_d = predicted one way distance (walking at 3 km per hour) beyond which field processing will increase the rate at which a forager can deliver edible flesh to a central locale
MTFD = maximum terminal foraging distance
* *Tridacna* spp. includes *T. squamosa* and *T. maxima*

cessing to reduce inedible bulk, or trade-off the benefits of field processing (increases in the utility of a load) by transporting more loads of whole shellfish back and forth from the reef to a central place. Metcalfe and Barlow (1992) provide a solution to this trade-off with the assumption that the goal of foraging is to maximize the rate at which high quality parts can be delivered to a central place where there are no home processing costs. Briefly, their model relates the *benefits* of field processing (increases in the proportion of high quality parts in a load if less useful parts are culled) with the *costs* of collecting and field processing (foraging time). The x-intercept of a line tangent with this cost/benefit function is the round trip travel time beyond which field processing will increase the delivery rate of high quality resource parts. This model is especially important for the analysis of processing and transport strategies where resources have few components and high proportions of low value parts (e.g., inedible shells of shellfish). If we assume that shell has no value and that foragers pay no post-delivery processing costs, measures of the costs (time), benefits (proportional utility), and constraints (load size) while collecting, field processing, and traveling can be used to predict whether different prey types should be transported in bulk or field processed to maximize the rate at which edible flesh can be delivered to a central place. Given that only the shells of molluscs survive in archaeological settings, if some resources are more likely than others to be field processed, this could have important implications for interpreting variability in archaeological shell assemblages.

Table 16.2 shows that the predictions of the Metcalfe-Barlow model quantitatively matched most of the major trends exhibited in the contemporary shell deposits (see Bird & Bliege Bird 1997 for detailed analysis). The round trip travel time beyond which field processing will maximize the delivery rate of *Hippopus* and *Tridacna squamosa/maxima* flesh for adult foragers is only 2.9 ± .3 min. (n = 43 loads) and 4.8 ± .8 min. (n = 30 loads) respectively. If we assume a travel speed of 3 km/hour, foragers need only walk 70-120 meters beyond a central place before field processing will on average increase delivery rate. This is because by field processing these tridacnids the utility of a load can be greatly increased in a very short amount of time. We calculated the field processing threshold for each resource during each adult follow and found that 88% of the *Hippopus* loads and 90% of the *Tridacna squamosa/maxima* loads were either field processed or transported whole as predicted by the model. As expected, the vast majority of the shells of these prey types were discarded at the procurement locale and, consequently, they are underrepresented in the residential deposits.

In contrast, if rocky shore resources are sometimes an effective alternative to reef flat collecting, for the most part foragers cannot walk far enough from a central place on these small islands to make it profitable to field process them. Foragers need to walk between about 2400 - 5400 meters beyond a central place before field processing will maximize the delivery rate of *Asaphis* and *Nerita* flesh (round trip travel time field processing threshold = 96.7 ± 23.9 min., n = 7 loads; and 214.2 ± 51.2 min., n = 6 loads respectively). Ninety-two percent (12 of 13) of rocky shore resource loads collected during focal follows were processed as predicted: foragers never field processed these resources and on only one follow did they walk beyond the point at which field processing could have increased delivery rate. As a result, relative to reef flat resources, rocky shore shells are overrepresented in the residential deposits.

Archaeological Implications

In brief, these results demonstrate the archaeological limitations and utility of simple foraging models. While Meriam foragers select prey on-encounter in a manner anticipated by the prey-choice model, the model does not predict the archaeological consequences of Meriam gathering. The patterns observed archaeologically are more consistent with the goal of maximizing resource *delivery* rate in patchy environments, rather than maximizing resource *capture* rate while foraging in a fine-grained setting. This finding has implications for (1) recent arguments about the role of transport in patterning archaeological bone assemblages, (2) recognizing home bases in the archaeological record, and (3) clarifying the utility of the prey-choice model in archaeological studies.

Interpreting bone assemblages: Much of the debate about Plio-Pleistocene hominid subsistence has centered around interpretations of anatomical part profiles of medium to large bodied mammals from East African faunal assemblages (see O'Connell 1995, 1997 for review). These arguments hinge primarily on analyses of the relationship between the "utility" of resources or resource parts and their probability of being transported from a procurement site to a residential locale. While investigators disagree about whether or not the patterns in many Plio-Pleistocene sites indicate early central place foraging (e.g., Binford 1983 vs. Bunn 1986), there is a general consensus that economic utility, in terms of the nutrients associated with particular anatomical parts (meat utility or modified general utility indices, Binford 1978), anticipates the probability of transport from a procurement locale. The inference is that assemblages dominated by the remains of parts with high proportional

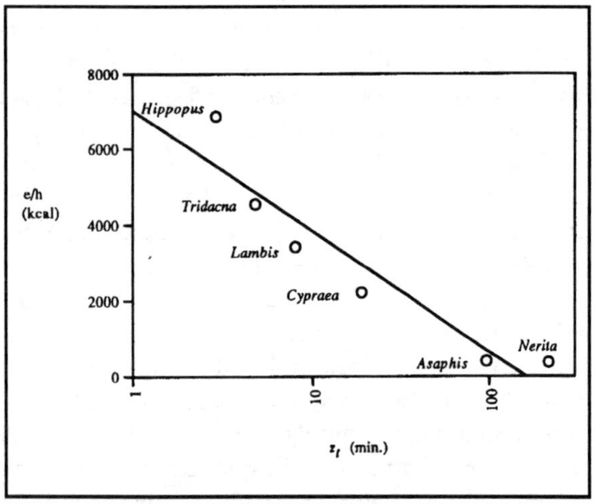

Figure 16.3 Graph of the relationship between post-encounter return rates (e/h, kcal/hr handling) and the round trip travel time beyond which field processing will increase delivery rate of edible flesh (z_t). y = .319LOG(x) + 7025.87, r^2 = 0.887

utility (high ratios of useful to less useful parts) are likely to indicate residential sites to which central place foragers transported resources procured elsewhere, with procurement locales reflecting the opposite pattern (cf. Binford 1981, Blumenschine & Madrigal 1993, Borrero 1990, Bunn & Kroll 1986, Klein & Scott 1986, Monahan 1996, Perkins & Daly 1968, Speth 1983). The results presented here call into question the assumption that such measures of utility, in and of themselves, will always be useful for interpreting variation in faunal assemblages.

The Meriam data demonstrate that, as hypothesized, the probability that less valuable parts will be culled before transport is a function of the relationship between proportional utility, field processing costs, travel distance, and load size (see Jones & Metcalfe 1988, Metcalfe & Jones 1988, and O'Connell et al. 1990 for similar argument). Of special importance for interpretations of faunal assemblage variability will be incorporating measures of the costs of field processing: all other things being equal (travel distance, proportional utility, load size), there will be an inverse relationship between the post-encounter profitability of a resource or resource part and its probability of ending up in a residential midden. Figure 16.3 demonstrates this with a comparison of the predicted point at which field processing will maximize the edible flesh delivery rate of each important Meriam shellfish resource and its post-encounter profitability. Where shellfish types have similar proportional utilities and load sizes are constant, a forager traveling from a central place should cull low quality parts of high ranked shellfish (high *e/h*) sooner than lower ranked shellfish. This is explains why certain high ranked resources are absent from Meriam residential shell deposits.

We would predict the same general relationship for mammal part transport: holding distance constant, resource loads whose proportional utility can be increased substantially with relatively *low* processing costs (high rank, often upper appendicular parts) would be more likely to be field processed and discarded at a butcher site than loads with high proportional utility and *high* field processing costs (often axial parts). These predictions are qualitatively matched by observations among Hadza, Alyawara, and Inuit large-game hunters (O'Connell et al. 1988, 1990, 1992; O'Connell & Marshall 1989; Savelle & Friesen 1996) and should be generally archaeologically applicable in situations where we can evaluate the variables of central place foraging models, especially differential processing costs.

Recognizing home bases: This analysis has further implications for future studies focused on recognizing 'home bases' in the archaeological record. Home bases should become distinct archaeologically when the foraging goal for individuals with load constraints is delivery rate maximization (e.g., delivery of energy to a central place per unit time foraging) rather than resource capture rate (e.g., energy acquired per unit time searching and handling). Foragers can always avoid transport/travel costs by processing and consuming resources on-encounter. If, however, the benefits of transporting resources to a home base make central place foraging an attractive option, resources with high field processing costs that remain in an optimal diet may be overrepresented in residential assemblages relative to more profitable and abundant resources with low field processing costs. Increasing transport distances will exaggerate this bias. It follows that assemblages dominated by high-quality/high-cost remains would be strong indicators of home bases, with procurement locales reflecting the opposite pattern. This hypothesis could be tested against other expectations about the archaeological reflection of central place foraging (e.g., the presence of storage or substantial structures) in regional site samples. The primary lesson here is that hypotheses that make a distinction between these two goals (capture vs. delivery rate) predict important differences in the patterns of subsistence remains.

Identifying patch differences: The predictions from the prey-choice model concerning the types of resources that are likely to be added to intensifying diets provide important insight into relative diachronic shifts in subsistence patterns (e.g., Broughton 1994, Edwards & O'Connell 1995). However, teasing out variability in intensification from differences in *patch* choice will require better control of variability in intra-patch efficiency. The Meriam data suggest that distinguishing between resource patches with mutually exclusive search times will be vital for archaeological interpretations of subsistence remains. For example, if we combine both rocky shore and reef flat resources in an analysis of Meriam intertidal gathering, the encounter-contingent prey-choice model predicts that rocky shore resources should never be pursued on-encounter while reef flat collecting because the mean return rate for reef flat collecting is so high. However, this does not necessarily mean that we should expect rocky shore resources to be absent from shell assemblages: there may be important occasions when foraging efficiency can be increased by avoiding the search costs of reef flat collecting by travel to and harvest prey from rocky shore patches (e.g., regular variability in tidal constraints and mother-child combined foraging opportunities).

Similar patch distinctions are likely to be important in a wide array of circumstances; and where they are, the archaeological signature of foraging will consistently violate the assumptions of encounter-contingent prey-choice models. For diachronic assessments of variability in diet breadth, this result emphasizes the importance of conducting separate analyses for sets of prey that have mutually exclusive search times (see also Smith 1991, Broughton 1995). Under such circumstances *marginal* return rates between patches or 'hunt types' (inclusive of travel costs) should be equalized to maximize foraging efficiency. In some cases, these return rates will be readily obtainable with experimental measures of patch variability (e.g., Barlow & Metcalfe 1996). Recognizing circumstances where these factors (differential transport and resource patchiness) condition variability in assemblage composition will be vital for more comprehensive applications of foraging theory to prehistoric contexts.

CONCLUSIONS

Important mismatches exist between predictions generated from the assumption that Meriam foragers will maximize their energetic capture rate while collecting shellfish in a fine-grained setting and the material consequences of intertidal gathering. These inconsistencies only become apparent when the predictions of quantifiable hypotheses are tested against observed behavior and its material signatures. By demonstrating these mismatches we can isolate more precisely what aspects of behavior and its material correlates require further explanation. Because the hypotheses tested here are generated from a general theory of human behavior, we would expect variation in archaeological products of behavior to be understandable in most circumstances where both behavior and ecological variables are measurable. While measures of prehistoric behavioral variables are never directly observable, in some cases they can be estimated through ethnoarchaeological or experimental quantifications (e.g., Barlow & Metcalfe 1996). For example, field processing and transport decisions should be contingent on travel distance, processing costs, and the proportional value of resources; and strategies that solve trade-offs in the relationship between these variables better than feasible alternatives should be common in a population. Figuring out useful techniques for measuring such variables, whether they are appropriate for anticipating foraging decisions, and whether material patterns reflect patterns in those decisions, all require explicit tests.

This research presents a case for both the importance of theoretically framed hypotheses and ethnoarchaeological observations designed to falsify them. Quantifiable ethnoarchaeological research permits observations of an entire pre-depositional behavioral sequence and its material consequences while theory can guide explanations of variation in the behavior. Models from behavioral ecology are important for archaeology not because they are accurate descriptions of behavioral outcomes, but because they generate predictions that are archaeologically testable under certain conditions. Future applications will move arguments about archaeological inference based on ethnographic observations beyond cautionary descriptions of how complex human behavior *can* be, and toward explanations that can account for its variability. Simply saying that archaeologically significant behavior happens in particular and previously unanticipated ways begs the real anthropological question about what conditions behavioral variability. Answers to this question that rely on explaining patterned behavior only as a product of cultural structure avoid a wealth of hypotheses that can assist in designing explanatory frameworks for archaeological investigation.

ACKNOWLEDGMENTS

I am most grateful for the generosity, kindness, and assistance of the Meriam people, especially the Passi family, our field assistant Edna Kabere, and the Mer Island Community Council. This study could not have been completed without Rebecca Bliege Bird's field expertise, lucid evaluation, and critical review. James O'Connell, Kristen Hawkes, Jack Broughton, Eric A. Smith, and Eric Fagan also enhanced this chapter with insightful comments and discussion. Finally, I thank John Beaton for support throughout this project and Ericka Luckel for assistance in designing and conducting the intertidal resource survey on Mer.

ENDNOTES

1. Energetic values for each prey type were obtained by averaging values of similar or identical species provided by Brand Miller et al. (1993), Sidwell (1981), Thomas et al. (in prep.), and the University of Salford

(1993). Tridacnids (*Hippopus* and *Tridacna* spp.) = 61.7 ± 7.3 kcal/100 gm edible flesh (n = 3 sample sets), *Trochus* = 97 ± 3.k kcal/100 gm (n = 3 sample sets), conch (*Lambis* and *Strombus*) = 82.5 ± 2.5 kcal/gm edible flesh (n = 2 sample sets), cowrie (*Cypraea*) = 80 kcal/100 gm edible flesh (n = 1 sample set), Venus clam (*Asaphis*) = 71.5 ± 3.9 kcal/100 gm edible flesh (n = 14 sample sets), nerites (*Nerita*) = 101.9 ± 4.4 kcal/100 gm edible flesh (n = 8 sample sets).

2. All means provided include only *adult* foragers 16 years or older and are given ± one standard error. Analyses of children's reef flat collecting efficiency are currently in progress. Total handling time and yields for each resource were summed for each follow. Intra-prey type variability in *e/h*, therefore, reflects differences across all follows during which processing took place. Each follow during which processing occurred includes no less than 4 and upwards of 50 processing sequences. For prey types that were rarely processed during follows, experimental processing sequences were used to estimate *e/h*.

3. Calculated as the ratio of *Hippopus* and *Tridacna squamosa/maxima* specimens predicted to be encountered per collecting hour (6.1 specimens, see Table 16.1) to total specimens in the optimal prey range predicted to be encountered per collecting hour (31.5 specimens).

4. Calculated as the ratio of *Hippopus* and *Tridacna squamosa/maxima* shell weight predicted to be encountered per collecting hour (10.13 kg) to total shell weight of all resources in the optimal prey range predicted to be encountered per collecting hour (16.47 kg).

REFERENCES CITED

Barlow, K. R. & Metcalfe, D.
1996 Plant utility indices: two Great Basin examples. *Journal of Archaeological Science* 23: 351-371.

Beaton, J. M.
1972 *The Nature of Aboriginal Exploitation of Mollusc Populations in Southern California.* Masters thesis, University of California, Los Angeles.

Beckett, J.
1987 *Torres Strait Islanders: Custom and Colonialism.* Melbourne: Cambridge University Press.

Bettinger, R., & M. Baumhoff
1982 The Numic spread: Great Basin cultures in competition. *American Antiquity* 47: 485-503.

Binford, L. R.
1978 *Nunamiut Ethnoarchaeology.* New York: Academic Press.
1981 *Bones: Ancient Men and Modern Myths.* New York: Academic Press.

Bird, D. W.
1996 *Intertidal Foraging Strategies Among the Meriam of the Torres Strait Islands, Australia: An Evolutionary Ecological Approach to the Ethnoarchaeology of Tropical Marine Subsistence.* Ph.D. dissertation, University of California, Davis.

Bird, D. W. & R. Bliege Bird
1997 Contemporary shellfish gathering strategies among the Meriam of the Torres Strait Islands, Australia: testing predictions of a central place foraging model. *Journal of Archaeological Science* 24: 39-63.

Bliege Bird, R.
1996 *The Behavioral Ecology of the Meriam Sexual Division of Labor: Gathering, Fishing, and Hunting in a Marine Ecosystem.* Ph.D. dissertation, University of California, Davis.

Bliege Bird, R. & D. W. Bird
1997 Delayed reciprocity and tolerated theft: the behavioral ecology of food sharing strategies. *Current Anthropology* 38: 49-78.

Blumenschine, R. J. & T. C. Madrigal
1993 Variability in long bone marrow yields of East African ungulates and its zooarchaeological implications. *Journal of Archaeological Science* 20: 555-587.

Borrero, L.
1990 Fuego-Patagonian bone assemblages and the problem of communal guanaco hunting. In *Hunters of the Recent Past* (L. B. Davis & B. O. K. Reeves, eds.), pp. 373-399. London: Unwin Hyman.

Botkin, S.
1980 Effects of human exploitation at Malibu Creek, California. In *Modeling Change In Prehistoric Subsistence Economies* (T. K. Earle & A. L. Christenson, eds.), pp. 121-139. New York: Academic Press.

Brand Miller, J., K. James, & P. Maggiore
1993 *Tables of Composition of Australian Aboriginal Foods.* Canberra: Aboriginal Studies Press.

Broughton, J. M.
1994 Declines in mammalian foraging efficiency during the late Holocene, San Francisco Bay, California. *Journal of Anthropological Archaeology* 13: 371-401.
1995 *Resource Depression and Intensification During the Late Holocene, San Francisco Bay: Evidence from the Emeryville Shellmound Vertebrate Fauna.* Ph.D. dissertation. University of Washington, Seattle.

Bunn, H. T.
1986 Patterns of skeletal representation and hominid subsistence activities at Olduvai Gorge, Tanzania, and Koobi Fora, Kenya. *Journal of Human Evolution* 15: 673-690.

Bunn, H. T. & E. M. Kroll
1986 Systematic butchery by Plio/Pleistocene hominids at Olduvai Gorge, Tanzania. *Current Anthropology* 27: 413-452.

Bunn, H. T., L. Bartram & E. Kroll
1988 Variability in bone assemblage formation from Hadza hunting, scavenging, and carcass proocessing. *Journal of Anthropological Archaeology* 7: 412-57.

Charnov, E. L.
1976 Optimal foraging: the marginal value theorem. *Theoretical Population Biology* 9: 129-136.

Cuthill, I. & A. Houston
1997 Managing time and energy. In *Behavioural Ecology: an Evolutionary Approach* (J. R. Krebs & N. B. Davies, eds.), pp. 97-120. Oxford: Blackwell.

Dawkins, R.
1982 *The Extended Phenotype.* Oxford: Oxford University Press.

Earle, T. K. & A. L. Christensen (editors)
1980 *Modeling Change in Prehistoric Subsistence Economies,* New York: Academic Press.

Edwards, D. & J. F. O'Connell
1995 Broad spectrum diets in arid Australia. In *Transitions: Pleistocene to Holocene in Australia and Papua New Guinea* (J. Allen & J. F. O'Connell, eds.). For *Antiquity* 69:769-783 (Special No. 265).

Emerson, A. M.
1993 The role of body part utility in small scale hunting under two strategies of carcass recovery. In *From Bones to Behavior: Ethnoarchaeological and Experimental Contributions to the Interpretation of Fauanal Remains* (J. Hudson, ed.), pp. 138-155. Carbondale: Occasional Paper No. 21, Center for Archaeological Investigations, Southern Illinois University.

Glassow, M. & D. Wilcoxon
1988 Coastal adaptation near Point Conception, California, with particular regard to shellfish exploitation. *American Antiquity* 53: 36-51.

Grafen, A.
1991 Modelling in behavioural ecology. In *Behavioural Ecology: An Evolutionary Approach* (J. R. Krebs & N. B. Davies, eds.), pp. 5-31. Oxford: Blackwell.

Haddon, A. C., editor
1908 and 1935 *Reports of the Cambridge Anthropological Expedition to Torres Straits. Sociology, Magic and Religion of the Eastern Islanders, vol. VI. General Ethnography, vol. I.* Cambridge: Cambridge University Press.

Hawkes, K.
1993 Why hunter-gatherers work: an ancient version of the problem of public goods. *Current Anthropology* 34: 341-361.

Hawkes, K., J. O'Connell & N. Blurton Jones
1995 Hadza children's foraging: juvenile dependency, social arrangements and mobility among hunter-gatherers. *Current Anthropology* 36: 688-700.

Isaac, G. L.
1984 The archaeology of human origins: studies of the Lower Pleistocene in East Africa, 1971-1981. *Advances in World Archaeology* 3: 1-87.

Jones, K. & D. Metcalfe
1988 Bare bones archaeology: bone marrow indices and efficiency. *Journal of Archaeological Science* 15: 415-423.

Jones, K. T. & Madsen, D. B.
1989 Calculating the cost of resource transportation: a Great Basin example. *Current Anthropology* 30: 529-534.

Klein, R. G. & K. Scott
1986 Re-analysis of faunal assemblages from the Haua Fteah and other late Quaternary archaeological sites in Cyrenaican Libya. *Journal of Archaeological Science* 13: 515-542.

Krebs, J. R. & N. B. Davies
1993 *An Introduction to Behavioural Ecology*. Oxford: Blackwell.

Krebs, J. R. & N. B. Davies (editors)
1997 *Behavioural Ecology: An Evolutionary Approach.* Oxford: Blackwell.

Krebs, J. R., A. I. Houston, & E. L. Charnov
1981 Some recent developments in optimal foraging. In *Foraging Behavior: Ecological, Ethological, and Psychological Approaches* (A. C. Kamil & T. D. Sargent, eds.), pp. 3-18. New York: Garland.

Levins, R.
1966 The strategy of model building in population biology. *American Scientist* 54: 421-431.

Maynard Smith, J.
1978 Optimization theory in evolution. *Annual Review of Ecology and Systematics* 9: 31-56.

Metcalfe, D. & K. R. Barlow
1992 A model for exploring the optimal trade-off between field processing and transport. *American Anthropologist* 94: 340-356.

Metcalfe, D. & K. Jones
1988 A reconstruction of animal body-part utility indeces. *American Antiquity* 53: 486-504.

Monahan, C. M.
1996 New zooarchaeological data from Bed II, Olduvai Gorge, Tanzania: implications for hominid behavior in the Early Pleistocene. *Journal of Human Evolution* 31: 93-128.

O'Connell, J. F.
1995 Ethnoarchaeology needs a general theory of behavior. *Journal of Archaeological Research* 3: 205-255.
1997 On Plio/Pleistocene archaeological sites and central places. *Current Anthropology* 38: 86-87.

O'Connell, J. F. & B. Marshall
1989 Analysis of kangaroo body part transport among the Alyawara of central Australia. *Journal of Archaeological Science* 16: 393-405.

O'Connell, J. F. & K. Hawkes
1981 Alyawara plant use and optimal foraging theory. In *Hunter-Gatherer Foraging Strategies: Ethnographic and Archaeological Analysis* (B. Winterhalder & E. A. Smith, eds.), pp. 99-125. Chicago: University of Chicago Press.

O'Connell, J. F., K. Hawkes, & N. Blurton Jones
1988 Hadza hunting, butchering, and bone transport and their archaeological implications. *Journal of Anthropological Research* 44: 113-61.
1990 Reanalysis of large animal body part transport among the Hadza. *Journal of Archaeological Science* 17: 301-316.
1997 Patterns in the distribution, site structure and asemblage composition of Hadza kill-butchering sites. *Journal of Archaeological Science* 19: 319-345.

Orians, G. H. & N. E. Pearson
1979 On the theory of central place foraging. In *Analysis of Ecological Systems* (D. J. Horn, R. D. Mitchell, & G. R. Stairs, eds.), pp. 154-177. Columbus: Ohio State University Press.

Perkins, D. & P. Daly
1968 A hunter's village in Neolithic Turkey. *Scientific American* 219: 97-106.

Perlman, S. M.
1980 An optimum diet model, coastal variability, and hunter-gatherer behavior. In *Advances in Archaeological Method and Theory, vol. 3* (M. B. Schiffer, ed.), pp. 257-310. New York: Academic Press.

Raab, L. M.
1992 An optimal foraging analysis of prehistoric shellfish collecting on San Clamente Island, California. *Journal of Ethnobiology* 12: 63-80.

Rose, L. & F. Marshall
1996 Meat eating, hominid sociality, and home bases revisited. *Current Anthropology* 37: 307-338.

Savelle, J. & T. Friesen
1996 An odontocete (Cetacea) meat utility index. *Journal of Archaeological Science* 23: 713-721.

Sharp, N.
1993 *Stars of Tagai: The Torres Strait Islanders.* Canberra: Aboriginal Studies Press.

Sidwell, V. D.
1981 *Chemical and Nutritional Composition of Finfishes, Whales, Crustaceans, Mollusks, and Their Products.* National Marine Fisheris Services, NOAA technical memorandum.

Simms, S. R.
1992 Ethnoarchaeology: obnoxious spectator, trivial pursuit, or the keys to a time machine. In *Quandries and Quests: Visions of Archaeology's Future* (L. Wandsnider, ed.), pp. 186-198. Carbondale: Center for Archaeological Investigations, Southern Illinois University.

Smith, E. A.
1987 Optimization theory in anthropology: applications and critiques. In *The Latest of the Best: Essays on Evolution and Optimality* (J. Dupré, ed.), pp. 201-249. Cambridge, MA: Bedford Books.

1991 *Inujjuamiut Foraging Strategies: Evolutionary Ecology of an Arctic Hunting Economy.* New York: Aldine de Gruyter.

Smith, E. A. & B. Winterhalder, eds.
1992 *Evolutionary Ecology and Human Behavior.* New York: Aldine de Gruyter.

Speth, J. D.
1983 *Bison Kill and Bone Counts: Decision Making by Ancient Hunters.* Chicago: University of Chicago Press.

Stephens, D. W. & J. R. Krebs
1986 *Foraging Theory.* Princeton: Princeton University Press.

Thomas, F. R.
1994 Of molluscs and women: foraging strategies in the central Gilbert Islands, Kiribati. *American Conchologist* 22: 21-23.

Thomas, F. R., H. Ako, D. Okuda, & R. Panguirigan
n.d. Prey choice and prey management: the case of mollusks in the Central Gilbert Islands, Kiribati. *Ecology of Food and Nutritio* (in prep.)..

University of Salford
1993 *Microdiet System.* London: Royal Society of Chemistry.

Part V
Epilogue

17

Rediscovering Darwin

G.A.Clark & C.M. Barton
Arizona State University

American social scientists fear and despise biology, although few of them have troubled to learn any (Daley & Wilson 1988: 154).

Stated most broadly, *Rediscovering Darwin* is about the largely unrealized potential for neo-Darwinian evolutionary theory (NDT) to serve as a paradigm for all kinds of archaeological research, from that of early hominids, to that of the relatively recent time frames with which New World workers are accustomed to deal. The book seeks to demonstrate at the levels of 'high' and 'middle range' theory (Parts II, III), and at the level of case studies and applications (Part IV), our shared conviction that NDT and its modern derivatives constitute the only generally acknowledged conceptual framework within which to understand human biological and cultural evolution, 'writ large' or 'small'. More precisely, evolutionary archeologists argue that Darwinian natural selection (and other NDT mechanisms and processes) act on human behavior and influence the material products of that behavior. By focusing on the requirements of NDT at the theoretical and methodological levels, the essays assembled here go some way toward creating a novel, coherent framework for explaining variation in the archaeological record.

An explicitly evolutionary archeology is the product of western science, and, at the level of the metaphysical paradigm, western science has subscribed to materialist biases, assumptions and preconceptions since it assumed its modern form in the last quarter of the 19th century (Clark 1993, 1997). Since prehistoric archaeology has aspired to be a 'science-like' endeavor over much of its existence as a relatively formalized intellectual discipline, and since NDT is the quintessential materialist paradigm, it would seem natural for archaeologists of whatever persuasion to incorporate Darwinian concepts and models in their research. So far, however, this has not happened, although there are indications that a synthetic evolutionary paradigm for archaeologists is beginning to emerge in the literature (see, e.g., Teltser 1995, Maschner 1996). *Why* it hasn't happened in the Americanist research tradition has been explored by Robert Dunnell (e.g., 1980, 1982, 1989; see also Lyman & O'Brien, Chapter 2) who suggested that modern efforts to make archaeology a materialist discipline must contend with the neo-evolutionary legacy of Leslie White (1959), Marshall Sahlins and Elman Service (1960, Service 1963), and with how this anthropological paradigm was interpreted, integrated and adopted by the emergent 'new archaeology' after the mid-1960s. Dunnell argued that these approaches derive from 'Spencerian' (or cultural) evolution, with its teleological, orthogenetic notions of progress, human intentionality, and essentialist biases. A Spencerian concept of evolution was adopted because of the prevailing misconception that humans, as cultural animals, were somehow 'unique' in the natural world, and that two kinds of evolutionary theory - one biological, the other cultural - were required to account for change in organisms and cultures respectively. It has become evident to even the most casual observer that humans have become a lot less 'unique' since the 1960s (e.g., Foley 1987, 1995).

DARWIN REVISITED

Darwin was aware of, and had to confront, these problems. In the England of the late Victorian era, the concept of 'evolution' was used in two ways: (1) as adaptation to local environments (via Darwin, explained by the theory of natural selection), and (2) as a notion of progressive complexity (via Herbert Spencer, not explained by natural selection). Darwin avoided the term because of the enormous influence of Spencer, for whom human behavior was rational and goal directed. Biologically, culturally, artistically, in terms of moral and ethical development etc., humans were supposedly striving for perfection, with the end result a "perfect human society" (1883). Although not yet achieved, the highest rung on the ladder of development so far attained was supposedly occupied by Victorian England. Darwin actively (and usually unsuccessfully) sought to avoid conflation of his ideas with those of Spencer, and seldom used the term evolution in either *The Origin of Species* (1859) or *The Descent of Man* (1871), preferring instead the more precise but cumbersome 'descent with modification'. This confounding of progressivism and the Darwinian notion of natural selection is with us yet and shows up in many aspects of modern public life (e.g., prescientific views of race and ethnicity, the abortion debate etc.).

In addition to Spencer, Darwin had to contend with the lingering influence of William Paley's 'Christian Evidences' movement, which, during the first half of the 19th century, aggressively sought to restore faith in an ordered and quasi-rational universe based on the concept of a divine creator and what is usually referred to as the 'argument from design'. Most fully elaborated in his *Natural Theology*, the argument from design was founded on the notion that the existence of god (and his omniscience, benevolence, intelligence etc.) is observable in the good design of organisms and the harmony of ecosystems, and that an orderly natural world could only be explained by invoking an omniscient creator (Paley 1802). The many varieties of plants and animals, including humans, were regarded as the direct, immutable and special products of divine creation. Darwin acknowledged the (relatively) good design of organisms and the fact that ecosystems were ordered phenomena, but, as early as 1842 (in *The Notebooks*), recognized that the theory of natural selection constituted an alternative and far more powerful explanation for the apparent order in the natural world than the argument from design, and that it undermined the latter's credibility. He proposed instead that the order apparent in the natural world arose as an accidental consequence of a universal striving for individual reproductive success - there was no principle or intelligence explicitly producing that order. Thus natural selection is both purposeless (relative to the notion of a divine creator or some quasi-absolute moral meaning), and directionless (relative to orthogenetic notions of progress).

Natural selection follows from three irrefutable facts: (1) all species produce more offspring than can possibly survive, (2) there is variation within a species, and (3) there is some principle of inheritance of that variation (which we now know to be genetics). From these facts, the theory of natural selection follows as a direct inference: if all organisms produce more offspring than can possibly survive, on average those organisms that are fortuitously better adapted to changing local environments will tend to produce more offspring than those that are not so well adapted, and the population average will tend to shift in their direction (Gould 1994).

Darwin got the idea of natural selection from Adam Smith's "invisible hand" transferred from the economic to the natural world. In *The Wealth of Nations*, Smith proposed that the way to get the best-functioning economy was not to attempt to plan or design it at all (whence *laissez-faire* capitalism). In a free economy, he argued, individuals would struggle to maximize personal profit, and those who did well would simply drive out the competition. The basic idea is that spontaneous order and harmony would result without intentional design (Smith 1776). Darwin was much influenced by Smith (1776) and by Thomas Malthus (1798), whose ideas - taken together - constituted a categorical refutation of the Enlightenment's faith in progress, but he didn't publish his theory earlier in part, at least, because he was reluctant to confront the enormous implications of natural selection for prevailing popular and scientific notions of humans and their place in the universe. Natural selection is a materialist theory that ran counter to the notion of dualism - the classical idea (Plato, Aristotle) remodelled and rearticulated in the context of a philosophy of scientific inquiry by the Enlightenment philosopher and mathematician René Descartes (1596-1650), that the universe was divided into the material stuff of everyday existence (matter) and intangible (mental or spiritual) stuff (of which the spirit, since it was more closely related to god, was the higher of the two forms). *Materialism holds that there is only matter,* and that what we tend to think of as 'spirit' or 'mind' is simply a manifestation of matter arranged in complex ways. Put another way, the mind is an accidental consequence of brain evolution (i.e., it is a product of the material sub-strate of the neurology of the brain). This is a radical idea even today, but it was so radical in the mid-19th century that Darwin resisted exploring its implications (although he was clearly aware of them) (Gould 1994).

THE EXTENDED PHENOTYPE

Basic to the position adopted in this book is the notion of the generalized or extended phenotype - the idea that the material remains of human behavior are as much aspects of the human phenotype as are the observable features of human biology (O'Brien & Holland 1995). Quantitatively, the archaeological record is immensely richer than that of fossil hominids, and in its later manifestations, effectively holds biological macroevolution 'constant', so it could be argued that it constitutes a better basis for building hypotheses about human evolutionary history than does the exceptionally 'coarse-grained' time-space grid of the human fossil record (even though the latter is relatively uncomplicated by 'culture'). The notion of the extended phenotype is a decidedly 'unradical' idea when it is noted that most evolutionary biologists routinely include things like spiderwebs and birds' nests in their concept of phenotype (e.g., Dawkins 1990). Of course, and without exception, New World archaeologists deal exclusively with *Homo sapiens sapiens*, so there is an enormous component of learned behavior which acts to generate phenotypic variety, manifest in the vast quantities of material remains humans manufacture, modify, lose and discard. Selection operates on the behaviors that produced this mountain of clutter, and culture, construed here as learned behavior, simply constitutes part of the human phenotype (just as it is part of the phenotypes of chimpanzees and bonobos). It is important to make clear, however, that selectionist approaches undertaken under the rubric of evolutionary archaeology have relatively little to do with ethnographic analogy, which was (and continues to be) the main focus of Schiffer's 'behavioral archaeology' (e.g., 1987).

Evolutionary Processes

Although the contributors to *Rediscovering Darwin* acknowledge the concept of the extended phenotype as the vehicle through which a Darwinian archaeology is operationalized, there are different construals of the mechanisms by which variation is transmitted, and a difference of opinion regarding the necessity for cultural selectionist approaches. The papers by Bettinger and Eerkens (Chapter 10) and Michael Shott (Chapter 11) adopt what is probably the best known of these, Boyd and Richerson's dual inheritance theory (1985), as a conceptual framework for modeling aspects of socially-transmitted behavior that affected aspects of form and style in Archaic projectile points. Adopting a Dunnell-like perspective, however, Lyman and O'Brien (Chapter 2) and Neff and Larson (Chapter 5) see no necessity for parallel conceptual frameworks like DIT, arguing that, since behavior is subsumed under the concept of the extended phenotype, selection operates on it just as directly as it does on somatic features. This issue is not resolved, although, guided by the conviction that selection is blind to the source of variation and even the mode of transmission, a tendency to subsume cultural under natural selection is perhaps discernible.

It is worth remarking that DIT has recently come under critical scrutiny by evolutionary biologists. Sober (1991) suggests that the adoption of evolutionary models by the social sciences involves borrowing the most unified and complete parts of evolutionary theory, leaving behind concepts about which there is less than full consensus. However problematic some of its applications, population genetics is really a unifying framework within evolutionary biology. It has no counterpart in DIT. The models of transmission systems upon which Boyd and Richerson (1985) focus their attention describe the quantitative *consequences* of systems of cultural influence, rather than the causes. Learning in a social context is a cultural universal (at least in regard to modern humans), and when patterns of social learning vary in the range and speed of transmission, it is the system of transmission (vertical, horizontal, oblique) that determines the rate of change and the end-state that the population achieves. To Sober, this means that models of cultural evolution have rather limited general applicability, and are suitable only for describing evolution *within* a cultural system, rather than the evolution of the cultural transmission system itself. A virtue is that they place the biological and cultural aspects of the phenotype in a common conceptual framework, so that the relative contributions of each to some eventual outcome can, in theory, be determined. Dual inheritance theory has also attracted the attention of a number of British workers interested in the archaeology of human ancestry (see, e.g., Steele & Shennan 1996).

In our opinion, the jury is still out on the necessity for and credibility of efforts to develop a cultural evolutionary theory under the aegis of NDT (e.g., Boyd and Richerson's dual inheritance theory [1985]). DIT combines aspects of biological and cultural evolution into a synthesis with considerable explanatory potential. It uses a generalized concept of information as a currency, and recognizes a fundamental distinction between genetic and cultural modes of information transmission. Although on the metaphysical level, it situates itself squarely within the conceptual framework of NDT, it integrates it with a parallel process of culturally-based learning that is argued, nevertheless, to be genetically mediated. The result is a theory with considerable predictive power that can be used to explain a wide variety of phenomena usually considered to be 'cultural' (e.g., the origins of symbol systems, stylistic variation, cooperation etc.).

DIT is based on three important principles: (1) there is a multiplicity of sources of information about culture, (2) cultural transmission of information can be 'horizontal', 'vertical' and 'oblique' (Boyd & Richerson 1992), and (3) cultural information is learned. Like biological transmission, cultural transmission (or inheritance) is conditioned by and responds to the environment. In addition to random variation in the content of culture generated by error and an analogue to genetic drift, three important forces generated by the interaction between humans and the environment are:

(1) *guided variation* - learning by trial-and-error, direct experience, rational evaluation and other means by

which individuals evaluate experience and change or maintain their behaviors.

(2) *biased transmission* - a learning process with three main variants:

(i) *direct bias* - when an individual adopts some cultural variant based on direct knowledge of the variant itself.

(ii) *frequency dependent bias* - doesn't require direct experience of a cultural variant for its adoption, but instead assumes that the trait will or will not be adopted based on its frequency (commonness or rarity) within the group.

(iii) *indirect bias* - learning in which cultural traits are adopted because they are perceived to be attractive on some criterion of 'attractiveness' (e.g., they connote prestige, status, wealth, power).

(3) *natural selection,* which can operate directly on culture.

Some cultural variants enhance an individual's survival within a particular environmental context - these variants, through time, can become 'fixed' within a group (see also Wilson 1997a, b). They may or may not be beneficial to individual survivorship, and can lead to behaviors that are maladaptive in terms of an individual's 'genetic' fitness while enhancing that individual's 'cultural' fitness.

Although not subjected to detailed critical scrutiny, we can detect nothing in DIT that stands in contradiction to NDT tenets as broadly construed. *Modern* humans, at least, are the quintessential cultural animals, so it would seem to make sense to model culturally-based learning as precisely as possible. So long as there is a rich archaeological record comprising the material remains of the extended phenotype, DIT might allow us to partition the sources of information about culture and the mechanisms for its transmission more effectively. However, as we suggested in Chapter 1, selection and other recognized evolutionary processes might well be blind to the mode of transmission in their fundamental operation. If so, this would obviate the need for DIT as conceptualized by Boyd and Richerson (1985). Processes they regard as distinct and specific to human cultural change, including guided variation and biased transmission, would be better subsumed under expanded concepts of selection, drift, information (including gene) flow, and so forth.

Finally, the appropriateness of DIT for tracking adaptation in 'deep time' is, in our view, deeply problematic. In addition to probably insurmountable problems with monitoring the extended phenotype in the remote past, there is no particular reason to think that the learned component of human behavior was especially significant until after the appearance of *Homo erectus,* early in the Lower Pleistocene. For practical reasons, then, it would make little sense to adopt an approach that depends upon an archaeological record with exceptionally high contextual resolution and integrity and then try to apply it to pre- and protocultural contexts wherein the record is also exceptionally 'coarse-grained'.

DUNNELL ON THE MATERIALIST PARADOX, UNITS AND SCALE

For almost 30 years, Robert Dunnell (e.g., 1971, 1980, 1986, 1995) has been concerned with the failure by archaeologists of all persuasions to address, make explicit, and/or rethink their observational languages or systematics (see also Clark [1991, 1993; Clark & Lindly 1991]). As most of us will readily acknowledge, there is a tendency to use analytical units either for convenience and/ or because of convention, without due consideration of how they are related to a phenomenon or process of interest (see Clark [1988, 1989] for a similar problem in human paleontology). While this fundamental epistemological issue is hardly confined to archaeology, it is perhaps more problematic in a discipline that has been little concerned with the inferential basis for its knowledge claims, and in which the peculiar constraints of CRM play such an important role. In any event, greater attention should clearly be directed toward the logic behind the analytical and conceptual units used in archaeological research, and how they articulate with one another. This is all the more worrisome in a nascent evolutionary archaeology, supposedly grounded in what is probably the most powerful explanatory paradigm ever developed in western science. If we want it to 'work', then we had better confront the fact that we have no underlying conceptual framework of the power and generality of genetic models in biology.

There is relatively little explicit consideration of, and certainly no consensus in *Rediscovering Darwin,* as to what should constitute the appropriate analytical and conceptual units of evolutionary archaeology (however, see Chapters 2, 5). Dunnell points out that archaeological research conducted under the aegis of neo-Darwinian evolutionary theory poses special problems in this regard (1995). On the one hand, NDT is the quintessential materialist paradigm.

> From a materialist perspective, everything is in the process of becoming - the relation we conceptualize as change . . . On the other hand, application of evolutionary theory requires units - things upon which selection can act, things that vary, evolve, or are adapted. It is thus apparent that materialism greatly complicates the construction of a language of observation (1995: 34).

But units are, by definition, 'kinds', and 'kinds' cannot possibly evolve. Units are essentialist constructs pressed into service in support of a materialist ontology, so adoption of an evolutionary perspective would seem to be at odds with their use. This conflict between paradigm at the level of the metaphysic (i.e., evolution's ontological materialism) and paradigm at the sociological and methodological levels (i.e., evolution's epistemological essentialism, which requires us to use constructs like types) is sometimes referred to as 'the materialist paradox'. Acknowledging that the materialist paradox is a problem in all science, how do we escape it here if we wish to develop an evolutionary archaeology that goes beyond the nominal invocation of evolutionary concepts and terms as "just another set of interpretive algorithms" (Dunnell 1995: 33)?

There is no easy answer to this question. Apart from Dunnell's work, there is so little in the literature that specifically addresses the nature of archaeological analytical and conceptual units that we can do little more than reiterate the main points of his arguments (see Dunnell [1980: 35-99] for an exegesis of the relationship between NDT and archaeology). Briefly, Dunnell contends that if we want to make evolutionary archaeology a truly scientific endeavor, we must pay greater attention to the systematics that underlie our logic of inference. Defined very broadly, evolution is a framework for explaining change as the differential persistence of variability. To operationalize a particular research program making full use of the powerful conceptual and methodological arsenal of NDT, we must distinguish units - things upon which selection can act etc. - along the dimensions of 'location' and 'scale' (Dunnell 1986).

Empirical and Theoretical Units

In regard to 'location', we should ascertain whether the unit is *empirical* (i.e., phenomenological, directly observable, lodged in the world of sense experience), or *theoretical* (i.e., ideational, relating to what goes on in the head of the investigator). Empirical units are conceptualized as naturally bounded phenomena which have specifiable, functional properties in relation to other such phenomena. Empirical units are the things we observe, measure and study. Examples are various kinds of objects (e.g., pots, points), and entities both above (e.g., activity areas, features, occupation surfaces, sites) and below (e.g., attributes like paste, temper; minerals, trace elements) the level of objects. Although in some cases intuitively obvious and therefore relatively uncontroversial (e.g., points), all empirical units require explicit criteria for their recognition - either a set of properties shared by group members or an algorithm for assigning group membership. In either case, the referent for group construction is observable and phenomenological - it pertains to the real world of sense experience. The strategy for group construction is inductive (Dunnell 1986). Dunnell (1978) would maintain that empirical units with functional properties are those that affect Darwinian fitness in the populations in which they occur, whereas stylistic features exhibit no detectable selective values. Although it is not always easy to distinguish one from the other, traits under selection should exhibit a distinctive frequency dependent distribution over time, showing an origin point, an increase in frequency, a plateau (reflective of some optimal value) and, eventually, a decline (indicating selection against the trait resulting from changes in the selective environment) (O'Brien & Holland 1992, 1995). Selectively neutral features would be subject to stochastic variation and would presumably have a more or less equal probability of increasing or decreasing in frequency over any time interval (Teltser 1995). In any event, they would not exhibit the characteristic sigmoid curves of variants under selective pressure. In short, empirical units expressing functional properties are our best bet if we hope to track selection trajectories and thus monitor adaptation over time. An interval or ratio scale of measurement would also seem to be required because nominal and ordinal scaling are probably too 'coarse-grained' to allow us to identify functional features accurately (see below).

Empirical units are constrasted with theoretical units, best conceptualized as measurement scales (nominal, ordinal, interval, ratio). Theoretical units - sometimes called 'instrumentalities' - exist only in the mind of the investigator; they have no empirical basis. Although use of higher scaling levels has increased (especially, and interestingly, in research oriented toward aspects of 'style' - Carr & Neitzel 1996), partly because of the quantification of research protocols in the 1970s, nominal scale measurement still dominates in archaeology, because it is the basis for the use of the other scales, because of time and money constraints (esp. evident in CRM contexts), and because the nearly universal inconsistencies in archaeological data recording force lower-order scaling on many archaeological data sets (Clark 1982, Clark & Stafford 1982).

Dunnell points out that while some measurement scales are "arbitrary dissections of a particular dimension," limited only by the precision of the measurement instrument (e.g., length), many others are more elaborate, theory-laden classifications, often misconstrued or confounded with empirical units (1986: 152). The important point here is that theoretical units like classification or measurement scales are *imposed upon*, rather than extracted from, empirical reality. They provide concepts by which we identify, describe, measure and compare phenomena; they allow us to model 'reality' in particular ways for particular purposes (Dunnell 1986). Theoretical units resolve the materialist paradox by accommodating continuity as change in the frequency distributions of variables recorded on empirical types. Using theoretical units above the level of nominal scaling allows us to treat variability in a truly quantitative fashion, rather than as a discrete set of nominal categories. The solution to the materialist paradox hinges, Dunnell would contend, on distinguishing empirical from theoretical units, and on employing the latter to describe and explain change (1995: 35). He suggests that archaeologists usually don't keep the two kinds of units conceptually separated, and that confusion almost inevitably arises because of formal similarities between some measurement scales (esp. nominal) and some empirical entities (1986: 153).

Issues of Scale

Dunnell's second preoccupation is with the 'goodness of fit' (or, more accurately, the lack thereof) between the scale at which units are formed and the scale at which evolutionary processes of interest operate (1980, 1995). In brief, a process operating at one scale can have effects at both higher and lower scales, and analytical units designed for one scale might not be appropriate if applied to another. This is an issue that has been much discussed in evolutionary biology. In the classical expression of new synthesis biology, selection and other evolutionary processes operate at the scale of the individual organism. Some evolutionary approaches to archaeology in 'deep time' also take a 'methodological individual' to be the unit upon which selection operates (e.g., Mithen 1990). Species, populations and other aggregates of organisms upon which the *effects* of selection can be detected are viewed as epiphenomenal. Although there are clearly problems with this (cf., e.g., Rensch [1959], Mishler & Donoghue [1982], Wilson [1989; 1997a, b]), most biologists would agree that the characteristics of populations, species and other higher order theoretical units must be reducible to, and understood in terms of, effects operating on the individual organism. Although it is a common circumlocution to speak of 'the evolution of species', what one is really talking about is not a process that operates directly on species, but rather the results of a process that operates on organisms. Individuals are subject to selection pressure, for example, but species evolve. This is one reason why, in discussions of species concepts, biologists feel more comfortable talking about the evolution of demes, or local breeding populations, in which organisms actually have some probability of interacting with one another, and can so be construed as analytical units, rather than species, where that assertion is more problematic (Dunnell 1995: 35,36).

Unlike the question of appropriate units, the correspondence between the scale of the processes or mechanisms of interest, and their empirical referents, has a long history in Americanist archaeology (was, in fact, a central tenet of processualism - e.g., Plog [1974]). A number of papers in *Rediscovering Darwin* specifically address this issue (e.g., and especially Neff and Larson's contribution [Chapter 5]). However, the time is long overdue to begin to take it more seriously.

POTENTIAL PROBLEM AREAS

Because of the fad-like character of the discipline, and the absence of a unifying framework like genetics in biology, there is a danger that an emergent evolutionary archaeology might go the way of the other discarded paradigm-like formulations that litter its intellectual landscape (e.g., culture history, processualism). There is also the likelihood that the approach will be more or less ignored by the overwhelming majority of American archaeologists who work in CRM contexts, typically little concerned with the inferential basis for knowledge claims. On the other hand, the explanatory power and reach of NDT as a truly scientific theory has been established for five decades in the life sciences. With time-honored but vacuous distinctions between humans and other animals collapsing every day, archaeologists would be well-advised to consider NDT, even if *how* it might be applied to particular kinds of archaeological problems is still a matter for discussion.

In an epilogue chapter analogous to this one, O'Brien and Holland (1995: 175-200) identify what they consider the major pitfalls that await the would-be evolutionary archaeologist. There are basically two of them - quantified essentialism and adaptationist accommodative argument. They suggest that even those evolutionary archaeologists who accept NDT at the level of the metaphysic can come to grief over these essentially methodological issues which arise from an incomplete understanding of evolutionary process, and the tendency of most workers to stick to the comfortable analytical categories and research protocols with which they are familiar.

Quantified Essentialism

By quantified essentialism, O'Brien and Holland (1995) mean the unthinking use of modal 'types' created on the basis of quantified attribute analyses but subsequently lumped together (and thereby nominalized or ordinalized) to produce new types predicated on measureable, albeit arbitrary, divisions of continuous frequency distributions. There is a lot of quantified essentialism in archaeology (and in paleoanthropology - see Clark [1988, 1989], Brace [1988, 1989]), and there is nothing inherently wrong with it so long as it is directed toward the analysis of the appropriate units. The appropriate units, they would argue, are stylistically-based types. There are 'types of types', of course, and stylistically-based types, while ill-suited for understanding the effects of selection, are well-suited to monitor rates of trait transmission (Teltser 1995). The appropriate use of quantification under an NDT conceptual framework presupposes a clear understanding of the differences between stylistic and functional features (see above) because it is the latter which allow us to track selection trajectories through time and which affect the Darwinian fitness of the populations in which they occur. Measurement at the level of interval or ratio scaling is a prerequisite for tracking change in functional traits, they argue, rather than nominal scaling built around modes.

While we are in substantial agreement with O'Brien and Holland's observations about quantified essentialism and the problems that it can cause in archaeology, it is worth remarking that an argument almost diametrically opposed to this one has been made recently by Foley and Lahr (1997) who, using a cladistic approach to Old World lithic technologies in 'deep time', suggest that there is a strong correlation between the evolution and spread of morphologically modern humans and Sir Grahame Clark's Mode 3 Technologies (Middle Stone Age/Paleolithic) (1977). These workers treat the emergence of Clark's modes as technological apomorphies (unique derived features) which can be arranged, according to the principles of phylogenetic systematics, into a branching sequence or cladogram (see also Foley [1987]). The function/style dichotomy is considered unimportant.[1] At each branching point in the cladogram a new trait is added (e.g., Levallois technique, blades), but at the same time the more ancient ones (e.g., bifaces) do not disappear. In cladistic terms they become plesiomorphies (primitive retentions); plesiomorphies do not provide phylogenetic information. Only the more derived traits (apomorphies - in this case, the most derived of Clark's modes) can give useful phylogenetic information (Foley & Lahr 1997: 11).

This approach, by workers who are both committed evolutionists and card-carrying cladists, is almost aggressively 'anti-Dunnellian' in terms of its biases and preconceptions about *how* to use NDT to track the evolution of technological systems. The point is simply that there are many construals of NDT, and while there is broad agreement at the level of the metaphysic, there can nevertheless be a very considerable 'parting of the ways' in respect of methodology (Clark 1993, 1994).

Adaptationist 'Just-So' Stories

Adaptationist 'just-so' stories constitute the second major kind of pitfall likely to arise in archaeological (in fact, all) applications of NDT (O'Brien & Holland 1995). Adaptationist 'just-so' stories are a form of post-hoc accommodative argument comprising explanations cooked up after an analysis has been completed to account for patterns in a data set. Post-hoc accommodative argument is a problem in all science, but is especially vexing in non-experimental disciplines like archaeology where long, complex inferential chains must be constructed to evaluate the plausibility of cause-and-effect relationships (e.g., Binford 1981, Clark 1982).

O'Brien and Holland (1995: 188) point out that the identification of functional features - adaptations - and the measurement of variation constitute only the first step in a two-step process. The second step is explanation. Unless variation is explained, they observe, why bother to quantify it in the first place? Archaeologists who subscribe to various construals of NDT frequently use selection to account for pattern in measureable variation in the functional properties of artifacts, but tend to treat this variation as evidence - in and of itself - for cause-and-effect relationships between those properties and some nebulous concept of adaptation or fitness. This tendency results in the construction of more or less plausible explanations for artifact functional attributes that are limited only by the ingenuity of the investigator (plausibility being determined by congruence - or lack thereof - between the adaptationist biases of the investigator and those of the reader), and that are, in any event, untestable.

The potential logical problems inherent in post hoc argument in adaptationist explanations for pattern were the targets of Gould and Lewontin's famous assault on biologists who identify adaptations ex post facto as solutions to problems posed by the environment, and then go about trying to 'atomize' the organisms in order to figure out how particular parts of it functioned (1978). Although, as O'Brien and Holland (1995) correctly observe, their critique centers on the fact that organisms are integrated wholes, and phylogenetic development and basic body plan (*bauplüne*) should be taken into account in identifying adaptations, Gould and Lewontin (1978) end up recommending decoupling 'adaptation' from 'selection' in efforts to explain aspects of organismic form, function and behavior (see Clark, this volume). In archaeology, the two are often conflated, with the emphasis on selection, although NDT clearly embraces various kinds of 'non-adaptive' evolution wherein natural selection, while of undeniable centrality, is not invoked as the cause of frequency shifts in the genotype. In light of parallels between models of biological and cultural evolution (e.g., Boyd & Richerson 1985, 1992), and whether or not the latter are necessary for an evolutionary archaeology (we are unconvinced), we would suggest that the holistic approach advocated by Gould and Lewontin (1978) is in fact operationally impossible. In practical terms, some degree of atomism is unavoidable in any effort to study adaptation, whether biological or cultural, and as the papers in *Rediscovering Darwin* make clear, a failure to atomize results in the practical impossibility of exploring *any* morphological or behavioral trait. It is worth remarking that Gould and Lewontin's strictures on *baupläne* and developmental constraints are probably intended to discourage inquiry into the adaptive basis for human behavior, essentially because their ideological biases render them unable to accept the possibility that cultural evolution is largely epiphenomenal, thinly covering the evolved genetic-developmental mechanisms that generate what some think of as 'uniquely human' (see Queller [1995] for a devastating critique of Gould and Lewontin's 'holism').

Acknowledging its existence, what do we do about this dilemma? In our opinion, there is no simple or programmatic answer to this question, except to 'get on with it'. Mayr is certainly correct when he notes that

> . . . the student of adaptation has to sail a perilous course between a pseudoexplanatory reductionist atomism and stultifying nonexplanatory holism (1988: 154).

In archaeology we can make informed, a priori judgements based on engineering design analysis (O'Brien & Holland 1990, 1992) as to what function postulated adaptive features might have had in order to sort out adaptive (functional) features from non-adaptive (stylistic) ones. In so doing, we can eventually build a corpus of credible inference that demonstrates both the power and versatility of the approach, and that avoids the more obvious examples of post hoc accommodative argument.

CONCLUSIONS

There is broad agreement that the selectionist approach embodied in NDT consists of two stages: (1) the identification and quantification of the appropriate kinds of variation, and (2) tracking the history of selective forces operation on that variation. *Rediscovering Darwin* challenges archaeologists with a commitment to strengthening our logic of inference to 'get on with it' and show by application how natural and sexual selection, and other NDT mechanisms act to shape the material residues of human behavior, resulting in the persistence of some varieties and the loss of others. As O'Brien and Holland (1995: 189) remark, there will be missteps of various kinds - methodological and conceptual errors, mistakes in judgment, and in the appropriate choice of analytical units - but we can hope that a corpus of solid work will eventually emerge that distinguishes between adaptive and non-adaptive features of the material record of human behavior, and that is founded on a better understanding of the proper function of adaptive features. We are of the opinion that archaeologists who envision the discipline as a 'science-like' endeavor can ill afford to ignore what is arguably the most powerful metaphysical paradigm ever to arise in western science. Nevertheless, we still must confront the fundamental ambiguity of the archaeological record, and the thorny issue of how to assign meaning to archaeological data under an evolutionary framework. We've got to show that NDT 'works' better than its competitors.

Finally, we should listen to Dunnell, and get rid of the orthogenetic notions of progress, human intentionality, and essentialist stage typologies that were implicit features of the cultural evolutionism of the 1940s and 1950s (see Chapters 2, 5). Dunnell has little use for cultural evolution, arguing that it had little in common with Darwinian evolution because it provided no mechanism by which to explain the 'history' of selection trajectories, but a great deal in common with the non-scientific, progressive, goal-directed evolutionary philosophy of Darwin's contemporary, Herbert Spencer. In retrospect, it seems pretty clear that some of these elements were unwittingly incorporated - wholly or in part - into the emergent 'new archaeology' of the 1960s. They've been there ever since. Modern attempts to make archaeology a materialist discipline must contend with the fact that the Darwinian paradigm fell from favor in the U.S. after World War II because of the misconception that humans, as cultural animals, were somehow 'unique', and that two kinds of evolutionary theory - one biological, the other cultural - were required to account for change in organisms and cultures respectively. Although this notion has fared poorly in recent years (see, e.g., Foley 1987, 1995; Sagan & Druyan 1992), it exhibits a remarkable tenacity, particularly in the archaeology of recent prehistory, where the learned component of human behavior is admittedly large. We are left with the suspicion, however, that underlying much behavior thought to be learned, conscious, goal-directed and rational, are ancient, primordial tendencies buried deep in our evolutionary past. Humans, after all, are nothing more (or less) than highly intelligent, technologically sophisticated, socially complex animals. It may be time - however bewildering, wrenching and innovative it will have to be - to stop using 19th century evolutionary concepts and terms as a basis for the archaeology of the 21st century.

Figure 17.1. A cartoon from the British social commentary and humor magazine *Punch* (1881) showing Darwin (seated, left) and the counterclockwise emergence of Victorian man (upper right) from earlier and more primitive forms of life arrayed around 'Time's Meter'. Note primordial CHAOS in the lower left (from the half-title page, *The Cambridge Encyclopedia of Human Evolution* (Jones et al., eds., Cambridge University Press, 1994; used with permission).

ENDNOTE

1. In fact, one of us, who has adopted a Dunnell-like perspective on NDT in modern human origins research (Clark 1992; Clark & Lindly 1989), is taken to task for arguing that technologies change for reasons related to adaptation!

REFERENCES CITED

Binford, L.R.
1981 *Ancient Men and Modern Myths.* New York: Academic Press.

Boyd, R. & P. Richerson
1985 *Culture and the Evolutionary Process.* Chicago: University of Chicago Press.
1992 Cultural inheritance and evolutionary ecology. In *Evolutionary Ecology and Human Behavior* (E. Smith and B. Winterhalder, eds.), pp. 61-92. New York: Aldine de Gruyter.

Brace, C.L.
1988 Punctuationism, cladistics and the legacy of Medieval Neoplatonism. *Human Evolution* 3: 121-138.
1989 Medieval thinking and the paradigms of paleoanthropology. *American Anthropologist* 91: 442-446.

Carr, C. & J. Neitzel (editors)
1996 *Style, Society and Person.* New York: Plenum.

Cavalli-Sforza, L. & M. Feldman
1981 *Cultural Transmission and Evolution: a Quantitative Approach.* Princeton: Princeton University Press.

Clark, G.A.
1982 Quantifying archaeological research. *Advances in Archaeological Method and Theory* 5: 217-273.
1987 Paradigms and paradoxes in contemporary archaeology. In *Quantitative Research in Archaeology* (M. Aldenderfer, ed.), pp. 30-60. Newbury Park, CA: Sage Publications.
1988 Some thoughts on the Black Skull: an archaeologist's assessment of WT-17000 *(A. boisei)* and systematics in human paleontology. *American Anthropologist* 90: 357-371.
1989 Paradigms and paradoxes in paleoanthropology. *American Anthropologist* 91: 446-450.
1991 Epilogue: Paradigms, realism, adaptation and evolution. In *Perspectives on the Past* (G. Clark, ed.), pp. 411-439. Philadelphia: University of Pennsylvania Press.
1992 Continuity or replacement? Putting modern human origins in an evolutionary context. In *The Middle Paleolithic: Adaptation, Behavior and Variability* (H. Dibble and P. Mellars, eds.), pp. 183-205. Philadelphia: University of Pennsylvania Museum.
1993 Paradigms in science and archaeology. *Journal of Archaeological Research* 1: 203-234.
1994 Origine de l'homme: le dialogue de sourds. *La Recherche* 25: 316-321.
1996 NAGPRA and the demon-haunted world. *SAA Bulletin* 14: 3, 15: 4.
1997 Through a glass darkly: conceptual issues in modern human origins research. In *Conceptual Issues in Modern Human Origins Research* (G.A.Clark and C.M. Willermet, eds.), pp. 60-76. New York: Aldine de Gruyter.

Clark, G.A. (editor)
1991 *Perspectives on the Past: Theoretical Biases in Mediterranean Hunter-Gatherer Research.* Philadelphia: University of Pennsylvania Press.

Clark, G.A. & J. Lindly
1989 Modern human origins in the Levant and western Asia: the fossil and archaeological evidence. *American Anthropologist* 91: 962-985.
1991 Paradigmatic biases and paleolithic research traditions. *Current Anthropology* 32: 577-587.

Clark, G.A. & C.R. Stafford
1982 Quantification in American archaeology: an historical perspective. *World Archaeology* 14: 98-119.

Clark, J.G.D.
1977 *World Prehistory: a New Outline.* Cambridge: Cambridge University Press.

Daley, M. & M. Wilson
1988 *Homicide.* New York: Aldine de Gruyter.

Darwin, C.
1845 *Journal of Researches into the Natural History and Geology of the Countries Visited during the Voyage of H.M.S. Beagle round the World.* London: John Murray.
1859 *On the Origin of Species by means of Natural Selection or the Preservation of Favoured Races in the Struggle for Life.* London: John Murray.
1871 *The Descent of Man and Selection in Relation to Sex.* London: John Murray.

Dawkins, R.
1990 *The Extended Phenotype: the Long Reach of the Gene.* Oxford: Oxford University Press.

Descartes, R.
1637 (1901) *Discours sur la Methode.* John Veitch translation *(The Philosophy of Descartes).* New York: Tudor Publishing Company.

Dunnell, R.
1971 *Systematics in Prehistory.* New York: Free Press.
1978 Style and function: a fundamental dichotomy. *American Antiquity* 43: 192-202.
1980 Evolutionary theory and archaeology. *Advances in Archaeological Method and Theory* 3: 35-99.
1982 Science, social science and common sense: the agonizing dilemma of modern archaeology. *Journal of Anthropological Research* 38: 1-25.
1986 Methodological issues in Americanist artifact classification. *Advances in Archaeological Method and Theory* 9: 149-207.
1989 Aspects of the application of evolutionary theory in archaeology. In *Archaeological Thought in America* (C.C. Lamberg-Karlowsky, ed.), pp. 35-49. Cambridge: Cambridge University Press.
1995 What is it that actually evolves? In *Evolutionary Archaeology* (P. Teltser, ed.), pp. 33-50. Tucson: University of Arizona Press.

Foley, R.
1987 *Another Unique Species?* London: Longman Scientific.
1995 *Humans before Humanity.* Oxford: Blackwell.

Foley, R. & M. Lahr
1997 Mode 3 technologies and the evolution of modern humans. *Cambridge Archaeological Journal* 7: 3-36.

Gould, S.J.
1994 *Darwin's Revolution in Thought.* Northampton, MA: Into the Classroom Video, Inc.

Gould, S.J. & R. Lewontin
1978 The spandrels of San Marco and the Panglossian paradigm: a critique of the adaptationist programme. *Proceedings of the Royal Society* 205: 581-598.

Jones, S., et al. (editors)
1994 *The Cambridge Encyclopedia of Human Evolution*. Cambridge: Cambridge University Press.

Malthus, T.
1798 *An Essay on the Principle of Population.* London: J. Johnston.

Maschner, H. (editor)
1996 *Darwinian Archaeologies.* New York: Plenum.

Mayr, E.
1988 *Toward a New Philosophy of Biology: Observations of an Evolutionist.* Cambridge, MA: Harvard University Press.

Mishler, B. & M. Donoghue
1982 Species concepts: a case for pluralism. *Systematic Zoology* 31: 491-503.

Mithen, S.
1990 *Thoughtful Foragers: a Study of Prehistoric Decision Making.* Cambridge: Cambridge University Press.

O'Brien, M. & T. Holland
1990 Variation, selection, and the archaeological record. In *Archaeological Method and Theory, Vol. 2* (M. Schiffer, ed.), pp. 31-79. Tucson: University of Arizona Press.
1992 The role of adaptation in archaeological explanation. *American Antiquity* 57: 36-59.
1995 The nature and premise of a selection-based archaeology. In *Evolutionary Archaeology* (P. Teltser, ed.), pp. 175-200. Tucson: University of Arizona Press.

Paley, W.
1802 (1963) *Natural Theology.* Indianapolis: Bobbs-Merrill.

Plog, F.
1974 *The Study of Prehistoric Change.* New York: Academic Press.

Queller, D.
1995 The spaniels of St. Marx and the Panglossian paradox: a critique of a rhetorical programme. *The Quarterly Review of Biology* 70: 485-489.

Rensch, B.
1959 *Evolution Above the Species Level.* New York: Columbia University Press.

Sagan, C. & A. Druyan
1992 *Shadows of Forgotten Ancestors.* New York: Random House.

Sahlins, M. & E. Service
1960 *Evolution and Culture.* Ann Arbor: University of Michigan Press.

Schiffer, M.
1987 *Formation Processes of the Archaeological Record.* Albuquerque: University of New Mexico Press.

Service, E.
1963 *Primitive Social Organization.* New York: Random House.

Smith, A.
1776 (1961) *An Inquiry into the Nature and Cause of the Wealth of Nations.* London: Methuen.

Sober, E.
1991 Models of cultural evolution. In *Trees of Life: Essays in the Philosophy of Biology* (P. Griffiths, ed.), pp. 17-38. Dordrecht: Kluwer.

Steele, J. & S. Shennan
1996 Introduction. In *The Archaeology of Human Ancestry* (J. Steele and S. Shennan, eds.), pp. 1-41. London: Routledge.

Teltser, P.
1995 Culture history, evolutionary theory, and frequency seriation. In *Evolutionary Archaeology* (P. Teltser, ed.), pp. 51-68. Tucson: University of Arizona Press.

Teltser, P. (editor)
1995 *Evolutionary Archaeology: Methodological Issues.* Tucson: University of Arizona Press.

White, L.
1959 *The Evolution of Culture: the Development of Civilization to the Fall of Rome.* New York: McGraw-Hill.

Wilson, D.
1989 Levels of selection: an alternative to individualism in biology and the human sciences. *Social Networks* 11: 257-272.
1997a Human groups as units of selection. *Science* 276: 1816-1817.
1997b (editor) Multilevel Selection. *American Naturalist* 150 (supplement).

List of Contributors

Douglas B. Bamforth
Anthropology Department, Campus Box 233
University of Colorado
Boulder, CO 80309-0233
email: bamforth@spot.colorado.edu

C. Michael Barton
Department of Anthropology
Box 872402, Arizona State University
Tempe, AZ 85287-2402
email: michael.barton@asu.edu

Robert L. Bettinger
Department of Anthropology
University of California, Davis
Davis, CA 95616
email: rlbettinger@ucdavis.edu

Douglas W. Bird
Department of Anthropology
Universtiy of Washington
Seattle, WA 98195-3100

Peter Bleed
Department of Anthropology
University of Nebraska
Lincoln, NB 68588-0368
email: pbleed@unlinfo.unl.edu

Geoffrey A. Clark
Department of Anthropology
Box 872402, Arizona State University
Tempe, AZ 85287-2402
email: gaclark@asu.edu

Michael W. Diehl
Department of Anthropology
University of Louisville
Louisville, KY 40292

Jelmer Eerkens
Department of Anthropology
University of California
Santa Barbara, CA 93106
email: 6500eerk@ucsbuxa.ucsb.edu

John A.J. Gowlett
Archaeological Sciences, The Hartley Building
University of Liverpool, P.O. Box 147
Liverpool L69 3BX, United Kingdom
email: gowlett@liv.ac.uk

Daniel O. Larson
Department of Anthropology
California State University
1250 Bellflower Blvd.
Long Beach, CA 90840-1003

R. Lee Lyman
Department of Anthropology
University of Missouri, Columbia
Columbia, MO 65211
email: anthrll@showme.missouri.edu

Steven Mithen
Department of Archaeology
University of Reading, Whiteknights
PO Box 218
Reading RG6 2AA, United Kingdom
email: s.j.mithen@reading.ac.uk

Hector Neff
230Z Research Reactor Center
University of Missouri
Columbia, MO 65211
email: neff@reactor.murr.missouri.edu

Fraser D. Neiman
Department of Architectural History
University of Virginia
and Department of Archaeology, Monticello
Box 316
Charlottesville, VA 22902
email: fraser.neiman@virginia.edu

Michael J. O'Brien
Department of Anthropology
University of Missouri, Columbia
Columbia, MO 65211
email: anthmjo@showme.missouri.edu

Dr. David Rindos
deceased

Michael J. Shott
Department of Sociology, Anthropology, and Criminology
University of Northern Iowa
Cedar Falls, IA 50614-0513
email: shott@uni.edu

R. Esmée Webb
Yamaji Language Centre
PO Box 433
Geraldton WA 6531, Australia

Polly Wiessner
Research Group for Human Ethology
in the Max-Planck-Society
Andech, 82346 , Germany

Additional Participants of the 1994 Society for American Archaeology Symposium

James L. Boone
Department of Anthropology
University of New Mexico
Albuquerque, NM 87131

C. Loring Brace
Museum of Anthropology
Ruthven Museums Building, 1109 Geddes Ave.
University of Michigan
Ann Arbor, MI 48109-1079

Robert C. Dunnell
Department of Anthropology
DH05, M32 Denny Hall
University of Washington
Seattle, WA 98195
email: dunnell@u.washington.edu

Kristen Hawkes
Department of Anthropology
University of Utah
Salt Lake City, UT 84112

Michael Jochim
Department of Anthropology
University of California, Santa Barbara
Santa Barbara, CA 93106
email: jochim@sscf.ucsb.edu

James O'Connell
Department of Anthropology,
University of Utah
Salt Lake City, UT 84112
email: oconnell@anthro.utah.edu

Patrice Teltser
Department of Sociology and Anthropology
Lehigh University
681 Taylor Street
Bethlehem, PA 18015-3169
email: pat6@lehigh.edu

Procedure for Approval of AP3A Volumes
An Annual Publication of the Archaeology Division

1. The editor(s) or author(s) should discuss the volume with the Series Editor, who evaluates its suitability for the series. The current Series Editor is Michael E. Whalen, Department of Anthropology, University of Tulsa, Tulsa, Ok 74104; e-mail: michael-whalen@utulsa.edu. The proposal is either declined or passes on to step 2.

2. If the proposed volume is an edited collection of papers, the editor(s) prepare four copies of a prospectus, which should consist of a statement of the volume's purpose or focus, a table of contents, and abstracts of each paper. Meeting abstracts are sufficient, if the proposed volume is the result of a symposium.

   If the proposed volume is an authored study, the author(s) should prepare four copies of a prospectus consisting of a statement of the volume's purpose or focus, a table of contents, and a short description of each chapter.

3. A copy of the prospectus should be sent to the Series Editor and to each of the other 3 members of the review committee, which will normally consist of the Series Editor, the Archaeology Division Chair-elect, and the two Archaeology Division Board Members-at-Large. The Series Editor will provide a mailing list.

4. The Series Editor and the Review Committee rank the candidate manuscripts in order of preference for publication, and the Series Editor presents the ranked manuscripts to the Archaeology Division Executive Board. The final decision for preliminary acceptance of one manuscript rests with the Executive Board.

5. The editor(s) or author(s) of the selected manuscript assemble four copies of a formal draft manuscript, with full versions of all papers or chapters, tables, and sketches of the illustrations. One copy is sent by the authors to the Series Editor and to each of the other 3 members of the Review Committee. A mailing list will be provided by the Series Editor.

6. The Review Committee members review the draft manuscript and vote on its final, formal acceptance.

7. Comments from the Review Committee members are collated and summarized by the Series Editor, who communicates with the volume editor(s) or volume author(s). Any necessary changes, additions, or deletions, are made to the satisfaction of the Series Editor.

8. The volume editor(s) or author(s) prepare(s) a final, camera-ready copy. This should be done in consultation with the Series Editor, who will provide a detailed style guide.

9. One complete copy of this camera-ready ms. is sent to the Series Editor for final approval of style and appearance.

10. The Series Editor's approval of the complete, camera-ready ms. completes the job of the volume editor(s) or author(s).

11. The complete, camera-ready ms. is sent by the Series Editor to the A.A.A. for printing and distribution.